Engineering Your Future

A Comprehensive Introduction to Engineering

Engineering Your Future

A Comprehensive Introduction to Engineering

NINTH EDITION

William C. Oakes, PE, PhD
Purdue University

Les L. Leone, PhD
Michigan State University

CONTRIBUTORS

David "Boz" Bowles, MFA
Louisiana State University

Frank M. Croft, Jr., PhD
Ohio State University

Toby Cumberbatch, PhD
The Cooper Union

John B. Dilworth, PhD
Western Michigan University

Heidi A. Diefes, PhD
Purdue University

Ralph E. Flori, PhD
University of Missouri-Rolla

Craig J. Gunn, MS
Michigan State University

Todd Hamrick, PhD
West Virginia University

Daniel F. Hartner, PhD
Rose-Hulman Institute of Technology

Neal A. Lewis, PhD
University of Bridgeport

Marybeth Lima, PhD
Louisiana State University

Melodee Moore, PhD
FAMU-FSU College of Engineering

Ahad Nasab, PhD, PE
Middle Tennessee State University

Merle C. Potter, PhD
Michigan State University

Yeow K. Siow, PhD
University of Illinois at Chicago

Michael F. Young, MS
Michigan Technological University

New York   Oxford
OXFORD UNIVERSITY PRESS

Oxford University Press is a department of the University of Oxford.
It furthers the University's objective of excellence in research,
scholarship, and education by publishing worldwide.
Oxford is a registered trade mark of Oxford University Press
in the UK and certain other countries.

Published in the United States of America by
Oxford University Press
198 Madison Avenue, New York, NY 10016,
United States of America.

For titles covered by Section 112 of the US Higher Education
Opportunity Act, please visit www.oup.com/us/he for the
latest information about pricing and alternate formats.

Library of Congress Cataloging-in-Publication Data

Names: Oakes, William C., 1962- author. | Leone, Les L., author.
Title: Engineering your future : a comprehensive introduction to engineering.
Description: Ninth edition. | New York, NY : Oxford University Press, [2017]
    | Includes bibliographical references and index.
Identifiers: LCCN 2016047099| ISBN 9780190279264 (pbk.) | ISBN
    9780190279288 (eISBN)
Subjects: LCSH: Engineering--Vocational guidance.
Classification: LCC TA157 .O223 2017 | DDC 620.0023--dc23 LC record
available at https://lccn.loc.gov/2016047099

978-0-19-020892-9
9 8 7 6 5 4 3 2 1
Printed by LSC Communications, United States of America

Contents

Preface

You can't make an educated decision about what career to pursue without adequate information. *Engineering Your Future* endeavors to give you a broad introduction to the study and practice of engineering. In addition to presenting vital information, we've tried to make it interesting and easy to read as well.

You might find Chapter 2, "Engineering Majors," to be a tremendous help to you in determining what areas of engineering sound most appealing to you as you begin your education. Our "Profiles of Engineers", available on the Companion Website, may also be of particular interest to you. It includes information from real people—engineers practicing in the field. They discuss their jobs, their lives, and the things they wish they had known going into the profession.

The rest of the book presents such things as the heritage of engineering; some thoughts about the future of the profession; some tips on how best to succeed in the classroom; advice on how to gain actual, hands-on experience; exposure to computer-aided design; and a nice introduction to several areas essential to the study and practice of engineering.

We have designed this book for modular use in a first-year engineering course that introduces students to the field of engineering. Such a course differs in content from university to university. Consequently, we have included many topics, too numerous to cover in one course. We anticipate that several of the topics will be selected for a particular course with the remaining topics available to you for outside reading and for future reference.

As you contemplate engineering, you should consider the dramatic impact engineers have had on our world. Note the eloquent words of American Association of Engineering Societies Chair Martha Sloan, a professor emeritus of electrical engineering at Michigan Technological University:

In an age when technology helps turn fantasy and fiction into reality, engineers have played a pivotal role in developing the technologies that maintain our nation's economic, environmental and national security. They revolutionized medicine with pacemakers and MRI scanners. They changed the world with the development of television and the transistor, computers and the Internet. They introduced new concepts in transportation, power, satellite communications, earthquake-resistant buildings, and strain-resistant crops by applying scientific discoveries to human needs.

Engineering is sometimes thought of as applied science, but engineering is far more. The essence of engineering is design and making things happen for the benefit of humanity.

Joseph Bordogna, former president of IEEE, adds:

Engineering will be one of the most significant forces in designing continued economic development and success for humankind in a manner that will sustain both the planet and its growing population. Engineers will develop the new processes and products. They will create and manage new systems for civil infrastructure, manufacturing, communications, health care delivery, information management, environmental conservation and monitoring, and everything else that makes modern society function.

We hope that you, too, will find the field of engineering to be attractive, meaningful, and exciting—one that promises to be both challenging and rewarding, and one that matches well with your skills and interests.

For the instructor's convenience, there is an Ancillary Resource Center site with support materials (PowerPoint figure slides and a test bank). This material may be found at http://oup-arc.com/oakes-engineering-9e/.

New to the Ninth Comprehensive Edition

- Chapter 1 "The Heritage of Engineering" replaces "The History of Engineering." This chapter was rewritten to move away from chronicling historical engineering achievements to describe engineering as a profession that has impacted so much of our daily lives and to appreciate the rich and inclusive heritage of engineering and engineers that contributed to what we see today. Diverse examples are used to discuss themes of the heritage of engineering that span genders and cultures with some discussion of the historical contexts to prompt ideas and allow for further research and discussions. Themes that are discussed include how engineers are making the world a better place and improving the human condition as well as the importance of teamwork and communication now and historically.
- Chapter 2, "Engineering Majors," was updated to reflect current technological advances, especially in the computer, electrical, and biological areas. Mobile computing is discussed as an example. Nanotechnology and its influence have also been reflected in the descriptions of the majors.
- Chapter 3, "A Statistical Profile of the Engineering Profession," provides the latest available data on the job market for engineers, recent starting salaries for the

different majors, and a variety of related information. This material includes updated college enrollment data trends, number of degrees awarded for the various engineering majors, and career-long projections of salaries by employer size and type, field of study, and geographical region. Updated information is also provided concerning the diversity of the profession, and engineering graduate school data.

- Chapter 5, "Future Challenges," was updated to include a list and description of the National Academy of Engineering's Grand Challenges. These descriptions, used with permission from the National Academy, are the result of the academy's study of the most significant technological challenges of the day. These have been added to the existing chapter and can be used as a standalone section or as part of the existing chapter.
- Previously called "Visualization and Graphics, Chapter 8 is now titled "Graphics and Orthographic Projection" and has been rewritten to be more concise and practical. The text has been refocused to concentrate on techniques applied by working engineers.
- Chapter 10, "Teamwork," has been completely updated with new examples and material. The chapter uses real examples from today's leading companies, including Netflix, Boeing, Tesla Motors, and Google.
- Chapter 11, "Project Management," has been completely rewritten with significant new material added. A sample student project is introduced and developed, showing how a project plan can be developed using project management tools. The application of Microsoft Project software is demonstrated.
- Chapter 12, "Engineering Design," was revised to help students gain insight into the more practical aspects of learning the engineering design process. The 10-stage process has been reduced to a more manageable five stages and includes an open-ended case study that can be used in the classroom as is or with modification.
- Chapter 14, "Ethics and Engineering," has been rewritten with the goal of introducing ethics to future professional engineers in a lively, more accessible way. In addition to systematically introducing the vocabulary and concepts needed to understand the nature of professional ethics and the difference between ethics and policy, the chapter now more directly confronts and clarifies some of the most common questions and confusions students have about ethics, including where professional ethical obligations come from, why the ethical obligations of engineers are not merely matters of subjective opinion and personal conscience, and why codes of professional ethics must be understood not as arbitrary lists of rules but rather as a reflection of rational, intuitive requirements on the practice of a learned profession. These insights about the nature of professional ethics are now also reinforced in the revised explanation and analysis of existing codes of engineering ethics as well as in the review questions.
- Chapter 15, "Units and Conversions," includes expanded sections on significant figures and unit conversion along with numerical examples. A new section on dimensionless numbers has been added. Several problems regarding dimensionless numbers have been added to the end-of-chapter problems.

- Chapter 16, "Mathematics Review," presents brief yet concise reviews of many of the mathematical concepts students will encounter in their engineering studies. Improvements to previous editions include "in line expansion" of select example problems, additional help with vector math, and a unit circle to accompany the trigonometry section of the chapter.
- Chapter 17, "Engineering Fundamentals," provides a review of specific math and science applications that are fundamental to engineering studies. Select example problems in this chapter also have more detailed "in line expansion" of solutions, designed to encourage good problem-solving skills and problem documentation. Included also in the revised chapter is a brief review of partial pressures in the thermodynamics section.
- Appendix A, "Nine Excel Skills Every Engineering Student Should Know," While the number of skills is retained, the skills themselves have been completely revised. Instead of focusing on "which button to click," the skills are now presented in a way that promotes everyday application as well as lifelong learning.
- Appendix B, "Impress Them: How to Make Presentations Effective," Given a complete overhaul, this appendix now offers guidelines for making a powerful presentation that will leave a lasting impression on the audience. The makeup of a presentation is dissected, and plenty of good and bad examples are included.
- Appendix C, "An Introduction to MATLAB," The programming section has been significantly expanded. Learning to code is an art, and making an efficient and elegant code is a lifelong pursuit—with this appendix serving as a starting point.

Acknowledgments

The authors are especially grateful to the reviewers whose opinions and comments directly influenced the development of this edition:

Anil Acharya, Alabama A&M University
Spyros Andreou, Savannah State University
Asad Azemi, Penn State University
Jerome Davis, University of North Texas
Chris Geiger, Florida Gulf Coast University
Nolides Guzman Zambrano, Lone Star College
Dr. Dominic M. Halsmer, Oral Roberts University
Todd Hamrick, West Virginia University
Matthew Jensen, Florida Institute of Technology
Benjamin S. Kelley, Baylor University
Mark Keshtvarz, Northern Kentucky University
Dr. Raghava R. Kommalapati, Prairie View A&M University
Tanya Kunberger, Florida Gulf Coast University
Andre Lau, Penn State University

Dean Lewis, Penn State University
Jennifer Light, Lewis-Clark State College
Dr. James McCusker, Wentworth Institute of Technology
Deepak Mehra, Potomac State College
Christopher Miller, University of Akron
Melodee Moore, Florida A&M University
Ahad Nasab, Middle Tennessee State University
Herbert Newman, Coastal Carolina University
Dr. John H. O'Haver, University of Mississippi
Olayinka Frank Oredeko, Central Georgia Technical College
Reginald Perry, FAMU-FSU College of Engineering
Cherish Qualls, University of North Texas
James Rantschler, Xavier University of Louisiana
Dr. Farhad Reza, Minnesota State University
Bernd F. Schliemann, University of Massachusetts at Amherst
Gary Scott, State of University of New York
Yeow Siow, Purdue University at Calumet
Yiheng Wang, Lone Star College

We would also like to thank those reviewers who provided feedback for previous editions:

Spyros Andreou, Savannah State University
Juan M. Caicedo, University of South Carolina
Matthew Cavalli, University of North Dakota
Rafael Fox, Texas A&M University–Corpus Christi
Keith Gardiner, Lehigh University
Chris Geiger, Florida Gulf Coast University
Yoon Kim, Virginia State University
Nikki Larson, Western Washington University
Keith Level, Las Positas College
Jennifer Light, Lewis-Clark State College
S. T. Mau, California State University at Northridge
Edgar Herbert Newman, Coastal Carolina University
John Nicklow, Southern Illinois University at Carbondale
Megan Piccus, Springfield Technical Community College
Charles E. Pierce, University of South Carolina
G. Albert Popson, Jr., West Virginia Wesleyan College
Ken Reid, Ohio Northern University
Nikki Strader, Ohio State University
Yiheng Wang, Danville Community College
Gregory Wight, Norwich University
David Willis, University of Massachusetts at Lowell
Shuming Zheng, Chicago State University

—The Authors

CHAPTER 1

The Heritage of Engineering

While writing this chapter, I was teaching a class over the Internet to engineering professors in India. The class was about how to integrate design experiences (addressing needs of underserved people and communities) into undergraduate engineering courses. I was excited when I finished that day's class as we had had a great conversation about how we can use engineering to meet human, community, and environmental needs in India and the United States. The same ideas could be applied to any country to make our world a better place. Today's technology has opened so many opportunities to make an impact in our communities, our countries, and our world. I ended the class thinking that this is really an exciting time to be an engineer or an engineering student—with all of the technological tools we have at our disposal and the exciting things we can do with them.

As I ended the class, I looked outside at the first snowfall of the year. Because of the time difference between India and the United States, I have to teach the class very early in the morning, so the sun was just coming up. The beautiful sunrise with the falling snow got me thinking. I had just been talking with about 40 colleagues who were literally on the other side of the world and spread out all over their country. I was in Indiana, and our course facilitator was from Massachusetts. The incredible technology that allowed us to discuss how to use technology to make a difference in the world was created by engineers who had come before us. A generation ago, we would have had to make a very expensive phone call to have that discussion. Earlier generations would have had to communicate with letters on actual paper that were physically carried from one place to the next. Technology has significantly changed the way we communicate, as well as so many other parts of our lives. Those changes were created and driven by engineers who started out a lot like you.

As I sat there in the warm house and watched the snow, I began to think about all of the other ways that engineers have impacted us. The materials to make the house to keep me warm were developed by engineers. The house is heated with an ultra-high-efficiency furnace that also protects the environment. The natural gas burning in the furnace was found, extracted, refined, and piped to the house using technology developed by engineers. The lights in the house were developed by engineers. The appliances in the house all have computers to make them more efficient and easier to use. Everywhere I looked I saw something that had been touched by engineers . . . with the exception of the snowflakes falling outside, of course.

There are so many engineers who have made an impact in our daily lives, and they came from many different places and backgrounds. I thought about them as I moved through the day. I had to pick up my daughter from a friend's house, and I was grateful for Mary Anderson, who had invented the windshield wiper to clear the snow from my car's windshield. When I got to the first intersection, I thought about Garrett Morgan, the African American inventor who developed the traffic light to keep us safe on the roads. I was grateful for the computer and electrical engineers who developed the technology in my hearing aids that allow me to have a conversation with my daughter when I picked her up.

1.1   Introduction

The impact of engineers on our everyday lives is incredible. Even our life expectancies are so much higher in large parts due to the technologies that engineers have developed to provide safe drinking water, sanitation, accessible medicines, and much more. Engineers have made an enormous impact on our world, and there are so many opportunities yet to come. Today's technology has given us the tools to address needs and opportunities to make a difference in our world.

The purpose of this first chapter is to give you a sense of the strong heritage of the engineering profession. We will provide a brief glimpse into some of those who have come before you and a feeling of the incredibly exciting profession you are exploring. This is not meant to be a comprehensive overview of the history of engineering, as that would be a book in itself. Instead we use history to illustrate some of the diversity and wondrous heritage of the engineering profession and highlight a few of the men and women who have developed the amazing world of technology we live in today.

Definition of Engineering

Even if you already have a general knowledge of what engineering involves, a look at the definition of the profession may give you some insight. The organization that accredits engineering programs is called ABET, and they define engineering as:

> *The profession in which knowledge of the mathematical and natural sciences, gained by study, experience, and practice, is applied with judgment to develop ways to use, economically, the materials and forces of nature for the benefit of mankind.*

This definition places three responsibilities on an engineer: (1) to develop judgment so that you can (2) help mankind in (3) economical ways. It places obligations on us to address needs that benefit others and to make sure we don't do harm. We seek to provide economical solutions because if they are too expensive, they are out of reach of people. Looking at case *histories* and *historical* overviews can help us see how others have applied these principles before us and understand more about the profession we are entering. Study of history can also give us a sense of belonging to the profession. There are engineers who come from the very kind of background you come from and look a lot like you—or did when they were your age.

Definitions are important, but they don't always inspire. The National Academy of Engineering is a body of outstanding engineers who advise the federal government on matters pertaining to engineering and technology. One has to be nominated and invited to become a member of the national academy. This body studied the perceptions of engineering and engineers in the United States and came to the conclusion that most people do not understand who we are and what great things we could do. They produced a report entitled *Changing the Conversation* to help us communicate the potential of engineering. Part of that report includes a positioning statement to help guide our conversations. It reads,

> *No profession unleashes the spirit of innovation like engineering. From research to real-world applications, engineers constantly discover how to improve our lives by creating bold new solutions that connect science to life in unexpected, forward-thinking ways. Few professions turn so many ideas into so many realities. Few have such a direct and positive effect on people's everyday lives. We are counting on engineers and their imaginations to help us meet the needs of the 21st century.*

We need this positioning statement because engineers and engineering are often misunderstood as a field. The contributions of engineers are not always seen, understood, or appreciated. As illustration, I think of a class I teach that engages about 500 students per semester in designs to meet community needs locally and globally. The students work together to develop designs, and they work with community partners. I often hear them describe themselves as "not a typical engineer." They like to work with others, have a social life, and want to make a difference in the world. I love that attitude, and I do wonder how I have 500 students who view themselves as "not typical." At least in our class they are typical and are very much more typical of engineers and the overall engineering profession, what it is and what is should be. It may not match the stereotypes, but it does match the heritage we have as engineers. We have a strong knowledge of math, science, and technology and have to work with many others to create solutions that can improve the human and environmental conditions. It takes many different people to do that, and it always has and always will. The following sections will explore history with examples of some of these diverse engineers who were real people who have helped make the world a better place.

1.2 The Beginnings of Engineering: The Earliest Days

The foundations of engineering were laid with our ancestors' efforts to survive and to improve their quality of life. From the beginning, they looked around their environments and saw areas where life could be made easier and more stable. They found improved ways to provide for food, through hunting and fishing. They discovered better methods for providing shelter for their families and ways to make clothing. Their main physical concern was day-to-day survival. As life became more complicated and small collections of families became larger communities, the need grew to look into new areas of concern and specialization.

If you look back at the definition of engineering given by ABET, you will notice a statement: "The profession in which knowledge of the mathematical and natural sciences . . . is applied." Prehistoric engineers applied problem solving and toolmaking but did not have a grasp of the same mathematical principles or knowledge of natural science *as we know it today*. They designed and built items more by trial and error, testing, and intuition. They built spears that worked and others that failed, but in the end they perfected weapons that allowed them to bring down game animals and feed their families. Although they couldn't describe it, they used principles of aerodynamics and mechanical advantage to develop more efficient tools to hunt.

Since written communication and transportation did not exist at that time, little information or innovation was exchanged with people from faraway places. Each group around the world moved ahead on its own. It is inspiring to see how people from all over the world developed innovations to improve the quality of life for their families and their communities.

Transportation was another area where early engineers made an impact. The designs of early boats, for example, inspire even today's engineers. Breakthroughs in transportation and exploration are being located ever earlier as we continue to make discoveries about various peoples traveling long before we thought they did— influencing others and bringing back knowledge. Transportation was used to hunt and fish, to move families, and to explore new areas. Polynesian boat designers, for example, developed crafts that could sail great distances and allowed people to settle many of the islands across the Pacific. Their use of mathematics and astronomy allowed them to navigate great distances on their vessels that were designed for long ocean voyages. Their vessels are still an engineering marvel today.

ACTIVITY 1.1	Prepare a brief report that focuses on engineering in a historical era and cultural area (for example, pre-Columbian Central America, Europe in the Industrial Revolution, Mesopotamia). Analyze the events that you consider to be engineering highlights and explain their importance to human progress.

1.3   Early Cities

As cities grew and the need to address the demands of the new fledgling societies increased, a significant change took place. People who showed special aptitude in certain areas were identified and assigned to ever more specialized tasks. This development gave toolmakers the time and resources to dedicate themselves to

building and innovation. This new social function created the first real engineers, and innovation flourished more rapidly.

Between 4000 and 2000 B.C., Egypt in Africa and Mesopotamia in the Middle East were two areas for early engineering activity. Stone tools were developed to help humans in their quest for food. Copper and bronze axes were perfected through smelting. These developments were not only aimed at hunting: The development of the plow was allowing humans to become farmers so that they could reside in one place and give up the nomadic life. Mesopotamia also made its mark on engineering by giving birth to the wheel, the sailing boat, and methods of writing. Engineering skills that were applied to the development of everyday items immediately improved life as they knew it.

During the construction of the pyramids (c. 2700–2500 B.C.) the number of engineers required was immense. They had to make sure that everything fit correctly, that stones were properly transported long distances, and that the tombs would be secure against robbery. Imhotep (chief engineer to King Zoser) was building the Step Pyramid at Sakkara (pictured in Fig. 1.1) in Egypt about 2700 B.C. The more elaborate Great Pyramid of Khufu (pictured in Fig. 1.2) would come about 200 years later. These early engineers, using simple tools, performed, with great acuity, insight, and technical rigor, tasks that even today give us a sense of pride in their achievements.

The Great Pyramid of Khufu is the largest masonry structure ever built. Its base measures 756 feet on each side. The 480-foot structure was constructed using over 2.3 million limestone blocks with a total weight of over 58 million tons. Casing blocks

Figure 1.1 The Step Pyramid of Sakkara
Source: © iStockPhoto

Figure 1.2 The Great Pyramid of Khufu
Source: © iStockPhoto

of fine limestone were attached to all four sides. These casing stones, some weighing as much as 15 tons, have been removed over the centuries for a wide variety of other uses. It is hard for us to imagine the engineering expertise needed to quarry and move these base and casing stones, and then piece them together so that they would form the pyramid and its covering.

Here are additional details about this pyramid given by Roland Turner and Steven Goulden in *Great Engineers and Pioneers in Technology, Volume 1: From Antiquity through the Industrial Revolution*:

> *Buried within the pyramid are passageways leading to a number of funeral chambers, only one of which was actually used to house Khufu's remains. The granite-lined King's Chamber, measuring 17 by 34 feet, is roofed with nine slabs of granite which weigh 50 tons each. To relieve the weight on this roof, located 300 feet below the apex of the pyramid, the builder stacked five hollow chambers at short intervals above it. Four of the relieving chambers are roofed with granite lintels, while the topmost has a corbelled roof. Although somewhat rough and ready in design and execution, the system effectively distributes the massive overlying weight to the sturdy walls of the King's Chamber.*
>
> *Sheer precision marks every other aspect of the pyramid's construction. The four sides of the base are practically identical in length—the error is a matter of*

inches—and the angles are equally accurate. Direct measurement from corner to corner must have been difficult, since the pyramid was built on the site of a rocky knoll (now completely enclosed in the structure). Moreover, it is an open question how the builder managed to align the pyramid almost exactly north-south. Still, many of the techniques used for raising the pyramid can be deduced.

After the base and every successive course was in place, it was leveled by flooding the surface with Nile water, no doubt retained by mud banks, and then marking reference points of equal depth to guide the final dressing. Complications were caused by the use of blocks of different heights in the same course.

The above excerpt mentions a few of the fascinating details of the monumental job undertaken to construct a pyramid with primitive tools and human labor. It was quite a feat for these early African engineers.

As civilizations grew around the world, the need for infrastructure increased, and it was the early civil engineers who met this challenge. Cities developed in many places, including India, China, and the Americas. Early engineering achievements can be seen even today in many places. For example, pyramids still stand in Latin America as a testament to the skill and expertise of early Native American engineers. Cities were constructed that included sophisticated infrastructure and building techniques. One extraordinary example of ingenuity and skill that inspires many visitors is the Incan city of Machu Picchu (Fig. 1.3) built on top of the Andes mountains in Peru. Constructed in the 15th century at the height of the Inca Empire, it is an

Figure 1.3 Machu Picchu in present-day Peru
Source: Damian Gil/Shutterstock.com

engineering marvel that used sophisticated techniques of dry-stone walls that fused huge blocks without the use of mortar. The design of the city itself is based on astronomical alignments that show mathematical and astronomical sophistication. The site at the top of the mountains would have created significant engineering challenges, as well as providing for incredible panoramic views that can be enjoyed today. Recreating that city would be a challenge even with today's technology.

Engineering the Temples of Greece

The Parthenon (Fig. 1.4) was constructed by Iktinos in Athens starting in 447 B.C. and was completed by 438 B.C. It is an extraordinary example of a religious temple. Engineers played a role in the religious aspects of societies all over the world. The Parthenon was to be built on the foundation of a previous temple using materials salvaged from its remains, making this an early example of recycling. The Parthenon was designed to house a statue of Athena that stood almost 40 feet tall. Iktinos performed the task that he was assigned, and the temple exists today as a monument to engineering capability.

Structural work on the Parthenon enlarged the existing limestone platform of the old temple to a width of 160 feet and a length of 360 feet. The building itself, constructed entirely of marble, measured 101 feet by 228 feet; it was the largest such temple on the Greek mainland. Around the body of the building Iktinos built a colonnade,

Figure 1.4 The Parthenon in Athens
Source: Rich Lynch/Shutterstock.com

customary in Greek temple architecture. The bases of the columns were 6 feet in diameter and were spaced 14 feet apart. Subtle harmonies were thus established, for these distances were all in the ratio of 4:9. Moreover, the combined height of the columns and entablatures (lintels) bore the same ratio to the width of the building.

Remember that this was the year 438 B.C. It would be a significant feat to replicate the Parthenon today.

Aqueducts and Roads

As cities and populations grew, additional needs had to be met, including the delivery of water. In Europe, the Romans developed sophisticated systems of aqueducts to deliver and distribute water into their cities. This was the work of early civil engineers who were using mathematics and an early understanding of sciences. One such aqueduct is shown in Figure 1.5. It is remarkable that these well-designed structures still stand.

Transportation, including the design and construction of roads, continues to be an active area of study for civil engineers, and the Romans were among the first great transportation engineers. Construction of the first great Roman road, the Appian

Figure 1.5 Roman aqueduct
Source: © iStockPhoto

Way, began around 312 B.C. It connected Rome and Capua, a distance of 142 miles. The Appian Way eventually stretched to Brundisium, at the very southernmost point in Italy, and covered 360 miles. The Roman engineers continued building roads until almost A.D. 200, when the entire empire was connected with a network of roads.

For those interested in civil engineering, the Roman roads followed elaborate principles of construction. A bedding of sand, 4 to 6 inches thick, or sometimes mortar 1 inch thick, was spread upon the foundation. The first course of large flat stones cemented together with lime mortar was placed upon this bedding of sand. If lime was not available, the stones (none smaller than a man's hand) were cemented together with clay. The largest were placed along the edge to form a retaining wall. This course varied from 10 inches thick on good ground to 24 inches on bad ground. A layer of concrete about 9 inches deep was placed on top of this, followed by a layer of rich gravel or sand concrete. The roadway would generally be 12 inches thick at the sides of the road and 18 inches in the middle, thus creating a crown that caused runoff. While this third course was still wet, the fourth or final course was laid. This was made of carefully cut hard stones. Upon completion these roads would be from 2 to 5 feet thick, quite a feat for hand labor.

It is interesting to note that after the fall of Rome, road building was no longer practiced by anyone in the world. It would be many hundreds of years before those who specialized in road building again took on the monumental task of linking the peoples of the world.

The Great Wall of China

In 220 B.C., during the Ch'in Dynasty, military general Meng T'ien led his troops along the borders of China. His primary role was that of a commander of troops charged with the task of repelling the nomadic hordes of Mongolians who occasionally surged across the Chinese border. The Ch'in emperor, Shih Huang Ti, commissioned him to begin building what would become known as the Great Wall of China (Fig. 1.6).

The emperor himself conceived the idea to link all the fortresses that guarded the northern borders of China. The general and the emperor functioned as engineers, even though this was not their profession. They solved a particular problem by applying the knowledge they possessed in order to make life better for their people. The ancient wall is estimated to have been 3,080 miles in length, while the modern wall runs about 1,700 miles. The original wall is believed to have passed Ninghsia, continuing north of a river and then running east through the southern steppes of Mongolia at a line north of the present Great Wall. It is believed to have reached the sea near the Shan-hal-huan River. After serving as a buffer against the nomadic hordes for six centuries, the wall was allowed to deteriorate until the sixth and seventh centuries A.D., when it underwent major reconstruction under the Wei, Ch'i, and Sui dynasties. Although the vast structure had lost military significance by the time of China's last dynasty, the Ch'ing, it never lost its significance as a wonder of the world and as a massive engineering undertaking.

Figure 1.6 The Great Wall of China
Source: © iStockPhoto

Agricultural Engineering

We have used a number of examples of civil engineering, and there were other branches of engineering that impacted people early in history, including agricultural engineering. The development of agricultural practices included many contributions by engineers. Earlier, we mentioned the plow as an example of a mechanism that made it easier and more productive to grow food. The Native Americans were very astute agricultural engineers. Today, we are still learning about the sophisticated ways that indigenous people incorporated an understanding of the land and the environment into their efforts to produce sustainable processes. They were truly the first sustainability engineers. Recent discoveries in the Amazon River basin show that native peoples had cultivated much of what is the Amazon jungle today, and it was done in a more environmentally friendly manner than our current practices. Researchers are studying the ancient methods to inform practices of today to develop a sustainable approach for protecting one of the most biodiverse places on the planet.

Like the Romans, Native Americans learned how to distribute water for drinking and for agriculture. The water systems were often very sophisticated for agriculture irrigation, drinking, and defense. The Spanish colonists learned from the indigenous people and their irrigation techniques. In Mexico today, many of the irrigation systems still derive their designs from the native ones.

Innovative ways of processing food were developed by Mayans, Incas, and others. For example, the Incas developed ways to freeze dry food, including potatoes, that could be stored for years. The technique was adapted by the Spanish to send fresh potatoes back to Europe.

Native Americans were some of the first genetic engineers, and corn is an example. There is not a wild form of corn that exists today, unlike most other crops. Scientists hypothesize that Native Americans cross-bred wild grains to produce what has become one of the largest agricultural crops today.

Early engineers from all over the world helped improve the quality of life of their fellow citizens. That tradition continues today, and we will discuss some examples of those engineers and their qualities in the sections that follow.

Industrial Age

The pace of technological change has increased as more technology has been developed. In the earliest centuries, advances were slow and developed over a long period of time. That changed significantly with the Industrial Revolution that began in the 1800s. Machines were created that performed tasks more efficiently than people or animals could. Transportation moved from relying on horses to locomotives and automobiles. Ships could power themselves instead of relying on wind or rowing. Machines were introduced to provide power and changed the way many industries were performed, including mining and agriculture. The Industrial Age produced machines that could replace the need for manual labor and also created new jobs for people to manufacture, operate, and repair these machines.

The invention of machines was significant, but engineers are also interested in how they are used and by whom. The invention of the automobile, for example, didn't change the lives of ordinary people until it was made affordable and thus accessible to more people. Making technology affordable and accessible to a broad and diverse section of people is, and always has been, an important aspect of engineering. In 1913, Henry Ford pioneered the moving assembly line for the automobile industry, which began to make the automobile affordable. The idea of mass production reduced the costs of cars and also provided jobs for people to earn the money to own one.

The advent of large-scale manufacturing created new challenges and areas of engineering related to manufacturing. Understanding the manufacturing processes

Figure 1.7 Dr. Lillian Gilbreth
Source: © Underwood & Underwood/Corbis

and how to make them more efficient and safe for the workers and the environment was important and opened the way for the modern-day field of industrial engineering. Dr. Lillian Gilbreth and her husband Frank were two of the pioneers in this new field. They introduced the use of time and motion studies to improve the efficiency of manufacturing processes. These techniques revolutionized the way things were made and are still making an impact in today's manufacturing arena. Lillian Gilbreth (Fig. 1.7) had a long and distinguished career, including becoming the first female engineering professor at Purdue University. She is also known for the book written about her family that included 12 children, entitled *Cheaper by the Dozen.*

Industrial engineers today still work with factories to optimize processes. Modern factories integrate robotics, automation, and people. Industrial engineers also use these techniques to work to optimize healthcare systems, warehouse and distribution networks for retail and grocery stores, and systems to address natural disasters. Industrial engineers also lay out amusement parks and the systems that allow you to optimize the flow of guests to the parks, including the reservation system to allow you to reserve a time for your favorite roller coaster.

The training we have as engineers can lead to many ___en applications, and an example from a friend who is an industrial engi___ strates this point. My daughters attended a school that made apple pies as ___ fundraising event. For several years, they used the same three-day proc___ thousands of pies, until my friend moved into town. He was a stay-at-h___and an industrial engineer. He looked at their process and made sugges___ modern manufacturing concepts. The result was that the three-day ___ cut to two days and produced more pies. The apple pie process was ___d. It struck me as meaningful as this was done in the same communi___an Gilbreth had lived and taught years earlier. I suspect she would b___

a) The Industrial Revolution changed the whole landscape of the world. How did the engineer fit into this revolution? What were some of the major contributions?

b) Research and compare the first rotary engine to the rotary engines of today.

c) Identify one aspect of our lives today that has been impacted by the Industrial Revolution. Describe how it has been impacted and how our quality of life has increased because of those inventions.

1.4   A Case Study of Two Historical Engineers

The important thing to realize about history is that it's about people. To fully understand the story of an invention, you need to investigate the people involved. An invention and its inventor are inextricably woven together. As you study the details of history, you'll likely find yourself most interested as you get to know more about the people involved and the challenges they were up against. We hope the two studies that follow will inspire you to further investigation.

Leonardo da Vinci

Leonardo da Vinci (Fig. 1.8) had an uncanny ability to envision mechanized innovations well ahead of his time. An example is seen in Figure 1.9 with his flying machine.

Figure 1.8 Leonardo da Vinci, born April 15, 1452, Vinci, Italy, and died May 2, 1519
Source: © Corbis

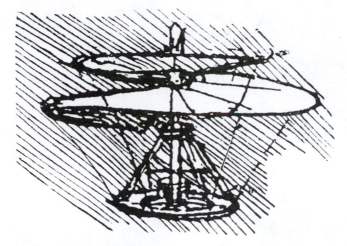

Figure 1.9 A flying machine
Source: © Bettmann/CORBIS

We want to discuss him for his engineering ability as well as for his amazing ability as an artist. Engineering has a heritage of diverse people with diverse interests and talents. As an artist, da Vinci created frescoes and sculptures. Figure 1.10 shows how he used his artistic talents and his knowledge of science and engineering to examine the anatomy of the human body. He was an architect too. He was a true Renaissance man and harnessed his immense genius to make improvements to almost every aspect of the lives of his contemporaries, and to greatly impact the lives of future generations. Da Vinci was disadvantaged as he couldn't read Latin. This prevented him from learning from the common scientific writings of the day, which focused on the works of Aristotle and other Greeks and their relation to the Bible. Instead, da Vinci was forced to make his assessments solely from his observations of the world around him. He was not interested in the thoughts of the ancients; he simply wanted to use his engineering skills to improve his environment.

He did indeed bring his genius to bear on an enormously wide range of subjects, but little of his work had any relationship to that of his contemporaries. He designed, he built, and he tested, but various essential aspects of development were always lacking from his innovations as he lacked a community of engineering peers with whom he could integrate his efforts. Even brilliant engineers need teams of people, as we will discuss later. da Vinci was so far ahead of his time relative to the development of the field of engineering in his day that it was centuries before many of his innovations came to fruition.

To obtain his first commission as an engineer he wrote a letter claiming that he could construct movable bridges; remove water from the moats of fortresses under siege; destroy any fortification not built of stone; make mortars, dart-throwers, flame-throwers and cannon capable of firing stones and making smoke; design ships and weapons for war at sea; dig tunnels without making any noise; make armored wagons to break up the enemy in advance of the infantry; design buildings; sculpt in

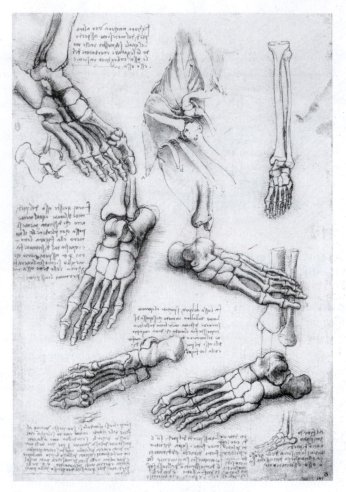

Figure 1.10 Foot anatomy by Leonardo da Vinci
Source: © Sheila Terry/Science Photo Library

any medium; and paint. With that résumé, he was hired with the title of Painter and Engineer to the Duke. Among his accomplishments was perfecting the canal system around Milan. Da Vinci's chief court function during the period he was employed by the Duke was to produce spectacular shows for the entertainment of the aristocrats who came to court. He produced musical events, designed floats, and delighted the court with processions that dazzled the eyes and flying devices that wowed the audience. The interesting thing is that during this time he wrote over 5,000 pages of notes, detailing every conceivable kind of invention. This truly was a great engineer at work.

There is so much you will encounter if you investigate Leonardo da Vinci. It is interesting to note that he seemingly examined every aspect of his world and left behind a significant record from which we can learn.

Gutenberg and His Printing Press

Another important innovator of this era developed one of the most history-altering inventions of all time—the printing press. In 1455, Johannes Gutenberg (Fig. 1.11) printed the first book, a Bible. He lived before Leonardo da Vinci and wasn't a generalist like da Vinci, but his invention changed society forever. With the invention of the printing press, humans were able to use, appreciate, and disseminate information as never before.

The development of printing came at a time when there was a growing need for the ability to spread information. There had been a long phase of general societal introversion from which humanity was prepared to emerge. Gutenberg's printing press was the spark that ignited the flame of widespread communication.

There were already printing presses when Gutenberg introduced his. As early as the 11th century, the Chinese had developed a set of movable type from a baked mixture of clay and glue. The type pieces were stuck onto a plate where impressions could be taken by pressing paper onto the stationary type. Since the type was glued to the plate, the plate could be heated and the type removed and resituated. The process was not used in any form of mass production, but it did form a basis for the future invention of the Gutenberg printing press. In Europe there were attempts at creating presses that would produce cheap playing cards and frivolous items, but it took Gutenberg to see the practical need for a press that could provide everyday humans with reading material (Fig. 1.12).

Johannes Gutenberg was born into a noble family of the city of Mainz, Germany. His early training is reputed to have been in goldsmithing. By 1428 he had moved to Strasbourg. It was there that he began to formulate the idea for his printing press. His first experiments with movable type stemmed from his notions of incorporating the techniques of metalworking—such as casting, punch-cutting, and stamping—for the mass production of books. Since all European books at this time were handwritten by scribes with elaborate script, Gutenberg decided to reproduce this writing style with

Figure 1.11 Johannes Gutenberg, creator of the first mass-producing printing press. He was born in 1394–1398, in Mainz, Germany, and died Feb. 3, 1468.

Source: © iStockPhoto

Figure 1.12 The printing press
Source: © iStockPhoto

a font of over 300 characters, far larger than the fonts of today. To make this possible, he invented the variable-width mold and perfected a rugged blend of lead, antimony, and tin used by type foundries up to the present century.

ACTIVITIES 1.3

a) Study some of the inventions of Leonardo da Vinci. Draw conclusions on how these early works influenced later inventions or innovations.
b) Research and build one of da Vinci's inventions, or a model of one, from his plans.
c) Where does Boyle's Law fit in your future studies?
d) How can the arts be integrated into engineering to make engineering better?
e) Explain why the printing press revolutionized the world and the life of the engineer.

1.5   Computers, Information, Networking, and People

The Industrial Revolution increased the pace of innovation, but it was nothing like what we see today with the invention of the computer. The computer itself has allowed engineers to perform analyses and create models that can test ideas without having to build an actual device. Thousands of ideas can be tested in a short time to find the optimal concept. As computing devices have become faster and smaller, applications for smart devices have permeated our society. Today, every car has a computer to control emissions, activate safety devices, and monitor the health of the car itself. The average appliance today has a computer to make it more efficient and user-friendly. The computers in today's kitchen appliances are more powerful than the computers on the Apollo spacecraft that carried astronauts to the moon and back.

When computers were invented, they were very large and took up whole rooms. Many of the early users greatly underestimated their potential and looked at computers in a very limited way. Fortunately, there were innovators who saw their potential. One of those was Bernard (Bernie) M. Gordon. Bernie is one of the most accomplished inventors of our time, with hundreds of patents. He helped design the first commercial digital computer in the United States, the UNIVAC. He saw the opportunity to use the computer to acquire and process data from measurement instruments. He invented the first analog to digital (A to D) converter. In the 1950s, this was revolutionary; when he first proposed this idea to start a company, it was turned down. He was told that there was a total market for perhaps 10 in the whole world. Initial rejection didn't stop him, but he needed to get someone to buy his invention.

I had the privilege to meet Bernie at a dinner at the National Academy of Engineering that gave an award in his name. Two of my daughters were also there, and he told them the story of how he found that first customer. He called Grace Hopper (Fig. 1.13), one of the pioneers in early computing. He said that Grace knew not only where all of the computers in the world were but also who owned and operated them. There were only a few dozen at that time. The result was that Bernie found someone who was interested, and it was a success that launched him on his amazing career. By the way, that initial estimate of 10 A to D converters in the world was a little off, as there are probably more than 10 million made every day.

Computers have opened so many doors, and Bernie Gordon is an example of someone who has continued to find new ways to apply technology to improve the lives of others. He was talking to a doctor about the need to get vital statistics on a baby during delivery. He knew that his expertise in data acquisition and signal processing could be of use, and he and his team invented the fetal monitor that is used today in hospitals all over the world. That invention has saved probably millions of lives of newborn babies. Bernie is also an example of someone who continued to invent, and in his 80s he worked with a team to develop brain scanning technologies that could save the lives of stroke victims in emergency rooms.

Hearing Bernie Gordon talk was memorable. He constantly was looking at how he could improve the human condition using his engineering skills. The way he talked with respect and admiration for Grace Hopper with my daughters was particularly moving. He described her as "the" person in the world who knew all of the computer users at that time. Whether she actually knew them all, I really don't know, but he saw her as the most personally networked person in the computing world at that time. Grace Hopper was and is an inspirational woman whose pioneering work in computer languages allowed many more people to participate in computing. She

Figure 1.13 Harvard Mark 1 computer team, 1944; Grace Hopper is second from the right in the front row.
Source: © US Air Force/Science Photo Library

created compiling languages that use English rather than machine language. She made computers more accessible to more people, which is what created the acceleration in technology today.

What also struck me from that conversation was how important networking was for the success of the innovation. Connecting people with the right people has been and still is an important part of engineering success. Bernie's network included Grace, and her network included the owners of the computers. He would not have been successful without those personal networks. Today social media allows us to have networks all around the world, with whom we can communicate almost instantaneously. Social media uses the technology that Grace Hopper and others created, and she did it before there were social media networks. Her tradition has continued with an annual conference for women in computing and technology that is named in her honor. This conference is a place where women come and learn from each other and build their own professional and personal networks. Male or female, engineers use professional and personal networks and relationships to make a difference in the world.

Teams and People

This idea of working together with other people is not always celebrated as one of the historical views of engineering. Engineers are people, and there is a diverse range of people in the profession. When I worked in industry, I saw the importance of being able to work with others. I worked on a development program for a new jet engine, and we had over 1,500 engineers working together. There is an enormous need in engineering for people who have diverse backgrounds and perspectives and who can work across these diverse groups. This has actually always been the case, but it is not the way we have always taught engineering and history.

An example comes from one of the greatest inventors in the United States, Thomas Edison. One of his most famous inventions was the lightbulb. When I think back on how I learned about Thomas Edison in school, I had the image of a hermit-like person working alone in his laboratory. That is not the kind of career or life I wanted. It also doesn't match what actually happened: Edison worked with teams of people who made many incremental advances that ultimately resulted in the lightbulb and many other inventions. Along the way, there were many setbacks, lessons learned, and discoveries. His team shared those frustrations and celebrations. His team that developed the lightbulb had dozens of members, including Lewis Latimer, who was the son of freed slaves. Latimer's contributions were to components that helped the team ultimately develop the first working lightbulb.

Lewis Latimer was a recognized inventor in his own right, and he worked on several other technical teams. His collaborators included Alexander Graham Bell, who led the development of the telephone. These inventors worked together on many projects and were colleagues. They also had professional and personal networks. Discussing Lewis Latimer is inspiring to me when I think about how much he gave to our society. It also gives me pause to think that that society allowed his parents to be owned as slaves. Social issues, including injustice, are part of our heritage and history as people and engineers.

The way we have learned history does not always recognize the idea of teams and shared recognition. We tend to want to assign a single name to an invention or accomplishment. There are many reasons why, but what this has done is to hide the contributions of others, and these were often people of color and women who were not socially recognized at the time. What this means is that to truly understand where we have come from, we need to work harder, research, and dig deeper. There were many people who were involved in the history of technology, and today you can use that technology, such as the Internet, to look at websites dedicated to people from different genders, ethnicities, nationalities, religions, and sexual orientations.

Another example of a team approach is the civil engineering accomplishments of the designing and building of the Brooklyn Bridge in New York City (Fig. 1.14). The original lead civil engineer was John Roebling, who died of tetanus while the bridge was being constructed. His son, Washington, assumed the leadership of the construction, but he became very ill and was bedridden. The bridge construction was completed with Emily Roebling (Fig. 1.15), his wife, as the leader. She had learned the

Figure 1.14 Brooklyn Bridge in New York
Source: © Batchelder/Alamy Stock Photo

Figure 1.15 Emily Warren Roebling (1843–1903),
field engineer for the Brooklyn Bridge
Source: © Everett Collection Historical/
Alamy Stock Photo

fundamentals of civil engineering for the bridge completion without attending a
formal college. Emily had been working on the project with her husband, but we may
never have known had he not fallen ill. She worked as the leader for 14 years until the
bridge was completed, but she worked to keep her husband listed as the lead engineer. He still had an important role as she regularly consulted with him, but she was

the one in the field taking the lead. Socially, women were not encouraged to become engineers, and they were often banned from engineering as a field of study, so they had to work as part of a team behind the scenes. The plaque on the Brooklyn Bridge recognizes the contributions of John, Washington, and Emily Roebling for their pioneering feat, but many historical accounts omit Emily. When I look at the Brooklyn Bridge, I see inspiration in many ways.

There are many other examples of teams where certain members may have remained unknown except for special circumstances that brought their contributions to light. One was the design of the city of Washington, D.C. The chief designer was Pierre Charles L'Enfant, but disagreements interrupted the development of the design and he was replaced by Andrew Ellicott and his team. What is less known is that a critical team member was Benjamin Banneker, a descendent of freed slaves, who is credited with saving the project. When L'Enfant departed, he apparently took the copies of the plans for the city. Banneker is credited with recreating the plans totally from memory in just a couple of days (Fig. 1.16). He subsequently worked with Ellicott until illness forced him to withdraw from the project. L'Enfant and Ellicott are credited with the designs, but they were part of a team that could not have been successful without contributions from Benjamin Banneker and others.

This has really always been the case. Think about the complexity of building Machu Picchu and the number of people and coordination it would have required. The pyramids in Egypt and the aqueducts of Rome were developed by large teams of

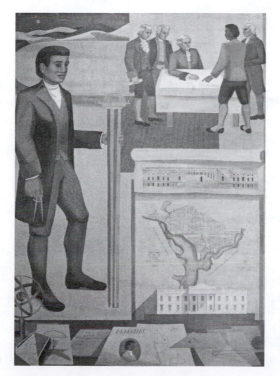

Figure 1.16 Benjamin Banneker, U.S. geographer and surveyor
Source: The George F. Landegger Collection of District of Columbia photographs in Carol M. Highsmith's America/Library of Congress/Science Photo Library

people working together on the designs and construction. These teams have included diverse people who at the times were discriminated against. Benjamin Banneker worked with slave owners, including some of the founders of our country. Emily Roebling would not have been allowed to study engineering formally, yet she learned the trade, and thousands drive over her accomplishment in New York every day. While there are still many struggles and obstacles to overcome, today's society is much more inclusive and there are many more opportunities.

1.6   The History of the Disciplines

As you investigate the many areas of engineering throughout this textbook, you will discover that there are a wide variety of career paths from which to choose. As a prelude to the in-depth look at engineering fields in Chapter 2, this section takes a brief look at the historical backgrounds of the following disciplines:

- Aerospace engineering
- Agricultural engineering
- Chemical engineering
- Civil engineering
- Computer engineering
- Electrical engineering
- Industrial engineering
- Mechanical engineering

From the core areas listed, many additional engineering specialties have evolved over time, for a total today of over 30 different fields in engineering. The information here touches only lightly on the backgrounds of some of these fields. We recommend that you research in greater detail the history of disciplines of interest to you.

The individual disciplines of engineering were actually named only a short time ago relative to the history of the world. What we present here is a short history of the major disciplines dating from after the time they were individually recognized.

Aerospace Engineering

Aerospace engineering is concerned with engineering applications in the areas of *aeronautics* (the science of air flight) and *astronautics* (the science of space flight). Aerospace engineering deals with flight of every kind: balloon flight, sailplanes, propeller- and jet-powered aircraft, missiles, rockets, satellites, and advanced interplanetary concepts such as ion-propulsion rockets and solar-wind vehicles. It is the field of future interplanetary travel. The challenge is to produce vehicles that can traverse the long distances of space in ever-shorter periods of time. New propulsion systems will require that engineers venture into areas never before imagined. By reading histories and biographies, you get a feel for the changes that aerospace pioneers brought about, which will help you to understand the mindset needed to make great leaps forward yourself.

ACTIVITY 1.4	Where do you think the next breakthrough will come for the aerospace engineer? How should this affect the approach we take in the education of the next generation of aerospace engineers? Do you think your generation will walk on Mars?

Agricultural Engineering

At the turn of the 20th century, a large majority of the working population was engaged in agriculture. The mass entry into industrial employment would come a few decades later. This movement to the factories of America lowered the number of workers in agriculture to under 5 percent. This drastic change occurred for a variety of reasons, the principal reason being the integration of technology and engineering into agriculture, which allowed modern farmers to feed approximately 10 times as many families per farmer as their ancestors did 100 years prior.

As you investigate the many facets of agricultural engineering, you will discover that America leads the world in agricultural technology. The world depends upon the United States to feed those in areas of famine, to supply agricultural implements to countries that do not have the resources to perfect such implements, and to continue to perfect technologies to feed more and more of the world's growing population. The agricultural engineer has one of the largest responsibilities in the engineering community. The modern world cannot survive without the efficiencies of a mass-produced food supply.

Agricultural engineering focuses on the following areas:

- Soil and water
- Structures and environment
- Electrical power and processing
- Food engineering
- Power and machinery

ACTIVITIES 1.5	a) Compare the growth of industry with the number of people leaving the farms of America. What innovations in farming helped to offset this exodus? How were we able to continue to feed the population with such a change?
	b) Inspect an early piece of farm machinery. Find a current model of that machine. How are they different? What has not changed in their design?
	c) Farm machinery can be highly dangerous if not used properly. Investigate the dangers posed by farm machinery and how the operators are protected from those dangers.

d) Follow a piece of farm machinery through its many changes from the 17th century to the present day.

e) Investigate where agriculture has gone in the late 20th century. What new aspects has it undertaken?

Chemical Engineering

Chemical engineering is one of the newer disciplines among engineering professions. A closely related field, chemistry, has been studied for centuries in its many forms from alchemy to molecular structure, but chemical engineering began its rise to prominence shortly after World War I. The number of activities involving chemical processing forced engineers and chemists to combine their efforts, bonding the two disciplines together. Chemical engineers needed to understand the way in which chemicals moved and how they could be joined together. This led chemical engineers to begin experimenting with the ways in which mechanisms were used to separate and combine chemicals. Experimenting with mass transfer, fluid flow, and heat transfer proved beneficial. Chemical engineering has become a critical engineering profession.

Chemical engineering applies chemistry to industrial processes that change the composition or properties of an original substance for useful purposes. Chemical engineering is involved in the manufacture of drugs, cements, paints, lubricants, pesticides, fertilizers, cosmetics, foods; in oil refining, combustion, extraction of metals from ores; and in the production of ceramics, brick, and glass. The petrochemical industry is a sizable market for chemical engineers. Chemical engineers can apply their skills in food engineering, process dynamics and control, environmental control, electrochemical engineering, polymer science technology, unit operations, and plant design and economics.

ACTIVITIES 1.6

a) Pick a product that you believe has had to be engineered by a chemical engineer. Explain the process that started with the raw materials and ended with the product you identified. Now look into the recent history of each step of this process. Where have the latest improvements occurred? Try to identify the oldest technique used today in the process. Any book covering the background of this discipline and its products should include such information.

b) Look at the history of the study of chemistry and try to pinpoint those times when the application of chemistry involved chemical engineers.

c) What inventions and innovations did the first chemical engineers develop?

d) Contact a chemical engineering society and collect information on their history, especially about their early members.

Civil Engineering

The 17th and 18th centuries gave birth to most of the major modern engineering disciplines in existence today, but civil engineering is the oldest. Civil engineering traces its roots to early 18th-century France, although the first man definitely to call himself a "civil engineer" was a keen Englishman, John Smeaton, in 1761 (Kirby 1956, p. xvi). Smeaton was a builder of lighthouses who helped distinguish the profession of civil engineering from architecture and military engineering. The surveying of property, the building of roads and canals to move goods and people, and the building of bridges to allow safe passage over raging waters all fall within the parameters of what we now call civil engineering. Construction was the primary focus of the civil engineer, while the military engineer focused on destruction. In America, construction of the earliest railroads began in 1827 and demonstrated the dedication of those early engineers in opening up the West to expansion.

Today, civil engineering focuses primarily on structural issues such as rapid transit systems, bridges, highway systems, skyscrapers, recreational facilities, houses, industrial plants, dams, nuclear power plants, boats, shipping facilities, railroad lines, tunnels, harbors, offshore oil and gas facilities, pipelines, and canals. Civil engineers are heavily involved in improving the movement of populations through their many physical systems. Mass transit development allows civil engineers to protect society and the environment, reducing traffic and toxic emissions from individual vehicles. Highway design provides motorists with safe motorways and access to recreation and work. The civil engineer is always at the center of discussion when it comes to transport and buildings.

ACTIVITIES 1.7

a) Trace the evolution of the first accounts of civil engineering in history. How has it changed? What new areas have entered the field and why did they become part of civil engineering?

b) Research a significant American civil engineering project and explain its importance in the history of the United States.

c) Investigate one of the following and describe in simple terms how it was constructed: a railroad, a bridge, a canal, or a sewer.

d) Research and discuss the connection between mining engineering and civil engineering.

Computer Engineering

Computers play such an important role in engineering design today that no major business can exist comfortably without access to computer technology. Large, medium, and small organizations alike all depend on computers to perform inventories, create billing, record sales, and order new stock. The storage of data alone makes the computer indispensable. The once-mammoth computers of the past have been downsized to the size of your fist, and computing speed today was unthinkable just years ago. Consumers vie with their neighbors to see who can purchase the fastest computer. The world has become almost completely computer-centered, and the computer's role should only increase.

ACTIVITY
1.8
It's time to use your imagination. What will computer engineers of the future be required to do in order to both lead and to keep up with innovations? What new things will computers do?

Electrical Engineering

The amount of historical perspective in this discipline is vast. "The records of magnetic effects date back to remotest antiquity . . . The whole of Electrical Engineering is based on magnetic and electrical phenomenon" (Dunsheath 1962, p. 21). In the late 1700s, experiments concerning electricity and its properties began. By the early 1800s, Volta began his studies of electric conduction through a liquid. Electric current and its movement became the topic of the day. As the need for electricity to power the many devices that were being fabricated by the mechanical engineers grew, the profession of electrical engineering prospered. Direct current, alternating current, the electric telegraph—all became areas to be investigated by early electrical engineers. The profession started with magnetism and electrical phenomena, but it has grown more recently in new directions as new requirements have been imposed by modern society.

Electrical engineering is the largest branch of engineering, employing over 400,000 engineers. Among the major specialty areas in electrical engineering are electronics and solid-state circuitry, communication systems, computers and automatic control, instrumentation and measurements, power generation and transmission, and industrial applications.

ACTIVITIES
1.9
a) Investigate early electrical phenomena and how they were perceived by early humans.
b) Study one pioneer in electrical engineering and show how his or her experiments aided the growth of electrical engineering.
c) What are some of the high points in the history of electrical engineering?

Industrial Engineering

Industrial engineering is a growing branch of the engineering family because of the awareness that the application of engineering principles and techniques can help create better working conditions. Industrial engineers must design, install, and improve systems that integrate people, materials, and equipment to provide efficient production of goods. They must coordinate their understanding of the physical and social sciences with the activities of workers to design areas in which the workers will produce the best results. The daily lives of the people with whom industrial engineers work are closely tied to the designs that they create.

ACTIVITY 1.10 Observe an area where people work. How would you as an industrial engineer improve their working conditions and productivity?

Mechanical Engineering

As coke replaced charcoal in the blast furnaces of England in the early 1700s, the modern mechanical engineering profession dawned. The introduction of coke allowed for larger blast furnaces and a greater ability to use iron. Higher-quality wrought iron could also be produced. These improvements laid the foundation for the Industrial Revolution, during which the production of great quantities of steel was made possible. As these materials became more readily available, mechanical engineers began to design improved lathes and milling and boring machines. Mechanical engineers realized that a wide variety of devices needed to be created to work with the quantities of iron and steel being produced. As the pace of manufacturing increased, the numbers of mechanical engineers also showed a marked increase. The profession would continue to grow in its size and importance. Steam power became vitally important during the Industrial Revolution. The skills of mechanical engineers were needed to create the tools to harness the power of steam. From the steam engine to the automobile, from the automobile to the airplane, from the airplane to the space shuttle, mechanical engineers have been and always will be in great demand for the development of new devices for the betterment of humanity.

Automobiles, engines, heating and air-conditioning systems, gas and steam turbines, air and space vehicles, trains, ships, servomechanisms, transmission mechanisms, radiators, mechatronics, and pumps are a few of the systems and devices requiring mechanical engineering knowledge. Mechanical engineering deals with power, its generation, and its application. Power affects the rate of change or motion of something. This can be a change of temperature or a change of motion due to an outside stimulus. Mechanical engineering is the broadest-based discipline in engineering. The breadth of study required to be a mechanical engineer allows these professionals to diversify into many of the other engineering areas. The major specialty areas of mechanical engineering are applied mechanics; control; design; engines and power plants; energy; fluids; lubrication; heating, ventilation, and air conditioning (HVAC); materials, pressure vessels and piping; and transportation and aerospace.

ACTIVITIES
1.11

a) Look around you. How many things can you find that have been influenced by the work of a mechanical engineer?
b) Investigate the early lathes and explain the improvements made in today's models.
c) Explore in greater depth the effect the iron industry had upon the Industrial Revolution.
d) Who are some of the important names in the early days of the Industrial Revolution, and what did they do?
e) Where would mechanical engineering be without the Industrial Revolution?
f) Pick a period between 1700 and 1999. Look closely at the contributions made by mechanical engineers.

Bioengineering

Bioengineering, the application of biotechnology, is also referred to as biological engineering or biosystems engineering. It involves the application of engineering principles to living things. The U.S. National Institutes of Health defines bioengineering as integrating "physical, chemical, or mathematical sciences and engineering principles for the study of biology, medicine, behavior, or health."

Bioengineers can work in a variety of fields, including medicine and medical device development, drug manufacturing, agriculture, and environmental science. They can work in industry, hospitals, academic institutions, and government agencies, among others. Biomechanical engineers, for instance, help design equipment for surgeons to perform arthroscopic procedures. Other bioengineers work to design better exercise equipment or therapeutic devices. Some work with nanotechnology to repair cell damage or control gene function. Still others build artificial limbs or organs.

Green Engineering

Green engineering is a general term used to refer to the application of engineering principles in an effort to conserve natural resources or to minimize adverse impact on the environment. Efficiency, renewability, and durability are some of the key aspects of green design.

Examples of green engineering include efforts by architects and engineers to develop structures that consume a minimum of energy, both in their construction and in their ongoing operation. This may involve the use of solar energy or other design elements. LEED (Leadership in Energy and Environmental Design) certification involves standards for construction related to energy conservation and can involve material selection, location of manufacture, water use efficiency, indoor air quality, and even landscaping and irrigation.

Green engineering principles are also applied to automobile manufacture, including hybrid vehicles and flex fuels. Other careers involve solar energy generation, thermal engineering, and providing clean water supplies.

1.7 Closing Thoughts

I cannot think of a more impactful profession to be entering today than engineering. I also cannot think of a better and more exciting time to do so. Bernard Amadei, a professor of civil engineering and the founder of Engineers Without Borders, opened a conference on assistive technology with the statement that he believed we were entering a "Golden Age of Engineering." His passion is to bring technology to improve the human condition, especially in the developing world. He sees so many opportunities to use today's technology to address human and environmental issues on a global scale. Technology allows us to team with anyone in the world as we address these challenges in any place in the world. It is an exciting time, but with significant challenges.

Today's exciting engineering opportunities are open to more people than ever before. You could say that "the sky's the limit," but that's not literally true. Engineers like Ellen Ochoa (Fig. 1.17), NASA's first Latina astronaut and the director of flight crew operations, would tell you that the sky is not the limit: She has been farther than that four times herself. Truly, the sky is not the limit for those who want to make a difference in the world.

As we move forward to address the many challenges and opportunities our world presents, we honor and learn from our predecessors individually and as a profession.

Figure 1.17 Ellen Ochoa, NASA's director of flight crew operations
Source: © NG Images / Alamy Stock Photo

They have given us a strong heritage that you can be part of. Part of this heritage is understanding and, sometimes, rediscovering the genius of the past. An example is in the area of sustainability, which has become an important part of engineering design and practice. Many of the sustainability approaches, practices, and philosophies build on practices and innovations from indigenous peoples who balanced advancement with care for the land. Lessons from the past with the technology of today make tomorrow exciting.

REFERENCES

American Society of Civil Engineers, *The Civil Engineer: His Origins*, New York, ASCE, 1970.

Burghardt, M. David, *Introduction to Engineering*, 2nd ed., New York, HarperCollins, 1995.

Burstall, A., *A History of Mechanical Engineering*, London, Faber, 1963.

De Camp, L. Sprague, *The Ancient Engineers*, Cambridge, MA, The MIT Press, 1970.

Dunsheath, P., *A History of Electrical Engineering*, London, Faber, 1962.

Gray, R. B., *The Agricultural Tractor 1855–1950*, St. Joseph, MI, American Society of Agricultural Engineers, 1975.

Greaves, W. F., and J. H. Carpenter, *A Short History of Electrical Engineering*, London, Longmans, Green, and Co. Ltd., 1969.

Kirby, R., et al., *Engineering in History*, New York, McGraw-Hill, 1956.

Miller, J. A., *Master Builders of Sixty Centuries*, Freeport, NY, Books for Libraries Press, 1972.

NASA, Johnson Space Center Director Dr. Ellen Ochoa biography. https://www.nasa.gov/centers/johnson/about/people/orgs/bios/ochoa.html, accessed 11/28/2015.

National Academy of Engineering (NAE), *Changing the Conversation: Messages for Improving Public Understanding of Engineering*, http://www.nae.edu/Publications/Reports/24985.aspx. 2008, accessed 1/13/13.

Progressive Engineer Profile, Bernard Gordon. http://www.progressiveengineer.com/profiles/bernardGordon.htm, accessed 11/22/2015.

Red, W. Edward, *Engineering—The Career and the Profession*, Monterey, CA, Brooks/Cole Engineering Division, 1982.

Rosenberg, S. H., *Rural America a Century Ago*, St. Joseph, MI, American Society of Agricultural Engineers, 1976.

Singer, Charles, ed., *A History of Technology, Volume II*, New York, Oxford University Press, 1956.

Turner, Roland, and Steven L. Goulden, eds., *Great Engineers and Pioneers in Technology, Volume 1: From Antiquity Through the Industrial Revolution*, New York, St. Martin's Press, 1981.

EXERCISES AND ACTIVITIES

1.1   The history of engineering is long and varied. It contains many interesting inventions and refinements. Select one of these inventions and discuss the details of its creation. For example, you might explain how the first printing presses came into being and what previous inventions were used to create the new device.

1.2   Build a simple model of one of the inventions mentioned in this chapter—a bridge, aqueduct, or submarine, for instance. Explain the difficulties of building these devices during the time they were invented.

1.3   Explain how easy it would be to create some inventions of the past using our present-day knowledge and capability.

1.4   Engineering history is filled with great individuals who have advanced the study and practice of engineering. Investigate an area of engineering that is interesting to you and write a detailed report on an individual who made significant contributions in that area.

1.5   Explain what it would have been like to have been an engineer during any particular historical era.

1.6   Compare the lives of any two engineers from the past. Are there similarities in their experiences, projects, and education?

1.7   What kind of education were engineers of old able to obtain?

1.8   What period of engineering history interests you most? Why? Explain why this period is so important in the history of engineering.

1.9   If you had to explain to a 7-year-old child why engineering is important to society, what information from the history of engineering would you relate? Why?

CHAPTER 2
Engineering Majors

2.1 Introduction

Engineers produce things that impact us every day. They invent, design, develop, manufacture, test, sell, and service products and services that improve the lives of people. The Accreditation Board for Engineering and Technology (ABET), which is the national board that establishes accreditation standards for all engineering programs, defines engineering as follows (Landis 1995):

> *Engineering is the profession in which a knowledge of the mathematical and natural sciences, gained by study, experience, and practice, is applied with judgment to develop ways to utilize, economically, the materials and forces of nature for the benefit of mankind.*

Frequently, students early in their educational careers find it difficult to understand exactly what engineers do, and often more to the point, where they fit best in the vast array of career opportunities available to engineers.

Common reasons for a student to be interested in engineering include:

1. Proficiency in math and science
2. Suggested by a high school counselor
3. Has a relative who is an engineer
4. Heard it's a field with tremendous job opportunity
5. Read that it has high starting salaries

While these can be valid reasons, they don't imply a firm understanding of engineering. What is really important is that a student embarking upon a degree program, and ultimately a career, understands what that career entails and the options it presents. We all have our own strengths and talents. Finding places to use those strengths and talents is the key to a rewarding career.

The purpose of this chapter is to provide information about some of the fields of engineering in order to help you decide if this is an area that you might enjoy. We'll explore the role of engineers, engineering job functions, and the various engineering disciplines.

The Engineer and the Scientist

To better understand what engineers do, let's contrast the roles of engineers with those of the closely related field of the scientist. Many students approach both fields for similar reasons: they were good at math and science in high school. While this is a prerequisite for both fields, it is not a sufficient discriminator to determine which is the right career for a given individual.

The main difference between the engineer and the scientist is in the object of each one's work. The scientist searches for answers to technological questions to obtain a knowledge of why a phenomenon occurs. The engineer also searches for answers to technological questions, but always with an application in mind.

Theodore Von Karman, one of the pioneers of America's aerospace industry, said, "Scientists explore what is; engineers create what has not been" (Wright 1994).

In general, science is about discovering things or acquiring new knowledge. Scientists are always asking, "Why?" They are interested in advancing the knowledge base that we have in a specific area. The answers they seek may be of an abstract nature, such as understanding the beginning of the universe, or more practical, such as the reaction of a virus to a new drug.

The engineer also asks, "Why?" but it is because of a problem that is preventing a product or service from being produced. The engineer is always thinking about the application when asking why. The engineer becomes concerned with issues such as the demand for a product, the cost of producing the product, and the impact on society and the environment of the product.

Scientists and engineers work in many of the same fields and industries but have different roles. Here are some examples:

- Scientists study the planets in our solar system to understand them; engineers study the planets so they can design a spacecraft to operate in the environment of that planet.
- Scientists study atomic structure to understand the nature of matter; engineers study the atomic structure in order to build smaller and faster microprocessors.
- Scientists study the human neurological system to understand the progression of neurological diseases; engineers study the human neurological system to design artificial limbs.
- Scientists create new chemical compounds in a laboratory; engineers create processes to mass-produce new chemical compounds for consumers.
- Scientists study the movement of tectonic plates to understand and predict earthquakes; engineers study the movement of tectonic plates to design safer buildings.

The Engineer and the Engineering Technologist

Another profession closely related to engineering is engineering technology. Engineering technology and engineering have similarities, yet there are differences;

they have different career opportunities. ABET, which accredits engineering technology programs as well as engineering programs, defines engineering technology as follows:

> *Engineering technology is that part of the technological field which requires the application of scientific and engineering knowledge and methods combined with technical skills in support of engineering activities; it lies in the occupational spectrum between the craftsman and engineering at the end of the spectrum closest to the engineer.*

Technologists work with existing technology to produce goods for society. Technology students spend time in their curricula working with actual machines and equipment that are used in the jobs they will accept after graduation. By doing this, technologists are equipped to be productive in their occupation from the first day of work.

Both engineers and technologists apply technology for the betterment of society. The main difference between the two fields is that the engineer is able to create new technology through research, design, and development. Rather than being trained to use specific machines or processes, engineering students study additional mathematics and engineering science subjects. This equips engineers to use these tools to advance the state of the art in their field and move technology forward.

There are areas where engineers and engineering technologists perform very similar jobs. For example, in manufacturing settings, engineers and technologists are employed as supervisors of assembly-line workers. Also, in technical service fields both are hired to work as technical support personnel supporting equipment purchased by customers. However, most opportunities are different for engineering and engineering technology graduates:

- The technologist identifies the computer networking equipment necessary for a business to meet its needs and oversees the installation of that equipment; the engineer designs new computer boards to transmit data faster.
- The technologist develops a procedure to manufacture a shaft for an aircraft engine using a newly developed welding technique; the engineer develops the new welding machine.
- The technologist analyzes a production line and identifies new robotic equipment to improve production; the engineer develops a computer simulation of the process to analyze the impact of the proposed equipment.
- The technologist identifies the equipment necessary to assemble a new CD player; the engineer designs the new CD player.
- The technologist identifies the proper building materials and oversees the construction of a new building; the engineer determines the proper support structures, taking into account the local soil, proposed usage, earthquake risks, and other design requirements.

What Do Engineers Do?

Engineering is an exciting field with a vast range of career opportunities. In trying to illustrate the wide range of possibilities, Professors Jane Daniels and Richard Grace from Purdue University constructed the cubic model of Figure 2.1. One edge of the cube represents the engineering disciplines that most students identify as their potential majors. A second edge of the cube represents the different job functions an engineer can have within a specific engineering discipline. The third edge of the cube represents industrial sectors where engineers work. A specific engineering position, such as a mechanical engineering design position in the transportation sector, is the intersection of these three axes. As one can see from the cube, there is a vast number of possible engineering positions.

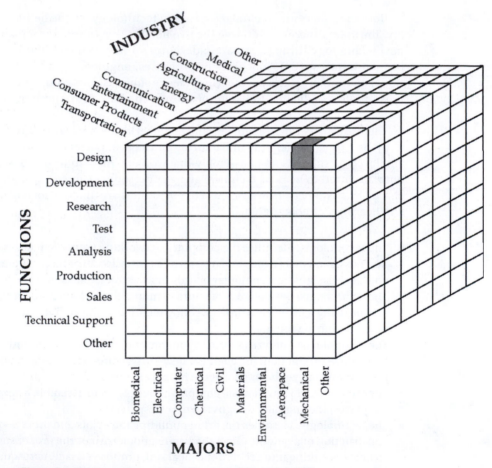

Figure 2.1 Engineering positions (Grace and Daniels 1992)

In the following sections of this chapter, the engineering functions and majors are described. The remaining axis, the industrial sectors, are dependent on the companies and governmental agencies that employ engineers.

To obtain more information about the various industrial sectors:

- Explore your school's placement center
- Visit job fairs
- Attend seminars on campus sponsored by various companies
- Search the Internet (visit websites describing career opportunities)
- Talk to faculty familiar with a certain industry
- Shadow a practicing engineer
- Work as an intern or co-op engineer
- Take an engineering elective course

Most engineering curricula include common courses within an engineering major for the first two years of study. It is not until the junior or senior year that students take technical electives that can be industry specific. Students are encouraged to explore various career opportunities so they can make better decisions when required to select their junior- and senior-level electives.

EXAMPLE 2.1

Let's consider a mechanical engineer (an ME) who performs a design function to illustrate how a specific job in one industrial sector can vary from that in another sector.

- Aerospace—Design of an aircraft engine fan blade: Detailed computer analyses and on-engine testing are required for certification by the Federal Aviation Administration (FAA). Reliability, efficiency, cost, and weight are all design constraints. The design engineer must push current design barriers to optimize the design constraints, potentially making tradeoffs between efficiency, cost, and weight.
- Biomedical—Design of an artificial leg and foot prosthesis giving additional mobility and control to the patient: Computer modeling is used to model the structure of the prosthesis and the natural movement of a leg and foot. Collaboration with medical personnel to understand the needs of the patient is a critical part of the design process. Reliability, durability, and functionality are the key design constraints.
- Power—Design of a heat recovery system in a power plant, increasing the plant productivity: Computer analyses are performed as part of the design process. Cost, efficiency, and reliability are the main design constraints. Key mechanical components can be designed with large factors of safety since weight is not a design concern.
- Consumer products—Design of a pump for toothpaste for easier dispensing: Much of the development work might be done with prototypes. Cost is a main design consideration. Consumer appeal is another consideration and necessitates extensive consumer testing as part of the development process.

- Computer—Design of a new ink-jet nozzle with resolution approaching laser printer quality: Computer analyses are performed to ensure that the ink application is properly modeled. Functionality, reliability, and cost are key design concerns.

2.2   Engineering Functions

Within engineering there are basic classifications of jobs that are common across the various engineering disciplines. What follows are brief descriptions of these different engineering job functions. A few examples are provided for each function. It is important to realize that all the fields of engineering have roles in each of the main functions described here.

Research

The role of the engineering researcher is the closest to that of a scientist of all the engineering functions. Research engineers explore fundamental principles of chemistry, physics, biology, and mathematics in order to overcome barriers preventing advancement in their field. Engineering researchers differ from scientists in that they are interested in the application of a breakthrough, whereas scientists are concerned with the knowledge that accompanies a breakthrough.

Research engineers conduct investigations to extend knowledge using various means. One of the means is conducting experiments. Research engineers may be involved in the design and implementation of experiments and the interpretation of the results. Typically, the research engineer does not perform the actual experiment. Technicians are usually called upon for the actual testing. Large-scale experiments may involve the coordination of additional supporting personnel, including other engineers, scientists, technologists, technicians, and craftspeople.

Research is also conducted using the computer. Computational techniques are developed to calculate solutions to complex problems without having to conduct costly and time-consuming experiments. Computational research requires the creation of mathematical models to simulate the naturally occurring phenomena under study. Research engineers also might develop the computational techniques to perform the complex calculations in a timely and cost-effective fashion.

Most research engineers work for some type of research center. A research center might be a university, a government laboratory such as the National Aeronautics and Space Administration (NASA), or an industrial research center. In most research positions an advanced degree is required, and often a doctoral degree (PhD) is needed. If research appeals to you, a great way to explore it is by doing an undergraduate research project with an engineering professor. This will allow you to observe the operation of a laboratory first hand and find out how well you enjoy being part of a research team.

Development

Development engineers bridge the gap between laboratory research and full-scale production. The development function is often coupled with research in research and development (R&D) divisions. Development engineers take the knowledge acquired by the researchers and apply it to a specific product or application. The researcher may prove something is possible in a laboratory setting; the development engineer shows that it will work on a large, production-size scale and under actual conditions encountered in the field. This is done in pilot manufacturing plants or by using prototypes.

Development engineers are continuously looking for ways to incorporate the findings of researchers into prototypes to test their feasibility for use in tomorrow's products. Often, an idea proven in a laboratory needs to be significantly altered before it can be introduced on a mass-production scale. It is the role of development engineers to identify these areas and work with the design engineers to correct them before full-scale production begins.

An example of a development process is the building of concept cars within the automotive industry. These are unique cars that incorporate advanced design concepts and technology. The cars are then used as a test case to see if the design ideas and technology actually perform as predicted. The concept cars are put through exhaustive tests to determine how well the new ideas enhance a vehicle's performance. Each year, new technology is introduced into production automobiles that was first proven in development groups using concept vehicles.

Testing

Test engineers are responsible for designing and implementing tests to verify the integrity, reliability, and quality of products before they are introduced to the public. The test engineer devises ways to simulate the conditions a product will be subjected to during its life. Test engineers work closely with development engineers in evaluating prototypes and pilot facilities. Data from these initial development tests are used to decide whether full production versions will be made or if significant changes are needed before a full-scale release. Test engineers work with design engineers to identify the changes in the product to ensure its integrity.

A challenge that engineers face is simulating the conditions a product will face during its lifespan, and doing so in a timely, cost-effective manner. Often the conditions the product will face are difficult to simulate in a laboratory. A constant problem for the test engineer is simulating the aging of a product. An example of such a testing challenge is the testing of a pacemaker for regulating a patient's heart, which is designed to last several decades. An effective test of this type cannot take 20 years or the product will be obsolete before it is introduced. The test engineer must also simulate conditions within the human body without exposing people to unnecessary risks.

Other challenges facing test engineers involve acquiring accurate and reliable data. The test engineer must produce data that show the product is functioning properly or that identify areas of concern. Test engineers develop data acquisition and instrumentation methods to achieve this. Techniques such as radiotelemetry may be used to transmit data from the inside of a device being tested. The measurement techniques must not interfere with the operation of the device, presenting a tremendous challenge when dealing with small, compact products. Test engineers must cooperate with design engineers to determine how the device being tested can be fitted with instrumentation yet still meet its design intent.

Test engineers must have a wide range of technical and problem-solving skills. They must also be able to work in teams involving a wide range of people. They work with design and development engineers, technicians, and craftspeople, as well as management.

EXAMPLE 2.2   Test engineers must understand the important parameters of their tests. The development of a certain European high-speed train provides an example of the potential consequences that may result when test engineers fail to understand these parameters. A test was needed to show that the windshield on the locomotive could withstand the high-velocity impacts of birds or other objects it might encounter. This is also a common design constraint encountered in airplane design. The train test engineers borrowed a chicken gun from an aerospace firm for this test. A chicken gun is a mechanism used to propel birds at a target, simulating in-flight impact. With modern laws governing cruelty to animals, the birds are humanely killed and frozen until the test.

On the day of the test, the test engineers aimed the gun at the locomotive windshield, inserted the bird, and fired. The bird not only shattered the windshield but put a hole through the engineer's seat.

The design engineers could not understand what went wrong. They double-checked their calculations and determined that the windshield should have held. The problem became clear after the test engineers reviewed their procedure with the engineers from the aerospace firm from whom they had borrowed the equipment. The aerospace test engineers asked how long they had let the bird thaw.

The response was, "Thaw the bird?"

There is a significant difference in impact force between a frozen eight-pound bird and a thawed bird. The test was successfully completed later with a properly thawed bird.

Design

The design function is what many people think of when they think of engineering, and this is where the largest number of engineers are employed. The design engineer is responsible for providing the detailed specifications of the products society uses.

Rather than being responsible for an entire product, most design engineers are responsible for a component or part of the product. The individual parts are then assembled into a product such as a computer, automobile, or airplane. Design engineers produce detailed dimensions and specifications of the part to ensure that the component fits properly with adjoining pieces. They use modern computer design tools and are often supported by technicians trained in computer drafting software.

The form of the part is also a consideration for the design engineer. Design engineers use their knowledge of scientific and mathematical laws, coupled with experience, to generate a shape to meets the specifications of the part. Often, there is a wide range of possibilities and considerations. In some fields, these considerations are ones that can be calculated. In others, such as in consumer products, the reaction of a potential customer to a shape may be as important as how well the product works.

The design engineer also must verify that the part meets the reliability and safety standards established for the product. The design engineer verifies the integrity of the product. This often requires coordination with analysis engineers to simulate complex products and field conditions, and with test engineers to gather data on the integrity of the product. The design engineer is responsible for making corrections to the design based on the results of the tests performed.

In today's world of ever-increasing competition, the design engineer must also involve manufacturing engineers in the design process. Cost is a critical factor in the design process and may be the difference between a successful product and one that fails. Communication with manufacturing engineers is therefore critical. Often, simple design changes can radically change a part's cost and affect the ease with which the part is made.

Design engineers also work with existing products. Their role includes redesigning parts to reduce manufacturing costs and time. They also work on redesigning products that have not lived up to expected lives or have suffered failure in the field. They also modify products for new uses. This usually requires additional analysis, minor redesigns, and significant communication between departments.

Analysis

Analysis is an engineering function performed in conjunction with design, development, and research. Analysis engineers use mathematical models and computational tools to provide the necessary information to design, development, or research engineers to help them perform their function.

Analysis engineers typically are specialists in a technology area important to the products or services being produced. Technical areas might include heat transfer, fluid flow, vibrations, dynamics, system modeling, and acoustics, among others. They work with computer models of products to make these assessments. Analysis engineers often possess an advanced degree and are experienced in their area of expertise.

To produce the information required of them, they must validate their computer programs or mathematical models. This may require comparing test data to their predictions. This job requires coordination with test engineers in order to design an appropriate test and to record the relevant data.

An example of the role of analysis is the prediction of temperatures in an aircraft engine. Material selection, component life estimates, and design decisions are based in large part on the temperature the parts attain and the duration of those temperatures. Heat transfer analyses are used to determine these temperatures. Engine test results are used to validate the temperature predictions. The permissible time between engine overhauls can depend on these temperatures. The design engineers then use these results to ensure reliable aircraft propulsion systems.

Systems

Systems engineers work with the overall design, development, manufacture, and operation of a complete system or product. Design engineers are involved in the design of individual components, but systems engineers are responsible for the integration of the components and systems into a functioning product.

Systems engineers are responsible for ensuring that the components interface properly and work as a complete unit. Systems engineers are also responsible for identifying the overall design requirements. This may involve working with customers or marketing personnel to accurately determine market needs. From a technical standpoint, systems engineers are responsible for meeting the overall design requirements.

Systems engineering is a field that most engineers enter only after becoming proficient in an area important to the systems, such as component design or development. Some graduate work often is required prior to taking on these assignments. However, there are some schools where an undergraduate degree in systems engineering is offered.

Manufacturing and Construction

Manufacturing engineers turn the specifications of the design engineer into a tangible reality. They develop processes to make the products we use every day. They work with diverse teams of individuals, from technicians on the assembly lines to management, in order to maintain the integrity and efficiency of the manufacturing process.

It is the responsibility of manufacturing engineers to develop the processes for taking raw materials and changing them into the finished pieces that the design engineers detailed. They use state-of-the-art machines and processes to accomplish this. As technology advances, new processes often must be developed for manufacturing the products.

The repeatability or quality of manufacturing processes is an area of increasing concern to modern manufacturing engineers. These engineers use statistical methods to determine the precision of a process. This is important, since a lack of precision in the manufacturing process may result in inferior parts that cannot be used or that may not meet the customer's needs. Manufacturing engineers are very concerned about the quality of the products they produce. High-quality manufacturing means lower costs since more parts are usable, resulting in less waste. Ensuring quality means having the right processes in place, understanding the processes, and working with the people involved to make sure the processes are being maintained at peak efficiency.

Manufacturing engineers also keep track of the equipment in a plant. They schedule and track required maintenance to keep the production line moving. They also must track the inventories of raw materials, partially finished parts, and completely finished parts. Excessive inventories tie up substantial amounts of cash that could be used in other parts of the company. The manufacturing engineer is also responsible for maintaining a safe and reliable workplace, including the safety of the workers at the facility and the environmental impact of the processes.

Manufacturing engineers must be able to work with diverse teams of people, including design engineers, trade workers, and management. Current "just in time" manufacturing practices reduce needed inventories in factories but require manufacturing engineers at one facility to coordinate their operation with manufacturing engineers at other facilities. They also must coordinate the work of the line workers who operate the manufacturing equipment. Manufacturing engineers must maintain a constructive relationship with their company's trade workers.

Manufacturing engineers play a critical role in modern design practices. Since manufacturing costs are such an important component in the success of a product, the design process must take into account manufacturing concerns. Manufacturing engineers identify high-cost or high-risk operations in the design phase of a product. When problems are identified, they work with the design engineers to generate alternatives.

In the production of large items such as buildings, dams, and roads, the production engineer is called a construction engineer rather than a manufacturing engineer. However, the role of the construction engineer is very similar to that of the manufacturing engineer. The main difference is that the construction engineer's production facility is typically outdoors while the manufacturing engineer's is inside a factory. The functions of construction engineers are the same as mentioned above when "assembly line" and "factory" are replaced with terms like "job site," reflecting the construction of a building, dam, or other large-scale project.

Operations and Maintenance

After a new production facility is brought online, it must be maintained. The operations engineer oversees the ongoing performance of the facility. Operations engineers must have a wide range of expertise dealing with the mechanical and electrical issues involved with maintaining a production line. They must be able to interact with manufacturing engineers, line workers, and technicians who service the equipment. They must coordinate the service schedule of the technicians to ensure efficient service of the machinery, minimizing its downtime impact on production.

Maintenance and operations engineers also work in non-manufacturing roles. Airlines have staffs of maintenance engineers who schedule and oversee safety inspections and repairs. These engineers must have expertise in sophisticated inspection techniques to identify all possible problems.

Large medical facilities and other service-sector businesses require operations or maintenance engineers to oversee the operation of their equipment. Obviously, it is critical that emergency medical equipment be maintained in peak working order.

Technical Support

A technical support engineer serves as the link between customer and product and assists with installation and setup. For large industrial purchases, technical support may be included in the purchase price. The engineer may visit the installation site and oversee a successful startup. For example, a new power station for irrigation might require that the technical support engineer supervises the installation and helps the customer solve problems to get the product operational. To be effective, the engineer must have good interpersonal and problem-solving skills as well as solid technical training.

The technical support engineer may also troubleshoot problems with a product. Serving on a computer company's helpline is one example. Diagnosing design flaws found in the field once the product is in use is another example.

Technical support engineers do not have to have in-depth knowledge of each aspect of the product. However, they must know how to tap into such knowledge at their company.

Modern technical support is being used as an added service. Technical support engineers work with customers to operate and manage their own company's equipment as well as others. For example, a medical equipment manufacturer might sell its services to a hospital to manage and operate its highly sophisticated equipment. The manufacturer's engineers would not only maintain the equipment, but also help the hospital use its facilities in the most efficient way.

Customer Support

Customer support functions are similar to those of technical support as a link between the manufacturer and the customer. However, customer support personnel also are involved in the business aspect of the customer relationship. Engineers are often used for this function because of their technical knowledge and problem-solving ability. Typically, these positions require experience with the products and customers and also some business training.

The customer support person works with technical support engineers to ensure proper customer satisfaction. Customer support is also concerned with actual or perceived value for customers. Are they getting what they paid for? Is it cost-effective to continue current practices? Customer support persons are involved in warranty issues, contractual agreements, and the value of trade-in or credits for existing equipment. They work very closely with the technical support engineers and also with management personnel.

Sales

Engineers are valuable members of the sales force in numerous companies. These engineers must have interpersonal skills conducive to effective selling. Sales engineers bring many assets to their positions.

Engineers have the technical background to answer customer questions and concerns. They are trained to identify which products are right for the customer and how they can be applied. Sales engineers can also identify other applications or other products that might benefit the customer once they become familiar with their customer's needs.

In some sales forces, engineers are hired because the customers are engineers themselves and have engineering-related questions. When airplane manufacturers market their aircraft to airlines, they send engineers. The airlines have engineers who have technical concerns overseeing the maintenance and operation of the aircraft. Sales engineers have the technical background to answer these questions.

As technology continues to advance, more and more products become technically sophisticated. This produces an ever-increasing demand for sales engineers.

Consulting

Consulting engineers either are self-employed or work for a firm that does not provide goods or services directly to consumers. Such firms provide technical expertise to organizations that do. Many large companies do not have technical experts on staff in all areas of operation. Instead, they use consultants to handle issues in those technical areas.

For example, a manufacturing facility in which a cooling tower is used in part of the operation might have engineers who are well versed in the manufacturing processes but not in cooling tower design. The manufacturer would hire a consulting firm to design the cooling tower and related systems. Such a consultant might oversee the installation of such a system or simply provide a report with recommendations. After the system is in place or the report is delivered, the consultant would move on to another project with another company.

Consulting engineers also might be asked to evaluate the effectiveness of an organization. In such a situation, a team of consultants might work with a customer and provide suggestions and guidelines for improving the company's processes. These might be design methods, manufacturing operations, or even business practices. While some consulting firms provide only engineering-related expertise, other firms provide both engineering and business support and require the consulting engineers to work on business-related issues as well as technical issues.

Consulting engineers interact with a wide range of companies on a broad scope of projects, and come from all engineering disciplines. Often a consultant needs to be registered as a professional engineer in the state where he or she does business.

Management

In many instances, engineers work themselves into project management positions, and eventually into full-time management. National surveys show that more than half of all engineers will be involved in some type of management responsibilities—supervisory or administrative—before their career is over. Engineers are chosen for their technical ability, their problem-solving ability, and their leadership skills.

Engineers may manage other engineers or support personnel, or they may rise to oversee the business aspects of a corporation. Often, prior to being promoted to this level of management, engineers acquire some business or management training. Some companies provide this training or offer incentives for employees to take management courses in the evening on their own time.

Other Fields

Some engineering graduates enter fields other than engineering, such as law, education, medicine, and business. Patent law is one area in which an engineering or science degree is almost essential. In patent law, lawyers research, write, and file patent applications. Patent lawyers must have the technical background to understand what an invention does so that they can properly describe the invention and legally protect it for the inventor.

Another area of law that has become popular for engineering graduates is corporate liability law. Corporations are faced with decisions every day over whether to introduce a new product, and they must weigh potential risks. Lawyers with technical backgrounds have the ability to weigh technical as well as legal risks. Such lawyers are also used in litigation for liability issues. Understanding what an expert witness is saying in a suit can be a tremendous advantage in a courtroom and can enable a lawyer to effectively cross-examine the expert witness. Often, when a corporation is sued over product liability, the lawyers defending the corporation have technical backgrounds.

Engineers are involved in several aspects of education. The one students are most familiar with is engineering professors. College professors usually have PhDs. Engineers with master's degrees can teach at community colleges and in some engineering technology programs. Engineering graduates also teach in high schools, middle schools, and elementary schools. To do this full-time usually requires additional training in educational methods, but the engineering background is a great start. Thousands of engineers are involved in part-time educational projects where they go to classes as guest speakers to show students what *real* engineers do. The American Society for Engineering Education (ASEE) is a great resource if you are interested in engineering education.

Engineers also find careers in medicine and business, on Wall Street, and in many other professions. In modern society, with its rapid expansion of technology, the combination of problem-solving ability and technical knowledge makes engineers very valuable and extremely versatile.

EXAMPLE
2.3
A dean of a New York engineering school passed along a story about her faculty, which was concerned that most of that year's graduates had not accepted traditional engineering positions. The largest employers were firms that hire stockbrokers. The school's engineering graduates were prized for their ability to look at data and make rational conclusions from the data. In other words, they were very effective problem solvers. They also had the ability to model the data mathematically. This combination made the engineering graduates a perfect fit for Wall Street. This is not an isolated story. Many engineering graduates find themselves in careers unrelated to their engineering education. All engineers are trained problem solvers, and in today's technically advanced society, such skills are highly regarded.

2.3 Engineering Majors

The following section is a partial listing of the various engineering disciplines in which engineering students can earn degrees. The list includes all of the most common majors. Some smaller institutions may offer programs that are not mentioned in this chapter. Be aware that some disciplines are referred to by different names at various institutions. Also, some programs may be offered as a subset of other disciplines. For instance, aeronautical and industrial engineering is sometimes combined with mechanical engineering. Environmental engineering might be offered as part of civil or chemical engineering.

These descriptions are not meant to be comprehensive. To describe a field such as electrical engineering completely would take an entire book by itself. These descriptions are meant to give you an overview of the fields and to answer some basic questions about the differences among the fields. It is meant to be a starting point. When selecting a major, a student might investigate several sources, including people actually working in the field.

An engineering student should keep in mind that the list of engineering fields is fluid. Areas such as aerospace, genetic, computer, and nuclear engineering did not even exist 50 years ago, yet men and women developed the technology to create these fields. In your lifetime, other new fields will be created. The objective is to gain the solid background and tools to handle future challenges. It is also important to find a field you enjoy.

EXAMPLE
2.4
"Show me the money!" As academic advisors, we see students regularly who are interested in engineering but not sure which discipline to pursue. Students are tempted to decide which field to enter by asking which offers the highest salary. There are two issues a student should consider when looking at salaries of engineers.

First, all engineers make a good starting salary—well over the average American household income. However, a high salary would not make up for a job that is hated. The salary spread between the engineering disciplines is not large, varying only by about 10 percent from the average. The career you embark on after graduation from college will span some 40 years—a long time to spend in a discipline you don't really enjoy.

Second, if you consider money a critical factor, you should consider earning potential, not starting salary. Earning potential in engineering is dependent, to a large extent, on job performance. Engineers who do well financially are those who excel professionally. Again, it is very rare for someone to excel in a field that he or she does not enjoy.

Aerospace Engineering

Aerospace engineering is also referred to as aeronautical or astronautical engineering. Technically, aeronautical engineering involves flight within the Earth's atmosphere and had its birth when two bicycle repairmen from Dayton, Ohio, made the first flight at Kitty Hawk, North Carolina, in 1903. Since that time, aerospace engineers have designed and produced aircraft that have broken the sound barrier, achieved near-orbit altitudes, and become the standard mode of transportation for long journeys. Astronautical engineering refers to flight outside of the Earth's atmosphere in space, which began on October 4, 1957, when the Soviet Union launched Sputnik into orbit. The United States achieved one of its proudest moments in the field of aerospace on July 20, 1969, when Neil Armstrong set foot on the moon's surface—the first person ever to do so. To describe what the broad field of aerospace engineering has become, it can be separated into the categories of aerodynamics, propulsion, structures, controls, orbital mechanics, and life sciences.

Aerodynamics is the study of the flow of air over a streamlined surface or body. The aerodynamicist is concerned with the lift that is produced by a body such as the wing of an aircraft. Similarly, there is a resistance or drag force when a body moves through a fluid. An engineer looks to optimize lift and minimize drag. While mechanical, civil, and chemical engineers also study flow over bodies, the aerospace engineer is interested especially in high-speed airflows. When the speed of the aircraft approaches or exceeds the speed of sound, the modeling of the airflow becomes much more complex. Aerodynamics plays an important role in power generation turbines, including wind turbines. Aerospace engineers are often involved in their designs.

Air-breathing propulsion systems that power airplanes and helicopters include propellers and gas turbine jets, as well as ramjets and scram-jets. Propulsion systems used to launch or operate spacecraft include liquid and solid rocket engines. Propulsion engineers are continually developing more efficient, quieter, and cleaner-burning conventional engines as well as new engine technologies. New engine concepts include wave rotors, electric propulsion, and nuclear-powered craft.

The structural support of an aircraft or spacecraft is critical. It must be both lightweight and durable. These conditions are often mutually exclusive and provide challenges for structural design engineers. Structural engineers use new alloys, composites, and other new materials to meet design requirements for more efficient and more maneuverable aircraft and spacecraft.

The control schemes for aircraft and spacecraft have evolved rapidly. The new commercial airplanes are completely digitally controlled. Aerospace engineers work with electrical engineers to design control systems and subsystems used to operate the craft. They must be able to understand the electrical and computational aspects of the control schemes as well as the physical systems that are being controlled.

Aerospace engineers are also interested in orbital mechanics. They calculate where to place a satellite to operate as a communication satellite or as a global positioning system (GPS). They might also determine how to use the gravity fields of the near-Earth planets to help propel a satellite to the outskirts of the solar system.

Aerospace engineers must be aware of human limitations and the effects of their craft on the human body. It would serve no useful purpose to design a fighter plane that was so maneuverable it incapacitated the pilot with a high G-force. Designing a plane and a system that keeps the pilot conscious is an obvious necessity. Understanding the physiological and psychological effects of lengthy exposure to weightlessness is important when designing a spacecraft or space station.

The American Institute of Aeronautics and Astronautics (AIAA) is one of the most prominent aerospace professional societies. It is composed of the following seven technical groups, which include 66 technical committees:

- Engineering and Technology Management
- Aircraft Technology Integration and Operations
- Propulsion and Energy
- Space and Missile Systems
- Aerospace Sciences
- Information and Logistics Systems
- Structures, Design, and Testing

Agricultural Engineering

Agricultural engineering traces its roots back thousands of years to the time when people began to examine ways to produce food and food products more efficiently. Today, the role of the agricultural engineer is critical to our ability to feed the ever-expanding population of the world. The production and processing of agricultural products is the primary concern of agricultural engineers. Areas within agricultural engineering include power machinery, food processing, soils and water, structures, electrical technologies, and bioengineering.

The mechanical equipment used on modern farms is highly sophisticated and specialized. Harvesting equipment not only removes the crop from the field but also begins to process it and provides information to the farmers on the quality of the harvest. This requires highly complicated mechanical and electrical systems that are designed and developed by the agricultural engineer.

Often, once the food is harvested it must be processed before it reaches the marketplace. Many different technologies are involved in the efficient and safe processing and delivery of food products. Food process engineers are concerned with providing healthier products to consumers who increasingly rely on processed food products. This often requires the agricultural engineer to develop new processes. Another modern-day concern is increased food safety. Agricultural engineers design and develop means by which food is produced free of contamination, such as the irradiation techniques used to kill potentially harmful bacteria in food.

The effective management of soil and water resources is a key aspect of productive agriculture. Agricultural engineers design and develop means to address effective land use, proper drainage, and erosion control. This includes such activities as designing and implementing an irrigation system for crop production and designing a terracing system to prevent erosion in a hilly region.

Agricultural structures are used to house harvested crops, livestock, and their feed. Agricultural engineers design structures including barns, silos, dryers, and processing centers. They look for ways to minimize waste or losses, optimize yields, and protect the environment.

Electrical and information technology development is another area important to the agriculture community due to the fact that farms are typically located in isolated regions. Agricultural engineers design systems that meet the needs of the rural communities.

Bioengineering has rapidly evolved into a field with wide uses in health products as well as agricultural products. Agricultural engineers are working to harness these rapidly developing technologies to further improve the quality and quantity of the agricultural products necessary to continue to feed the world's population.

The American Society of Agricultural Engineers (ASAE) is one of the most prominent professional societies for agricultural engineers. It has eight technical divisions that address areas of agricultural engineering:

- Food Processing
- Information and Electrical Technologies
- Power and Machinery
- Structures and Environmental
- Soil and Water
- Forest
- Bioengineering
- Aquaculture

Architectural Engineering

In ancient times, major building projects required a master builder who was responsible for the design and appearance of the structure, for selecting the appropriate materials, and for personally overseeing construction. With the advent of steel construction in the 19th century, it became necessary for master builders to consult with specialists in steel construction in order to complete their designs. As projects became more complex, other specialists were needed in such areas as mechanical and electrical systems. Modern architectural engineers facilitate the coordination between the creativity of the architect and the technical competencies of a variety of technology specialists.

Architectural engineers are well grounded in the engineering fundamentals of structural, mechanical, and electrical systems and have at least a foundational understanding of aesthetic design. They know how to design a building and how the various technical systems are interwoven within that design. Their strengths are both creative and pragmatic.

The National Society of Architectural Engineers, which is now the Architectural Engineering Institute, defined architectural engineering as

the profession in which a knowledge of mathematics and natural sciences, gained by study, experience, and practice, is applied with judgment to the development of ways to use, economically and safely, the materials and forces of nature in the engineering design and construction of buildings and their environmental systems. (Belcher 1993)

There are four main divisions within architectural engineering: structural; electrical and lighting; mechanical systems; and construction engineering and management.

Structural is primarily concerned with the integrity of the structure of buildings. Determining the integrity of a building's structure involves the analysis of loads and forces resulting from normal usage and operation as well as from earthquakes, wind forces, snow loads, or wave impacts. The structural engineer takes the information from the analysis and designs the structural elements that will support the building while at the same time meeting the aesthetic and functional needs involved.

The area of electrical and lighting systems is concerned with the distribution of utilities and power throughout the building. This requires knowledge of electricity and power distribution as well as lighting concerns. Lighting restrictions may require energy-efficient lighting systems that are sufficient to meet the functional requirements related to use of the building. The architectural engineer also must ensure that the lighting will complement the architectural designs of the building and the rooms within that structure.

Mechanical systems control the climate of a building, which includes cooling and heating the air in rooms as well as controlling humidity and air quality. Mechanical systems are also used to distribute water through plumbing systems. Other mechanical systems include transportation by way of elevators and escalators within a building. These systems must be integrated to complement the architectural features of the building.

The fourth area is construction engineering and management, which combine the technical requirements with the given financial and legal requirements to meet project deadlines. The architectural engineer is responsible for implementing the design in a way that both ensures the quality of the construction and meets the cost and schedule of the project. Modern computer tools and project-management skills are implemented to manage such complex projects.

Other areas that are emerging for architectural engineers include energy management, computerized controls, and new building materials, including plastics and composites and acoustics. Many programs have added explicit areas of emphasis in the application of alternative energy, energy efficiency, and sustainability into architectural engineering programs. Designing structures to reduce energy use, material consumption, and environmental impact in the design, construction, and operation phases of buildings is a growing area of interest and a topic of a great deal of research within the field.

Architectural engineers are employed by consulting firms, contractors, and government agencies. They may work on complex high-rise office buildings, factories, stadiums, research labs, or educational facilities. They also may work on renovating historic structures or developing affordable low-income housing. As the construction industry continues to grow and as projects continue to become more complex, the outlook for architectural engineers is bright.

There are fewer than 20 ABET-accredited programs in architectural engineering in the United States. Some of these are four-year programs, while others provide the combined engineering foundation and architectural insights in a five-year program.

The Architectural Engineering Institute (AEI) was formed in 1998 with the merger of the National Society of Architectural Engineers and the American Society of Civil Engineers Architectural Engineering Division. AEI was created to be the home for professionals in the building industry. It is organized into divisions, which include:

- Commercial Buildings
- Industrial Buildings
- Residential Buildings
- Institutional Buildings
- Military Facilities
- Program Management
- Building Systems
- Education
- Architectural Systems
- Fully Integrated and Automated Project Process
- Glass as an Engineered Material
- Mitigation of the Effects of Terrorism
- Sick/Healthy Buildings
- Designing for Facilities for the Aging (Belcher 1993)

Biological Engineering

Biological engineering, one of the newer fields of engineering, spans the fields of biology and engineering. The integration of biology and engineering is one of the new frontiers for engineering, and biological engineers are at the leading edge of those efforts. Traditionally, engineering has applied concepts from mathematics and the physical sciences to design, analyze, and manufacture inanimate products and processes to address human and environmental needs. Biological engineering, also called bioengineering or biotechnological engineering, integrates the study of living systems into engineering, opening opportunities to adapt and even create living products. A search of ABET's website (http://main.abet.org/aps/accreditedprogram-search.aspx, accessed July 2016) for accredited biological engineering shows growing number of accredited programs. One example is at the Massachusetts Institute of Technology, whose website defines this new discipline as follows:

> *The goal of this biological engineering discipline is to advance fundamental understanding of how biological systems operate and to develop effective biology-based technologies for applications across a wide spectrum of societal needs including breakthroughs in diagnosis, treatment, and prevention of disease, in design of novel materials, devices, and processes, and in enhancing environmental health.* (http://be.mit.edu/, accessed July 2016)

This definition shows the breadth of and opportunities for this emerging field. Biological engineering is being practiced as separate programs as well as integrated with more traditional engineering majors at many schools. Some schools, for example, include biological engineering as a complement to their agricultural engineering programs to differentiate the parts of their curriculum and research programs that address new opportunities in biological areas that have been opened via modern technology. Another example is the inclusion of bioengineering as part of biomedical engineering programs. Similar to other fields of engineering that have grown out of the more traditional disciplines, biological engineering is evolving rapidly. At this point in the discipline's development, this means that the term "biological engineering" may mean slightly different things at different institutions. Students are encouraged to research individual programs to understand what this exciting and emerging field means at their respective institution.

Biomedical Engineering

Biomedical engineering is one of the newer fields of engineering, first recognized as a discipline in the 1940s. Its origins, however, date back to the first artificial limbs, which were made of wood or other materials. Captain Ahab in *Moby Dick* probably didn't think of the person who made his wooden leg as a biomedical engineer, yet he was a predecessor of today's biomedical engineers who design modern prosthetics.

Biomedical engineering is a very broad field that overlaps with several other engineering disciplines. In some institutions, it may be a specialization within another discipline. Biomedical engineering applies the fundamentals of engineering to meet the needs of the medical community. Because of the wide range of skills needed in the fields of engineering and medicine, biomedical engineering often requires graduate work. The broad field of biomedical engineering encompasses the three basic categories of bioengineering, medical, and clinical.

Bioengineering is the application of engineering principles to biological systems. This can be seen in the production of food or in genetic manipulation to produce a disease-resistant strain of a plant or animal. Bioengineers work with geneticists to produce bioengineered products. New medical treatments are produced using genetically altered bacteria to produce human proteins needed to cure diseases. Customized drugs and delivery methods can also be developed for individual patients and diseases. Nanotechnology is important in these endeavors to create customized molecules and structures. Bioengineers may work with geneticists in producing these new products in the mass quantities needed by consumers.

The medical aspect of biomedical engineering involves areas such as the development of the chemical processes necessary to make an artificial kidney function and the electrical challenges in designing a new pacemaker. It also involves the design and development of devices to solve medical challenges, such as a prosthesis or artificial joints to replace arthritic joints, allowing individuals increased mobility and reduced pain. Bioengineers are also developing new ways to grow new cartilage that can eliminate traumatic surgeries required to insert artificial joints or new skin to treat burn victims.

Medical engineers develop instrumentation for medical uses, including noninvasive surgical instruments. Much of the trauma of surgery results from the incisions made to gain access to the area of concern. Procedures allowing a surgeon to make a small incision, such as orthoscopic surgery, have greatly reduced risk and recovery time for patients. New imaging and sensor technology is being used to identify and treat many diseases and conditions such as brain trauma and concussions.

Rehabilitation is another area in which biomedical engineers work. It involves designing and developing devices for the physically impaired. Such devices can expand the capabilities of impaired individuals, thereby improving their quality of life or shortening recovery times.

Clinical engineering involves the development of systems to serve hospitals and clinics. Such systems exhaust anesthetic gases from an operating room without impairing the surgical team. Air lines in ventilating systems must be decontaminated to prevent microorganisms from spreading throughout the hospital. Rehabilitation centers must be designed to meet the needs of patients and staff.

Chemical Engineering

Chemical engineering differs from most of the other fields of engineering in its emphasis on chemistry and the chemical nature of products and processes. Chemical engineers take what chemists do in a laboratory and, applying fundamental engineering, chemistry, and physics principles, design and develop processes to mass produce products for use in our society. These products include detergents, paints, plastics, fertilizers, petroleum products, food products, pharmaceuticals, electronic circuit boards, and many others.

The most common employment of chemical engineers is in the design, development, and operation of large-scale chemical production facilities. In this area, the design function involves the design of the processes needed to safely and reliably produce the final product. This may involve controlling and using chemical reactions, separation processes, or heat and mass transfer. While the chemist might develop a new compound in the laboratory, the chemical engineer would develop a new process to make the compound in a pilot plant, which is a small-scale version of a full-size production facility. An example would be the design of a process to produce a lower-saturated-fat product with the same nutritional value yet still affordable. Another example would be the development of a process to produce a higher-strength plastic used for automobile bumpers.

With respect to energy sources, chemical engineers are involved in the development of processes to extract and refine crude oil and natural gas. Petroleum engineering grew out of chemical engineering and is described in a separate section. Chemical engineers are also involved in alternative fuel development and production.

The processing of petroleum into plastics by chemical engineers has created a host of consumer products that are used every day. Chemical engineers are involved in adapting these products to meet new needs. For example, a research house in Massachusetts is made entirely from plastic. It is used as a research facility to develop new building materials that are cheaper and more beneficial to the consumer and the environment.

Many of the chemicals and their byproducts used in industry can be dangerous to people and/or the environment. Chemical engineers must develop processes that minimize harmful waste. They work with both new and traditional processes to treat hazardous byproducts and reduce harmful emissions.

Chemical engineers are also very active in the bioproducts arena. This includes the pharmaceutical industry, where chemical engineers design processes to manufacture new lines of affordable drugs. Geneticists have developed the means to artificially produce human proteins that can be used to treat many diseases. They use genetically altered bacteria to produce these proteins. It is the job of the chemical engineer to take the process from the laboratory and apply it to a larger-scale production process to produce the quantities needed by society.

Chemical engineers are also involved in bioprocesses such as dialysis, where chemical engineers work along with other biomedical engineers to develop new ways to treat people. Chemical engineers might be involved in the development of an artificial kidney. They could also be involved in new ways to deliver medicines, such as through skin implants or patches.

Chemical engineers have become very active in the production of circuit boards, such as those used in computers. To manufacture the very small circuits required in modern electronic devices, material must be removed and deposited very precisely along the path of the circuit. This is done using chemical techniques developed by chemical engineers. With the demand for smaller and faster electronics, challenges continue for chemical engineers to develop the processes to make them possible.

Chemical engineers are also involved in the modeling of systems. Computer models of manufacturing processes are made so that modifications can be examined without having to build expensive new facilities. Managing large processes involving facilities that might be located around the globe requires sophisticated scheduling capabilities. Chemical engineers who developed these capabilities for chemical plants have found that other large industries with scheduling challenges, such as airlines, can use these same tools.

The American Institute of Chemical Engineering (AIChE) is one of the most prominent professional organizations for chemical engineers. It is organized into 13 technical divisions that represent the diverse areas of the chemical engineering field:

- Catalysis and Reaction Engineering
- Computing and Systems Technology
- Engineering and Construction Contracting
- Environmental
- Food, Pharmaceutical, and Bioengineering
- Forest Products
- Fuels and Petrochemicals
- Heat Transfer and Energy Conversion
- Management
- Materials Engineering and Sciences
- Nuclear Engineering
- Safety and Health
- Separations

Civil Engineering

Ancient examples of early civil engineering can be seen in the pyramids of Egypt, the Roman roads, bridges, and aqueducts of Europe, and the Great Wall in China. These all were designed and built under the direction of the predecessors to today's civil engineers. Modern civil engineering continues to face the challenges of meeting the needs of society. The broad field of civil engineering includes these categories: structural, environmental, transportation, water resources, surveying, urban planning, and construction engineering.

Structural engineers are the most common type of civil engineer. They are primarily concerned with the integrity of the structure of buildings, bridges, dams, and highways. Structural engineers evaluate the loads and forces to which a structure will be subjected. They analyze the structural design in regard to earthquakes, wind forces, snow loads, or wave impacts, depending on the area in which the building will be constructed. Modern structures incorporate sensors that can detect earthquakes and change the characteristics of the building to increase its tolerance to the motion of an earthquake.

A related field of structural engineering is architectural engineering, which is concerned with the form, function, and appearance of a structure. The architectural engineer works alongside the architect to ensure the structural integrity of a building. The architectural engineer combines analytical ability with the concerns of the architect.

Civil engineers in the environmental area may be concerned with the proper disposal of wastes—residential and industrial. They may design and adapt landfills and waste treatment facilities to meet community needs. Industrial waste often presents a greater challenge because it may contain heavy metals or other toxins that require special disposal procedures. Environmental engineering has come to encompass much more and is detailed in a later section of this chapter.

Transportation engineers are concerned with the design and construction of highways, railroads, and mass transit systems. They are also involved in the optimization and operation of the systems. An example of this is traffic engineering. Civil engineers develop the tools to measure the need for traffic control devices such as signal lights and to optimize these devices to allow proper traffic flow. This can become very complex in cities where the road systems were designed and built long ago but the areas around those roads have changed significantly with time. Modern designs use sensors to detect vehicles and traffic patterns. Cellphones emit signals to communicate with the cellular towers, and these produce an enormous amount of data that transportation engineers can use to take real-time traffic measurements to allow systems to adjust to traffic patterns, congestion, and accidents.

Civil engineers also work with water resources as they construct and maintain dams, aqueducts, canals, and reservoirs. Water resource engineers are charged with providing safe and reliable supplies of water for communities. This includes the design and operation of purification systems and testing procedures. As communities continue to grow, so do the challenges for civil engineers to produce safe and reliable water supplies.

Before any large construction project can be started, the construction site and surrounding area must be mapped or surveyed. Surveyors locate property lines and establish alignment and proper placement of engineering projects. Modern surveyors use satellite technology as well as aerial and terrestrial photogrammetry. They also rely on computer processing of photographic data.

A city is much more than just a collection of buildings and roads. It is a home and working place for people. As such, the needs of the inhabitants must be taken into account in the design of the city's infrastructure. Urban planning engineers are involved in this process. They incorporate the components of a city (buildings, roads, schools, airports, etc.) to meet the overall needs of the population. These needs include adequate housing, efficient transportation, and open spaces. The urban planning engineer is always looking toward the future to fix problems before they reach a critical stage.

Another civil engineering area is construction engineering. In some institutions, this may even be a separate program. Construction engineers are concerned with the management and operation of construction projects. They are also interested in the improvement of construction methods and materials. Construction engineers design and develop building techniques and building materials that are safe, reliable, cost-effective, and environmentally friendly. There are many technical challenges that construction engineers face and will continue to face in the coming decades.

An example of one such construction challenge is the rebuilding of a city's infrastructure, such as its sewers. While the construction of a sewer may not seem glamorous or state of the art, consider the difficulty of rebuilding a sewer under an existing city. A real problem facing many cities is that their infrastructure was built decades or centuries ago, but it has a finite life. As these infrastructures near the end of their expected lives, the construction engineer must refurbish or reconstruct these systems without totally disrupting the city.

The American Society of Civil Engineers (ASCE) is one of the most prominent professional organizations for civil engineers. It is organized into the following 16 technical divisions covering the breadth of civil engineering:

- Aerospace
- Air Transport
- Architectural Engineering
- Construction Division
- Energy
- Engineering Mechanics
- Environmental Engineering
- Geomatics
- Highway
- Materials Engineering
- Pipeline
- Urban Planning and Development
- Urban Transportation
- Water Resources Engineering
- Water Resources Planning and Management
- Waterways, Ports, Coastal, and Ocean Engineering

Computer Engineering

Much of what computer and electrical engineers do overlaps. Many computer engineering programs are part of electrical engineering or computer science programs. However, computer technology and development have progressed so rapidly that specialization has been required in this field, and thus computer engineering is treated separately. This is one of the fastest-evolving fields as computer technology drives more innovation and what is defined as a computer changes. "Computer" once meant a desktop, a laptop, and/or large servers, but the term has evolved to include mobile phones and the integration of computing into everything from household appliances to smart buildings. Given the wide range of computer applications and their continued growth, computer engineering has a very dynamic future.

Computer engineering is similar to computer science, yet distinct. Both fields are extensively involved with the design and development of software. The main difference is that the computer scientist focuses primarily on the software and its optimization. The computer engineer, by contrast, focuses primarily on computer hardware—the machine itself. Software written by the computer engineer is often designed to control or to interface more efficiently with the hardware of the computer and its components.

Computer engineers are involved in the design and development of operating systems, compilers, and other software that requires efficient interfacing with the components of the computer. These attributes are important whether the device is a traditional computer, a tablet, a smartphone, or some other computing device. Computer engineers also work to improve computing performance by optimizing the software and hardware in applications such as graphic displays.

Computer engineers work on the design of computer architecture. Designing faster and more efficient computing systems is a tremendous challenge as faster and smaller processors are developed. Applications for even faster processors continue to increase in traditional computers and in a host of other applications, including mobile devices, automobile technologies, household appliances, and the healthcare field. The continual quest for faster and smaller microprocessors involves overcoming barriers introduced by the speed of light and by circuitry so small that the molecular properties of components become important. Computer engineers are at the forefront of nanotechnology to produce modern microprocessors and microelectronics.

Computer engineers develop and design electronics to interface with computers. These include Wi-Fi systems, Bluetooth Ethernet connections, and other means of data transmission. Computer engineers created the devices that made the Internet possible as well as mobile communication and data transmission.

In addition to having computers communicate with each other, the computer engineer is also interested in having computers work together. This may involve increasing the communication speed of computer networks between each other or between other devices. The cloud computing concept integrates and uses multiple devices that are connected and coordinated. Modern computing devices have chips or processors that are actually multiple processors linked together so that they can perform calculations in parallel to increase their speed.

Security is an enormous issue as more and more information is transferred using computers. Computer engineers are developing new means of commercial and personal security to protect the integrity of electronic communications, including advanced encryption and biometric security.

Artificial intelligence, voice recognition systems, and touchscreens are examples of other technologies with which computer engineers are involved. The applications that computer engineers work on today range from your mobile phone, to your car, to your home and much more. The devices that we use for computing today look very different than those from even a decade ago, and they will continue to change and evolve—and it is the computer engineer who will make this happen.

Electrical Engineering

Considering the wide range of electronic devices people use every day, it is not surprising that electrical engineering has become the most populated of the engineering disciplines. Electrical engineers have a wide range of career opportunities in almost all of the industrial sectors. To provide a brief discussion of such a broad field, we will divide electrical engineering into eight areas: computers, communications, circuits and solid-state devices, control, instrumentation, signal processing, bioengineering, and power.

Engineers specializing in computer technology are in such high demand that numerous institutions offer a separate major for computer engineering. Please refer back to the earlier section that described computer engineering separately.

Electrical engineers are responsible for the explosion in communication technologies, including mobile devices and smartphones. Cellular networks have become a dominant form of communication that has changed the way we live, work, and communicate with each other. Wireless technologies have allowed developing countries to bypass traditional phone lines and set up national communication systems with mobile networks. Satellites provide nearly instantaneous global communication. Anyone with a GPS-connected device can pinpoint precisely where he or she is located anywhere in the world. Widespread access to GPS through mobile devices has changed the way we navigate our planes, boats, and cars. Fiberoptics and lasers are rapidly improving the reliability and speed with which information can be exchanged. Wireless communication allows people to communicate anywhere, with anyone. Communication between large servers on cloud systems and smaller mobile devices dramatically increases the ability to perform computing over a wide network. Future breakthroughs in this field will have a tremendous impact on how we live in tomorrow's society.

Electrical engineers also design and develop electronic circuits. Circuit design has changed rapidly with the advent of microelectronics and nanotechnology. Nanotechnology plays an important part of modern circuit designs as the thickness of the wires, now called transfer points, within circuits has shrunk to the point where their thickness is measured in atoms. This significantly changes the physics of the materials and the properties of the systems. As the limits of current technology are approached, there will be incentives to develop new and faster ways to accomplish the same tasks.

Almost all modern machines and systems are digitally controlled. Digital controls allow for safer and more efficient operation. Electronic systems monitor processes and make corrections faster and more effectively than human operators can. This improves reliability, efficiency, and safety. The electrical engineer is involved in the design, development, and operation of these control systems. The engineer must determine the kind of control required, what parameters to monitor, the speed of the correction, and many other factors. Control systems are used in chemical plants, power plants, automotive engines, airplanes, healthcare systems, and a variety of other applications.

For a control system to operate correctly, it must be able to measure the important parameters of whatever it is controlling. Doing so requires accurate instrumentation, another area for the electrical engineer. Electrical engineers who work in this area develop electrical devices to measure quantities such as pressure, temperature, flow rate, speed, heart rate, and blood pressure. Often, the electrical devices convert the measured quantity to an electrical signal that can be read by a control system or a computer. Instrumentation engineers also design systems to transmit the measured information to a recording device, using telemetry. For example, such systems are needed for transmitting a satellite's measurements to the recording computers back on Earth as it orbits a distant planet.

Signal processing is another area where electrical engineers are needed. In many instances the electrical signals coming from instrumentation or other sources must be conditioned before the information can be used. Signals may need to be electronically filtered, amplified, or modified. An example is the active noise control system on a stethoscope that allows a paramedic to listen to a patient's heart while in a helicopter so the paramedic can make a quick, accurate assessment of the patient. Similar technology is used in modern hearing aids to allow users to be able to hear in crowded and noisy rooms. Another example is your ability to talk to your smartphone.

Image processing has had a significant impact on our world. Modern medical imaging has become an important tool in the diagnosis and treatment of a wide range of diseases. It has dramatically increased the ability of doctors to examine, identify, and diagnose conditions without invasive surgeries. Facial recognition is used in many applications from airports to border crossings. As computing capabilities continue to advance, image recognition and processing will become more important to more aspects of our lives.

Electrical engineers also work in biomedical or bioengineering applications, as described earlier. Electrical engineers work with medical personnel to design and develop devices used in the diagnosis and treatment of patients. Examples include noninvasive techniques for detecting tumors through magnetic resonance imaging (MRI) or computed tomography (CT) scans. Other examples include pacemakers, cardiac monitors, and controllable prosthetic devices.

The generation, transmission, and distribution of electric power are, in some ways, the most traditional aspects of electrical engineering, but they are also areas of innovation for the future, as power demands increase and the need for smart electrical grids increases. Electrical engineers work closely with mechanical engineers in the production of electrical power. They also oversee the distribution of power through electrical networks and must ensure reliable supplies of electricity to our communities. With modern society's dependence on electricity, interruptions in the flow of electricity can be catastrophic. Smart grid systems are used to effectively distribute electricity and to shift supply to meet changing demands. Power engineers are at the forefront of alternative energy in the development of solar energy and the conversion of wind, hydroelectric, wave, and other sources of energy into electricity. Smart grid systems are needed to integrate these sources into the overall electrical system. Smart systems allow producers as small as individual homeowners to provide energy back into the main electrical grid when they have excess capacity. Electrical engineers also work with materials engineers to incorporate superconductivity and other technology in more efficient power transmission.

The Institute of Electrical and Electronics Engineers (IEEE) is the largest and most prominent professional organization for electrical engineers. It is organized into 37 technical divisions, which indicates the breadth of the field of electrical engineering:

- Aerospace and Electronic Systems
- Antennas and Propagation
- Broadcast Technology
- Circuits and Systems
- Communications

- Components Packaging and Manufacturing Technology
- Computer
- Consumer Electronics
- Control Systems
- Dielectrics and Electrical Insulation
- Education
- Electromagnetic Compatibility
- Electron Devices
- Engineering in Medicine and Biology
- Engineering Management
- Instrumentation and Measurement
- Lasers and Electro-Optics
- Magnetics
- Microwave Theory and Techniques
- Neural Networks
- Nuclear and Plasma Sciences
- Oceanic Engineering
- Power Electronics
- Power Engineering
- Professional Communication
- Reliability
- Robotics and Automation
- Signal Processing
- Social Implications of Technology
- Solid-State Circuits
- Geoscience and Remote Sensing
- Industrial Electronics
- Industrial Applications
- Information Theory
- Systems, Man, and Cybernetics
- Ultrasonics, Ferroelectrics, and Frequency Control
- Vehicular Technology

Environmental Engineering

Environmental engineering is a field that has evolved to improve and protect the environment while maintaining the rapid pace of industrial activity. This challenging task has three parts to it: disposal, remediation, and prevention.

Disposal is similar to that covered under civil engineering. Environmental engineers are concerned with disposal and processing of both industrial and residential waste. Landfills and waste treatment facilities are designed for residential waste concerns. The heavy metals and other toxins found in industrial wastes require special disposal procedures. The environmental engineer develops the techniques to properly dispose of such waste.

Remediation involves the cleaning up of a contaminated site. Such a site may contain waste that was improperly disposed of, requiring the ground and/or water to be removed or decontaminated. The environmental engineer develops the means to remove the contamination and return the area to a usable form.

Prevention is an area that environmental engineers are becoming more involved in. Environmental engineers work with manufacturing engineers to design processes that reduce or eliminate harmful waste. One example is in the cleaning of machined parts. Coolant is sprayed on metal parts as they are cut to extend the life of the cutting tools. The oily fluid clings to the parts and has to be removed before the parts are assembled. An extremely toxic substance had been used in the past to clean the parts because there was not a suitable alternative. Environmental engineers discovered that the oil from orange peels works just as well and is perfectly safe (even edible). Since this new degreasing fluid did not need to be disposed of in any special way, manufacturing costs were reduced, and the environment of the workers improved.

Prevention does not just mean avoiding disasters. It also means reducing the environmental impact (footprint) of products and processes. Environmental engineers may be engaged in improving the sustainability of processes, including manufacturing and construction as well as the design and operation of new products. Environmental engineers study the environmental impact of the processes or products, identify alternatives, and work to implement these alternatives to reduce the impact on our environment. Reducing the environmental impact of the products we use every day is vital to protecting the global environment as the world's population continues to grow.

Environmental engineers must be well grounded in engineering fundamentals and current on environmental regulations. Within their companies, they are the experts on compliance with the ever-changing environmental laws. Environmental engineers must be able to understand the regulations and know how to apply them to the various processes they encounter.

Industrial Engineering

Industrial engineering is described by the Institute of Industrial Engineers (IIE) as the design, improvement, and installation of integrated systems of people, material, and energy. Industrial engineering is an interdisciplinary field that involves the integration of technology, mathematical models, and management practices. Traditional industrial engineering is done on a factory floor. However, the skills of an industrial engineer are transferable to a host of other applications. As a result, industrial engineers find themselves working within a wide variety of industries. Four of the main areas of emphasis for industrial engineers are production, manufacturing, human factors, and operations research.

The production area includes functions such as plant layout, material handling, scheduling, and quality and reliability control. An industrial engineer would examine the entire process involved in making a product and optimize it by reducing cost and production time and by increasing quality and reliability. In addition to factory layout, industrial engineers apply their expertise in other ways. For example, an industrial engineer might analyze the flow of people through an amusement park to reduce bottlenecks and provide a more pleasant experience for the patrons.

Manufacturing differs from production in that it addresses the components of the production process. While production concerns are on a global scale, manufacturing concerns address the individual production station. The actual material processing, such as machining, is optimized by the industrial engineer.

The human factors area involves the placement of people into the production system. An industrial engineer in this area studies the interfaces between people and machines in the system. The machines may include production machinery, computers, or even office chairs and desks. Industrial engineers consider ergonomics in finding ways to improve the interfaces. They look for ways to improve productivity while providing a safe environment for workers.

Operations research is concerned with the optimization of systems. This involves mathematically modeling systems to identify ways to improve them. Project management techniques such as critical path identification fall under operations research. Often, computer simulations are required either to model the system or to study the effects of changes to the system. These systems may be manufacturing systems or other organizations. The optimizing of sales territories for a pharmaceutical sales force provides a non-manufacturing example.

The IIE is one of the most prominent professional organizations for industrial engineers. It is organized into three societies, 10 technical divisions, and eight interest groups. The following technical divisions show the breadth of industrial engineering:

- Aerospace and Defense
- Energy, Environment, and Plant Engineering
- Engineering Economy
- Facilities Planning and Design
- Financial Services
- Logistics Transportation and Distribution
- Manufacturing
- Operations Research
- Quality Control and Engineering Reliability
- Utilities

Marine and Ocean Engineering

Nearly 80 percent of the Earth's surface is covered with water. Engineers concerned with the exploration of the oceans, the transportation of products over water, and the use of resources in the world's oceans, seas, and lakes are involved in marine and ocean engineering.

Marine engineers focus on the design, development, and operation of ships and boats. They work together with naval architects in this capacity. Naval architects are concerned with the overall design of the ship. They focus on the shape of the hull in order to provide the appropriate hydrodynamic characteristics. They are also concerned with the usefulness of the vessel for its intended purpose, and with the design of the ship's subsystems, including ventilation, water, and sanitary systems, to allow the crew to work efficiently.

The marine engineer is primarily concerned with the subsystems of the ship that allow the ship to serve its purpose. These include the propulsion, steering, and navigation systems. The marine engineer might analyze the ship for vibrations or stability in the water. The ship's electrical power distribution and air conditioning fall under the responsibility of marine engineers. They also might be involved in the analysis and design of the cargo handling systems of the ship.

The responsibilities of an ocean engineer involve the design, development, and operation of vehicles and devices other than boats or ships. These include submersible vehicles used in exploring oceans and obtaining resources from the ocean depths. He or she might be involved in designing underwater pipelines or cables, offshore drilling platforms, and offshore harbor facilities.

Ocean engineers also are involved with the interaction of the oceans and things with which oceans come in contact. They study wave action on beaches, docks, buoys, moorings, and harbors. Ocean engineers design ways to reduce erosion while protecting the marine environment. They study ways to protect and maintain marine areas that are critical to our food supply. Ocean engineers become involved with pollution control and treatment in the sea and alternative sources of energy from the ocean.

One of the professional societies in which these engineers may be involved is the Society of Naval Architects and Marine Engineers. The society is subdivided into the following nine technical and research committees:

- Hull Structure
- Hydrodynamics
- Ship's Machinery
- Ship Technical Operations
- Offshore
- Ship Production
- Ship Design
- Ship Repair and Conversion
- Small Craft

Materials Engineering

The origins of materials engineering can be traced to around 3000 B.C. when people began to produce bronze for use in creating superior hunting tools. Since that time, many new materials have been developed to meet the needs of society. Materials engineers develop these new materials and the processes to create them. The materials may be metals or non-metals: ceramics, plastics, and composites. Materials are at the forefront of many technologies, such as nanotechnology, in the development of new materials for a large range of applications. Materials engineers are generally concerned with four areas of materials: structure, properties, processes, and performance.

Materials engineers study the structure and composition of materials on a scale ranging from the microscopic to the macroscopic. They are interested in the molecular bonding and chemical composition of materials. They are also concerned with the effect of grain size and structure on the material properties.

The properties in question might include strength, crack growth rates, hardness, and durability. Numerous technological advances are impeded by a lack of materials possessing the properties required by the design engineers. Materials engineers seek to develop materials to meet these demands.

A given material may have very different properties depending on how it is processed. Steel is a good example. Cooling can affect its properties drastically. Steel that is allowed to cool in air will have different properties than steel that is cooled through immersion in a liquid. The composition of a material also can affect its properties. Materials such as metallic alloys contain trace elements that must be evenly distributed throughout the alloy to achieve the desired properties. If the trace elements are not well distributed or form clumps in the metal, the material will have very different properties than the desired alloy. This could cause a part made with the alloy to fail prematurely. Materials engineers design processes and testing procedures to ensure that the material has the desired properties.

The materials engineer also works to ensure that a material meets the performance needs of its application by designing testing procedures to ensure that these requirements are met. Both destructive and nondestructive testing techniques are used to serve this process.

Materials engineers develop new materials, improve traditional materials, and produce materials reliably and economically through synthesis and processing. Subspecialties of materials engineering, such as metallurgy and ceramics engineering, focus on classes of materials with similar properties.

Metallurgy involves the extraction of metals from naturally occurring ore for the development of alloys for engineering purposes. The metallurgical engineer is concerned with the composition, properties, and performance of an alloy. Detailed investigation of a component failure often identifies design flaws in the system. The materials engineer can provide useful information regarding the condition of materials to the design engineer.

Ceramics is another area of materials engineering. In ceramic engineering, the naturally occurring materials of interest are clay and silicates, rather than an ore. These nonmetallic minerals are employed in the production of materials that are used in a wide range of applications, including the aerospace, computer, and electronic industries.

Other areas of materials engineering focus on polymers, plastics, and composites. Composites are composed of different kinds of materials that are synthesized to create a new material to meet some specific demands. Materials engineers are also involved in biomedical applications. Examples include the development of artificial tissue for skin grafts, or bone replacement materials for artificial joints.

One of the professional societies to which materials engineers may belong is the Minerals, Metals, and Materials Society. It is organized into the following five technical divisions:

- Electronic, Magnetic, and Photonic Materials
- Extraction and Processing
- Light Metals
- Materials Processing and Manufacturing
- Structural Materials

Mechanical Engineering

Mechanical engineering is one of the largest and broadest of the engineering disciplines. Fundamentally, mechanical engineering is concerned with machines and mechanical devices. Mechanical engineers are involved in the design, development, production, control, operation, and service of these devices. Mechanical engineering is composed of two main divisions: design and controls, and thermal sciences.

The design function is the most common function of mechanical engineering. It involves the detailed layout and assembly of the components of machines and devices. Mechanical engineers are concerned about the strength of parts and the stresses the parts will need to endure. They work closely with materials engineers to ensure that correct materials are chosen. Mechanical engineers must also ensure that the parts fit together by specifying detailed dimensions.

Another aspect of the design function is the design process itself. Mechanical engineers develop computational tools to aid the design engineer in optimizing a design. These tools speed the design process by automating time-intensive analyses.

Mechanical engineers are also interested in controlling the mechanical devices they design. Control of mechanical devices can involve mechanical or hydraulic controls. However, most modern control systems incorporate digital control schemes. The mechanical engineer models controls for the system and programs or designs the control algorithm.

The noise generated from mechanical devices is often a concern, so mechanical engineers are often involved in acoustics—the study of noise. The mechanical engineer works to minimize unwanted noise by identifying the source and designing ways to minimize it without sacrificing a machine's performance.

In the thermal sciences, mechanical engineers study the flow of fluids and the flow of energy between systems. Mechanical engineers deal with liquids, gases, and two-phase flows, which are combinations of liquids and non-liquids. Mechanical engineers might be concerned about how much power is required to supply water through piping systems in buildings. They might also be concerned with aerodynamic drag on automobiles.

The flow of energy due to a temperature difference is called **heat** transfer, another thermal science area in which mechanical engineers are involved. They predict and study the temperature of components in environments of operation. Modern personal computers have microprocessors that require cooling. Mechanical engineers design the cooling devices to allow the electronics to function properly.

Mechanical engineers design and develop engines. An engine is a device that produces mechanical work. Examples include internal combustion engines used in automobiles and gas turbine engines used in airplanes. Mechanical engineers are involved in the design of the mechanical components of the engines as well as the overall cycles and efficiencies of these devices.

Performance and efficiency are also concerns for mechanical engineers involved in the production of power in large power generation systems. Steam turbines, boilers, water pumps, and condensers are often used to generate electricity. Mechanical engineers design these mechanical components needed to produce the power that operates the generators. Mechanical engineers also are involved in alternative energy sources including solar and hydroelectric power, and alternative fuel engines and fuel cells.

Another area in the thermal sciences is heating, ventilating, and air conditioning (HVAC). Mechanical engineers are involved in the climate control of buildings, which includes cooling and heating the air in buildings as well as controlling humidity. In doing so, they work closely with civil engineers in designing buildings to optimize the efficiency of these systems.

Mechanical engineers are involved in the manufacturing processes of many different industries. They design and develop the machines used in these processes and develop more efficient processes. Often, this involves automating time-consuming or expensive procedures within a manufacturing process. Mechanical engineers also are involved in the development and use of robotics and other automated processes.

In the area of biomedical engineering, mechanical engineers help develop artificial limbs and joints that provide mobility to physically impaired individuals. They also develop mechanical devices used to aid in the diagnosis and treatment of patients. Their knowledge of mechanics and forces has made them an important player in the diagnosis and prevention of concussions, working with medical professionals and neuroscientists.

The American Society of Mechanical Engineering (ASME) is one of the most prominent professional societies for mechanical engineers. It is divided into 35 technical divisions, indicating the diversity of this field:

- Advanced Energy Systems
- Aerospace Engineering
- Applied Mechanics
- Basic Engineering Technical Group
- Bioengineering
- Design Engineering
- Dynamic Systems and Control
- Electrical and Electronic Packaging
- FACT
- Fluids Engineering
- Fluids Power Systems and Technology Systems
- Heat Transfer

- Information Storage/Processing
- Internal Combustion Engine
- Gas Turbine
- Manufacturing Engineering
- Materials
- Materials Handling Engineering
- Noise Control and Acoustics
- Non-Destructive Evaluation Engineering
- Nuclear Engineering
- Ocean Engineering
- Offshore Mechanics/Arctic Engineering
- Petroleum
- Plant Engineering and Maintenance
- Power
- Pressure Vessels and Piping
- Process Industries
- Rail Transportation
- Safety Engineering and Risk Analysis
- Solar Energy
- Solid Waste Processing
- Technology and Society
- Textile Engineering
- Tribology

Mining Engineering

Modern society requires a vast amount of products made from raw materials such as minerals. The continued production of these raw materials helps to keep society functioning. Mining engineers are responsible for maintaining the flow of these raw materials by discovering, removing, and processing minerals into the products society requires.

Discovering the ore involves exploration in conjunction with geologists and geophysicists. The engineers combine the use of seismic, satellite, and other technological data, using their knowledge of rocks and soils. The exploration may focus on land areas, the ocean floor, or even below the ocean floor. In the future, mining engineers may also explore asteroids, which are rich in mineral deposits.

Once mineral deposits are identified, they may be removed. One way minerals are removed is by way of mining tunnels. The engineers design and maintain the tunnels and the required support systems, including ventilation and drainage. Other times, minerals are removed from open-pit mines. Again, the engineers analyze the removal site and design the procedure for removing the material. Engineers also develop plans for returning the site to a natural state. Mining engineers use boring, tunneling, and blasting techniques to create a mine. Regardless of the removal technique, the environmental impact of the mining operation is taken into account and minimized.

The mining engineer is also involved in the processing of the raw minerals into usable forms. Purifying and separating minerals involves chemical and mechanical processes. While mining engineers may not be involved in producing a finished product that consumers recognize, they must understand the form their customers can use and design processes to transform the raw materials into usable forms.

Mining engineers also become involved in designing the specialized equipment required for use in the mining industry. The design of automated equipment capable of performing the most dangerous mining jobs helps to increase the safety and productivity of a mining operation. Since mines typically are established in remote areas, mining engineers are involved in the transportation of minerals to the processing facility.

The expertise of mining engineers is not used exclusively by the mining industry. The same boring technology used in developing mines is used to create subway systems and railroad tunnels, such as the one under the English Channel.

Nuclear Engineering

Nuclear engineers are concerned primarily with the use and control of energy from nuclear sources. This involves electricity production, propulsion systems, waste disposal, and radiation applications.

The production of electricity from nuclear energy is one of the most visible applications of nuclear engineering. Nuclear engineers focus on the design, development, and operation of nuclear power facilities. This involves using current fission technology as well as the development of fusion, which would allow seawater to be used as fuel. Nuclear energy offers an environmentally friendly alternative to fossil fuels. A current barrier to the use of nuclear energy is the high cost of plant construction. This barrier provides a challenge for design engineers to overcome. Research is currently being performed on the viability of smaller, more efficient nuclear reactors.

Nuclear power is also used in propulsion systems. It provides a power source for ships and submarines, allowing them to go years without refueling. It is also used as a power source for satellites. Nuclear-powered engines are being examined as an alternative to conventional fossil-fueled engines, making interplanetary travel possible.

One of the main drawbacks to nuclear power is the production of radioactive waste. This also creates opportunities for nuclear engineers to develop safe and reliable means to dispose of spent fuel. Nuclear engineers develop ways to reprocess the waste into less hazardous forms.

Another area in which nuclear engineers are involved is the use of radiation for medical or agricultural purposes. Radiation therapy has proven effective in treating cancers, and radioactive isotopes are also used in diagnosing diseases. Irradiation of foods can eliminate harmful bacteria and help ensure a safer food supply.

Due to the complex nature of nuclear reactions, nuclear engineers are at the forefront of advanced computing methods. High-performance computing techniques, such as parallel processing, constitute research areas vital to nuclear engineering.

The American Nuclear Society (ANS) is one of the professional societies to which nuclear engineers belong. It is divided into these 16 technical divisions:

- Biology and Medicine
- Decommissioning, Decontamination, and Reutilization
- Education and Training
- Environmental Sciences
- Fuel Cycle and Waste Management
- Fusion Energy
- Human Factors
- Isotopes and Radiation
- Materials Science and Technology
- Mathematics and Computations
- Nuclear Criticality Safety
- Nuclear Operations
- Nuclear Installations Safety
- Power
- Radiation Protection and Shielding
- Reactor Physics

Petroleum Engineering

Petroleum and petroleum products are essential components in today's society. Petroleum engineers maintain the flow of petroleum in a safe and reliable manner. They are involved in the exploration for crude oil deposits, the removal of oil, and the transporting and refining of oil.

Petroleum engineers work with geologists and geophysicists to identify potential oil and gas reserves. They combine satellite information, seismic techniques, and geological information to locate deposits of gas or oil. Once a deposit has been identified, it can be removed. The petroleum engineer designs, develops, and operates the needed drilling equipment and facilities. Such facilities may be located on land or on offshore platforms. The engineer is interested in removing the oil or gas in a safe and reliable manner—safe for the people involved as well as for the environment. Removal of oil is done in stages, with the first stage being the easiest and using conventional means. Oil deposits are often located in sand. A significant amount of oil remains coating the sand after the initial oil removal. Recovery of this additional reserve requires the use of secondary and tertiary extraction techniques using water, steam, or chemical means.

Transporting the oil or gas to a processing facility is another challenge for the petroleum engineer. At times this requires the design of a heated pipeline such as the one in Alaska to carry oil hundreds of miles over frozen tundra. In other instances this requires transporting oil in double-hulled tankers from an offshore platform near a wildlife refuge. Such situations necessitate extra precautions to ensure that the wildlife is not endangered.

Once the oil or gas arrives at the processing facility, it must be refined into usable products. The petroleum engineer designs, develops, and operates the equipment to chemically process the gas or oil into such end products. Petroleum is made into various grades of gasoline, diesel fuel, aircraft fuel, home heating oil, motor oils, and a host of consumer products from lubricants to plastics.

Other Fields

The most common engineering majors have been described in this chapter. However, there are other specialized engineering programs at some institutions. Here is a partial listing of some of these other programs:

- Automotive Engineering
- Acoustical Engineering
- Applied Mathematics
- Bioengineering
- Engineering Science
- Engineering Management
- Excavation Engineering
- Fire Engineering
- Forest Engineering
- General Engineering
- Genetic Engineering
- Geological Engineering
- Inventive Design
- Manufacturing Engineering
- Packaging Engineering
- Pharmaceutical Engineering
- Plastics Engineering
- Power Engineering
- Systems Engineering
- Theater Engineering
- Transportation Engineering
- Welding Engineering

2.4　Emerging Fields

The fields of engineering have been, and will continue to be, dynamic. As new technologies emerge, new definitions are needed to classify disciplines. The boundaries will continue to shift and new areas will emerge. The explosion of technological advances means that there is a good chance you will work in a field that is not currently defined. Technological advances are also blurring the traditional delineations between fields. Areas not traditionally linked are coming together and providing cross-disciplinary opportunities for engineers. A good example of this is in the development of smart buildings, which sense the onset of an earthquake, adapt their structures to survive the shaking, and then return to normal status afterward. Research is being conducted on these technologies that bridge computer engineering with structural (civil) engineering to produce such adaptable buildings.

The incredible breakthroughs in biology have opened many new possibilities and will continue to impact most of the fields of engineering. Historically, biology has not been as integrated with engineering as has physics and chemistry, but that is rapidly changing and will have an enormous impact on the future of engineering. The ability to modify genetic codes has implications in a wide range of engineering applications, including the production of pharmaceuticals that are customized for individual patients, alternative energy sources, and environmental reclamation.

Nanotechnology is an area that is receiving a great deal of attention and resources and is blurring the boundaries of the fields of engineering. Nanotechnology is an emerging field in which new materials and tiny structures are built atom by atom, or molecule by molecule, instead of the more conventional approach of sculpting parts from preexisting materials. The possibilities for nano-applications include the following:

- The creation of entirely new materials with superior strength, electrical conductivity, resistance to heat, and other properties
- Microscopic machines for a variety of uses, including probes that could aid diagnostics and repair
- A new class of ultra-small, super-powerful computers and other electronic devices, including spacecraft
- A technology in which biology and electronics are merged, creating gene chips that instantly detect foodborne contamination, dangerous substances in the blood, or chemical warfare agents in the air
- The development of molecular electronics and devices that self-assemble, similar to the growth of complex organic structures in living organisms

Nanotechnology requires specialized laboratory and production facilities, providing further challenges and opportunities for future engineers.

The explosion of information technology with the Internet and wireless communication has produced a melding of disciplines to form new fields in information science and technology. Information management and transfer is an important and emerging issue in all disciplines of engineering and has opened opportunities for engineers who want to bridge the gaps between the traditional fields and information technologies.

Sustainability is another important concept that spans many engineering disciplines. Sustainable development is a term that was defined by a report to the United Nations, known as the Brundtland Report, in 1987 as "development that meets the needs of the present without compromising the ability of future generations to meet their own needs" (*Our Common Future*). Designing products and processes that contribute to sustainable development is important for all engineering disciplines and is becoming a larger part of undergraduate and graduate education as well as professional development programs. As the global population continues to rise, so too do the pressures on the global environment. Sustainability will have to play a large and significant role in product development, manufacturing, and construction if we are to meet these challenges.

Undoubtedly, more new fields will open and be discovered as technology continues to advance. As the boundaries of the genetic code and molecular-level devices are crossed, new frontiers will open. As an engineer, you will have the exciting opportunity to be part of the discovery and definition of these emerging fields, which will have a tremendous impact on society's future.

2.5   Closing Thoughts

The information presented in this chapter is meant to provide a starting point on the road to choosing a career. There may have been aspects of one or more of these engineering fields that appealed to you, and that's great. The goal, however, is not to persuade you that engineering is for everyone; it is not. The goal is to provide information to help you decide if engineering would be an enjoyable career for *you*.

As a student, it is important to choose a career path that will be both enjoyable and rewarding. For many, an engineering degree is a gateway to just such a career. Each person has a unique set of talents, abilities, and gifts that are well matched for a particular career. In general, people find more rewarding careers in occupations where their gifts and talents are well used. Does engineering match your talents, abilities, and interests?

Right now, choosing a career may seem overwhelming. But at this point, you don't have to. What you are embarking on is an education that will provide the base for such decisions. Think about where a degree in engineering could lead. One of the exciting aspects of an engineering education is that it opens up a wide range of jobs after leaving college. Most students will have several different careers before they retire, and the important objective in college is obtaining a solid background that will allow you to move into areas that you enjoy later in life.

As you try to make the right choice for you, seek out additional information from faculty, career centers, professional societies, placement services, and industrial representatives. Ask a lot of questions. Consider what you would enjoy studying for four years in college. What kind of entry-level job would you be able to get with a specific degree? What doors would such a degree open for you later in life? Remember, your decision is unique to you. No one can make it for you.

2.6   Engineering and Technical Organizations

The following is a list of many of the engineering technical societies available to engineers. They can be a tremendous source of information for you. Many have student branches that allow you to meet both engineering students and practicing engineers in the same field. Join one or more of these organizations during your freshman or sophomore year.

American Association for the Advancement of Science
1200 New York Avenue, NW
Washington, DC 20005
(202) 326–6400
www.aaas.org

American Association of Engineering Societies
1111 19th Street, NW
Suite 403
Washington, DC 20036
(202) 296–2237
www.aaes.org

American Ceramic Society
735 Ceramic Place
Westerville, OH 43081-8720
(614) 890–4700
www.acers.org

American Chemical Society
1155 16th Street, NW
Room 1209
Washington, DC 20036-1807
(202) 872–4600
www.acs.org

American Concrete Institute
38800 Country Club Drive
Farmington Hills, MI 48331
(248) 848–3700
www.aci-int.org

American Congress on Surveying and Mapping
5410 Grosvenor Lane
Suite 100
Bethesda, MD 20814–2122
(301) 493-0200

American Consulting Engineers Council
1015 15th St., NW, Suite 802
Washington, DC 20005
(202) 347–7474
www.acec.org

American Gas Association
1515 Wilson Blvd.
Arlington, VA 22209
(703) 841–8400
www.aga.com

American Indian Science and Engineering Society
5661 Airport Blvd.
Boulder, CO 80301–2339
(303) 939-0023
www.colorado.edu/AISES

American Institute of Aeronautics and Astronautics
1801 Alexander Bell Drive, Suite 500
Reston, VA 20191–4344
(800) NEW-AIAA or (703) 264–7500
www.aiaa.org

American Institute of Chemical Engineers
345 East 47th Street
New York, NY 10017-2395
(212) 705–7000 or (800) 242-4363
www.aiche.org

American Institute of Mining, Metallurgical and Petroleum Engineers
345 East 47th Street
New York, NY 10017
(212) 705-7695

American Nuclear Society
555 North Kensington Avenue
La Grange Park, IL 60526
(708) 352–6611
www.ans.org

American Oil Chemists' Society
1608 Broadmoor Drive
Champaign, IL 61821-5930
(217) 359–2344
www.aocs.org

**American Railway
Engineering Association**
8201 Corporate Drive
Landover, MD
(301) 459–3200

**American Society for
Engineering Education**
1818 N St., NW, Suite 600
Washington, DC 20036–2479
(202) 331–3500
www.asee.org

**American Society for Heating,
Refrigeration and
Air Conditioning Engineers**
1791 Tulie Circle NE
Atlanta, GA 30329
(404) 636–8400
www.ashrae.org

American Society for Quality
611 East Wisconsin Avenue
Milwaukee, WI 53202
(414) 272–8575
www.asq.org

**American Society of
Agricultural Engineers**
2950 Niles Road
St. Joseph, MI 49085-9659
(616) 429–0300
www.asae.org

**American Society of Civil
Engineers**
1801 Alexander Bell Drive
Reston, VA 20191-4400
(800) 548-2723 or (703) 295-6000
www.asce.org

**American Society of
Mechanical Engineers**
345 East 47th Street
New York, NY 10017-2392
(212) 705-7722
www.asme.org

**American Society of Naval
Engineers**
1452 Duke Street
Alexandria, VA 22314
(703) 836-6727
www.jhuapl.edu/ASNE

**American Society of
Nondestructive Testing**
1711 Arlingate Lane
P.O. Box 28518
Columbus, OH 43228-0518
(614) 274-6003
www.asnt.org

**American Society of
Plumbing Engineers**
3617 Thousand Oaks Blvd.
Suite 210
Westlake Village, CA 91362-3649
(805) 495-7120
www.aspe.org

**American Water Works
Association**
6666 West Quincy Avenue
Denver, CO 80235
(303) 443-9353
www.aws.org

**Architectural Engineering
Institute**
1801 Alexander Bell Drive,
1st Floor
Reston, VA 20191–4400
(703) 295–6370, Fax (703)
295–6132
www.aeinstitute.org

**Association for the
Advancement of Cost
Engineering International**
209 Prairie Avenue
Suite 100
Morgantown, WV 26505
(304) 296-8444
www.aacei.org

Board of Certified Safety Professionals
208 Burwash Avenue
Savoy, IL 61874–9571
(217) 359–9263
www.bcsp.com

Construction Specifications Institute
601 Madison Street
Alexandria, VA 22314–1791
(703) 684–0300
www.csinet.org

Information Technology Association of America
1616 N. Fort Myer Drive,
Suite 1300
Arlington, VA 22209
(703) 522–5055
www.itaa.org

Institute of Electrical and Electronics Engineers
1828 L Street NW, Suite 1202
Washington, DC 20036
(202) 785–0017
www.ieee.org

Institute of Industrial Engineers
25 Technology Park
Norcross, GA 30092
(770) 449–0461
www.iienet.org

Iron and Steel Society
410 Commonwealth Drive
Warrendale, PA 15086–7512
(412) 776–1535
www.issource.org

Laser Institute of America
12424 Research Parkway, Suite 125
Orlando, FL 32826
(407) 380–1553
www.laserinstitute.org

Mathematical Association of America
1529 18th Street, NW
Washington, DC 20036
(202) 387–5200
www.maa.org

Minerals, Metals and Materials Society
420 Commonwealth Drive
Warrendale, PA 15086
(412) 776–9000
www.tms.org

NACE International
1440 South Creek Drive
Houston, TX 77084–4906
(281) 492–0535
www.nace.org

National Academy of Engineering
2101 Constitution Avenue, NW
Washington, DC 20418
(202) 334–3200

National Action Council for Minorities in Engineering, Inc.
The Empire State Building
350 Fifth Avenue, Suite 2212
New York, NY 10118–2299
(212) 279–2626
www.naof cme.org

National Association of Minority Engineering Program Administrators, Inc.
1133 West Morse Blvd.,
Suite 201
Winter Park, FL 32789
(407) 647–8839, Fax (407) 629–2502

National Association of Power Engineers
1 Springfield Street
Chicopee, MA 01013
(413) 592-6273
www.powerengineers.com

National Conference of Standards Laboratories International
2995 Wilderness Place, Ste. 101
Boulder, CO 80301-5404
(303) 440-3339
www.ncsli.org

National Institute of Standards and Technology
Publications and Programs Inquiries
Public and Business Affairs
Gaithersburg, MD 20899
(301) 975-3058
www.nist.gov

National Science Foundation
4201 Wilson Blvd.
Arlington, VA 22230
(703) 306-1234
www.nsf.gov

National Society of Black Engineers
1454 Duke Street
Alexandria, VA 22314
(703) 549-2207
www.nsbe.org

National Society of Professional Engineers
1420 King Street
Alexandria, VA 22314
(888) 285-6773
www.nspe.org

Society of Allied Weight Engineers
5530 Aztec Drive
La Mesa, CA 91942
(619) 465-1367

Society of American Military Engineers
607 Prince Street
Alexandria, VA 22314
(703) 549-3800 or (800) 336-3097
www.same.org

Society of Automotive Engineers
400 Commonwealth Drive
Warrendale, PA 15096
(412) 776-4841
www.sae.org

Society of Fire Protection Engineers
7315 Wisconsin Avenue
Suite 1225W
Bethesda, MD 20814
(301) 718-2910

Society of Hispanic Professional Engineers
5400 East Olympic Blvd.
Suite 210
Los Angeles, CA 90022
(213) 725-3970
www.engr.umd.edu/
organizations/shpe

Society of Manufacturing Engineers
One SME Drive
P.O. Box 930
Dearborn, MI 48121-0930
(313) 271-1500
www.sme.org

Society for Mining, Metallurgy and Exploration, Inc.
8307 Shaffer Parkway
Littleton, CO 80127
(303) 973-9550
www.smenet.org

Society of Naval Architects and Marine Engineers
601 Pavonia Avenue
Jersey City, NJ 07306
(800) 798-2188
www.sname.org

Society of Petroleum Engineers
P.O. Box 833836
Richardson, TX 75083-3836

Society of Plastics Engineers
14 Fairfield Drive
Brookfield, CT 06804-0403
(203) 775-0471
www.4spe.org

Society of Women Engineers
120 Wall Street
11th Floor
New York, NY 10005
(212) 509-9577
www.swe.org

SPIE—International Society for Optical Engineering
P.O. Box 10
Bellingham, WA 98227-0010
(360) 676-3290
www.spie.org

Tau Beta PI
508 Dougherty Engineering Hall
P.O. Box 2697
Knoxville, TN 37901-2697
(423) 546-4578
www.tbp.org

Women in Engineering Initiative
University of Washington
101 Wilson Annex
P.O. Box 352135
Seattle, WA 98195-2135
(206) 543-4810
www.engr.washington.
edu/~wieweb

REFERENCES

Belcher, M. C., *Magill's Survey of Science: Applied Science*, Pasadena, CA, Salem Press, 1993.

Burghardt, M. D., *Introduction to the Engineering Profession*, 2nd ed., New York, Harper Collins College Publishers, 1995.

Garcia, J., *Majoring {Engineering}*, New York, Noonday Press, 1995.

Grace, R., and J. Daniels, *Guide to 150 Popular College Majors*, New York: College Entrance Examination Board, 1992, pp. 175–178.

Irwin, J. D., *On Becoming an Engineer*, New York, IEEE Press, 1997.

Kemper, J. D., *Engineers and Their Profession*, 4th ed., New York, Oxford University Press, 1990.

Landis, R., *Studying Engineering, A Road Map to a Rewarding Career*, Burbank, CA, Discovery Press, 1995.

Smith, R. J., B. R. Butler, and W. K. LeBold, *Engineering as a Career*, 4th ed., New York, McGraw-Hill Book Company, 1983.

World Commission on Environment and Development, *Our Common Future*, New York, Oxford University Press, 1987.

Wright, Paul H., *Introduction to Engineering*, 2nd ed., New York, John Wiley and Sons, Inc., 1994.

EXERCISES AND ACTIVITIES

2.1   Contact an engineer practicing in a field of engineering that interests you. Write a brief report on his or her activities and compare them to the description of that field of engineering in this chapter.

2.2   For a field of engineering that interests you, make a list of potential employers that hire graduates from your campus, and list the cities in which they are located.

2.3   Visit a job fair on your campus and briefly interview a company representative. Prepare a brief report on what that company does, what the engineer you spoke to does, and the type of engineers they are looking to hire.

2.4   Make a list of companies that hire co-op and/or intern students from your campus. Write a brief report on what these companies are looking for and their locations.

2.5   Select a company that employs engineers in a discipline that interests you, and visit their web page. Prepare a brief report on what the company does, hiring prospects, engineering jobs in that organization, and where the company is located.

2.6   Contact a person with an engineering degree who is not currently employed in a traditional engineering capacity. Write a one-page paper on how that person uses his or her engineering background on the job.

2.7   Write a one-page paper on an engineering field that will likely emerge during your lifetime (a field that does not currently exist). Consider what background an engineering student should obtain in preparation for this emerging field.

2.8   Draft a sample letter requesting a co-op or intern position.

2.9   Identify a modern technological problem. Write a brief paper on the role of engineers and technology in solving this problem.

2.10   Select an engineering discipline that interests you and a particular job function within that discipline. Write a brief paper contrasting the different experiences an engineer in this discipline would encounter in each of three different industries.

2.11   Make a list of your own strengths and talents. Write a brief report on how these strengths are well matched with a specific engineering discipline.

2.12   Pick an engineering discipline that interests you. List and briefly describe the technical and design electives available in that discipline for undergraduates.

2.13   Select a consumer product you are familiar with (smartphone, stereo, automobile, food product, etc.). List and briefly describe the role of all the engineering disciplines involved in producing the product.

2.14   Select an engineering discipline that interests you. Write a brief paper on how the global marketplace has altered this discipline.

2.15   Write a brief paper listing two similarities and two differences for each of the following engineering functions:

   a) Research and development
   b) Development and design
   c) Design and manufacturing
   d) Manufacturing and operations
   e) Sales and customer support
   f) Management and consulting

2.16   Find out how many engineering programs your school offers. How many students graduate each year in each discipline?

2.17   Which of the job functions described in this chapter is most appealing to you? Write a brief paper discussing why it is appealing.

2.18   Write a paper about an engineer who made a significant technical contribution to society.

2.19   Report on the requirements for becoming a registered professional engineer in your state. Also report on how registration would be beneficial to your engineering career.

2.20   Make a list of general responsibilities and obligations you would have as an engineer.

2.21   Write a brief paper on the importance of ethical conduct as an engineer.

2.22   Select one of the industrial sectors within engineering and list five companies that do business in that sector. Briefly describe job opportunities in each.

2.23   Prepare a report on how the following items work and what engineering disciplines are involved in their design and manufacture:

a) Cell phone
b) CT scan machine
c) DVD player
d) Dialysis machine
e) Flat-screen TV

2.24   Answer the following questions and look for themes or commonality in your answers. Comment on how engineering might fit into these themes.

a) If I could snap my fingers and do whatever I wanted, knowing that I wouldn't fail, what would I do?
b) At the end of my life, I'd love to be able to look back and know that I had done something about _____.
c) What would my friends say I'm really interested in or passionate about?
d) What conversation could keep me talking late into the night?
e) What were the top five positive experiences I've had in my life, and why were they meaningful to me?

2.25   Write a short paper describing your dream job, regardless of pay or geographical location.

2.26   Write a short paper describing how engineering might fit into your answer to 2.25.

2.27   List the characteristics a job must have for you to be excited to go to work every morning. How does engineering fit with those characteristics?

2.28   List the top 10 reasons why a student should study engineering.

2.29   List 10 inappropriate reasons for a student to choose to study engineering.

2.30   Select an industrial sector and describe what you suppose a typical day is like for a:

a) Sales engineer
b) Research engineer
c) Test engineer
d) Manufacturing engineer
e) Design engineer

2.31   Make a list of five non-engineering careers an engineering graduate could have and describe each one briefly.

2.32   Select one of the professional societies that was listed at the end of this chapter. Prepare a report on what the organization does and how it could benefit you as a practicing engineer.

2.33   Select one of the professional societies that was listed at the end of this chapter. Identify one of its technical divisions and prepare a report on that division's activities.

2.34   Identify one of the student branches of an engineering professional society on your campus and prepare a report on the benefits of involvement with that organization for an engineering student.

2.35   Visit the website of one of the professional societies listed in this chapter. Prepare a brief presentation summarizing the material.

2.36   Write a letter to a ninth-grade class explaining the exciting opportunities in your chosen major within engineering. Include a short discussion of the classes they should be taking to be successful in that same major.

2.37   Write a letter to a ninth-grade class describing the opportunities a bachelor's degree in engineering provides in today's society.

2.38   Write a letter to your parents detailing why you are going to major in the field you have chosen.

2.39   Write a one-page paper describing how your chosen field of engineering will be different in 25 years.

2.40   Select one field of engineering and write a one-page paper on how the advances in biology have influenced that field.

2.41   Select one field of engineering and write a one-page paper on how the advances in nanotechnology have influenced that field.

2.42   Select one field of engineering and write a one-page paper on how the advances in information technology have influenced that field.

2.43   Select two fields of engineering and describe problems that span these two disciplines.

2.44   Write a brief paper on an emerging area within engineering. Relate the area to your chosen engineering major.

2.45   Research the current spending priorities of the U.S. government in the area of technical research. How will these priorities impact your chosen major?

2.46   For each grouping of engineering disciplines, describe applications or problems that span the disciplines:

a) Aerospace, Materials, and Civil
b) Mechanical, Agricultural, and Computer

    c) Biological and Environmental
    d) Industrial, Chemical, and Electrical
    e) Biomedical and Nuclear
    f) Agricultural and Aerospace
    g) Materials and Biomedical
    h) Civil and Computer

2.47   Select one professional organization and find out how they handle new and emerging technologies within their society. Where do they put them and where can people working in emerging areas find colleagues?

2.48   Interview a practicing engineer and a faculty member from the same discipline about their field. Compare and contrast their views of the discipline.

2.49   Select two engineering majors and compare and contrast the opportunities available in the two.

2.50   Identify how your skills as an engineer can be used within your local community, either as full-time work or as a volunteer. Share your findings with the class by preparing a short oral presentation.

CHAPTER 3

A Statistical Profile of the Engineering Profession

3.1 Statistical Overview

Many students who choose a major in engineering know little about the profession or the individuals who work in the field. This chapter will answer such questions as: How many people study engineering? What are their majors? What is the job market for engineers like? How much do engineers earn? How many women and minorities are studying engineering? How many practicing engineers are there in the U.S.?

The picture of the engineering profession presented in this chapter will assist you in better understanding the field. This information has been gathered from a variety of surveys and reports that are published on a regular basis by various organizations. As with many types of data, some are very current and others not as current. We have endeavored to provide commentary that explains the information presented.

3.2 College Enrollment Trends of Engineering Students

As shown in Figure 3.1, the number of first-year students enrolled in engineering programs has fluctuated greatly during the period from 1955 to 2013, according to the 2014 Engineering Workforce Commission's annual survey of enrollment patterns. During the 1950s and 1960s the number of first-year students pursuing bachelor's degrees in engineering programs ranged between 60,000 and 80,000. During the 1970s there was a significant drop in student interest, and freshman enrollment fell to a low of 43,000. This was followed by the largest long-term climb in student enrollment, which eventually peaked in the early 1980s at about 118,000 students. From that point until 1996 there was a steady decline in student enrollment, with a low of about 85,000 first-year students. This was followed in 1997 to 2002 with a modest increase in student enrollment each year, which was projected to last for several years. From 2003 to 2006 there was a decline in first-year students enrolling in engineering programs, even though total undergraduate engineering enrollment remained high.

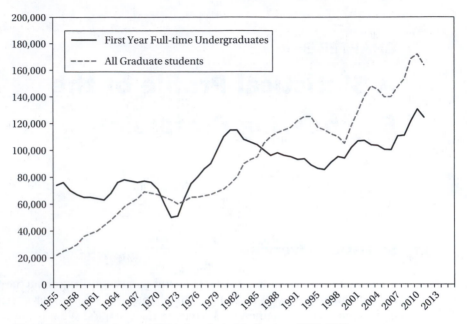

Figure 3.1 Engineering enrollments: selected indicators, 1955–2013.
Source: Engineering Workforce Commission (EWC), 2014.

2007 and 2008 showed a modest increase in first-year students to 111,000. However, beginning in 2009, there has been a steady, significant increase: 2013 data show over 124,000 first-year engineering students, although there was a slight decline compared with 2012.

Why has enrollment in engineering (and higher education in general) fluctuated so greatly over the last few decades? The primary factor influencing enrollment is the number of high school graduates in a given year. Of course, those numbers are directly correlated to the number of births 18 years previously. Given these facts and the information in Figure 3.2, it can be determined how the birth rates from 1960 to 2014 have influenced, and will influence, the enrollment of college-bound students. Likewise, Figure 3.3 provides data concerning the number of high school graduates from 1988 to 2022. This shows that the number of graduates started to climb in 1992 and increased in 1999 through 2005. There were modest increases in 2006 through 2009, and then projections show a slight decline followed by a rise in the number of graduates through 2022.

Obviously, there are other factors that influence a student's choice of a major, including the general economy and the relative popularity of various fields in any given year.

It also should be noted that the total number of students pursuing graduate engineering degrees increased significantly from the late 1970s to 1992. Since then, there was a relatively steady decline until 1997, when enrollment started increasing again, with 2004 numbers reaching an all-time high of over 145,000. 2005 and 2006 numbers show a decline to about 140,000, but 2008 figures rose to over 153,000 graduate students and the number continued to climb to almost 164,000 in 2013.

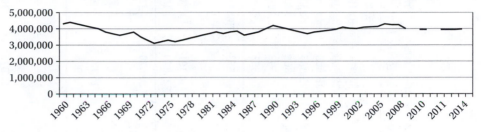

Figure 3.2 Births in the U.S., 1960–2014
Source: U.S. Dept. of Health & Human Services, CDC National Center for Health Statistics, 2015.

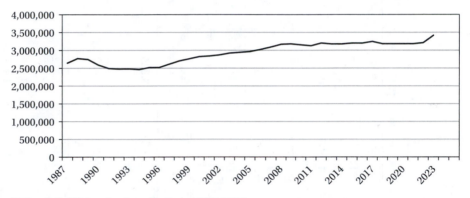

Figure 3.3 High school graduations, 1987–2022
Source: U.S. Dept. of Education, National Center for Education Statistics, 2013.

3.3 College Majors of Recent Engineering Students

Table 3.1 provides specific 2013 enrollment data for undergraduate and graduate students in several broad engineering fields. This information shows that over 540,000 students were majoring in one of the undergraduate engineering disciplines, with another 164,000 pursuing graduate engineering degrees. The largest field of study at all levels is the electrical and computer engineering disciplines, followed by mechanical and aerospace. It is interesting to note that over 106,000 students are in "other engineering disciplines" (which includes the many smaller, specialized fields of engineering) and almost 14,000 are in pre-engineering programs.

3.4 Degrees in Engineering

When examining the number of degrees awarded to engineering graduates during the years 1983 to 2014 (Fig. 3.4), there is a close correlation between enrollment and degrees. Since 1986, when over 78,000 engineering graduates were produced, to

Table 3.1 Fall 2013 Engineering Enrollments by Broad Disciplinary Groups

	All Engineering Disciplines	Electrical and Computer	Mechanical and Aerospace	Civil and Environmental	Chemical and Petroleum	Industrial, Manufacturing, & Management	All Other Disciplines	Pre-Engineering
Full-Time Undergrads								
First Year	123,877	28,518	28,512	11,643	9,923	2,182	34,984	8,115
Second Year	108,304	27,637	28,157	11,835	10,380	3,485	22,909	3,901
Third Year	108,197	30,939	29,345	13,100	10,653	4,641	18,505	1,014
Fourth Year	144,262	40,398	39,420	20,326	14,590	6,761	22,535	232
Fifth Year	11,897	3,403	3,255	1,588	1,261	706	1,665	19
Full-Time Undergrad	496,537	130,895	128,689	58,492	46,807	17,775	100,598	13,281
Part-Time Undergrad	43,624	16,594	11,015	5,246	2,672	1,581	5,955	561
Total: All Undergrad	540,161	147,489	139,704	63,738	49,479	19,356	106,553	13,842
Candidates for M.S.								
Full Time	61,307	26,368	9,992	7,205	2,422	5,128	10,191	1
Part Time	37,299	12,472	5,556	3,942	903	4,537	9,879	10
Total: All M.S.	98,606	38,840	15,548	11,147	3,325	9,665	20,070	11
Candidates for Ph.D.								
Full Time	56,632	18,894	9,407	5,461	5,823	1,909	15,138	–
Part Time	8,541	3,306	1,348	905	389	597	1,996	–
Total: All Ph.D.	65,173	22,200	10,755	6,366	6,212	2,506	17,134	–
Full-Time Grads	117,939	45,262	19,399	12,666	8,245	7,037	25,329	1
Part-Time Grads	45,840	15,778	6,904	4,847	1,292	5,134	11,875	10
Total: All Grads	163,779	61,040	26,303	17,513	9,537	12,171	37,204	11

Source: Engineering Workforce Commission of the American Association of Engineering Societies.

1997, when approximately 65,000 degrees were awarded, there was a steady decline (with the exception of 1993) until 1995, and then a slight increase until 1997. However, there were declines in 1998 and 1999, and then a gradual increase between 2000 and 2006, to a total of 76,000. 2007, however, shows a slight decline in degrees awarded, followed by an increase in 2008, a decline in 2009, and a significant increase beginning in 2010 through 2014. However, this same period does show a slight decline in graduate degrees in 2013 and 2014, perhaps reflecting an improving job market for recent bachelor's degree graduates. Table 3.2 documents the number of bachelor's degrees awarded by field of study for the years 2005 to 2014. It is interesting to note that the greatest number were earned by students in mechanical engineering, which in 2005 surpassed electrical engineering and the closely related fields of computer engineering and computer science (which at many schools are in the same department or are closely aligned with electrical engineering programs).

3.5 Job Placement Trends

The Collegiate Employment Research Institute at Michigan State University conducts an annual hiring survey of employers and is able to track the labor market for new college graduates. From 1994 to mid-2001, the job market for graduating engineers was excellent; experts agree that it was probably the hottest job market in over 30 years.

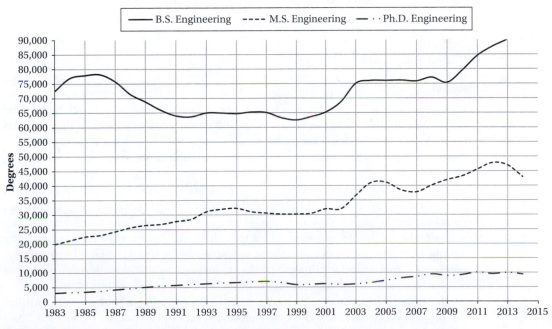

Figure 3.4 Engineering degrees, 1983–2014, by degree level

Source: EWC, 2015.

Table 3.2 Bachelor's Degrees in Engineering, by Discipline, 2005–2014

Curriculum	2005	2006	2007	2008	2009	2010	2011	2012	2013	2014
Aerospace	2,395	2,681	2,763	2,897	2,989	3,150	3,286	3,318	3,587	3,262
Agricultural	302	343	351	355	383	418	427	440	336	423
Bioengineering	2,638	3,028	3,055	3,478	3,787	3,951	4,293	4,629	5,079	5,204
Chemical	4,621	4,590	4,607	4,915	5,151	5,849	6,297	6,812	7,270	7,638
Civil	8,857	9,432	9,875	10,862	11,274	11,835	13,175	13,186	13,232	12,389
Computer	16,019	14,282	13,171	11,471	10,370	10,764	11,610	12,355	12,996	14,674
Electrical	14,742	14,329	13,783	13,113	11,978	11,968	12,005	12,201	11,710	12,590
Engr. Science	1,019	1,015	1,097	1,104	1,104	1,212	1,233	1,306	1,089	1,071
Industrial	3,220	3,079	3,150	3,064	3,077	3,293	3,423	3,660	3,897	4,174
Materials & Metallurgical	815	919	924	1,027	985	1,111	1,134	1,265	1,335	1,186
Mechanical	14,835	15,698	16,172	17,865	17,016	18,040	19,016	19,516	20,433	22,106
Mining	224	243	345	345	341	380	417	443	457	471
Nuclear	239	327	324	400	394	412	463	536	584	507
Petroleum	298	381	468	541	665	784	994	1,005	1,171	1,322
Other Engineering. Disciplines	5,779	5,756	5,738	5,670	5,806	6,361	6,826	7,061	7,007	6,866
TOTAL	76,003	76,103	75,823	77,107	75,320	79,528	84,599	87,733	90,233	93,883

Source: Engineering Workforce Commission of the American Association of Engineering Societies, 2015.

Unfortunately, overall hiring contracted nearly 50 percent in 2002–2003, due in part to the aftershock of 9/11. Hiring trends slowly recovered each year from 2004 to 2008, but then the economic downturn of 2008 began to negatively influence employment opportunities for new college graduates, including engineers, as available positions became highly competitive. It was not until 2010–2011 that a gradual shift in hiring patterns began to emerge. Students were once again finding that employers were returning to college campuses, job prospects were improving, and more graduates were finding employment.

The most recent survey of over 4,700 employers for 2015–2016 shows that the job market has exploded during the last three years. The overall hiring of new college graduates has increased by the largest amount in the last decade. If this trend continues, engineering students should devote increased efforts to finding and choosing the right opportunity among several options they should have available. Also, it should be noted that employment opportunities and starting salaries for engineers have traditionally surpassed those of most other majors on college campuses, and this trend continues into 2016.

The Bureau of Labor Statistics, in its 2015 report, shows that over 2.4 million people work as engineers and other closely related professions. Figure 3.5 illustrates the recent trends in engineering employment over the last decade. These trends involve some modest fluctuation, perhaps due to corporate downsizing, and some engineering job losses. However, the overall picture is one of fairly stable employment.

Figure 3.6 illustrates the relatively low unemployment in the engineering profession. In 1994 the unemployment rate of engineers was about 3.5 percent. Since that time, there was a steady decrease, until 2001, when the unemployment rate was around 1.7 percent. Since 2001, engineering unemployment climbed over 3.0 percent in 2003, and then steadily dropped to 1.3 percent in 2007. However, 2008 numbers show engineering unemployment at 3.3 percent; the rate climbed to 6 percent in mid-2010. While the 2010 engineering unemployment rate was the highest since

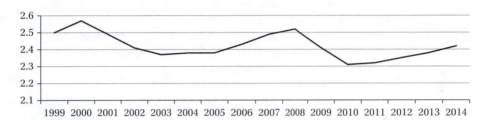

Figure 3.5 Numbers of employed engineers and other closely related occupations

Source: Bureau of Labor Statistics, 2015

Figure 3.6 U.S. General population unemployment compared to engineering unemployment: recent trends, 1990–2015

Source: Bureau of Labor Statistics, 2015.

1994, the unemployment rate of engineers continues to be well below that of the general population. Data from 2011 to 2016 show unemployment rates steadily dropping for the entire workforce. The engineering figures show a similar decline.

3.6   Salaries of Engineers

Salaries of engineering graduates consistently have been among the highest for all college graduates over the past several decades. To some extent, this has been due to the shortage of engineering candidates to fill available jobs. However, even in down years and leaner economic periods, engineers tend to do better in the job market than do their counterparts from other majors. Table 3.3 compares the starting salaries of graduates in a broad range of disciplines for the years 2016 versus 2015. Once again, engineers and computer scientists have the highest starting salaries as well as the biggest percentage gain from 2015 to 2016.

As reflected in Table 3.4, recent data indicate that average starting salaries for 2016 graduates in selected engineering fields have varied significantly. Starting salary offers in many disciplines have increased while a few have dropped slightly. Petroleum engineering graduates with bachelor's degrees are at the top of the salary list. This trend may reflect the shortage of trained graduates in that field.

Table 3.5 provides starting salary data for 2016 graduates by engineering curriculum and some of the types of employers that hire students in those fields.

One question often raised by engineering students concerns the long-term earning potential for engineers. In other words, does the high starting salary hold up over the length of one's engineering career? Figure 3.7 presents 2015 median salaries for

Table 3.3 Average Starting Salary Offers in Broad Area Disciplines Reported, January 2016 vs. January 2015 (for graduates with bachelor's degrees)

Discipline	2016 Average	2015 Average	Percent Average Change in Starting Salaries
Engineering	$64,891	$62,998	3.0%
Computer Science	$61,321	$61,287	0.06%
Math & Sciences	$55,087	$56,171	–1.93%
Business	$52,236	$51,508	1.41%
Agriculture & Natural Resources	$48,729	$51,220	–4.86%
Healthcare	$48,712	$50,839	–4.18%
Communications	$47,047	$49,395	–4.75%
Social Sciences	$46,585	$49,047	–5.10%
Humanities	$46,065	$45,042	2.27%
Education	$34,891	$34,498	1.14%

Source: Reprinted from the January 2016 Salary Survey, with permission of the National Association of Colleges and Employers, copyright holder.

Table 3.4 Starting Salary Offers in Engineering Reported, January 2016

Curriculum	2016 Average	2013 Average	3-Year Percent Change in Average Starting Salaries	25th Percentile (2016)	75th Percentile (2016)
For graduates with bachelor's degrees					
Aerospace	$63,117	$64,400	–1.99%	$62,000	$68,000
Bioengineering / Biomedical	$61,108	$47,300	29.19%	$58,000	$65,000
Chemical	$69,196	$67,600	2.36%	$64,000	$73,000
Civil	$61,734	$57,300	7.73%	$57,800	$65,000
Computer Engr.	$65,606	$71,700	–8.50%	$61,500	$69,500
Computer Sci.	$62,675	$64,800	–3.27%	$56,000	$67,000
Electrical	$66,269	$63,400	4.53%	$63,000	$70,000
Environmental	$60,375	*	–	$55,000	$66,500
Industrial / Manufacturing	$62,242	$56,300	10.55%	$60,000	$65,000
Materials	$63,478	*	–	$60,000	$68,000
Mechanical	$65,593	$64,000	2.49%	$61,000	$68,500
Nuclear	$62,625	*	–	$61,000	$68,000
Petroleum	$89,563	$93,500	–4.21%	$81,500	$103,500
For graduates with master's degrees					
Aerospace	$72,887	*	–	$70,490	$77,500
Bioengineering / Biomedical	$71,140	*	–	$64,000	$80,700
Chemical	$79,894	*	–	$75,000	$81,000
Civil	$67,962	*	–	$60,000	$72,350
Computer Engr.	$74,513	*	–	$68,000	$77,000
Computer Sci.	$73,563	$74,200	–0.86%	$64,000	$82,000
Electrical	$74,896	$67,200	11.45%	$69,700	$80,000
Industrial / Manufacturing	$73,756	*	–	$66,000	$81,350
Materials	$74,721	*	–	$71,280	$80,700
Mechanical	$75,391	$67,900	11.03%	$68,000	$78,000

* Data not reported.

Source: Reprinted from the January 2016 Salary Survey, with permission of the National Association of Colleges and Employers, copyright holder.

all engineers based on the number of years since their bachelor's degree. It is apparent that an engineer's relative earnings hold up well for many years following graduation.

One significant difference in salaries is shown in Figures 3.8 and 3.9. These data relate the 2015 salary difference between engineers who are placed into supervisory

Table 3.5 Average 2016 Starting Salary by Curriculum and Some Common Types of Employers for Each Field.

Curriculum/Type of Employer*	2016
For graduates with bachelor's degrees	
▪ Aerospace / Aeronautical Engineering	
Miscellaneous Manufacturing	$62,600
▪ Bioengineering / Biomedical Engr.	
Computer & Electronics Manufacturing	$62,333
▪ Chemical Engineering	
Chemical & Pharmaceutical Manufacturing	$73,318
Computer & Electronics Manufacturing	$67,600
Engineering Services	$65,750
Food & Beverage Manufacturing	$66,750
Miscellaneous Manufacturing	$65,786
▪ Civil Engineering	
Construction	$59,875
Engineering Services	$59,595
Miscellaneous Manufacturing	$61,600
Utilities	$66,250
▪ Computer Engineering	
Computer & Electronics Manufacturing	$69,833
Finance, Insurance & Real Estate	$70,000
Information	$61,600
Miscellaneous Manufacturing	$62,250
Miscellaneous Professional Services	$72,167
▪ Computer Science	
Computer & Electronics Manufacturing	$75,250
Finance, Insurance & Real Estate	$59,893
Information	$67,000
Miscellaneous Manufacturing	$61,917
Miscellaneous Professional Services	$67,340

Curriculum/Type of Employer*	2016
▪ Electrical / Electronics Engineering	
Chemical & Pharmaceutical Manufacturing	$76,917
Computer & Electronics Manufacturing	$71,417
Engineering Services	$61,287
Miscellaneous Manufacturing	$64,667
Miscellaneous Professional Services	$61,233
Utilities	$66,400
▪ Industrial / Manufacturing Engineering	
Chemical & Pharmaceutical Manufacturing	$62,667
Computer & Electronics Manufacturing	$63,857
Finance, Insurance & Real Estate	$69,500
Miscellaneous Manufacturing	$61,571
Wholesale Trade	$58,133
▪ Materials Engineering	
Computer & Electronics Manufacturing	$68,000
Finance, Insurance, & Real Estate	$71,000
Miscellaneous Manufacturing	$61,750
▪ Mechanical Engineering	
Chemical & Pharmaceutical Manufacturing	$72,864
Computer & Electronics Manufacturing	$66,636
Construction	$63,200
Engineering Services	$60,625
Finance, Insurance, & Real Estate	$64,700
Miscellaneous Manufacturing	$62,988

Curriculum/Type of Employer*	2016
For graduates with master's degrees	
▪ Chemical Engineering	
Chemical & Pharmaceutical Manufacturing	$76,125
Computer & Electronics Manufacturing	$78,333
Oil & Gas Extraction	$86,333

- **Civil Engineering**
  - Construction — $61,000
  - Engineering Services — $60,908
  - Utilities — $73,000
- **Computer Engineering**
  - Computer & Electronics Manufacturing — $88,500
  - Information — $66,000
  - Miscellaneous Professional Services — $82,250
- **Computer Science**
  - Computer & Electronics Manufacturing — $93,000
  - Finance, Insurance, & Real Estate — $61,000

- Information — $75,667
- **Electrical / Electronics Engineering**
  - Computer & Electronics Manufacturing — $88,714
  - Engineering Services — $70,090
  - Miscellaneous Professional Services — $75,625
- **Mechanical Engineering**
  - Chemical & Pharmaceutical Manufacturing — $75,667
  - Computer & Electronics Manufacturing — $83,500
  - Miscellaneous Manufacturing — $72,397
  - Utilities — $73,000

Source: From the January 2016 Salary Survey, with permission of the National Association of Colleges and Employers, copyright holder.

* Definitions of Employer Types:

Construction: The construction sector comprises establishments primarily engaged in the construction of buildings or engineering products (e.g., highways and utility systems.)

Engineering Services: This industry comprises establishments primarily engaged in applying physical laws and principles of engineering in the design, development, and utilization of machines, materials, instruments, structures, processes, and systems.

Finance, Insurance, and Real Estate: This sector includes establishments primarily engaged in financial transactions and/or in facilitating financial transactions, and consulting in fields such as insurance and real estate.

The *Information* sector comprises establishments engaged in the following processes: (a) producing and distributing information and cultural products, (b) providing the means to transmit or distribute these products as well as data or communications, and (c) processing data.

Management of Companies and Enterprises: This sector comprises 1) establishments that hold the securities of companies and enterprises for the purpose of owning a controlling interest or influencing management decisions, or 2) establishments that administer, oversee, and manage establishments of the company or enterprise.

Manufacturing: The manufacturing sector comprises establishments engaged in the mechanical, physical, or chemical transformation of materials, substances, or components into new products.

Mining, Quarrying, and Oil and Gas Extraction: This sector comprises establishments that extract naturally occurring mineral solids, such as coal and ores; liquid minerals, such as crude petroleum; and gases, such as natural gas.

Professional Services: This sector comprises establishments that specialize in performing professional, scientific, and technical activities for others. These activities require a high degree of expertise and training. (Examples: accounting, bookkeeping, payroll services, architectural, engineering, and computer services.)

The *Utilities* sector comprises establishments engaged in the provision of the following utility services: electric power, natural gas, steam supply, water supply, and sewage removal. Within this sector, the specific activities associated with the utility services provided vary by utility: electric power includes generation, transmission, and distribution; natural gas includes distribution; steam supply includes provision and/or distribution; water supply includes treatment and distribution; and sewage removal includes collection, treatment, and disposal of waste through sewer systems and sewage treatment facilities.

The *Wholesale* Trade sector comprises establishments engaged in wholesaling merchandise, generally without transformation, and rendering services incidental to the sale of merchandise. The merchandise described in this sector includes the outputs of agriculture, mining, manufacturing, and certain information industries, such as publishing.

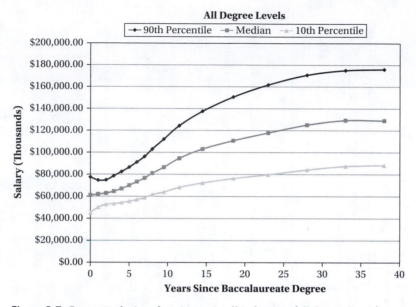

Figure 3.7 Current salaries of engineers in all industries (all degree levels)
Source: EWC, 2015 Engineering Salary Survey.

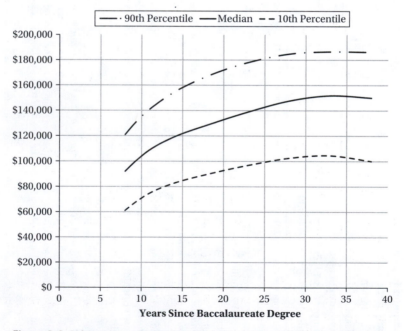

Figure 3.8 Salary curves for engineering supervisors, all degree levels
Source: EWC, 2015 Engineering Salary Survey.

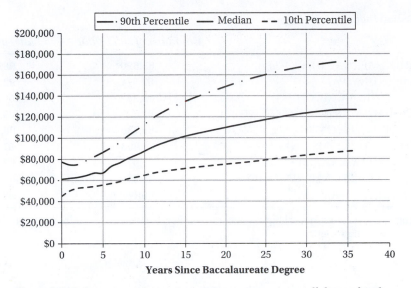

Figure 3.9 Salary curves for engineering non-supervisors, all degree levels
Source: EWC, 2015 Engineering Salary Survey.

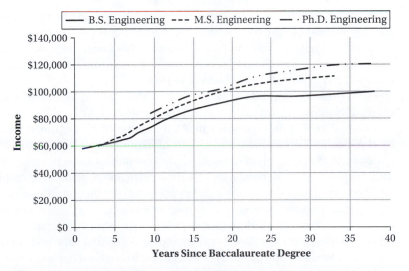

Figure 3.10 Current median income of all engineers in all industries by
highest degree earned and length of experience
Source: EWC, 2015 Engineering Salary Survey.

positions and those who remain in non-supervisory roles. Over the course of their career, supervisors tend to earn significantly higher salaries than those in non-supervisory positions.

Figure 3.10 shows how an engineer's income is influenced by the level of education achieved. As might be expected, those with master's and doctoral degrees have

Table 3.6 Median Income ($) by Region

Region	Graduates of 2000–2001	Graduates of 1990–1993
New England (CT, ME, MA, NH, RI, VT)	85,690	98,422
Middle Atlantic (NJ, NY, PA)	82,474	92,379
East North Central (IL, IN, MI, OH, WI)	85,282	100,374
West North Central (IA, KS, MN, MO, NE, ND, SD)	81,679	105,461
South Atlantic (DE, DC, FL, GA, MD, NC, SC, VA, WV)	84,736	98,979
East South Central (AL, KY, MS, TN)	74,963	91,901
West South Central (AR, LA, OK, TX)	80,972	82,501
Mountain (AZ, CO, ID, MT, NM, NV, UT, WY)	88,829	105,981
Pacific (AK, CA, HI, OR, WA)	85,177	99,172

Source: 2012 Engineers' Salaries: Special Industry Report. Engineering Workforce Commission of the American Association of Engineering Societies, Inc.

higher compensation levels. Length of experience in the engineering field is also an income determinant. Figure 3.10 demonstrates a consistent, regular growth in an engineer's income level as experience is gained.

Table 3.6 demonstrates 2012 median salaries by region of employment. This compares graduates of 1990–1993 with those of 2000–2001. For 10-year-out graduates, salaries are highest in the Mountain Region. The lowest reported salaries are for engineers in the East South Central Region. (Geographic strongpoints have shifted dramatically and repeatedly in recent years.)

For 20-year-out graduates, salaries are highest in the Mountain and West North Central Regions. Lowest salaries are in the West South Central Region.

Figure 3.11 shows the current median income of all engineers in all industries by the size of the company and length of experience. Large companies are those with over 5,000 total employees. Medium companies have between 500 and 5,000 total employees. Small companies are those with 500 or fewer total employees.

Size of company does make a difference in career income. Those with large companies maintain the highest salaries over a 38-year career. Those with medium companies stay second highest, while those with small companies remain third throughout their career. However, it should be noted that salaries in companies of all sizes maintain a steady upward growth curve throughout one's career.

Figure 3.12 provides the 2015 median annual income by major branch of engineering for all working engineers. Petroleum engineers show the highest income levels, with agricultural engineers at the lowest levels.

It is also interesting to examine median annual income by industry or service. Table 3.7 compares the 2012 median salaries of graduates from 1990–1993 with those of 2000–2001. The data show those involved in certain manufacturing areas are at the top of the salary list. The lowest from these classes are involved in public administration.

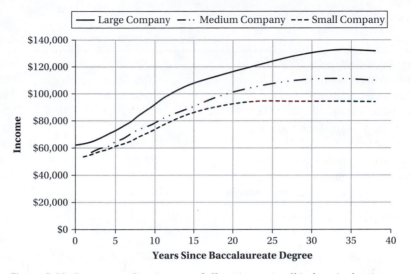

Figure 3.11 Current median income of all engineers in all industries by size of company and length of experience

Source: EWC, 2015 Engineering Salary Survey.

Figure 3.12 Income by engineering discipline, 2014 (median annual wage)

Source: Bureau of Labor Statistics, Division of Occupational Employment Statistics, 2015.

Table 3.7 Median Income ($) by Industry Sector

Industry Sector	Graduates of 2000–2001	Graduates of 1990–1993
All Manufacturing Industries	87,929	100,659
Metal Manufacturing Industries	N/A	88,751
Metal Products Manufacturing	73,931	81,631
Machinery Manufacturing	74,932	89,979
Transportation Equipment Manufacturing	94,540	100,130
Motor Vehicle Manufacturing	96,014	100,754
All Non-Manufacturing Industries	78,433	93,281
Transportation and Warehousing	73,231	84,196
Architectural, Engineering, and Related Services	82,246	106,204
Construction	82,174	100,082
Mining	86,471	107,730
Electric Power Generation	75,655	103,106
Scientific Research and Development Services	84,951	104,269
Professional, Scientific, and Technical Services	84,083	103,682
Public Administration	70,583	88,875
Utilities	79,161	103,900

Source: 2012 Engineers' Salaries: Special Industry Report. Engineering Workforce Commission of the American Association of Engineering Societies, Inc.

3.7 The Diversity of the Profession

For many years, engineering was a profession dominated by white males. In recent years, this has been changing as more women and minority students have found engineering to be an excellent career choice. Unfortunately, the rate of change in both enrollment and degrees awarded to these students has been slower than expected. This is especially discouraging at a time when population statistics show a significant increase in the number of college-bound women and minority students.

Table 3.8 provides information concerning degrees awarded to women and minority groups over a recent 10-year period. While the numbers have been increasing, the growth rate has been slower than expected. Since recent demographic data show that the number of minority students attending college is growing, we would expect to see a larger percentage of these students earning engineering degrees. So far, this is not the case—especially at the graduate level, where increases have been steady, but modest.

Table 3.9 on page 104 shows enrollment data for women and minorities in all undergraduate engineering programs. The percentage of women in engineering is almost 20 percent, while the percentage of under-represented minorities is about 15 percent. These figures have dropped during the past few years, despite increased recruitment efforts by colleges and universities.

Table 3.8 Degrees in Engineering by Level and Type of Student, 2005–2014

Level and Type	2005	2006	2007	2008	2009	2010	2011	2012	2013	2014
Bachelor's Degrees	76,003	76,103	75,823	77,107	75,320	79,528	84,599	87,761	90,233	93,883
Women	14,868	14,654	14,101	13,865	13,432	14,478	15,437	16,367	17,110	18,535
African Americans	3,756	3,673	3,735	3,470	3,392	3,635	3,457	3,587	3,940	4,936
Hispanic Americans	4,890	4,957	5,133	5,486	5,488	6,053	6,574	7,203	7,901	12,155
Native Americans	378	456	426	427	328	378	363	355	376	311
Asian Americans	10,033	9,719	9,466	9,143	8,743	9,174	9,680	10,129	10,881	10,807
Foreign Nationals	5,644	5,354	5,152	4,787	4,519	4,948	5,532	6,325	6,866	7,129
Master's Degrees	41,087	38,451	37,805	40,122	41,967	43,257	45,589	47,780	46,949	42,966
Women	9,212	8,731	8,393	9,237	9,596	9,798	10,310	10,889	11,190	10,356
African Americans	1,072	1,009	1,078	1,116	1,137	1,219	1,246	1,324	1,351	1,212
Hispanic Americans	1,194	1,185	1,292	1,318	1,393	1,543	1,617	1,872	2,028	1,931
Native Americans	123	128	81	78	94	98	105	98	83	69
Asian Americans	3,994	3,990	3,995	3,863	3,629	4,097	4,316	4,033	4,004	3,236
Foreign Nationals	17,536	15,441	14,604	16,408	18,426	18,360	18,868	19,837	20,412	19,766
Doctoral Degrees	7,276	8,116	8,614	9,449	8,907	9,243	10,086	9,558	9,957	9,373
Women	1,322	1,592	1,689	1,976	1,896	2,073	2,185	2,055	2,193	2,106
African Americans	111	121	122	127	139	165	169	188	189	175
Hispanic Americans	107	99	120	152	163	224	206	190	192	238
Native Americans	15	11	15	15	16	14	12	12	4	14
Asian Americans	391	496	432	508	582	678	839	641	594	580
Foreign Nationals	4,405	5,048	5,180	5,564	4,869	4,871	5,334	5,063	5,554	5,117

Source: Engineering Workforce Commission of the American Association of Engineering Societies, 2005–2014.

Table 3.9 Women and Minorities in the Engineering Pipeline, 2014 (% of all full-time undergraduates)

	First-Year Full-Time B.S. Students	*All Full-Time B.S. Students*
Women	19.66	19.63
African Americans	5.68	4.80
Hispanic Americans	10.48	10.38
Native Americans	0.43	0.42
Asian Americans	9.54	10.31

Source: Engineering Workforce Commission of the American Association of Engineering Societies.

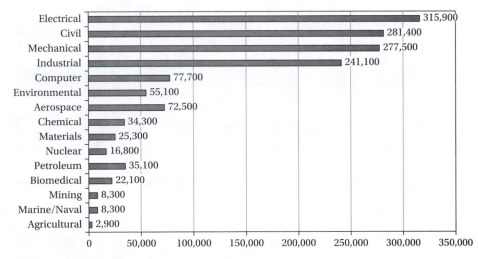

Figure 3.13 Practicing engineers in the U.S., 2015
Source: Bureau of Labor Statistics, 2015.

3.8   Distribution of Engineers by Field of Study

According to the Bureau of Labor Statistics (Fig. 3.13), almost 316,000 of all practicing engineers are employed in the electrical engineering discipline. This is followed by over 281,000 civil engineers and 277,000 mechanical engineers. The smallest numbers of practicing engineers are in the agricultural (2,900), marine/naval (8,300), and mining (8,300) sectors. Comparing this information to previously presented data on enrollment and degrees, one would expect these numbers to remain somewhat steady.

3.9   Engineering Employment by Type of Employer

According to the Bureau of Labor Statistics, in 2015 the largest sector of engineering and related science employment, 34 percent, was in Professional, Scientific, and Technical Services. This includes areas such as engineering and architectural services, research and testing services, and business services in which firms designed construction projects or did other engineering work on a contractual basis. Another 22 percent were involved in information-related occupations.

Government agencies at the federal, state, and local levels employed approximately 8 percent of engineers in 2015. The largest number were employed by the federal government in the Departments of Defense, Transportation, Agriculture, Interior, and Energy and the National Aeronautics and Space Administration (NASA). The majority of state and local workers are usually involved in local transportation, highway, and public works projects. Engineers were also working in manufacturing, communications, utilities, and computer industries.

3.10 Percent of Students Unemployed or in Graduate School

How does engineering unemployment compare to the number of students who go on to graduate school? As shown in Table 3.10 and referred to earlier in Figure 3.1, the total number of students pursuing graduate engineering degrees increased significantly between the late 1970s and 1992. Since then, there was a steady decline until 2000. From 2001 to 2011 enrollment in graduate engineering programs soared to 168,000. The recent trend is similar to the early period covered by Table 3.10 when jobs for bachelor-level graduates were not as plentiful, and many were opting for graduate school, which may explain the rapid rise in graduate school enrollments during that period.

Table 3.10 also shows the lower unemployment rate for graduating engineers between 1990 and 2013. The figures for 1992 and 1994 and 2002–2011 reflect a tighter job market than that of recent years. It should also be noted that, in general, the unemployment rate for new engineering graduates is far lower than for any other major.

As the engineering job market has improved in recent years, one might expect that the unemployment rate for engineers will continue to decrease, and the number going to graduate school may also start to decline.

3.11 A Word from Employers

As discussed earlier, the current job market for graduating engineers is encouraging. However, there are also important issues that employers want students to consider. Even though this study is over 16 years old, the message is still important for today's engineering college student. In a 1999 survey of 450 employers done by the Collegiate Employment Research Institute at Michigan State University, recruiters were asked to provide commentary on any special concerns they had about new engineering graduates. One interesting conclusion to the study found that

> *Employers want the total package when they hire their next engineering graduates. Not satisfied with academically well-prepared graduates, employers want individuals who possess and can demonstrate excellent communication and*

Table 3.10 Selected Disciplines, Engineering Graduate School Full-Time and Part-Time Enrollment Trends, and Percentage of Unemployed Engineers

	1990	1992	1994	1998	2000	2002	2004	2005	2007	2008	2011	2013
Aerospace	3,563	3,935	3,550	2,780	3,042	3,171	3,750	3,758	3,990	4,291	5,250	5,056
Chemical	6,657	6,926	7,292	6,683	6,714	6,963	7,282	6,878	6,949	7,268	8,073	7,884
Civil	11,625	13,900	14,164	11,009	11,151	12,546	12,942	12,561	12,787	13,430	16,545	14,967
Electrical	32,095	32,635	29,524	25,302	27,764	33,851	34,127	32,691	35,306	35,778	35,750	34,576
Industrial	9,713	11,552	11,401	9,428	5,234	6,330	5,789	5,340	5,677	6,302	6,913	6,550
Mechanical	17,115	18,566	17,310	14,032	15,201	17,085	17,945	17,554	18,349	18,880	21,860	21,247
TOTAL	80,750	87,510	83,230	69,220	69,106	79,946	81,835	78,782	83,058	85,949	94,391	90,280
% Unemployed Engineers	2.9%	5.2%	4.3%	1.7%	1.5%	3.0%	2.3%	2.1%	3.3%	3.1%	5.1%	3.5%

Sources: Engineering Enrollments and Degrees, Engineering Manpower Commission of the American Association of Engineering Societies, Inc. 1990, 1992, 1994, and Engineering Workforce Commission of the American Association of Engineering Societies, Inc., 1998, 2000, 2002, 2004, 2005, 2007, 2008, 2011, 2013.

interpersonal skills, teamwork, leadership, and computer/technical proficiency. A willingness to learn quickly and continuously, to problem solve effectively, and to use their common sense is also desired. New employees must be hard-working, take initiative, and be able to handle multiple tasks. . . . In addition, employers are looking for students with new emerging skills and aptitudes such as the ability to understand e-commerce, computer capabilities that include programming skills, and the ability to adapt to constant change. In the new global economy, increasing competition requires a strategy to respond quickly.

Just earning an engineering degree is no longer sufficient to prepare students for the fast-paced, technologically changing global economy. Students must devote time in college to developing those critical competencies that employers are seeking in order to be well prepared for the challenges that await them.

EXERCISES AND ACTIVITIES

3.1 If an administrator used only the data from 1975 to 1980 to predict future engineering enrollment, how many first-year, full-time students would have been expected in 2010? What would the percentage error have been?

3.2 What is the percentage increase in projected high school graduates in 2021 compared to those in 2010?

3.3 Using the data from Figure 3.5, estimate the number of employed engineers in three years.

3.4 Using the starting salary data from Table 3.4, and assuming an annual increase at the current cost of living (currently about 2 percent), calculate what you could expect to receive as a starting salary in your chosen field of study if you graduate three years from now.

3.5 Using the information from Figure 3.10, estimate the engineering salary after 30 years assuming (a) a bachelor's degree; (b) a master's degree; and (c) a doctorate degree.

3.6 Using the information from Figures 3.8 and 3.9, estimate the median salary after 30 years assuming the engineer is (a) a supervisor and (b) a non-supervisor.

3.7 If you were a supervisor in the upper decile rather than a non-supervisor in the upper decile, what would be your percentage increase in salary 20 years after graduation?

3.8 Calculate the percentage increase of the highest-paid bachelor's-degree engineer in 2016 compared to the lowest-paid bachelor's-degree engineer in 2016.

3.9   Suggest at least one reason for the relatively low income in the East South Central Region of Table 3.6.

3.10   Suggest at least one reason for the relatively low income of government or public administration workers compared to those in the private sector.

3.11   Review the data in Tables 3.8 and 3.9. As the enrollment of women and minority students is rising on college campuses, the relative growth for these groups studying engineering has not been increasing as quickly. Discuss some reasons, and provide examples, why you feel that more women and minorities are not pursuing engineering degrees. Explore some of the student support programs on your campus. What are some of the things your school is specifically doing to encourage and support these populations to pursue engineering degrees?

3.12   Estimate the percentage of engineers that are chemical engineers using the data in Figure 3.13.

3.13   Based on the current economic situation, do you expect the employment demand for graduating engineers to increase or decrease? Explain the basis for your answer.

3.14   If we experience a significant economic recovery, what do you think will happen to future enrollments in graduate engineering programs?

3.15   Review the information presented in Figures 3.2 and 3.3 and Table 3.8. Make a prediction about the future enrollment of women and minorities in engineering. Explain your answer.

3.16   What factors do you think influence how employers determine hiring needs in any given year?

3.17   Attend a campus career fair. Talk with at least three recruiters from different firms. Make a list and discuss those factors that employers are stressing in their hiring decisions.

3.18   Based on the current economic situation, do you expect the salaries of engineers to continue to increase at the recent pace? Why or why not? Using other tables in this chapter, explain the basis for your answer.

3.19   Review the enrollment data in this chapter. What do you think high schools, colleges, and the engineering profession can do to increase the number of students choosing to pursue an engineering career?

3.20   What factors do you think contribute to the fact that entry-level engineering salaries are usually higher than any other major on campus?

CHAPTER 4

Global and International Engineering

4.1 Introduction

Many students who choose to study engineering make this decision primarily because of their aptitude in mathematics and science. As a result, most enter college thinking that they will no longer be required to take courses in the liberal arts and humanities. Many are especially pleased to avoid further foreign language study. Once they arrive on campus, they learn that some study in the liberal arts is still required, but the fact remains that few, if any, engineering students are required to complete a foreign language requirement. For beginning students, the concept of using and applying their background in an environment requiring the use of a foreign language seems terribly remote to them. Most envision their career as remaining fairly stable, with one employer, in one location, for many years, perhaps even a lifetime. They have probably never considered the opportunity to "go global" with their engineering career.

Perhaps no other factor has more dramatically shaped an engineering career than the rapid globalization of the engineering profession. This globalization evolved initially from the post–World War II movement toward international markets, sustained by large multinational corporations providing goods and services to many regions of the world. In more recent years the globalization of the engineering profession has increased rapidly due to corporate mergers and acquisitions, the availability of high-speed, low-cost communication systems, and cheap labor costs in many nations.

In the Old World marketplace, countries competed by trading goods and services that were primarily developed and produced using capital and labor from within their own borders. However, as global restrictions eased and trade barriers changed, companies found they could maximize profits by investing capital resources in a variety of countries. Coupled with new policies that enable the movement of labor across political boundaries, many organizations have added incentives to globalize their business. The current market climate encourages firms to shift factories, jobs, research facilities, distribution centers, and even corporate management operations in order to maximize market share, develop new products, and access needed raw materials and less costly skilled and unskilled labor, while taking advantage of incentives that lower costs and maximize profits.

Since most of the products and services that are generated by these corporations are developed, designed, and manufactured by engineers, these engineers now find themselves in a unique position as critical players in this rapidly expanding global economy. Today's engineers must be better educated and properly trained to meet the challenges of this role. They must have an understanding of the world community around them, world cultures, and the world marketplace. Strong technical skills by themselves are often insufficient. Unfortunately, in many cases engineers are the problem solvers but are weak in the cultural and language skills required in today's global marketplace.

This chapter will focus on several factors that drive the global economy. It also will explore many of the international opportunities available to engineers, and discuss the preparation needed for an international engineering career.

4.2   The Evolving Global Marketplace

Many factors have been instrumental in the rapid acceleration of the global marketplace.

Changing World Maps and Political Alliances

One need only examine a map from 1990 and compare it to a current map to see the many changes that have occurred in that short period of time. New countries have been created by the breakup of the former Soviet Union. Organizational and governmental changes have occurred in Eastern Europe (the reunification of Germany, the breakup of Czechoslovakia, etc.). New nationalistic tendencies have taken root in the Far and Middle East, Asia, and Africa. The effects of September 11, 2001, and world terrorism and the wars in Iraq and Afghanistan have also changed the world's political and economic climate. Each of these situations has had both positive and negative impacts on the economic conditions that influence market demand and world production and distribution of goods and services.

Additionally, ever-evolving political alliances, the changing of governments and political leaders, and the development of new laws, regulations, and policies have greatly affected the global market. In recent years, some former enemies have become allies, while some former friends are now embroiled in turbulent relationships. Russia, for decades seen as America's chief adversary, is now a partner in exciting new ventures in space exploration and industrial development. Former enemies in the Middle East and Asia are now economic partners in many joint endeavors with U.S. business and industry. As these changes unfold, new opportunities emerge for American firms to expand markets, establish new production sites, and enhance their global presence.

The Role of Mergers, Acquisitions, and International Partnerships

Chrysler and Daimler-Benz, Pfizer, Pharmacia, and Upjohn, Boeing and McDonnell Douglas, Dow Chemical and Union Carbide, General Motors and Saab are but a few examples of the thousands of mergers, acquisitions, and industrial partnerships that have become commonplace in the global economy. Almost on a daily basis the media inform us of yet another joint venture, buyout, or corporate spinoff with international implications. No industry has been immune from these activities, as computer, electronic, pharmaceutical, chemical, communication, automotive, and banking firms have joined forces in an attempt to combine technologies, increase revenues, and manage costs.

The net effect of these activities has accelerated global opportunities for engineers, regardless of their initial interest in international careers. If engineers working for a U.S.-based corporation suddenly learn that their employer has been acquired by an international firm, by default those individuals become global engineers. Following such an acquisition, management decisions may now originate from an overseas site. Product development, design, and manufacturing may now be performed from a different cultural perspective. Expectations and goals may change significantly. Travel, communications, and interactions with others in the firm may now take on a new perspective.

NAFTA

One of the most dramatic, controversial, and significant influences on the globalization of the engineering profession was the 1994 North American Free Trade Agreement (NAFTA) between the United States, Canada, and Mexico. NAFTA was designed to lower tariffs and increase international competitiveness. Based on original projections it was forecast that NAFTA would eventually create the world's largest market, comprising 370 million people and $6.5 trillion of production—from "the Yukon to the Yucatan." At present, NAFTA has created the second largest free trade market in the world, after the European Economic Area, which has a population of 375 million and a gross domestic product of $7 trillion. Large and small corporations around the world have felt the impacts of NAFTA.

NAFTA's primary objective was to liberalize trade regulations as a way to stimulate economic growth in this region, and to give NAFTA countries equal access to each other's markets. Virtually all tariff and non-tariff barriers to trade and investment between NAFTA partners were eliminated in 2009. However, as an important side feature, each NAFTA member can still establish its own external tariffs for trade with third countries.

Other important outcomes for this agreement include an increased market access for trade, greater mobility for professional and business travelers, legal protection for copyrights and patents, and a mechanism for tariffs to reemerge if an import surge hurts a domestic industry. Almost all economic sectors are covered by NAFTA, though special rules apply to particularly sensitive areas such as agriculture, the automotive industry, financial services, and textiles and apparel. Side agreements cover cooperation on labor issues and environmental protection.

Since its inception, NAFTA has been very controversial in all participating countries. Many in the U.S. felt that it would result in the move of critical manufacturing jobs to cheaper labor markets (Mexico), the relocation of new production facilities, and a rise in unemployment and inflation. In actuality, more U.S. jobs have been lost to locales with cheaper labor than Mexico, notably Central America and Asia.

According to the NAFTA Free Trade Commission of trade representatives of the three participating countries, NAFTA has achieved many positive results. Based on the data in this report, the success of NAFTA is illustrated by the following:

- Since 1994, trade between the U.S., Canada, and Mexico has more than doubled. From less than $297 billion in 1993, trilateral trade has surpassed $1.6 trillion in less than 20 years.
- Investment in the three economies has increased significantly; total foreign direct investment in the NAFTA countries has increased by over US$1.7 trillion.
- As a result of this growth in trade and investment, 40 million new jobs have been created in all three countries, meaning lower costs and more choice for consumers.

The report summarizes: "The evidence is clear—NAFTA has been a great success for all three parties. It is an outstanding demonstration of the rewards that flow to outward-looking, confident countries that implement policies of trade liberalization as a way to increase wealth, improve competitiveness and expand benefits to consumers, workers and businesses."

Many displaced U.S. workers would strongly disagree with this assessment. In several areas, notably the automotive and manufacturing sectors, union workers are fighting to preserve jobs, worker security, and plant projects as more activities are being moved to non-U.S. locales.

The opportunities, challenges, and experiences available to engineers as a result of NAFTA are exciting. Engineers now find that their assignments may take them to Mexico or Canada, perhaps even for extended periods of time. Interactions between work colleagues may actually take place from remote distances via e-mails or Internet-based teleconferences. Today's engineer may be planning production implementation for a plant opening in Mexico or Canada, with the engineering done in the U.S. Or a Canadian firm may be producing goods in a U.S. or Mexican plant. The world of engineering has truly changed in North America and for the North American engineer.

The European Union

Another factor that has played a significant role in the globalization of the engineering profession has been the development of the European Union (EU). Previously known as the European Community, it is an institutional framework for the construction of a united Europe. It was created after World War II in order to unite the nations of Europe economically, with the intent that such an alliance would make another war among them unthinkable. In May 2004, the number of member countries grew from 15 to 25, and in 2007 it grew again to 27 with the addition of Romania and Belgium. They share the common economic institutions and policies that have brought an unprecedented era of peace and prosperity to Western and Eastern Europe. The first steps toward European integration were made when the six founding member countries pooled their coal and steel industries. They then set about creating a single market in which goods, services, people, and capital would move as freely as within one country. The single market came into being in January 1993. The original treaties gave the EU authority in a number of areas, including foreign trade, agriculture, competition, and transport. Over the years, in response to economic developments, formal authority was extended to new areas such as research and technology, energy, the environment, development, and education and training.

Of great relevance to world trade and economic competition, the EU cleared the way for the introduction of a single currency, which was implemented in January 2002, and the creation of a European Central Bank. As a result, the United States and Europe are now more economically interdependent than ever. According to the Delegation of the European Commission to the United States Report, the EU and the United States are involved in a thriving economic relationship. Their trade and investment relationship is the largest in the world. As each other's main trading partners, they are also in position as the largest trade and investment partners for nearly all other countries, therefore shaping the picture of the global economy as a whole.

In 2011, the EU generated an estimated gross domestic product (GDP) of US$15.04 trillion, which represents 31% of the world's total. It is the largest exporter of goods and the second largest importer. The investment trends are even more impressive as both the EU and the U.S. are each other's most important source of foreign direct investment, with a total two-way investment amount of $1.8 trillion. Direct employment due to these investments, along with indirect employment such as joint ventures and other trade-related job creation, indicates that 7 million U.S. workers are in jobs connected to EU companies.

As one of the world's largest trading powers, and as a leading economic partner for most countries, the EU is a major player on the world scene. Its scope for action extends increasingly beyond trade and economic questions. More than 130 countries maintain diplomatic relations with the EU, and the EU has over 100 delegations around the world.

The expansion of EU countries from 15 to 27 includes many in Central and Eastern Europe. Since the independence of these Central and Eastern European countries in 1989, the EU has drafted trade and cooperation accords with most of them and has been at the forefront of the international effort to assist them in the process of economic and political reform. The EU is also strengthening its links with the Mediterranean countries, which will receive some $6 billion in EU assistance over the next few years. The EU also maintains special trade and aid relationships with many developing countries. Under a special agreement, virtually all products originating from 70 African, Caribbean, and Pacific countries enjoy tariff-free access to the EU market.

It is obvious that the EU is, and will continue to be, a formidable ally as well as competitor to U.S. business interests. Many of the excellent international opportunities available to U.S. engineers have emerged and will continue to increase as a result of the partnerships, the competition, and the presence of the EU.

4.3   International Opportunities for Engineers

Automobile Industry

From the time the assembly-line process was developed by Henry Ford in the early 1900s until the late 1970s, the automobile industry was pretty much dominated by the Big Three in Detroit (General Motors, Ford, and Chrysler). Products from Germany, Japan, and Korea were present in the United States, but their quality was considered suspect. In the mid-1970s, after Japanese automakers began to implement quality improvement processes, their products started to become more popular in the world marketplace. The U.S. was recovering from a major escalation in gasoline prices, and high-mileage, low-cost cars were suddenly in demand. The foreign automakers' products fit this niche quite nicely. From that point in time, sales of U.S.-made vehicles dropped and those of foreign competitors gained overall market share. General Motors, which once commanded over half the market, is now fighting to achieve a 20% share, under the auspices of federal government subsidies.

Engineering students interested in a career in the automobile industry now have many exciting choices as a result of the changes in this dynamic industry. The traditional Big Three have expanded their territories by opening new plants and operating facilities in many regions of the world. Each of the major U.S. carmakers has established joint ventures, or bought or merged, with former competitors with the intent of increasing their presence and sales in new markets. It is not uncommon for their engineers to be involved in the planning and design of new products or facilities that are being developed for international customers. These individuals commonly interact with foreign colleagues.

Just as the U.S. automakers have positioned themselves strategically in important world markets, so has the foreign competition. Corporations such as Honda, Toyota, Nissan, Mazda, BMW, Mercedes Benz, and Volkswagen have all developed design, testing, and manufacturing facilities in the United States and other key locales. This allows them to be in closer contact with their foreign customer base, community, and workforce. While many of their employees have been transplanted from their home country, each company has implemented hiring processes to employ more American workers in the U.S., including engineers. It is now common for these firms to recruit full-time employees as well as co-op and internship candidates at many U.S. colleges.

Concurrent with this movement to locate facilities in foreign countries has been the increase in the development and expansion of firms that specialize in supplying products and parts used by the automakers. When some foreign carmakers began to establish their U.S.-based facilities, many of their foreign-based suppliers did the same. Now it is common to see many international suppliers located in close proximity to international automobile corporations. Firms that specialize in such things as electronics, engine technology, interior and exterior car products, and manufacturing equipment are expanding their American presence. Many that emerged solely as suppliers to foreign automakers have now found a strong customer base among the traditional Detroit Big Three as well. The auto supply business has become a competitive, lucrative market as all of the auto producers seek to keep costs down and maximize efficiency and profits.

As a result of this massive globalization of the automobile industry, many engineering graduates are finding rewarding opportunities with U.S.-based foreign automotive divisions and the rapidly growing base of suppliers, in addition to the Detroit Big Three. This industry has clearly emerged as one that provides many great opportunities for today's globally minded engineer.

Manufacturing

The philosophy associated with producing the world's goods and services has changed dramatically in recent years. As multinational corporations struggle to minimize costs while maximizing efficiency, there has been a significant increase in the location of manufacturing facilities and jobs in foreign lands. Many firms have found it more efficient and less costly to distribute products from a facility that is in close proximity to their customer base. Others have found it more cost-effective to produce their products in places where labor is cheaper, the quality is higher, and the government policies are more favorable for doing business. Therefore, it is no longer unusual in the realm of global production for a firm to have dozens of production facilities in many different countries.

Typically, much of today's research, design, and development is still being done in central locations, though actual production may be occurring in many different parts of the world. Engineers may find that their work will involve communication and input from colleagues around the globe. In addition to international telephone and e-mail communication, production problems, quality control issues, and meaningful design changes may necessitate frequent international travel for some engineers.

The typical foreign-based manufacturing facility is usually staffed by foreign national employees. Almost all of the production staff and skilled and unskilled labor are hired from the particular region. The corporate staff is usually involved in the initial startup of the facility, the hiring and training of the employees, and the monitoring of the manufacturing process. Often, it is most efficient for a corporation to use local engineering talent, so many firms will recruit international students attending U.S. colleges and universities to fill these roles. The overall objective is to provide the technical expertise to the local labor force so they will be in a position to run the operation. The number of actual U.S. employees permanently assigned to a particular facility at any one time will probably be small. Their role will be to manage the operation, deal with immediate issues that arise, and serve as the liaison between the home office and the production facility.

A similar situation exists for foreign firms that have established manufacturing facilities in the United States. Most of the onsite corporate staff will be from the home country, but the objective is to try to hire and train as many U.S. employees as possible. Foreign firms will invest time and money to train these individuals to adapt to their particular style of doing business. Many employees at these facilities will need to gain not only a technical knowledge of the engineering process, but also a proper cultural perspective. Certain methods or procedures may actually arise from a particular cultural background, belief system, or age-old methodology. This can be quite an adjustment for U.S. workers hired by these firms. Often, newly trained technical employees may be recruited from American colleges and universities to take advantage of their solid technical education and a broader understanding of cultural issues.

Many of the ideas and concepts that have revolutionized the manufacturing industry have been a result of the globalization of world production processes. Engineers and managers are adopting some of the best manufacturing technologies that have been developed in many different countries. Today's world of manufacturing is often a blend of the best practices and procedures gained from the integration of systems and concepts from engineers around the world. Manufacturing is truly a global enterprise in the life of today's engineer.

Construction Industry

The construction of roads, bridges, dams, airports, power plants, tunnels, buildings, factories, even whole cities, has created a worldwide boom for engineers involved in construction. While many U.S. firms have a long history of international construction activities, the many economic, political, and social changes around the globe have created a broad range of new opportunities for them and their engineers. Over the last 10 years, U.S. construction firms have recorded a significant increase in new foreign contracts. It is not uncommon for large construction firms to have corporate and field offices in several regions of the world. In fact, Bechtel, one of the largest global construction firms, lists on its website that it has offices in 140 countries.

The rapid growth of worldwide construction activity has been enhanced by the expansion of global economies, a loosening of trade restrictions due to NAFTA, increased competition from the EU, and a desire by many countries to improve productivity and living conditions after years of decay and neglect of their major facilities. Many of these countries look to U.S. engineering firms for assistance with these projects, since they have demonstrated the necessary technical and managerial expertise for large-scale construction jobs. One of the most impressive projects was the construction of the city of Jubail in Saudi Arabia by Bechtel in the 1970s. Recognized by the *Guinness Book of World Records* as the largest construction project in history, this project involved the construction of 19 industrial plants, almost 300 miles of roads, over 200,000 telephone lines, an airport, nine hospitals, a zoo, and an aquarium.

However, as the world's economic power base is changing, these U.S. construction firms find that they are now in direct competition for new building projects with firms from such countries as Germany and Japan. In fact, many building projects are actually joint ventures between U.S. firms and those of other places. A prime example was the construction of the Chunnel, the tunnel constructed to link Great Britain and France, which was actually developed by a team of construction companies, including several from Europe and the United States.

Many republics of the former Soviet Union and many Eastern European countries are relying on global engineering construction expertise. In addition, the rebuilding of Iraq will involve a worldwide construction effort. The end of the Cold War and the fall of the Berlin Wall have revealed massive needs for infrastructure improvement and industrial revitalization. The updating of transportation systems, the restoration of buildings, the renovation of power plant facilities (including nuclear-powered ones), and the conversion of industrial facilities from defense-oriented products to peace-time goods and services will require huge construction assistance. However, political instability coupled with uncertain economies may impede the actual rebuilding process.

As the world population grows, the need for the construction of new facilities to satisfy the demands of global customers should remain strong. Engineers should have ample opportunities to compete for many exciting global projects that will require a unique understanding of foreign cultures and world regions, and the ability to work as a team with others from across the globe—in addition to strong technical skills. However, competition will be strong from many new and existing construction firms around the world. Those who are prepared to compete in this environment will likely be successful.

Pharmaceutical Industry

Over time, the American pharmaceutical industry has been one of the most profitable and fastest-growing segments of the U.S. economy. The nation's pharmaceutical companies are generally recognized as global leaders in the discovery and development of new medicines and products. In addition, the U.S. is the world's largest single market, accounting for an estimated two-thirds of world pharmaceutical sales. For engineering students, the pharmaceutical industry can provide some interesting international opportunities in research and development, testing and manufacturing of equipment, construction of production facilities, and technical sales and management around the globe.

However, this field currently faces many challenges. The growing concerns and intensifying pressure from governments and the private sector to control costs have forced many manufacturers to restrict operations. In reaction to escalating drug prices, which grew at nearly three times the overall rate of inflation during the past 10 years, government and private managed-care providers have instituted new policies to hold down drug costs.

For those considering a career in this industry, there are many favorable conditions that should create interesting international opportunities. However, increasing competition from firms around the globe has provided new challenges as well. While some companies have shown only modest growth, the highly successful firms have been those whose research and development efforts have generated lucrative, innovative treatments. Long-term growth for the industry is supported by a number of favorable conditions: the recession-resistant nature of the business, the aging population, and ambitious research and development efforts aimed at generating new and improved drugs (especially for the treatment of chronic, long-term conditions).

Responding to an increasingly competitive global pharmaceutical market and a more restrictive pricing environment, drug companies have tried to solidify their operations through advanced product development as well as a variety of mergers, acquisitions, and alliances with international firms. Acquisitions are often viewed as the most efficient means of obtaining desirable products, opening up new market areas, and acquiring manufacturing facilities in various regions of the world. Acquiring existing products and facilities with established consumer bases is usually less expensive than developing a product from scratch. Additionally, competition in many foreign regions is not as regulated as in the United States. In the U.S., the Food and Drug Administration regulates the pharmaceutical industry. To gain commercial approval for new drugs, modified dosages, and new delivery forms, a manufacturer must show proof that the medical substance is safe and effective for human consumption. Many of the competing international firms are not so tightly regulated.

Despite these concerns, the pharmaceutical industry has clearly become a global industry, with an abundance of exciting opportunities for engineers in a variety of countries.

Food Industry

One of the basic maxims of human existence is that we all have to eat to survive. Probably few engineering-related industries are as essential as the food industry. From the earliest times, humans have had to design and develop methods to plant, grow, harvest, and process food for consumption. As machinery and technology evolved, the process of food production and distribution grew more efficient. Today's food industry is involved not only with providing better products but also with developing new products that have a longer shelf life. This industry provides many global opportunities for engineers since many of the largest food corporations are involved in worldwide product distribution. Pizza Hut in Russia, Coca-Cola in China, American breakfast cereals in Africa, and M&Ms in Europe are just a few examples of the expanding global food industry.

General Mills, one of the largest consumer food companies in the United States, has been active in developing new global products and entering joint ventures, which earned it worldwide revenues totaling $17 billion in 2012. Its employees work in plants and offices throughout the U.S., as well as Canada, Mexico, Central and South America, Europe, and Asia. International exports represented a significant portion of General Mills' sales from its very beginning back in 1928. A more aggressive approach to the international marketplace began in the 1990s with the creation of four new strategic alliances: Cereal Partners Worldwide, Snack Ventures Europe, International

Dessert Partners, and Tong Want—a joint venture in China with Want Want Holdings Ltd. These joint ventures are building a strong foundation for General Mills' future growth worldwide. In recent years, General Mills has acquired companies such as Pillsbury, Haagen Dazs, and Old El Paso Mexican Foods, which continue to enhance its global presence. Sales in their international businesses grew to $4.2 billion, with sales in more than 100 global markets.

Engineers from virtually every field of study are in demand in this growing and dynamic global industry. Technical expertise is needed for all stages of worldwide food production and distribution. Since many of the raw materials are grown throughout the world, engineering knowledge is needed to ensure that the food being grown meets the highest quality specifications. Scientists and engineers research and develop new products, while others are involved in the design of production and processing systems. Engineers also are used in the layout of production plants and facilities and are often involved in the startup, training, and management of the local workforce. Ensuring consistently high quality throughout the processing system is another essential function for engineers. Distribution in a timely and efficient manner is critical to the success of a corporation, and engineers can help ensure that the products are reaching customers safely and quickly.

Engineers entering this industry will find many interesting and challenging opportunities. Food has truly become a global commodity and the U.S. is an important contributor to the feeding of the world. Students with interest in virtually any area of this industry should be able to find ample opportunity to apply their engineering talents throughout the world.

Petroleum Industry

While many of the industries discussed in this chapter are relatively new to the global marketplace, the oil industry has long been involved on a worldwide scale. Historically the vast majority of the world's oil and gas reserves were located outside of the United States, the world's largest consumer, so the petroleum industry has been a global operation out of necessity for many decades. However, U.S.-based petroleum output is increasing significantly, putting the U.S. on pace to surpass Saudi Arabia as the world's largest producer by 2020. The increasing demand and price structures of recent years have made an engineering career in this field quite challenging and demanding.

Since oil supplies in certain areas have become depleted, there is a strong demand for individuals interested in exploration of new sites. Oil companies need engineers for the design and construction of new facilities, production, processing, and storage. Their expertise is used primarily to make a facility operational, but the customary objective is to train and develop a local workforce to run the plant once it is fully functional. Once the facility is fully operational, the engineers involved at the site function primarily as managers and troubleshooters. While new designs and operational concepts continue to be developed, they generally are being handled at U.S.-based R&D centers, with the technology adapted to the facility, wherever it may be needed.

With the strong emphasis on international operations, engineers involved in this industry must have an appreciation for cultural differences and must be able to work effectively with people of different backgrounds. As worldwide demand and increasing price fluctuations become more acute, it will become increasingly more important for individuals in this field to possess strong managerial skills to help their companies succeed in a very competitive market. While petroleum production will continue to be critical, the exploration and development of new sites will become ever more essential.

Chemical Processing Industry

Throughout the world, thousands of products have been created by the chemical processing industry. Industries across the globe rely on a vast array of chemicals, fibers, films, finishes, petroleum, plastics, pharmaceuticals, biotechnology, and composite material products developed and produced by this industry. While these firms often are not involved in the actual production of the end product, they are responsible for the key ingredients that are used by a wide range of other industries. A review of the literature from some of the leading chemical processing manufacturers shows that their products are being used worldwide in such industries as aerospace, agriculture, automotive, chemical, computer, construction, electronics, energy, environment, food, packaging, pharmaceutical, printing and publishing, pulp and paper, textile, and transportation fields, among others.

Since these materials are used by such a diverse group of industries, it is important for the manufacturing facilities to be located in close proximity to the customers incorporating them into their products. Therefore, it is typical for these chemical processing corporations to have facilities in many regions of the world. For example, according to one of its reports, Dow Chemical Company provides chemicals, plastics, energy, agricultural products, consumer goods, and environmental services to customers in 160 countries around the world. The company operates 197 manufacturing sites in 36 countries and employs more than 52,000 people who are dedicated to applying chemistry to benefit customers, employees, shareholders, and society.

DuPont is another worldwide chemical company, with 70,000 employees operating in more than 90 countries. The corporation is involved in a wide range of product markets, including agriculture, nutrition, electronics, communications, safety and protection, home and construction, transportation, and apparel.

DuPont expanded its global operations significantly in the 1990s, with new plants in Spain, Singapore, Korea, Taiwan, and China, and a major technical service center opening in Japan. In 1994, a Conoco joint venture began producing oil from the Ardalin Field in the Russian Arctic—the first major oilfield brought into production by a Russian/Western partnership since the dissolution of the Soviet Union.

BASF is the world's largest chemical processing company, with 111,000 employees at 370 production sites on five continents. Core businesses are in chemicals, plastics, agricultural products, and oil and gas. Even though Europe is their home market they have significantly increased their presence with NAFTA countries, Asia, and South America.

As indicated in these examples, engineers who work for such companies must have a global perspective. Many of the pilot-scale projects they work on could possibly be implemented as full production processes in several global locales. For the engineer, this may require active involvement with people from different cultures, using different languages and different methods of operation. Being successful in their endeavors will require an understanding of current global issues in addition to the technical skills needed to implement their product.

Computer and Electronics Industry

Probably no engineering-based industry has undergone more rapid, dynamic changes than the computer and electronics industry. New products are being developed, manufactured, and distributed so rapidly that in many cases last year's product is quickly out of date. This is an industry where U.S. businesses must share world leadership in some product areas. Among the wide range of specialties included in this field are communications, computers, solid state materials, consumer and industrial electronics, robotics, power and energy, and biomedical applications.

Besides U.S. corporations, the world leaders in many of these technologies are companies in the Far East, notably Japan, Taiwan, and South Korea. Companies in the U.S. have been engaged in a competitive battle to maintain their position in product markets they once dominated. Engineers in this industry are often employed by a firm whose home base is outside the U.S. Those working for a foreign company at a U.S. site may face an adjustment in their method of operation. In addition to learning new applications of their technology, engineers in the global marketplace may need to learn much about foreign business culture. For many, this can be an exciting new experience.

While the U.S. no longer dominates many sectors of the electronics and computer industry, one area where the U.S. remains dominant is in the field of computer software development. U.S. computer specialists have the technical expertise that is in demand by firms throughout the world. Many software companies are now specializing in the development of products to serve clients in a multitude of foreign languages to help them in their business operations. It is not unusual for U.S. computer specialists to work on products exclusively geared for the foreign market.

Regardless of the home base of the computer or electronics manufacturer, much of the actual production of the products is now being done in Third World and developing countries. The intent is to keep costs down and profit margins up by making use of foreign labor and production. This requires engineering involvement not only in product development, but often in facility planning, plant startup, and hiring and training of a local workforce as well. For those interested in an evolving field with a variety of global applications, a career in the dynamic, challenging computer and electronics industry may be ideal.

Telecommunications Industry

Engineers have revolutionized the telecommunications industry and, as a result, have dramatically changed the world of engineering. The world around the U.S. has grown smaller, and the interaction of people worldwide has become faster and easier due to the rapid developments in this industry. Engineering developments in cellular and wireless telephone communication, voice and data transmission, video communication, satellite technology, and electronic mail have all had tremendous global impact. Engineers from around the globe now interact daily on routine matters such as program development, product design, plant facilities, operations, distribution, and quality-control issues. The developments in this industry have made it possible to bring fast, efficient, reliable, and affordable communication and information to new markets of the world that had previously been considered unreachable.

Some common international engineering employment opportunities in this field include designing, manufacturing, and distributing products such as integrated semiconductors, networks and networked computers, wireless telephone products and networks, and advanced electronic and satellite communications. Other telecommunication products include two-way voice and radio products for global applications, onsite to wide-area communication systems, messaging products (pagers and paging systems), handwriting-recognition products, image communications products, and Internet software products for worldwide distribution.

Alcatel-Lucent Technologies is an example of one major organization involved in the telecommunications field. They design, build, and service optical and wireless networks, broadband access, and voice enhancement equipment to assist the world's communication service providers in building future networks combining voice, data, and video. Many of the world's telephone calls and Internet sessions flow through their equipment. They employ about 76,000 employees who work with service providers on five continents.

Verizon Communications was formed by the merger of Bell Atlantic and GTE and is now one of the world's largest providers of wireline and wireless communication services for 94 million customers. Verizon had over 188,000 employees and over $110 billion in revenues in 2011. Their global operations are present in Europe, Asia, the Pacific Rim, and the Americas. Primary growth areas are projected in wireless and data transmission.

With the fast pace of telecommunications development and the increasing worldwide demand, there should be ample opportunities for global engineering in traditional markets as well as in emerging markets that have traditionally presented significant communications challenges.

Environmental Industry

One field where the technical expertise of U.S. engineers remains dominant is in the environmental arena. Given the excellent environmental training programs in the U.S., coupled with strict domestic environmental regulations, the American environmental industry is far ahead of most every other region of the world. There are many new and exciting global challenges waiting for the engineer with interest in this field.

Since many countries of the world lag far behind the U.S. in environmental policy, serious problems are surfacing at a rapid rate. Many countries are now looking to U.S. environmental firms to assist them with a wide range of projects. Many regions are in dire need of assistance with water purification systems, sanitation facilities, air pollution remediation, waste management, hazardous and toxic waste cleanup, pest control, landfill and recycling issues, and issues related to nuclear safety. In some areas of the world, the population is increasing at such a rapid rate that these problems are accelerating at a pace faster than the pace of technological development. In other areas, the funds, facilities, and resources required to meet environmental challenges are not available.

Engineers who work in this field will encounter many global challenges. One industrial employer is Walsh Environmental, an environmental consulting firm providing services in the U.S. and overseas. Its engineers work on such global projects as site assessment; remediation of soil, water, and air; environmental impact assessments; underground storage tank removal; and industrial hygiene, health, and safety issues. It is also involved in projects with energy and mining companies dealing with exploration, production, and power generation in remote locations around the world. It has particular expertise in the rainforest environments of South America and Southeast Asia, as well as with bioremediation and the remediation of chlorinated solvents. It has project locations in such diverse areas as South America, Japan, China, Australia, Canada, Botswana, Bangladesh, Bahrain, Gabon, South Africa, Indonesia, Ireland, New Guinea, Pakistan, and the Philippines.

AMEC is an international network of engineering and environmental professionals. Tapping the expertise of its engineers, AMEC has undertaken several large industrial construction projects throughout the world. These projects include coastal structures and power generation and distribution projects throughout North America, South America, Africa, and Asia. The global experience of AMEC's engineers enables them to effectively understand the needs of clients and to develop innovative solutions to the engineering and environmental challenges.

For environmental engineers, technical comprehension of environmental problems is not enough. They must also have cultural experience and sensitivity, and motivation to work with difficult challenges in many different locales.

Consulting

Countries, governments, and firms around the globe need the expertise of American engineers. Thus there is a huge global demand for consultants with a wide range of technical backgrounds. Over time, consulting firms have evolved as a source of technical assistance for entities that do not have the time, resources, or capability to address specific engineering issues from within their organization. Some of the large international consulting firms maintain a staff of engineers from diverse backgrounds who are technically and culturally prepared to assist global clients with a wide range of engineering problems. A consulting firm may be hired for a very short period of time (weeks or months) or may be involved with a project for several years, to completion.

One example of a large international consulting firm is Accenture. To its clients Accenture offers a combination of services that include strategic technical planning, business management, and engineering consulting. Its mission statement emphasizes its global commitment: "Accenture is a leading global management and technology consulting firm whose mission is to help its clients change to be more successful." The firm works with thousands of client organizations worldwide. Its clients represent a wide range of industries such as automotive and industrial equipment, chemicals, communications, computers, electronics, energy, financial services, food and packaged goods, media and entertainment, natural resources, pharmaceuticals and medical products, transportation and travel services, and utilities. To maintain competence in so many areas requires over 259,000 people and 200 offices in 120 countries. Its business in 2012 grew at an average annual rate of 10%. In this challenging field Accenture must have the ability to tap global resources and quickly deploy them to address customer needs. International teams are often employed to serve a rapidly growing number of global clients. Accenture's consultants often have the opportunity to work in more than one of the world's major markets during the course of their careers. Additionally, they encounter many opportunities to exchange knowledge and experiences with colleagues around the world, from diverse backgrounds, competencies, and industry expertise.

For engineering students interested in a wide variety of experiences involving a broad range of global industries, working in the consulting industry can provide the unique and interesting careers they seek.

Technical Sales

Technical sales is an important function for all engineering firms, both domestic and abroad. All firms need engineers who can competently present their products or services to potential clients. These specialists articulate the technical capabilities of the product and demonstrate how it can meet the needs of a prospective customer. In addition, skilled technical sales people are trained to assist with the installation and startup of the product, as well as troubleshooting onsite problems. In some cases, technical sales engineers work concurrently with their clients and their home-office design team to develop or modify existing products for the customer. Once the product has been implemented, the sales engineer also may be responsible for field testing and operator training. Another important responsibility is maintaining and ensuring customer satisfaction throughout the entire process, from initial purchase to installation to product implementation, and perhaps even to product phase-out.

On a global scale, technical sales takes on an even more critical role. The rapid increase in the demand for global technology translates into a critical need for trained technical people to handle international sales, installation, field engineering, and customer liaison roles with clients across the globe. These engineers must not only be technically skilled but also must have appropriate training in language, culture, customs, and sensitivity to effectively interact with foreign customers. They must appropriately represent themselves, their company, the product, and the culture, and most significantly, they must be able to demonstrate that their firm has the motivation and talent to adequately serve the global customer in a manner consistent with the customer's cultural and technical needs.

4.4 Preparing for a Global Career

We hope that the variety of exciting global opportunities available to engineers described in this chapter will get you thinking about your particular interests and motivation. While some engineers may drift unintentionally into an international career, most careers are the result of careful planning, experience, and preparation. There are several things that engineering students can do during their college years to better position themselves for a global engineering career. The remainder of this chapter will explore some of these factors.

Language and Cultural Proficiency

As stated earlier in this chapter, many engineering students have a difficult time seeing the value of studying a foreign language and culture. We hope that some of the material presented in this chapter will give you a different perspective. Even if you are studying engineering and have no specific plans for an international career, it would still be wise to anticipate that the global economy is likely, at some point, to impact your work as an engineer.

So the question remains: Are foreign language proficiency and cultural study necessary to communicate in the global workforce? The answer is, "probably not." English is commonly used as the international language among companies and project teams around the globe. Many countries now teach English as a second language to better prepare their students and workers for communication in the global workplace. It is not uncommon for U.S. engineers to deal with foreign colleagues who are fluent in several languages. However, despite its growing worldwide usage, an engineer cannot assume that English will be understood everywhere.

Therefore, it is important that engineering students take advantage of any language and cultural training available to them. Most students have had some foreign language education in high school, and this provides a good foundation for further study. Being able to speak foreign associates' language and knowing about their region's culture demonstrates a sensitivity and willingness to work together in the global workplace as an equal partner.

Unfortunately, the structure of most engineering curricula does not provide much room for the study of foreign language and culture. Most students find that they can satisfy their humanities requirement with a liberal arts course, but foreign language study generally only provides elective credit. Therefore, students who wish to add language proficiency to their education must do so by adding to their course load, or by learning the language outside their college studies.

Some schools and companies offer intensive language training designed to develop foreign language proficiency in a relatively short period of time. Other programs concentrate on teaching basic terminology, phrases, and concepts that prove useful for getting around in a foreign country. Some companies also provide training programs that emphasize cultural protocol to assist those working with foreign colleagues.

While in college, there are many things you can do on your own to become more culturally sensitive. Getting to know professors and classmates from different countries, and taking part in any special cultural activity on campus, can be enlightening. Attending meetings of various international clubs in your community can also be of benefit. Try to experience anything international that you come across: seminars, lectures, and other events. On most campuses, you will find that you don't have to travel very far to broaden your international understanding. While knowledge of a foreign language and culture may not be required for success as a global engineer, individuals who make the effort to develop some competency in this area may find more opportunities available to them.

Study Abroad and Exchange Programs

One of the very best ways to prepare for a global engineering career is to participate in an international study program. While not all schools offer a broad selection of such opportunities, most offer some programs for their students. If the opportunities at your school are limited, you may be able to participate in such a program with a nearby college or university as a guest or temporary student. Usually you can transfer the credits you earn back to your own institution. Some schools offer exchange programs that allow a student from one institution to exchange places with a foreign student from a partner school under an established agreement.

A **foreign study** experience can provide many benefits to you. It is an opportunity to develop and expand your problem-solving skills in a different environment with new perspectives and challenges. And an international study experience can greatly facilitate the learning and application of a foreign language while enhancing your geographical and historical knowledge. You will find yourself exposed to individuals and groups who may process information differently than you. An overseas study program provides you with new professional contacts, and can give you direction with your career. It can also give you a new level of confidence with a strengthened sense of personal identity, flexibility, and creativity.

There usually are a variety of foreign study options available to students. The full-year program provides students the opportunity to study at a foreign university for the equivalent of two semesters. Typically, students are enrolled in regular courses (preapproved as equivalent to those required in their program) that are taught and graded by faculty from the host institution. The credits are then transferred back and applied to the students' engineering program requirements. The semester-length program is structured similarly to the full-year program except that this program is limited to the equivalent of one semester. The intent of both programs is to immerse students in a foreign setting, having them live with nationals instead of Americans, and to teach them to incorporate their knowledge of the language and culture of the host country into their daily activities. This is the best way for students to develop language and cultural skills that can be applied to their future career.

In a variation of the above model, the instruction and many of the onsite activities are under the direct supervision of faculty from the student's home institution. Typically, the U.S. institution offers an academic program, usually a semester or less in length, with regular university courses, taught by its own faculty, in rented or leased facilities that are exclusively designated for this program. The advantage of this approach is that students of varying levels of language and cultural preparation can be accommodated, and all courses are treated as though they are offered by the home institution, thereby eliminating any potential problems with course transfer or application to degree requirements.

The **exchange program model** is founded on special agreements between two institutions in which they agree to exchange equal numbers of students for semester-length programs. Common course, credit, and transfer issues are generally resolved in advance. Typically, most of these programs are designed so that students from each country pay standard tuition and fees at their home institution, with those funds then set aside to cover the costs of hosting students from the other school.

The **short-term specialty program** provides an opportunity for students to visit and see, firsthand, specialized facilities, operations, or research projects in a foreign location. Typically, these programs are for a short period of time, and the program is often incorporated as part of an existing class, almost like a special field trip. The focus of these programs is to expose students to a specialty unavailable to them through normal on-campus teaching. While this program provides little opportunity for students to experience and reflect on the local language and culture, it does provide a means of experiencing a foreign study program on a limited basis.

The **study tour model** provides an academic experience that is usually oriented more toward tourism than study. Generally, these short-term programs are designed to help students learn about global technical areas and topics through nontraditional forms of instruction at an international location. This may include taking field trips to several international production facilities, writing journals in which students record the various global engineering design and development strategies they observe, or making a series of visits to research facilities and reporting back to their classes. The intent of this program is to provide students with a short, but educational, focus on different forms of global engineering. Unfortunately, this type of opportunity is limited in its ability to incorporate the cultural and language experience into the program model.

International Work Experience: Co-ops and Internships

Another excellent way for you to prepare for a global engineering career is by obtaining an international cooperative education (co-op) or internship. This can be a very complicated task, but for students motivated enough to make the effort, it could be very beneficial. Typically, these positions emphasize the integration of responsibilities that are directly related to your particular field of study. A co-op is a paid position, monitored by your college to ensure that the experience is meeting certain established learning objectives. On the other hand, internships may or may not be paid and may or may not be monitored by your school. The awarding of credit varies from school to school.

The benefits of obtaining an international work experience include the following:

- **Valuable work experience.** By participating in an international work experience, you have the opportunity to experience firsthand the opportunities, challenges, and frustrations of working in a different cultural environment. The total experience is bound to make you more aware of, and sensitive to, the critical issues of the global workplace.
- **Guaranteed immersion in a new cultural setting.** By participating in a well-planned international program, you should enjoy a comprehensive cultural learning experience. Interacting with others in an international work setting can help you determine if the factors that initially attracted you to a certain culture and language in a particular area of the world are truly of interest to you. These experiences can help you focus your career planning.
- **Greater foreign language competency.** There is no question that a person's foreign language skills improve significantly when used regularly in the society in which it is the native language.
- **The challenges of living and learning in an unfamiliar work environment.** Many students go through a period of culture shock, which can impact their work performance. They must deal with differences in work ethics, performance expectations, and culturally dictated ways of doing business. These can have a significant impact on a student's global career planning.
- **Developing an international network.** International work experience can provide you with a variety of new friends, professional contacts, and mentors. These individuals can serve as valuable resources to you as you pursue your career.

Special Considerations

While we have discussed the positive aspects of working in an international setting, there are other issues that also should be considered:

- **Language and cultural skills.** If you cannot demonstrate language proficiency and cultural sensitivity, many foreign organizations will be hesitant to hire you. You must be able to demonstrate that you will be an asset, not an impediment, to the daily operations of the firm.
- **Time and effort.** Obtaining an international work opportunity can be a time-consuming process that usually is more complicated than landing one in the U.S. A great deal of effort may be expended writing letters of inquiry; customizing

your résumé (perhaps in a different language); and completing application materials, visa forms, travel arrangements, and other documents.

- **Confusion over job responsibilities.** It is not uncommon for students to have misconceptions about the exact nature of their work responsibilities. Many have high expectations for challenging work and a significant level of responsibility and are disappointed when the actual assignments fail to meet their expectations.
- **Cost.** There are many additional costs associated with an international work assignment that may not be readily apparent. Expenses for airfare, local and regional travel, housing, and possible administrative fees are often overlooked when planning a budget. While most co-ops and a few internships are paid positions, many others are not. Generally, any salary that is provided is scaled down to reflect local wages and the economic pay rates of the region. For most students, this will probably be much less than they expect.
- **Housing issues.** Many students find that international housing can be difficult to find, expensive, and generally not what they are accustomed to in the U.S. Students should try to work with their foreign employer to secure decent and affordable housing, if possible.

Most students feel that the positive benefits of an international work experience far outweigh the challenges and effort. The technical experience gained and the opportunity to develop language and cultural skills provide precisely the type of background that employers seek when selecting candidates to fill global positions.

To obtain more information about an international work experience, talk with your campus placement office, your co-op or internship coordinator, and your academic advisor. Some excellent organizations involved in providing technical students with international work assignments and general information include Cultural Vistas (www.culturalvistas.org), the Council on International Exchange (www.ciee.org), and the NAFSA Association for International Education (www.nafsa.org).

Choosing the Right Employer or Opportunity

Most engineers fresh out of college are not likely to find a long line of employers at the campus placement office waiting to hire them just because of their interest in pursuing a global engineering career. You will probably have to thoroughly research many companies, industries, and government agencies to find the right situation for you. You must identify organizations that have a global commitment that is consistent with your background, skills, experiences, and interests.

Once you have identified employers involved in the global marketplace, you should try to contact them directly through on-campus contacts or by correspondence. Since most companies do not immediately thrust new graduates into the global workforce, you must not limit your job campaign to the narrow objective of obtaining an international career. Rather, the international interests, experiences, and competencies that you have developed over time should be viewed as valuable commodities that are supplemental to your core technical skills and abilities. The goal of your initial job campaign should be to obtain a position with an organization that will be able to provide international experiences once you have been trained and are experienced in the company's culture. It is important to realize that this process takes time; it may take several years before a global opportunity emerges.

In the meantime, there are several things you can do to position yourself for a global opportunity. For example, keeping co-workers and supervisors informed of your interest of being involved in an international project could pay off in the future. It is also important to maintain your technical capabilities as well as your language and cultural skills. One never knows when an opportunity may develop. Building relationships with those who can help you achieve your objectives can be very helpful. It is important to establish a solid work reputation as someone who can be trusted and is prepared for the responsibilities of a global assignment. You must continually demonstrate that you are a team player who can work well with others, that you are sensitive to cultural differences, and that you have strong communication skills. It is important to remember that sometimes the best opportunities emerge for those who demonstrate flexibility. Your goal may be to work on a project in the Far East, but if an opportunity comes up in Europe, you may want to take it as a way of gaining additional international experience that may eventually lead to an assignment in your preferred location.

The world of global engineering is here to stay. The work of engineers is reaching into all regions of the world. Companies need professionals who are ready to accept the challenges of the global marketplace. This world economy will provide excellent opportunities for those who show the interest, motivation, capability, and leadership to accept these challenges.

EXERCISES AND ACTIVITIES

4.1   Research some current articles about NAFTA. Write an essay about the current pros and cons of this agreement.

4.2   Research some current articles about the evolving European Union. In an essay, discuss whether or not you feel that the EU has been a successful venture.

4.3   Find some current information about the global economy (stock markets, status of world currencies, general economic conditions, etc.). What is the state of the current world economy? Which regions of the world are doing well and which are struggling? What industries are prospering and which are having difficulty?

4.4   Visit your campus career planning or placement center and read the recruiting brochures of some employers in which you have an interest. Make a list of those who offer international opportunities in your field of interest. Write an essay describing which industry is of most interest to you in your career, and why.

4.4 Develop a list of your present international skills and abilities. Suggest how you can strengthen these skills in the next few years.

4.6 Attend a foreign study orientation session, if available, or visit your campus international center, if one is available, to find out more about some of the opportunities that are available. List some of the advantages and disadvantages of these opportunities as they relate to your personal priorities.

4.7 Attend your campus career/job fair. Talk with at least three recruiters from different firms. Write an essay comparing their views concerning the opportunities for international work experiences such as co-ops and internships.

4.8 Visit with some upper-class engineering students who have studied abroad. Write an essay that compares and contrasts their experiences. Which students do you feel gained the most from their experience, and why?

4.9 Meet with your academic advisor and develop a long-range course plan. Modify this plan to include an opportunity for international experience. Write a short essay discussing the advantages and disadvantages of including this experience in your academic program.

4.10 Browse the web, exploring the sites of employers that are of interest to you. Many will have copies of annual reports and other operations information available for review. Develop a list of those firms that are expanding their global operations. What factors are cited for this expansion? Does there seem to be a pattern among organizations? Does it appear that any of these organizations are reducing their global operations? If so, why?

4.11 Select a world region that is of interest to you. Using information from your foreign study office, career center, or web resources, research the study abroad and/or internship opportunities that are available through your school or another college in this area. Which programs appeal to you and why? Visit with your academic adviser or professor to discuss the feasibility of this opportunity in your academic program.

4.12 Foreign language study is generally not required in most engineering schools. Given the increasing expansion of global engineering, do you think the study of a foreign language should be a required part of the engineering curriculum? Why, or why not?

4.13 One of the current criteria that ABET uses in accrediting engineering programs states that engineering graduates should have "the broad education necessary to understand the impact of engineering solutions in a global and societal context." What courses in your engineering program are designed to provide this exposure? If you are not sure, discuss this with your academic adviser or a professor in your program. Do you feel that the offered coursework will adequately prepare you to meet this criterion?

CHAPTER 5

Future Challenges

Throughout history, engineers have been problem-solvers at the forefront of change—change that has left its imprint on world history, politics, economics, and global development. Now that we are well into a new millennium, it is important not only to reflect on the successful accomplishments of engineers, but also to focus on the exciting challenges and opportunities that lie ahead. Today's engineers are confronted with a world in which technology is exploding at a breakneck pace, and technical expertise is needed to solve many global issues. Some believe that the next generation of engineers will be responsible for more developments during their lifetime than all the technological advancements since the construction of the pyramids. There are many issues and challenges that could be addressed, but we will touch on just a few in this chapter.

5.1 Expanding World Population

As can be observed in Table 5.1 and Figure 5.1, it took from the beginning of time until 1804 to reach a world population of 1 billion people. World population statistics show that there were approximately 1.6 billion people in 1900, and this figure reached 6 billion before the end of 1999—one century to add 4.4 billion people! According to the United Nations Population Division, this number could climb to almost 9 billion people by 2050. The concern for engineers, as well as others, is that this rapid increase in population, coupled with fast-paced industrial growth, places a tremendous strain on land, air, and water resources that are essential for human survival.

Nearly all world population growth is occurring in developing countries. The 49 least developed countries will almost triple in population over the next 50 years. World population is growing at a rate of 1.3% annually, or 77 million people per year. Half the growth is in six countries: India, China, Pakistan, Nigeria, Bangladesh, and Indonesia. Africa is projected to double its population to 1.9 billion by 2050. This massive growth cycle, coupled with extreme poverty and environmental devastation, has left millions of people without adequate food and water. In contrast, the U.S. population has been projected to grow to 335 million in 2025 and 422 million in 2050. The population should stabilize or decline in Japan, Europe, Russia, and many other

Table 5.1 Number of Years Needed to Add a Billion People to the World's Population

- 1st billion: from beginning of time to 1804
- 2nd billion: 123 years (reached in 1927)
- 3rd billion: 33 years (reached in 1960)
- 4th billion: 14 years (reached in 1974)
- 5th billion: 13 years (reached in 1987)
- 6th billion: 12 years (reached in 1999)
- 7th billion: 12 years (reached in 2011)
- 9th billion: projected by 2050

Source: United Nations Population Division, Department of Economic and Social Affairs

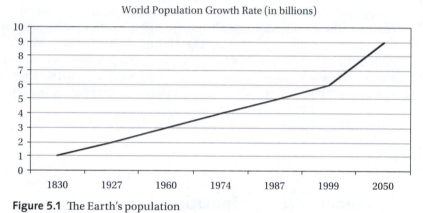

Figure 5.1 The Earth's population

countries of the established industrial world. Recent studies indicate, however, that the overall population growth rate has slowed from these earlier projections due to increased education and use of family planning, combined with increased mortality rates due to the spread of disease, including AIDS. Despite these recent projections, the overall trend of rapid population growth remains, creating numerous challenges for engineers.

The increasing global population will have significant impact on land use, air and water resources, and food production and distribution processes. In many areas of the world, water is often used faster than it can be replaced, and land suited for food production is dwindling. Forests are being destroyed for fuel and to make way for housing. More and more industrial plants are being built, which contributes to the pollution of air and water. In rapidly growing cities, the demands for housing, food, water, energy, and waste disposal are taxing available resources at an alarming rate.

Engineers will need to use all their talents in an effort to improve these situations. As the population grows, technological solutions will be needed to maintain a proper ecological balance. Many believe that this population challenge has implications for the other challenges outlined in this chapter.

5.2   Pollution

Engineers must always be aware of the impact of their work on the environment. Sometimes small, seemingly harmless actions can have long-term, far-reaching effects. Two critical areas of pollution concern for engineers are air pollution and fresh water resources.

Air Pollution

Air pollution results when the atmosphere absorbs gases, solids, or liquids. The sources of pollution can be either natural or manmade. When these contaminants enter the atmosphere, they can endanger the health of humans, animals, fish, and plants. They can damage materials, create visibility problems, and cause unpleasant odors.

The major cause of air pollution is burning gasoline, oil, or coal in automobiles, factories, power plants, and residences, without proper controls to remove the offending pollutants. Once the pollutants have entered the atmosphere, they can travel great distances and can cause problems such as acid rain in areas far from the original emissions site. Many nations of the world are burning coal and oil at record rates, which has increased atmospheric levels of carbon dioxide. Some scientists feel that this has contributed to a greenhouse effect in which the release of infrared radiation from the Earth is inhibited; in theory this could contribute to a global warming trend. Some evidence seems to suggest that certain airborne pollutants are damaging the ozone layer, although this is still a scientific controversy.

While acid rain has long been recognized as a problem in the industrialized countries, there is now evidence of the increasing danger of acid rain in Southeast Asian nations. According to the United Nations System-Wide Earthwatch 2005 Report, emissions of sulfur dioxide have declined significantly in Europe and North America with reduced coal use, and with the application of new emission cleanup techniques, further progress is expected. This has reduced the sulfur contribution to acid rain. However, the same level of improvement has not been achieved in nitrogen oxides and other pollutants from vehicles, where the reduction in emissions due to catalytic converters has been offset by the growing number of motor vehicles. Therefore, urban air quality in many areas has continued to deteriorate.

Some of the worst pollution problems actually appear in unexpected places, such as the Arctic, where high levels of toxins such as PCBs, DDT, mercury, and dioxin have been found. There appears to be a global process of distillation where pollutants evaporate in warmer areas, are transported by winds to the Arctic, and then condense to become concentrated in Arctic food chains.

The 2007 Energy Information Administration report "Greenhouse Gases, Climate Change, and the Economy" reports that world carbon dioxide emissions are expected to increase by 1.8% per year from 2004 levels through 2030. As indicated in Figure 5.2, total world carbon emissions were 33 billion metric tons in 2010 and were projected to reach 37 billion by 2020 and 43 billion by 2030. A significant portion of this increase is expected to occur in the developing countries, where emerging economies, such as China and India, spur their economic development using fossil energies. Emissions from developing world countries are expected to increase greater than the world average by 2.6% per year from 2004 levels through 2030.

In the United States, several recent laws call for monitoring air quality and the ozone layer. The Clean Air Act, enforced by the Environmental Protection Agency, requires the monitoring of air quality levels. Additionally, many member countries of the United Nations have joined forces to monitor and phase out many ozone-destroying chemicals and materials. These are some beginning steps that will have to continue on a worldwide basis for many years. Engineers have the technical expertise to develop and implement the necessary controls and devices to protect delicate air resources, but they must have the authority and financial support to meet these challenges.

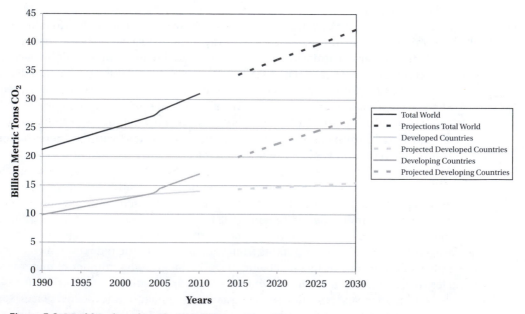

Figure 5.2 World carbon dioxide emissions

Source: U.S. Energy Information Administration, 2011.

Water Pollution and Freshwater Resources

Water is a most precious resource, vital for the existence of life. Unfortunately, the pollution of lakes, streams, ponds, wells, and reservoirs has become a serious challenge for engineers. Water pollution is the contamination of a water source by any agent that negatively affects the quality of the water. The major sources of water pollution include industrial wastewater, chemicals, detergents, pesticides and weed killers, municipal sewage, acid rain, residue from metals, bacteria and viruses, and even excessive, artificially enriched plant growth.

When these wastes infiltrate a water supply, they upset the delicate balance in the ecosystem and can cause human and environmental health problems. Water quality legislation enacted in recent decades has improved conditions significantly. Modern water treatment plants and purification systems have been useful in improving the overall situation, but more comprehensive systems still are needed to remove the most elusive pollutants resulting from such substances as herbicides and pesticides.

Freshwater is clearly becoming a major constraint on global development. According to the United Nations System-Wide Earthwatch 2005 Report, 40% of the world's population already faces chronic water shortages. One recent estimate suggests that more than half of available freshwater resources are already being used to meet human needs, and this could rise to 70% in 30 years. Water supplies could therefore run out in the next century if per capita consumption and excessive use in agriculture are not controlled. Another projection shows that between 1 and 2.4 billion people will live in water-scarce countries by 2050. The conflicts over sharing water in international river basins are increasing.

There are growing concerns about major regional water scarcities. Countries in North Africa and the Middle East already have 45 million people without adequate drinking water. Per capita water availability has shrunk by more than half in 30 years, and could be halved again in the next 30 years, requiring an investment of $50 billion. Three-fifths of Chinese cities are short of water and 80 million Chinese do not have adequate drinking water.

Water quality is gradually improving in most parts of Europe as a result of a massive investment program. However, there is still a problem with groundwater and diffuse sources of pollution, particularly from agriculture.

The challenges for engineers of the future will be to develop more sophisticated treatment systems that can more thoroughly treat water pollution at its initial sources, while continuing to cleanse polluted water sites and facilities in order to provide freshwater resources to growing populations.

Solid Waste

The safe and efficient disposal of solid waste materials has become one of the most serious challenges facing engineers. Solid waste is usually defined as any trash, garbage, or refuse made of solid or semisolid materials that are useless, outdated, unwanted, or hazardous. Generation of solid waste has increased significantly as the population has expanded and consumers have adopted a use-it-and-toss-it attitude.

Estimates indicate that in 1920, the U.S. was generating about 2.75 pounds of solid waste per person per day. By 1970, this figure had increased to about 5.5 pounds per person per day. As indicated in Figure 5.3, in 2010 the U.S. was generating over 4.43 pounds of trash per person per day. While recycling efforts have made some impact, U.S. waste production is still significantly above the world average and some have called U.S. consumers the "throwaway society."

Generally, the establishment of solid waste management and disposal programs has been the responsibility of state and local governments. They typically develop policies and procedures concerning how waste will be managed, using such options as landfills, incineration, recycling, and composting methods. Engineers are, and will continue to be, actively involved at all levels of governmental policy implementation as well as in roles as consultants and providers of the services needed by state and local agencies.

Until the mid-1980s, approximately 90% of the disposal of solid waste involved sanitary landfills—specially constructed facilities used to bury the nation's refuse. In a typical landfill, deep holes are dug and lined with special protective materials. The refuse is spread in thin, compacted layers and then covered by a layer of clean earth. Pollution of surface water and groundwater is minimized by lining and sloping the fill, by compacting and planting the top cover layer, and by diverting drainage. Landfills are a fairly cost-effective method of solid waste disposal, assuming there is an adequate supply of available land close to the sources of the waste. Unfortunately, this has become a critical issue in recent years as land appropriate for use as landfills is disappearing.

Since 1988, 79% of U.S. landfills have closed because they became filled, or due to new restrictive disposal legislation. Many landfills have been augmented by new methodologies, including waste-to-energy plants. In this process, refuse (which has

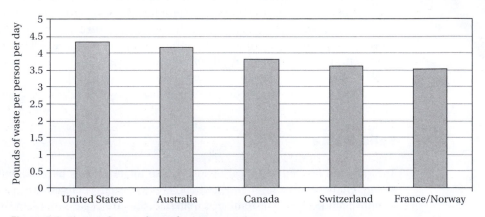

Figure 5.3 The top five trash-producing countries

about 50% of the energy content of coal) is burned in special incinerators and special thermal processes are used to recover the energy from the solid waste; the volume of the waste is reduced by 90%. This method now accounts for almost 20% of the solid waste management programs used across the country. However, in recent years, concerns over increasing costs and environmental problems have slowed the use of these processes.

In the last decade, recycling and composting have become the fastest-growing methods of solid waste management, and in 2010 they made up about 34% of waste management programs. An increasing number of local governments have implemented recycling systems for solid waste materials and have established composting programs for yard waste.

According to the United Nations System-Wide Earthwatch 2005 Report, there are several new areas of concern emerging as challenges related to solid waste issues. As military tensions change and disarmament agreements have been implemented, there has been a growing recognition of the enormous problem with the disposal of obsolete weapons, particularly nerve gas and chemical and nuclear weapons, which were never designed with safe disposal in mind. The combination of explosives and highly dangerous chemicals, often deteriorating and becoming increasingly unstable, makes dismantling such weapons, neutralizing their contents, and even transporting them to disposal facilities extremely expensive and environmentally risky. The U.S. alone has over 30,000 tons of chemical weapons, whose disposal could cost at least $12 billion. More than 50 ocean and inland lake sites across the U.S. contain explosive items, and harbors and beaches at the site of old battles throughout the world are riddled with unexploded bombs.

Another issue of concern is the cross-national transportation and dumping of waste. This problem has significantly increased in areas such as South Asia, Eastern and Central Europe, and the former Soviet Union. In 1992, Poland intercepted 1,332 improper waste shipments from Western Europe alone, and similar cases grew by 35% in the first half of 1993, which illustrates the scale of the problem in less developed countries. There are about 100,000 tons of obsolete and unused pesticides in developing countries, with 20,000 tons in Africa alone that will cost $80 million to clean up. The threats to health, water supplies, and the environment from these and other dumped toxic chemicals are serious. Worldwide legislation was initiated in 1995 to ban waste exports among some participating countries, but it will take time, resources, and strong governmental commitment to achieve effective implementation.

Based on the U.N. Earthwatch Report, as the world economy grows so does its production of wastes. For example, U.S. production of hazardous and toxic waste rose from 9 million tons in 1970 to 40 million tons in 2010. Europe produces more than 2.5 billion tons of solid waste a year, and every day the inhabitants of New York throw away approximately 26,000 tons.

So, what possible solutions exist for the challenges of solid waste disposal? In recent years recycling has become a preferred choice of waste disposal. The British government set a target of recycling 25% of all household waste within the next few years. Likewise, the proportion of household waste recycled in Germany increased

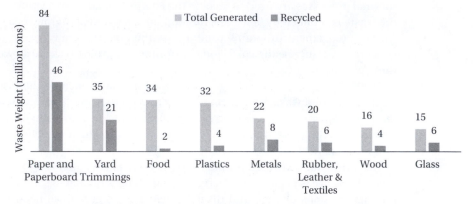

Figure 5.4 Amount of waste generated and recycled by selected categories
Source: Data from U.S. Environmental Protection Agency.

from 12% to 30% between 1992 and 1995. Around 75% of the average European car is already recycled, largely because the metal can be sold as scrap. However, electrical scrap accounts for merely 2% of waste produced in the European Union, and car scrap even less. While U.S. efforts in recycling have improved in recent years, as illustrated in Figure 5.4, the U.S. is recycling only small amounts of products generated.

Each method of waste disposal has drawbacks. Reusing glass bottles can require more energy than their initial manufacture as they have to be sterilized. Incineration is a source of greenhouse gases and toxic chemicals like dioxins and lead. Landfill sites are a possible source of toxic chemicals and produce large quantities of methane gas. They must be managed so that pollutants do not seep into groundwater and should therefore be kept dry, but this slows down the rate of decomposition.

Tires constitute another problem, as their resilience and indestructible nature become distinct disadvantages when it comes to disposal. Tires are virtually nondegradable and produce noxious fumes when burned. Western Europe, the U.S., and Japan produce around 580 million tires a year. Possible short-term solutions might include banning whole and shredded tires in any landfill. However, long-term acceptable disposal solutions continue to present significant challenges to engineers.

Further research needs to be carried out on the effects of various waste management options to determine the extent to which they benefit the environment. There are complex tradeoffs between costs, energy consumption, transportation, pollution, greenhouse gas production, and toxic byproducts, which require the expertise of engineers.

The challenge for engineers will be to continue to develop processes for the recycling of parts and components, and to solve all the intricate problems related to hazardous materials. These green engineers will need to incorporate the types of methodologies that have been successful in the recycling of such things as aluminum and paper into future designs for products such as automobiles, computers, factory equipment, and machinery.

5.3 **Energy**

The 2011 International Energy Outlook report prepared by the U.S. Energy Informa-tion Administration predicts that world energy demand will continue to grow steadily, and liquid fuels will account for a substantial portion of world energy demand by 2035. (For these purposes, liquid fuels are defined as petroleum and petroleum-derived fuels such as ethanol and biodiesel, coal to liquids, and gas to liquids. Petroleum coke, which is a solid, is included, as are natural gas liquids, crude oil consumed as fuel, and liquid hydrogen.) There will be a shift in energy demand by world regions as many developing and Third World countries increasingly demand significantly greater energy resources.

Other specific highlights of the report include the following:

- World energy consumption is projected to increase by 53% from 2008 levels to 2035. Projections show that world energy consumption could increase to 770 quadrillion British thermal units (Btu) by 2035 (Fig. 5.5).
- Although high prices for oil and natural gas, which are expected to continue throughout the period, are likely to slow the demand for energy in the long term, world energy consumption is projected to increase steadily as a result of strong economic growth and expanding populations in the world's developing countries.
- China and India will be major contributors to world energy consumption in the future. Over the recent past decades, their energy consumption as a share of total world energy use has increased significantly. In 1990, China and India together accounted for about 10% of the world's total energy consumption; in 2008 their share had grown to 21%. Even stronger growth is projected over the next 25 years, with their combined energy use more than doubling and their share increasing to

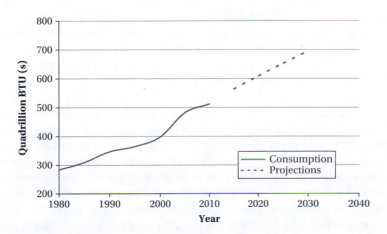

Figure 5.5 World energy consumption

Source: U.S. Energy Information Administration, 2011.

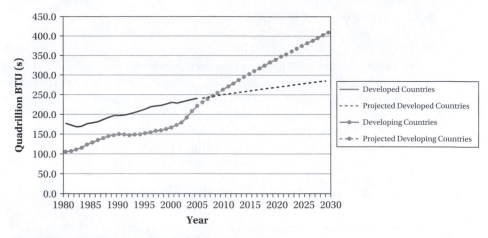

Figure 5.6 World energy consumption by region
Source: U.S. Energy Information Administration, 2011.

31% of world energy consumption in 2035. In contrast, the U.S. share of total world energy consumption is projected to contract from 22% in 2008 to about 17% in 2035. China's energy demand is projected to be 68% higher than the U.S. in 2035.

- Energy consumption in other regions also is expected to grow strongly from 2008 to 2035, with increases of around 60% projected for the Middle East, Africa, and Central and South America. A smaller increase, about 16%, is expected for developing countries in Europe and Eurasia (including Russia and the other former Soviet republics) (Fig. 5.6).

- In a dramatic shift toward energy independence, the U.S., which historically has relied heavily on foreign oil sources to meet its total energy needs, has reduced its international dependence down to 20% in 2012 (Fig. 5.7) In fact, it is now projected that the U.S. will be virtually self-sufficient by 2035. This is due to rising production of oil, shale gas, and bioenergy and significantly improved fuel efficiency in transportation. Evaluating the data associated with this decline in U.S. oil imports now indicates that North America will become a net oil exporter by 2030. The U.S. is now projected to become the world's largest global oil producer by 2020, exceeding Saudi Arabia by 2025.

- The use of all energy sources increases over the projected timeframe. Given expectations that world oil prices will remain relatively high throughout the projection period, liquid fuels are the world's slowest-growing source of energy; liquids consumption increases at an average annual rate of 1.0% from 2005 levels to 2035. Renewable energy and coal are the fastest-growing energy sources, with consumption increasing by 1.5% and 2.8%, respectively. Projected high prices for oil and natural gas, as well as rising concern about the environmental impacts of fossil fuel use, improve prospects for renewable energy sources. Coal's costs are comparatively low relative to the costs of liquids and natural gas, and abundant resources in large energy-consuming countries (including China, India, and the United States) make coal an economical fuel choice (Fig. 5.8).

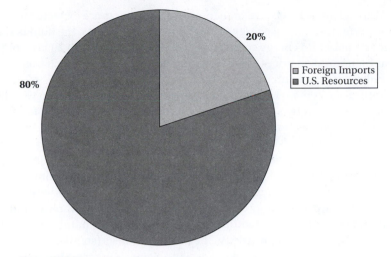

Figure 5.7 Total U.S. oil consumption
Source: U.S. Energy Information Administration 2012.

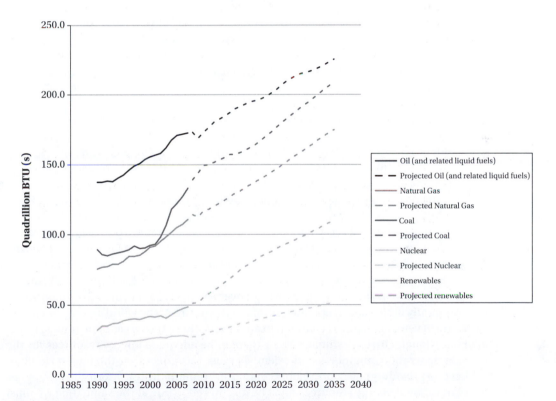

Figure 5.8 World energy consumption by fuel types
Source: U.S. Energy Information Administration, 2011.

- Although liquid fuels and other petroleum products are expected to remain important sources of energy throughout the projections, the liquids share of marketed world energy consumption declines from 37% in 2005 to 29% in 2035 as high world oil prices could force many consumers to switch from liquid fuels and other petroleum products when feasible.
- Natural gas remains an important fuel for electricity generation worldwide, because it is more efficient and less carbon intensive than other fossil fuels. In the 2005–2035 projections, total natural gas consumption increases by 1.7% per year on average, from 104 trillion cubic feet to 169 trillion cubic feet, while its share of world electricity generation increases from 20% in 2005 to 25% in 2035.
- Renewable energy refers to those energy sources that are continually replenished naturally, such as sunlight, wind, rain, tides, waves, and geothermal heat. About 16% of the world's energy consumption comes from renewable resources, with 10% of all energy from traditional sources of biomass, mainly used for heating, and 3.4% from hydroelectricity. The new renewables (smaller-scale hydroelectric power, modern sources of biomass, wind, solar, geothermal, and biofuels) account for another 3% and are growing very rapidly. The share of renewables in electricity generation is about 19%, with 16% of electricity coming from hydroelectricity and 3% from new renewables.
- Coal consumption is projected to increase by 1.5% per year from 2005 to 2030 and to account for 29% of total world energy consumption in 2030. In the absence of policies or legislation that would limit the growth of coal use, China, especially, and to a lesser extent India and the United States are expected to turn to coal in place of more expensive fuels. Together, these three nations account for 90% of the projected increase from 2005 to 2030 (Fig. 5.9).
- There is still considerable uncertainty about the future of nuclear power, and a number of issues could slow the development of new nuclear power plants. Plant safety, radioactive waste disposal, and the proliferation of nuclear weapons, which continue to raise public concerns in many countries, may hinder plans for new installations, and high capital and maintenance costs may keep some countries from expanding their nuclear power programs.

Despite rising levels of consumption, estimates of available world energy resources have become more optimistic in recent years. According to the January 2003 *Journal of Petroleum Technology*, "if energy consumption were to remain constant at current levels, proved reserves would supply world petroleum needs for 40 years, natural gas needs for 60 years and coal needs for well over 200 years." There are other reports that are equally optimistic, if not more so, about the future of the world's energy sources.

However, one theme is common to all reports: these resources appear to be finite. Even though current estimates show there to be adequate energy resources for the next generation, engineers and scientists must recognize that the long-term challenge of providing energy to a global community remains a high priority. New supplies of traditional energy resources must be discovered, while at the same time we must develop and expand options for alternative energy sources (such as wind, bio-based fuels, solar, photovoltaics, geothermal, and low-head hydro).

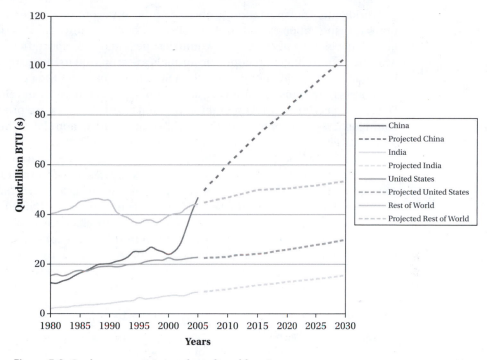

Figure 5.9 Coal consumption in selected world regions
Source: U.S. Energy Information Administration, 2011.

5.4 **Transportation**

Americans, as well as a growing worldwide population, continue to rely heavily on highway systems as their primary means of transportation. The ever-increasing use of the automobile has created significant problems such as crowded highways, inadequate parking, and overall congestion in major urban areas of the world. Transportation problems in larger cities have increased as the number of cars and trucks on the roads has surpassed the capacities of streets and bridges built to accommodate the traffic flow of earlier decades.

According to a 2013 study by the Texas Transportation Institute, it has been estimated that traffic delay and congestion cause motorists in the U.S. urban areas studied to spend the equivalent of more than one work week per year in traffic jams. Researchers tried to estimate the annual economic impact of traffic congestion and determined that the value of motorists' time, coupled with wasted fuel, resulted in combined congestion costs of $121 billion in the areas studied. Twenty-eight percent of the energy used in the U.S. goes to transporting people and goods from one place to another. In essence, trips take longer with more of the day spent in traffic congestion. This impacts our work days, weekend and personal travel, and commercial and freight shipments. Trip travel times have become increasingly unreliable.

Worldwide, transportation problems are also becoming serious issues. A nation's transportation system is generally an excellent indicator of its level of economic development. In many countries personal transportation still means walking or bicycling, and domestic animals are still used to move freight. According to the 2002 World Energy Projection System Report, over the next two decades, rapid growth of transportation infrastructure is expected in the developing world. Developing countries in Asia and Central and South America are predicted to account for 52% of the increase in the world's motor vehicle population between 1996 and 2020 (Fig. 5.10).

According to the 2011 U.S. Energy Information Administration projections, energy use for transportation in developing countries is projected to grow at an average annual rate of 2.6%, nearly triple the rate of growth in industrialized countries. Growth in the transportation sector in the industrialized countries—where modern transportation systems have been in place for many decades—is expected to average only 0.3% per year. Even in the most economically advanced countries, however, transportation energy consumption per capita continues to increase over the projection period, as rising per capita incomes are accompanied by purchases of larger personal vehicles and by increased travel for business and vacations.

The challenge facing engineers is to continue the process of building new highways, bridges, and parking facilities that can meet world demand. However, new construction should be viewed as only one solution to a very complicated problem. Engineers will need to provide their expertise to develop new ways to better control traffic flow, improve the use of intelligent traffic communication systems, change usage patterns, develop efficient methods for quicker accident identification and removal, encourage

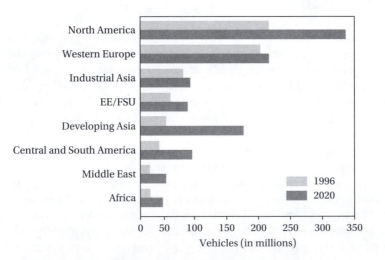

Figure 5.10 Road vehicle numbers by region

Sources: American Automobile Manufacturers Association, World
Motor Vehicle Data, Detroit, MI 1997. 2020: EIA, World Energy Project System, 2002.

ride-sharing and flex-time programs that reduce peak-hour travel, and design and implement imaginative, appealing mass transportation systems.

5.5 Infrastructure

Infrastructure is a term referring to entities designed to support human activities, such as roads, bridges, transportation services (including railways, airports, and waterways), public service facilities, and buildings. The current state of the world's infrastructure is causing great concern as many aging facilities and systems rapidly deteriorate. This, too, presents a significant challenge for engineers.

Many of these problems have become more visible in the U.S. in recent years. Roads are deteriorating faster than they can be repaired due to a lack of available funding. Bridges and highway overpasses are slowly falling apart and pose safety hazards due to traffic loads and volume. Miles of railway tracks and beds are in dire need of maintenance or replacement. Municipal water and wastewater systems cannot handle the increasing public demands, and new environmental safeguards are needed to meet evolving legislative guidelines.

In 2001, the American Society of Civil Engineers first issued a report card for a variety of American infrastructure categories. The report gave an overall grade of D+, with total investment needs of $1.3 trillion over five years to remediate the problems. In 2005, the report was re-evaluated; the grade dropped to a D, and total dollar investment needs climbed to $1.6 trillion. In the 2009 report, the overall grade remained a D, but the total dollar needs climbed to $2.2 trillion. The most recent report, released in 2013, showed a slight overall improvement in a few areas, which resulted in the cumulative GPA rising to D+. However, the total investment needed by 2020 has climbed to $3.6 trillion (Table 5.2).

Because of the increased visibility of many of these problems in the U.S., local, state, and federal government agencies are working diligently to secure adequate funding for repairs. Once secured, it will then be a challenge for engineers to effectively design, develop, and implement the necessary programs and systems to address these critical issues. It is encouraging that in some areas of the country efforts to deal with this massive problem are under way. However, funding must increase and the ingenuity, creativity, and expertise of the engineering community must be put to work to solve these problems.

On a global scale, infrastructure will continue to be a critical factor in economic competitiveness. Developing nations need to, and are, investing in highways, railroads, and utilities to allow their economies to grow. The linkage between economic activity and infrastructure continues to grow stronger and more critical as economic activity becomes increasingly more complicated and global in scope.

Historically, as engineering infrastructure improvement has enhanced the transportation of people, goods, commodities, water, waste, energy, and information, this increased mobility has translated directly into better choices for people across the globe: choices of where to work, what to buy, how to communicate, and in some ways

Table 5.2 U.S. Infrastructure Report Card

	2009 Grade	2013 Grade
Aviation	D	D
Bridges	C	C+
Dams	D	D
Drinking Water	D−	D
Energy	D+	D+
Hazardous Waste	D	D
Levees	D−	D−
Navigable Inland Waterways	D−	D−
Ports	N/A	C
Public Parks and Recreation	N/A	C
Rail	N/A	C+
Roads	D−	D
Schools	D	D
Solid Waste	C+	B
Transit	D	D
Wastewater	D−	D

2013 U.S. Infrastructure GPA = D+

Total Investment Needs (by 2020) = $3.6 trillion

Source: American Society Of Civil Engineers, 2013

how to live. In an era of accelerating change, the infrastructure challenge for the global engineer will continue to be to expand that range of choices, and in so doing, to improve the quality of people's lives.

5.6 Aerospace and Defense

The U.S. space and defense industries, which have provided so many career opportunities for engineers since the 1960s, have been in a period of dramatic change and transition over the past several years. The Challenger disaster in 1986, coupled with the loss of the Columbia space shuttle in 2003, caused a total reevaluation of the future of NASA and the U.S. space program. The breakup of the former Soviet Union in the early 1990s, the wars in Iraq and Afghanistan, and the continuing tensions in many parts of the world have necessitated a major refocusing of defense spending priorities. All of these issues have created an uneven job market for engineers interested in careers in space and defense during the last 20 years.

Worldwide terrorist activity and localized conflicts in Iraq, the Middle East, Eastern Europe, and parts of the former Soviet Union have demonstrated that the U.S. must continue to maintain a prepared and technically current military. This has resulted in increased spending and business for many sectors of the defense industry and homeland security.

- President Obama's shifting of space efforts to include private companies as well as NASA is designed to ensure U.S. leadership in space science and exploration, support the development of new space capabilities, make air travel safer and more affordable, and answer critical scientific questions about Earth, the solar system, and the universe. In conjunction with American industry, NASA is developing new, lower-cost, safer approaches to spaceflight. This will allow the development of more efficient methods to transport people and cargo to locations such as the International Space Station, and to reduce the reliance on Russian capabilities for these functions. It also promotes U.S. jobs and accelerates the growth of the American aerospace industry.
- This policy promotes innovation and advances understanding of the universe: NASA will continue to operate satellites and aircraft to better understand the Earth and improve the ability to forecast climate change and natural disasters. The program will also support space probes and the development of new super-telescopes to further understand of the solar system.
- It fosters research and development breakthroughs in innovative technologies: New programs will allow the development of a new broad spectrum of space technology research grants to promote high-priority technologies such as laser space communications and unmanned aerial and in-space transportation systems. The overall goal will be to create the innovations necessary to keep the U.S. aerospace industry as the world leader in creating leading-edge new technologies.

Another key component in the future of American aerospace and defense technologies will be a greater emphasis and reliance on partnerships with U.S. industry.

President Obama also proposed a sweeping upheaval of NASA's human space-flight program, canceling the current program that would send astronauts back to the moon and instead investing in commercial companies to provide transportation to orbit. Rather than operate its own shuttles, NASA would buy space for its astronauts on commercial space taxis. NASA would shift its focus to unmanned exploration of mysterious deep outer space. The plan has been lauded by many so-called New Space advocates, who assert that traditional NASA programs have been too big, too expensive, and too slow.

Here are a few examples of efforts being made in the burgeoning commercial space industry:

- SpaceX became the first commercial company to successfully send supplies to the International Space Station. SpaceX also launches rockets that carry satellites into orbit and returns the rockets to earth for reuse.
- Bigelow Aerospace, in North Las Vegas, Nevada, is assembling the solar system's first private space station, expandable modules that could hold 36 people. Paying customers would primarily be nations that do not have the funding or expertise

to build their own station. A 30-day trip to a Bigelow station would cost about $25 million a person.

- Two major companies now offer space tourism. Virgin Galactic, of Mojave, California, takes passengers 50,000 feet up on a 90-minute flight aboard its mother ship, then launches them into outer space at four times the speed of sound, for four to five minutes, before returning to Earth. More than 500 people have already signed up, at $250,000 a seat. Another company, XCOR Aerospace, in Midland, Texas, offers a trip up to 100,000 feet—the engines turn off after reaching Mach 3.5, and then passengers coast for several minutes before gliding back down.

Other corporations are also developing plans to launch much smaller and lower-flying satellites, capable of a variety of tasks such as quickly mapping and measuring all the agricultural acreage on Earth.

The net effect of all these increased opportunities in the space and defense industries could provide many new challenging positions for engineers. Employment in aerospace-related industries was about 1.3 million workers in the late 1980s. This figure dropped to 800,000 by the mid-1990s but rebounded to over 1 million in 2010. The U.S. aerospace and defense industries also affect the jobs of 3.5 million other workers in the American economy, including those in large and small nongovernmental firms and related private firms, according to a 2012 study produced by Deloitte.

There are many challenges ahead for engineers in all aspects of the aerospace and defense industries. Since the nature of these industries cuts across virtually all areas of engineering, there should be opportunities for engineering students from a variety of majors and backgrounds. It appears that once the uncertainty in these industries ends, funding and demand for work in these areas will increase. There should be more stabilized opportunities and career challenges for those interested in these dynamic fields.

5.7　Competitiveness and Productivity

The health of a nation's economy in the global marketplace can often be measured by its relative productivity and competitiveness. Global competitiveness is the ability of a nation to successfully compete in the world marketplace while improving the real income of its citizens. Measures of competitiveness include savings rates (which ultimately provide the resources for facilities and equipment needed to improve efficiency); financial investments in areas such as research, development, and education; and the productivity of its industries, which generate economic growth and increased wages.

During recent decades, there has been increasing concern over the declining U.S. position in global competitiveness, especially in certain vital technological areas. There is specific concern regarding those firms that once dominated world markets and are gradually losing significant market share to foreign competitors.

Advances by foreign countries in basic science and research now often surpass those of the U.S., and this decline has implications for jobs, industry, national security, and the strength of the nation's intellectual and cultural life.

One area of intense global competition involves patents. High-quality patents are seen as strong indicators of a nation's future prosperity because they signify the emergence of important new technologies that will be under the patent holder's control for many years. While U.S. researchers still earn a significant number of new patents, the percentage has been falling as international scholars, particularly in Asia, have become more active. The U.S. corporate share of patents has fallen from over 60% in 1980 to 49% in 2008.

Another area of concern is in published scientific papers. According to the National Science Foundation, works by Americans peaked in 1992 and have fallen by 10% since. A 2012 study by Science Watch, which studies yearly scientific publishing activity, indicates that the entire European Union surpassed the United States in the mid-1990s as the world's largest producer of scientific literature. However, the U.S. is still number one when rankings are examined by individual country only.

The number of new U.S. doctorate degrees in science and engineering has been increasing in the past several years, influenced by large numbers of foreign students. However, increasing numbers of doctoral students from countries such as China, India, and Taiwan are electing to return to their home countries, causing a reverse brain drain. These declines are critical because new scientific knowledge is a significant factor in the American economy in a wide range of industries and the development of new products and technologies.

The World Economic Forum, in its annual Global Competitiveness Report for 2012–2013, ranked the top 10 countries based on a variety of factors. After many years at the top of the world rankings, the United States fell to seventh place behind number-one Switzerland. The report indicated that the U.S. continues to be strengthened by many structural factors that can make the economy extremely productive and in a position to survive business cycle shifts and economic turbulence. However, a number of escalating issues hindered the U.S. ranking in 2012–2013. Specifically, the report outlined several areas of strength and some issues of concern.

Strengths were as follows:

- Highly sophisticated and innovative companies that operate in efficient market areas
- An excellent higher education system with strong collaborations with the research and development sectors of business and industry
- Good labor markets with the ease and affordability of hiring with significant wage flexibility
- The world's largest overall domestic economy, which provides many opportunities for capitalizing on the points listed above

Issues of concern were as follows:

- Sometimes turbulent government interactions with the business community, with the perception that government spends its resources wastefully

- An overall concern with the functioning of private institutions and issues related to recent turmoil and scandals in the financial sector
- Weakening assessments of the U.S. financial market capabilities
- Large economic imbalances that have mounted up in recent years, with continuing fiscal deficits and increasing levels of public indebtedness, magnified by significant stimulus spending

In partial response to previous years' reports, and as part of its legislative agenda, the U.S. Congress authorized the America COMPETES Reauthorization Act. The intent was to continue funding programs that invest in modernizing manufacturing, to help spur American innovation through basic research and development, to invest in high-risk/high-reward clean energy research, and to strengthen math and science education to prepare students for the good jobs of the 21st century. The overall goals of the act were to do the following:

- Keep the United States on a path to double funding for basic scientific research, crucial to some of the most innovative breakthroughs, over the next several years
- Create jobs with innovative technology loan guarantees for small and midsized manufacturers and Regional Innovation Clusters to expand scientific and economic collaboration
- Promote high-risk/high-reward research to pioneer cutting-edge discoveries through the Advanced Research Projects Agency for Energy (APRA-E)
- Create the next generation of entrepreneurs by improving science, math, technology, and engineering education at all levels

The goals are to create prosperity through science and innovation, reassert the necessary economic and technological leadership over the next several years, and give future generations greater opportunities. Boosting U.S. competitiveness is essential to the economy, as almost half of the growth in the gross domestic product (GDP) since World War II is related to the development and adoption of new technology.

5.8 Engineering's Grand Challenges

The National Academy of Engineering brought together diverse teams of engineers and scientists from around the world and asked them what the most significant engineering challenges of today are. The results of that work are the 14 "Grand Challenges" listed in Table 5.3. These topics are not ranked in any order; they represent a view of the most significant challenges our world faces to which engineers can make a significant contribution. You will see an overlap in these challenges with the topics discussed earlier in this chapter, as they touch on many aspects of our world. These challenges cut across all of the engineering disciplines. Most of them address complex social issues that require collaboration with

Table 5.3 National Academy of Engineering's Grand Challenges

Grand Challenge	Description
Advance Personalized Learning	A growing appreciation of individual preferences and aptitudes has led toward more personalized learning, in which instruction is tailored to a student's individual needs. Given the diversity of individual preferences, and the complexity of each human brain, developing teaching methods that optimize learning will require engineering solutions of the future.
Make Solar Energy Economical	Currently, solar energy provides less than 1 percent of the world's total energy, but it has the potential to provide much, much more.
Enhance Virtual Reality	Within many specialized fields, from psychiatry to education, virtual reality is becoming a powerful new tool for training practitioners and treating patients, in addition to its growing use in various forms of entertainment.
Reverse-Engineer the Brain	A lot of research has been focused on creating thinking machines—computers capable of emulating human intelligence— however, reverse-engineering the brain could have multiple impacts that go far beyond artificial intelligence and will promise great advances in health care, manufacturing, and communication.
Engineer Better Medicines	Engineering can enable the development of new systems to use genetic information, sense small changes in the body, assess new drugs, and deliver vaccines to provide health care directly tailored to each person.
Advance Health Informatics	As computers have become available for all aspects of human endeavors, there is now a consensus that a systematic approach to health informatics— the acquisition, management, and use of information in health—can greatly enhance the quality and efficiency of medical care and the response to widespread public health emergencies.
Restore and Improve Urban Infrastructure	Infrastructure is the combination of fundamental systems that support a community, region, or country. Society faces the formidable challenge of modernizing the fundamental structures that will support our civilization in centuries ahead.
Secure Cyberspace	Computer systems are involved in the management of almost all areas of our lives, from electronic communications, and data systems, to controlling traffic lights to routing airplanes. It is clear that engineering needs to develop innovations for addressing a long list of cybersecurity priorities.
Provide Access to Clean Water	About 1 out of every 6 people living today do not have adequate access to water, and more than double that number lack basic sanitation, for which water is needed. It's not that the world does not possess enough water—it is just not always located where it is needed.
Provide Energy from Fusion	Fusion is the energy source for the sun. The challenges facing the engineering community are to find ways to scale up the fusion process to commercial proportions, in an efficient, economical, and environmentally benign way.
Prevent Nuclear Terror	Engineering shares the formidable challenges of finding the dangerous nuclear material in the world, keeping track of it, securing it, and detecting its diversion or transport for terrorist use.

(continued)

Table 5.3 National Academy of Engineering's Grand Challenges *(continued)*

Grand Challenge	*Description*
Manage the Nitrogen Cycle	It doesn't offer as catchy a label as "global warming," but human-induced changes in the global nitrogen cycle pose engineering challenges just as critical as coping with the environmental consequences of burning fossil fuels for energy.
Develop Carbon Sequestration Methods	The growth in emissions of carbon dioxide, implicated as a prime contributor to global warming, is a problem that can no longer be swept under the rug. But perhaps it can be buried deep underground or beneath the ocean.
Engineer the Tools for Scientific Discovery	Grand experiments and missions of exploration always need engineering expertise to design the tools, instruments, and systems that make it possible to acquire new knowledge about the physical and biological worlds.

Source: http://www.engineeringchallenges.org/challenges.aspx (used with permission, accessed 12/1/2015)

people outside of engineering to understand the social context and broader implications, as well as innovative technology. These challenges have been a focus area for the National Academy, whose function is to advise the federal government. As a result, they have been a priority of activity and investment at the federal level. More information can be found about the Grand Challenges at the NAE website, www.engineeringchallenges.org/.

EXERCISES AND ACTIVITIES

5.1 Discuss how population growth is related to most of the other challenges outlined in this chapter. Use examples to support your comments.

5.2 Further explore some of the causes of air pollution discussed in this chapter. Write an essay on new engineering developments that might address these problem areas.

5.3 Experts claim that the water supply system in the U.S. is the safest and most drinkable in the world. Write a research report on factors that have adversely influenced water quality in other parts of the world.

5.4 Research some of the alternatives to current methods of waste treatment and waste storage. What might be some more desirable solutions to the waste management problem?

5.5 Explore some of the long-term projections for world energy usage over the next 25 years. Develop a list of factors that could increase the demand for energy resources. Develop another list of variables that could reduce energy consumption during this period.

5.6   Propose some alternatives to the construction of new highways, bridges, and parking facilities that would help to reduce traffic congestion in urban areas.

5.7   Suggest some alternatives that could provide the necessary funding to improve the nation's infrastructure.

5.8   What issues do you feel are of most importance if the U.S. is to improve its competitive position in the global economy? Provide specific examples to support your comments.

5.9   Research one aspect of the U.S. space or defense programs. Write an essay that outlines your assessment of where this part of the industry will be in the next 20 years, and explain why.

5.10   Select one of the challenge topics presented. Research the problem and prepare a paper that discusses additional negative impacts such as unemployment, economic growth, health, and global stability issues related to your topic area.

5.11   Review recent articles on nuclear power. Prepare a report that explores the advantages and disadvantages of developing additional nuclear power facilities.

5.12   With a group of four or five students, discuss the issues presented in this chapter. Brainstorm and consider what new majors, programs, or courses might evolve during the next 20 years in an attempt to meet these future challenges. Present your results to the class.

5.13   Some legislators have recommended further budget reductions for space and defense programs. Review recent articles on this topic. Do you agree or disagree with these recommendations? Explain your position.

5.14   Develop a list of other challenges that could be added to this chapter. Select one and write an essay that outlines the challenge in greater detail, using recent articles as resources.

5.15   Select one of the NAE Grand Challenges (www.engineeringchallenges.org/) that would relate to the field of engineering you are considering. Explore potential solutions. How could your future branch of engineering find solutions?

CHAPTER 6

Succeeding in the Classroom

6.1 Introduction

As an engineering student, an important goal is to succeed in the classes you need to attain an engineering degree. How you do this depends to a great deal on you and your style, temperament, and strengths. This chapter presents strategies proven to be effective in helping students succeed. As with any tips on success, look at the techniques being outlined and decide if they can be applied to you and your own individual style. If you are unsure about some of the suggestions, give them a try. If those don't work, move on to others until you find something that fits with your own style. You may find yourself in the same situation as one freshman engineering student after being given a book on study skills.

The student enters a professor's office very excited and asks for a short conference with the professor:

> *Student: Professor! It worked!*
> *Professor: That's great. What worked?*
> *Student: The book!*
> *Professor: That's great. What book?*
> *Student: That study skills book.*
> *Professor: So you read it?*
> *Student: Yes, I read it like you suggested. It sure sounded like a lot of unworkable stuff at first. But then I decided, what the heck. I'll try it for a week. Couldn't hurt, I figured. Well, it really did work! It even worked for physics lab. I got an A on the last lab. I still can't believe it.*
> *Professor: That's great. I am glad it helped.*
> *Student: Could I get another copy of that book? I have some friends who want to read it now . . .*

Just like that student, we are often reluctant to try new things. We are comfortable in what we do. You may have been very successful in high school with your study methods and wonder why you need to develop different study methods and techniques. The answer is that your environment has changed. College is very different from high school. The courses are structured differently and require significantly more preparation outside of class. The expectations of the instructors are different and the grading is different. Modifying your habits to become more effective is a key to succeeding in engineering studies. Learning to adapt to meet new challenges is a skill that will serve you well not only in college, but for the rest of your life.

The three components to succeeding in your academics are your ability, your attitude, and your effort. There is no book that can address the first item. While there is unquestionably a level of ability that is needed to succeed in engineering, there are numerous students who had the ability but failed at engineering because of poor attitudes or poor or inefficient effort. There are also many students who lacked natural ability but because of their positive attitude and extraordinary effort are now excellent engineers. This chapter focuses on techniques to improve these two components for success that are totally under your control.

6.2 Attitude

> *Ability is what you're capable of doing. Motivation determines what you do. Attitude determines how well you do it.*
>
> —Lou Holtz, South Carolina football coach

Your attitude is the first thing that you have control over when beginning any challenge, including studying engineering. You can expect success or expect failure. You can look for the positives or dwell on the negatives.

Approaching your classes, professors, and teaching assistants with a positive attitude is the first key to succeeding in your engineering studies. Look at each class as an opportunity to succeed. If there are difficulties along the way, learn to deal with them and move ahead. Many students will decide that a certain class is too difficult or that the professor is a poor teacher and will expect to do poorly. Many fulfill their own prophecy and actually fail. Other students look at the same situation and overcome the hurdles and excel.

It is a distinct possibility that you will have professors you don't like or whom you think teach poorly. You may even have an experience like the following example.

EXAMPLE 6.1 A physics professor of mine discovered the overhead projector during a semester and moved his lecture from the blackboard to overheads. The problem was that he was very tall and stood so that he blocked the overhead. It shined on his coat and tie, but we couldn't see what he was writing. When we told him in class that we couldn't see, he moved out of the way and looked at the overhead. He then adjusted the focus and asked if we could see now. After we responded yes, he moved back to write more and blocked the overhead again. When another student tried to explain that he was blocking his view of the overhead, the professor explained that no matter where he stood, he would block someone's view. He suggested that the student move to the other side of the room. A third student from the other side of the room chimed in that he couldn't see from that side either. Unfortunately he was in the back of the lecture hall, so the professor thought that was the reason. He moved out of the way of the projector and asked one of the students in the front row if she could see. She said, "Yes, but . . ." The professor cut her off before she could finish and asked the guy in the back of the room if he had glasses. When he said no, the professor recommended he get his eyes checked because the woman in front said she could see. At this point we all gave up taking notes for the rest of the lecture.

Later, I complained to an advisor who was the director of the university honors program. Rather than a strategy for communicating with the professor, I received a story. It seems that he had an English professor who was about 108 years old (or seemed that old). He couldn't stand up without leaning on something, so he used the blackboard. His shoulder made contact with the blackboard, which wasn't a problem in itself, except that he wrote at shoulder level. The result is that he erased what he wrote as he moved along the board. So if you weren't in a seat where you could see the three or four words he wrote before he erased them with his shoulder, you couldn't take notes.

The point he was making was that there are challenges in learning as well as later in life. What we have to do is to find ways to overcome obstacles, not to dwell on them. If your attitude is that it is a professor's fault that you won't learn, you won't. If you keep a positive attitude, you can turn challenges into opportunities. He shared that his class became much closer and studied more together because they had to piece the lecture together from their collective notes. They learned a lot, and the positive outcome was the closeness the class developed. He challenged us to go and talk to the professor outside of class, during his office hours, and try to explain our difficulty with the lecture. Instead of just complaining, we should take action with a positive attitude to solve our problem. It was a real lesson for life.

I am still not sure what lesson I was supposed to learn from that story. It struck me, however, that even with professors like that, the advisor earned his PhD, became a professor, and eventually ran the university's honors program. He had developed strategies to cope and kept a good attitude. Attitude is totally under your control.

6.3 Goals

Do you feel the world is treating you well?

If your attitude toward the world is excellent, you will receive excellent results. If you feel so-so about the world, your response from that world will be average. Feel badly about your world and you will seem to have only negative results from life.

—Dr. John Maxwell, founder of INJOY, Inc.

As freshmen, there were a number of us who were in an honors program. From the first day of his freshman year, one in the group set a goal to be a Rhodes Scholar. The rest of us were too busy adjusting to college and taking it all in. In our senior year, when we were interviewing for jobs and deciding whether to go to graduate school, the friend who had set the goal was being toured around the state as the newest Rhodes Scholar from our university. The difference in our college careers was astounding, simply because he had established clear, elevated goals. The rest of us did okay, but he excelled, in large part because he had defined a standard and kept striving for it.

The lesson that I took away from that friendship in college was that setting clearly defined goals early is a key to success in school, in an engineering career, or in life as a whole.

In setting goals, there are two essential elements—height and time. You might ask, "How high do I set my goals?" This will vary from individual to individual. Goals should be set high, but attainable. A common management practice is for a team to decide on a reasonable goal to accomplish. Then the bar is raised and a stretch goal is defined that encourages the team to produce more. This stretch goal is designed to push the group farther than they would have gone with only the original goal. Sometimes this stretch goal looks reasonable and sometimes it looks totally unreachable. What often happens, though, is that the stretch goal is met, realistic or not. As an individual, setting stretch goals helps us grow. When setting goals for yourself, look at what you think you can do and set base goals. Then establish stretch goals a little higher than the base goals. You may just find yourself meeting those higher goals.

The second key element in goal setting is time. When will the goal be accomplished? It is important to have long-term goals and short-term goals. Long-term goals help guide where you are headed. For example, what do you want to do with your engineering degree? An appropriate long-term goal as a freshman could involve determining the kind of entry-level job you desire. This will help answer the question "Why am I doing this?" when your classes get tough.

Short-term or intermediate goals should be stepping-stones to the achievement of your long-term goals. An advantage of an academic setting is that there are natural breaks, semesters, or quarters, for the establishment of goals within specific time frames. Goals can be set for the academic year, and for each semester or quarter. These provide intermediate goals.

Short-term goals need to be set and rewarded. Rewarding short-term goals is essential to keeping motivation high in the quest for long-term goals. Many students lose sight of their long-term goals because they didn't set short-term, rewardable goals. Short-term goals may be an "A" on an upcoming quiz or test, or completing a homework set before the weekend.

EXAMPLE 6.2	Long-term goal	Job as a design engineer for a major manufacturer of microprocessors
	Intermediate goals	3.7 GPA after freshman year 3.5 GPA at graduation Hold an office in the local chapter of IEEE Internships after sophomore and junior years

Short-term goals   B on first calculus test
A in calculus
3.7 GPA after first semester
Join IEEE and attend meetings
Complete chemistry lab by Friday

In the list of short-term goals above, there are opportunities for rewards. Perhaps there's a CD you want. Set a goal of a certain grade on that calculus test. If you meet the goal, you can go buy the CD.

Clearly define your goals by writing them down. It is much more powerful to write down your goals than to simply think about them. Some people post them where they can see them daily. Others put them in a place where they can retrieve and review them regularly. Either method is effective. For short-term goals, establish weekly goals. Select a day, probably Sunday or Monday, and establish the goals for the week. This is a great time to establish the rewards for your weekly goals. Plan something fun to do on the weekend if you meet those goals.

At the end of each semester, examine your progress toward your long-term goals. This is an advantage to being in school. Every semester, there is an opportunity to assess your progress and start fresh in new classes. Take advantage of this. Examine your long-term goals and make any changes if necessary.

Determining Goals

It can be intimidating or difficult for students early in their careers to decide on appropriate goals. Fortunately, there are numerous resources on campus to help. Academic advisors can provide criteria for grade-point goals. Things to consider include:

1. What is the minimum GPA to stay enrolled in school?
2. What grades do you need to continue in engineering or to enter engineering?
3. What grades are needed to be eligible for co-op or intern positions?
4. What GPA is required to make the dean's list or honor list?
5. What GPA is needed to be eligible for the engineering honorary organizations?
6. What grades are needed for admittance to graduate school?
7. What grades are needed for scholarships?

There are also people on campus who work with the placement of graduates. They are excellent resources to get information on what employers are looking for. What credentials do you need to be able to get the kind of job you desire? If you know this as a freshman, you can work toward it. You can also go directly to the employers and ask these same questions of them. If companies come to your campus, talk to them about what credentials are needed to get a job. Job fairs are great times to do this. Approaching company representatives about establishing goals is also a great way to show initiative and distinguish yourself from your classmates.

Goals are important for outside the classroom too. It is important to have goals in all areas of your life. Later in this chapter we will discuss such goals.

6.4　Keys to Effectiveness

Once goals have been set regarding grades, the next step is to achieve those grades. Effort and effectiveness may be the most important components of your success as a student. There are numerous cases of well-equipped students who end up failing. There are also students who come to the university poorly prepared or start off poorly but eventually succeed. The difference is in their effort and their effectiveness. The following suggestions contain strategies to improve personal effectiveness in your studies.

> A general rule is that it takes a minimum of two hours of study outside of class for each hour of class lecture.

Take Time to Study

One of the strongest correlations with a student's performance is the time he or she spends studying.

According to the reminder above, if you are taking 15 hours, that means that you need to spend 30 effective hours a week studying outside of class. In a typical high school, the time requirement is much less than this. Developing the study habits and discipline to spend the needed time studying is a prime factor in separating successful and unsuccessful students.

Go to Class

There is also a high correlation between class attendance and performance. This may seem obvious, but you will run into numerous students who will swear that it doesn't matter if you go to class. Many instructors in college do not take attendance, so no one but you knows if you were there, which is different from high school. The classroom is where important information will be presented or discussed. If you skip class, you may miss this information. Also, by skipping class, you miss an opportunity to be exposed to the material you will need to know, and you will have to make up that time on your own. Missing class only delays the expenditure of time you will need to master the material. Most often, it may take much more than an hour of independent study to make up an hour's worth of missed class time. It is possible for you to never miss a class in your undergraduate years, except for illness or injury, serious family problems, or trips required by groups to which you belong. Never missing a class could be one of your goals.

Also, you may want to schedule classes at 8 a.m. and 4 p.m. This will get you up and going and keep you there! You will seldom, if ever, study at 8 a.m. and at 4 p.m. Do your studying in the middle of the day. This will help you to use your time effectively.

> After class, reread the assigned sections in your textbook. Concentrate on the sections highlighted by the lecture. By doing so, you will not waste time trying to understand parts of the book that are not critical to the course.

Make Class Effective

The first component of making class effective is sitting where you can get involved. If the classroom is large, can you see and hear in the back? If not, move up front. Do you need to feel involved in the class? Sit in the front row. Do you fall asleep in class? Identify why. Ask yourself, "Am I sleeping enough at night?" If the answer is no, get more sleep.

The second step for making class time effective is to prepare for class. Learning is a process of reinforcing ideas and concepts. As such, use the class time to reinforce the course material by reviewing material for the lecture beforehand. Most classes will have a textbook and assigned reading for each lecture. Take time before the lecture to skim over the relevant material. Skimming means reading it through but not taking time to understand it in depth. This will make the class time more interesting and more understandable, and it will make note taking easier.

By following the above advice, you will have had three exposures to the material, which helps you remember and understand the material better.

Keep Up with the Class

Class is most effective if you keep up. An excellent short-term goal is to master the lecture material of each course before the next lecture. A course is structured in such a way that you master one concept and then move on to the next during the next lecture. Often in science and engineering classes the concepts build on each other. The problem with falling behind is that you will be trying to master concepts that depend on previous lectures, and you are going to hear lectures that you will not understand.

There is a very practical reason that a three-credit class meets three times per week. During a 15-week semester, there are only 45 hours of class meeting time. That amount of class time could be held in a week of consecutive nine-hour days. Arranged that way, a whole semester could be taken in a month. Then it would only take a year to get a bachelor's degree. So why don't we do it that way?

The reason is that the current educational model is designed to introduce a concept and give you, the learner, time to digest it. It is an educationally sound model. In engineering study, digesting means reviewing notes, reading the text, and working several problems related to the topic. If you are still confused about an aspect of the previous lecture, the next class provides a great opportunity to bring it up with the professor. Get your questions answered before going on to the next topic.

A note of caution should be given here. Friends of yours may be studying subjects where they can study from test to test rather than from class to class. They may insist that you don't need to keep up so diligently because they don't. Some majors cover material that is more conceptually based. That is, once you understand the concept, you have it. Engineering, math, and science courses are not this way. To be mastered they require extensive study and preparation over an extended period of time. In some courses there are only a few basic concepts, yet the entire course is dedicated to the application of these few concepts. An example is Statics, a sophomore-level course, in which there are only two fundamental equations germane to the course: the sum of forces equals zero, and the sum of moments equals zero. The entire course involves applying these two equations to various situations, and using the information in engineering applications.

Take Effective Notes

A main component in keeping up with your classes is taking effective notes for each lecture. Effective notes capture the key points of the lecture in a way that allows you to understand them when you are reviewing for the final exam three months later. Suggested note-taking strategies follow:

1. Skim the assigned reading prior to class to help identify key points.
2. Take enough notes to capture key points but don't write so much that you fail to adequately listen to the presentation in class.
3. Review your notes after class to annotate them, filling in gaps so they will still make sense later in the semester.
4. Review your notes with other students to ensure that you captured the key points.
5. Review your notes early in the semester with your professor to be certain you are capturing the key points.

In class, your job is to record enough information to allow you to annotate your notes later. You don't have to write everything down. Getting together with classmates after class is a terrific way to annotate your notes. That way you have different perspectives on what the main points were. If you really want to make sure you captured the key points, go and ask your professor. Doing so early in the semester will not only set you up for successful note taking throughout the semester, it will also allow you to get to know your professor.

Work Lots of Problems

Because the applications of the concepts are the core of most of your classes, the more problems you can work, the better prepared you will be. In math, science, or engineering courses, doing the assigned homework problems is a minimum. Search for additional problems to work. These may be from the text or from old exams. Your professor should be able to steer you to appropriate problems to supplement the homework.

Use Caution With Solution Manuals and Files

For some classes, there are homework files or solution manuals available. While these tools may help get the homework done quicker, they very often adversely affect exam performance. The problem is that most professors don't allow the solutions to be used during the test. Using them becomes a crutch and can impair your ability to really learn the material and excel on exams. If they are used at all, use them only after exhausting all other possibilities of working out the answer yourself. If it takes a lot of work to figure out a homework problem, you will learn that concept well and you will have a higher probability of demonstrating your knowledge on the next test. Taking the easy way out on homework has a very consistent way of showing up on test results. Minimal learning results from copying down a solution.

Group Studying

A better model for studying than using solutions is to study in a group. Numerous studies have shown that more learning takes place in groups than when students study by themselves. Retention is higher if a subject is discussed, rather than just listened to or read. If you find yourself doing most of the explaining of the ideas to your study partners, take heart! The most effective way to learn a subject is to teach it. Anyone who has taught can confirm this anecdotally; they really learned the subject the first time they taught it.

> *The most important single ingredient to the formula of success is knowing how to get along with people.*
>
> —Theodore Roosevelt

If you aren't totally convinced of the academic benefits of group studying, consider it a part of your engineering education. Engineers today work in groups more than ever before. Being able to work with others effectively is essential to being an effective engineer. If you spend your college years studying in groups, working in groups will be second nature when you enter the workforce.

Studying with others will also make it more bearable, and even fun, to study. There are numerous stories about students preparing for exams and spending the whole day on a subject. It is difficult to stick to one subject for an entire day by yourself. With study partners you will find yourself sticking to it and maybe even enjoying it.

Studying with others makes it easier to maintain your study commitments. Something we all struggle with is discipline. It is much easier to keep a commitment to study a certain subject if there are others who are depending on you. They will notice if you are not there and studying.

I will pay more for the ability to deal with people than any other ability under the sun.

—John D. Rockefeller

Group studying is more efficient if properly used. Chances are that not everyone in the group will be stuck on the same problems, so right away, you will have someone who can explain the solution. The problem areas common to everyone before the group convenes can be tackled with the collective knowledge and perspectives of the group. Solutions are found more quickly if they can be discussed. More efficient study time will make more time for other areas of your life!

Choosing a group or partner may be hard at first. Try different study partners and different-size groups. Ray Landis, Dean of Engineering at California State University at Los Angeles and a leading expert on student success, suggests studying in pairs. "That way each gets to be the teacher about half the time." Larger groups also can be effective. The trick is to have a group that can work efficiently. Too large a group will degenerate into a social gathering and will not be productive in studying. Whichever group size you choose, here are some basic tips for group studying.

1. Prepare individually before getting together.
2. Set expectations for how much preparation should be done before getting together.
3. Set expectations on what will be done during each group meeting.
4. Find a good place to convene the group that will allow for good discussion without too many distractions.
5. Hold each other accountable. If a group member is not carrying his or her weight, discuss it with that person and try to get him or her to comply with the group's rules.
6. If a member continues to fail to carry his or her own weight or comply with the group's rules, remove that person from the group.

Select a Good Study Spot

In a subsequent section, differences in learning styles will be discussed. Depending on your own learning style and personality, you may need a certain kind of environment to study efficiently. Some resources attempt to describe the *perfect* study environment. In reality, there is no one perfect study environment or method of studying. What is crucial is that you find what you need and a place that meets your needs. For some, it involves total quiet, sitting at a desk. Others may prefer some noise, such as background music, and prefer to spread out on the floor. Whatever you decide is the best environment, pick one where you can be effective.

Cynthia Tobias describes the conflict she has with her husband's view of a proper study environment in her book *The Way They Learn:*

I have always favored working on the floor, both as a student and as an adult. Even if I'm dressed in a business suit, I close my office door and spread out on the floor before commencing my work. At home, my husband will often find me hunched over books and papers on the floor, lost in thought. He is concerned.

"The light is terrible in here!" he exclaims. "And you're going to ruin your back sitting on the floor like that. Here, here! We have a perfectly clean and wonderful rolltop desk." He sweeps up my papers and neatly places them on the desk, helps me into the chair, turns on the high-intensity lamp, and pats my shoulder.

"Now, isn't that better?" he asks.

I nod and wait until he is down the hall and out of sight. Then, I gather all my papers and go back down on the floor. . . . It does not occur to him that anyone in their right mind could actually work better on the floor than at a desk, or concentrate better in 10-minute spurts with music or noise in the background than in a silent 60-minute block of time.

Different people have different ways in which they can be most efficient. You need to discover yours so that you can be efficient. If you are unsure of what works best for you, test some options. Try different study environments and keep track of how much you get done. Stick with the one that works best for you.

If you and your roommate have different styles and needs, then you will need to negotiate on how the room will be used. Both of you will have to be considerate and compromise, because it is a shared space.

One item that is a distraction for almost everyone is television. Because it is visual, it is almost impossible to concentrate on homework with a TV on, although many students swear they are still "productive." A few may be, but most are not. Watching TV is one of the biggest time-wasters for students. It is hard enough for first-year students to adjust to college life and the rigors of pre-engineering classes. We strongly recommend ***not*** having a television set in your residence hall room or apartment for at least your first year. It is not even good for study breaks, which should be shorter than 30- or 60-minute programs. This may be one temporary sacrifice you have to make to be successful in your studies.

6.5 Test-Taking

Most courses use written, timed exams as the main method of evaluating your performance. To excel in a course, you need to know the material well and be able to apply it in a test situation. The early part of this chapter provided tips on how to improve your understanding of the course material. The second part is test-taking skills. Ask any student who has been away from school for a few years. Taking tests is a skill and needs to be practiced. Just like a basketball player who will spend hours practicing lay-ups, it is important to practice taking tests. A great way to do this is to obtain past exams. A couple of days before the test, block out an hour and sit down and take one of the old tests. Use only the materials allowed during the actual test. For instance, if it is a closed-book test, don't use your book. In this way, you are doing two things. The first is assessing your preparation. A good performance on the practice test indicates that you are on the right track. The second thing you are doing is practicing the mechanics of taking the test. Again, with the basketball analogy, players will scrimmage in practice to simulate game conditions. You should simulate test conditions. What do you do when you reach a problem you can't do? Most people will start to panic, at least a little bit. It is critical that you stay calm and reason your way beyond whatever is blocking you from doing the problem. This may mean skipping the problem and coming back to it, or looking at it in another way. In any case, these are the types of things you have to do on tests, and by practicing them under test conditions, you will do better on the actual tests.

Before the first test in the course, visit your professor and ask him or her how to assemble a simulated test. It may be that old exams are the way to go, or possibly certain types of problems will be suggested. Professors don't like the question "What is going to be on the test?" but are much more receptive to requests for guidance in your preparation. The professor's suggestions will also help you focus your studying, to make it more efficient.

Taking the Test

There are some general guidelines for taking any test. The first one is to come prepared to take the test. Do you have extra pencils just in case? Can you use your calculator efficiently? Breaking your only pencil or having your calculator fail during the test can be very stressful and prevent you from doing your best, even if you are given a replacement. Proper planning for the test is the first step toward succeeding.

The second step is to skim over the entire test before beginning to work the problems. This is partly to make sure that you have the complete test. It also gives you an overview of what the test is going to be like so you can plan to attack it efficiently. Take note of the weighting of the points for each problem and the relative difficulty.

The next step is to look for an easy problem or one that you know you can do. Do this problem first. It will get you into the flow of doing problems and give you confidence for the rest of the test. Also, doing the problems you know you can do will ensure that you have made time for the problems for which you should get full credit.

Always keep track of the time during a test. Tests are timed, so you need to use the time efficiently. Many students will say they "just ran out of time." While this may prevent you from finishing the test, it should not prevent you from showing what you know. Look at the number of points on the test and the time you have to take the test. This will tell you how long to spend on each problem. If there are 100 points possible on the test and 50 minutes to complete it, each point is allotted 30 seconds, so a 10-point problem should only take five minutes. A common mistake is wasting too much time struggling through a difficult problem and not even getting to others on the test. Pace yourself so you can get to every problem even if you don't complete them. Most instructors will give partial credit. If they do, write down how you would have finished the problem and go on to the next when the allotted time for that one is up. If there is time at the end, you can go back to the unfinished problem. But if not, the instructor can see that you understood the material and can award partial credit.

Remember that the goal of a test is to get as many points as possible. A professor can't give you any credit if you don't write anything down. Make sure you write down what you know and don't leave questions blank. Another common mistake occurs when students realize they have made a mistake and erase much of their work. Often, the mistake they made was minor and partial credit could have been given, but none is awarded because everything was erased.

Concentrate on the problems that will produce the most points. These are either problems you can do relatively quickly and get full credit for, or ones worth a large portion of the total points. For instance, if there are four problems, three worth 20 points and one worth 40 points, concentrate on the 40-point problem.

Leave time at the end, if possible, to review your answers. If you find a mistake and there is not enough time to fix it, write down that you found it and would have fixed it if you had time. If you are taking a multiple-choice test, be careful about changing your answers. Many studies find that students will change right answers as often as they change wrong answers. If you make a correction on a multiple-choice test, make certain you know that it was wrong. Otherwise, your initial answer may have been more likely to have been correct.

Think. This may sound basic, but it is important. What is the question asking? What does the professor want to see you do? When you get an answer, ask yourself if it makes sense. If you calculate an unrealistic answer, comment on it. Often a simple math error produces a ridiculous answer. Show that you know it is ridiculous but that you just ran out of time to find the error.

After the Test

After you get your test back, look it over. It is a great idea to correct the problems you missed. It is much easier to fill in the holes in what you missed as you go along in the semester than to wait until you are studying for the final exam. Most final exams cover the material from all the semester tests. If you can't correct the problems yourself, go and see the professor and ask for help. Remember, you *will* see the material again!

6.6 Making the Most of Your Professors

One of the most underused resources at a university is the faculty. Many students do not take full advantage of the faculty that they are paying to educate them. The professor is the one who decides on course material, what is important and what is not, and how you will be evaluated. Yet few students take the time to get to know their professors. Here are some reasons to get to know your professors:

1. They are professionals and have experience in the fields you are studying and, therefore, have perspectives that are valuable to you.
2. Every student will need references (for scholarships, job applications, or graduate school applications) and professors can provide them.
3. They are the ones in charge of the course and can help you focus your studying to be more effective.
4. They are the experts in the field and can answer the hard questions that you don't understand.
5. They assign your grades, and at some point in your college career, you may be on the borderline between grades. If they know you and know that you are working hard, they may be more likely to give you the higher grade.
6. They are likely to know employers and can provide job leads for full-time positions or internships.
7. They may be aware of scholarship opportunities or other sources of money for which you could apply.
8. They can provide opportunities for undergraduates to work in labs, which is great experience. You may even get paid!

It is interesting to examine the history of the university. In centuries past, students did not go to a university to get trained to do a job; rather, they went to be mentored by great scholars (the faculty). In modern universities, many students come looking primarily for the training needed to get a good job, and getting mentored by faculty is low among their priorities. Consider this: A school could send you the course materials at home and have you show up for one day to take your exams. Yet students still come to a university to study. Why? A main reason is for the interaction with the people at the university—faculty, staff, and other students.

If there are so many reasons to get to know professors, why don't most students do so? Many professors don't encourage students to come see them, or don't present a welcoming appearance. Professors may remind students of their parents, people who seem to be out of touch with college students. It may also be that many students don't really understand the advantages of getting to know their professors.

Here are some things to keep in mind when getting to know professors. The first is that they have spent their careers studying the material you are covering in your courses and, therefore, they enjoy it. So it would be a very bad idea to start by telling your calculus professor that you find math disgusting. This is insulting and suggests he has wasted his career.

The second is that professors teach by choice, especially in engineering. Industry salaries are higher than academic salaries for engineering PhDs, so professors teach because they want to. They also probably think that they are good at it. If you are having a problem with an instructor's teaching style, approaching him or her with a problem-solving strategy such as the following can be very constructive: "I am having trouble understanding the concepts in class. I find that if I can associate the concepts with applications or examples I understand them better. Can you help me with identifying applications?"

As a group, professors also love to talk, especially about themselves and their area of expertise. They have valuable experience that you can benefit from. Ask them why they chose the field they did. Ask them what lessons they learned as students. You will get some valuable insights and also get to know someone who can help you succeed in your class, and possibly in your career. Steven Douglass, in his book *How to Get Better Grades and Have More Fun*, suggests asking each professor, "What is the main objective in the course?"

Faculty members are very busy. They have many demands on their time (research, securing grants, writing, committee assignments, etc.) in addition to your class. These other demands are what keep the university running. So respect their time. It is okay to get to know them but don't keep popping in just to shoot the breeze. Also, when you come with questions, show them the work you have been doing on your own; you only need to get past a hurdle. A busy faculty member won't mind helping a hard-working student clear up an idea or concept. However, showing up and asking how to do the homework can get interpreted as "Do my homework for me." This would not make the positive impression you want.

6.7 **Learning Styles**

In the past several decades scientists have made many discoveries regarding the complexities of the human brain. While the brain is an important organ in each person's body, we now know that each person's brain is unique to him or her. Both our genetic code before birth and our environment after birth shape our developing brain. Nutrition, stress, and drugs and alcohol are some of the factors that can affect a developing brain.

Each person is born with all the brain cells, or neurons, that he or she will ever have (estimates reach as high as 180 billion neurons). Neurons differ from other cells in that they do not reproduce. However, they do form connections with other neurons. As they are used, neurons can grow tentacle-like protrusions called dendrites that can receive signals from other neurons. In the first five years of life, with normal stimulation, a multitude of connections are created as the brain learns to move the body, speak, analyze, create, and control. However, people can continue to create more neurological connections if they continue to use their brain and learn new material. In the absence of diseases like Alzheimer's or physical injury, the brain is capable of creating new connections throughout our lifetime.

This is great news! None of us is ever too old or too dumb to learn something new. Another benefit of brain research is that it has identified several different ways that people think and memorize. Memorizing refers to how people assimilate, or add, new material to existing knowledge and experience. It also includes how we accommodate, or change our previous way of organizing material. Thinking refers to how we see the world, approach problems, and use the different parts of our brain. One method does not fit all when it comes to how we learn best, so spending time to identify your own strengths and weaknesses will prove very beneficial. Once you identify the aspects of your own preferred learning style, you can manipulate information so that you learn more information in less time, with higher rates of recall.

Memory Languages

Place a check mark by all the statements that strongly describe your preferences.

Auditory

_____ I enjoy listening to tapes of lectures.

_____ I often need to talk through a problem aloud in order to solve it.

_____ I memorize best by repeating information aloud or explaining it to others.

_____ I use music and/or jingles to memorize.

_____ I remember best when information fits into a rhythmic pattern.

_____ I would rather listen to the recording of a book than read it.

_____ I remember names of people easily.

Visual

_____ I follow pictures or diagrams when assembling something.

_____ I am drawn to flashy, colorful, visually stimulating objects.

_____ I prefer books that include pictures or illustrations with the text.

_____ I create mental pictures to help me remember.

_____ I usually remember better when I can see the person talking.

_____ I remember faces of people easily.

_____ I prefer to follow a map rather than ask for directions.

Kinesthetic

_____ I can memorize well while exercising.

_____ I usually learn best by physically participating in a task.

_____ I almost always have some part of my body in motion.

_____ I prefer to read books or hear stories that are full of action.

_____ I remember best when I can do something with the information.

_____ I solve problems best when I can act them out.

You probably checked some in each of the three areas; however, most people have a preferred modality for learning new material. See which area you checked most often and develop ways to study within that area. For example, if you are an auditory learner, record your lectures to listen to over and over again—perhaps while going to and from class. In lecture, sit where you can hear the professor well. Try focusing on what is being said during the class and taking notes on the material from your recording later. Ask the professor questions during class and office hours. Read your assignments out loud to yourself, or into your recorder. Minimize visual distractions in your study place so you can focus on active listening, to your own voice or to a recording. Create jingles, songs, poems, or stories to help you remember. When studying with others, have them drill you on what you're learning, or spend time explaining it to them.

If you are a visual learner, you will benefit from a different set of strategies. During class make sure you sit where you can see the professor and the board or screen clearly. Write notes during lectures with plenty of pictures and meaningful doodles. Rewrite notes later in a more organized fashion, highlighting the main ideas emphasized in class. Create diagrams, lists, charts, and pictures to help clarify material. Write out questions that you have to ask the professor, either in class or during office hours. Highlight main ideas in your books as you read, and don't be afraid to write notes in the margins. Use more than one color to emphasize different things. Use mental images to help you remember how to solve problems. When studying with others, have them give you written practice problems and tests. Use mental maps to illustrate how ideas fit together or to plan upcoming papers or projects. Use video material to reinforce concepts and events.

EXAMPLE 6.3   I sat down next to a PhD student for a lecture. As we got ready for class, she prepared a little differently. Instead of taking out one writing instrument, she had four—she held four different-colored pens in her left hand. She would take one at a time and use them with her right hand to write. Different ideas had different colors. As I watched, she drew diagrams between points with the different colors. Her notes looked like a work of art. She was one of the most visual learners I have ever seen. Early in her academic career she struggled until she developed methods for succeeding. And she has succeeded, all the way through her PhD program in engineering.

Kinesthetic learners may have the biggest disadvantage, because most formal college learning environments lack kinesthetic activities. The exception to this would be classes with labs. Take advantage of these as often as possible, since doing something with the information you are learning is easiest for you. During lectures, make connections between what is being said and similar things you have done in the past. Talk with your professor during office hours about ways to be more hands-on with the material. Perhaps you could volunteer in your professor's lab, or use models and experiments on your own at home. Try memorizing while exercising, especially during repetitive activities like jogging, running stairs, jumping rope, or swimming laps. When studying with others, try acting out the material. Shorter, more frequent study times may be more productive than a single marathon session.

Thinking Skills

The way the brain works is much more complex than simply adding new material to old. Thinking refers to how we see the world, approach problems, and use the different parts of our brain. As science has rapidly improved its ability to measure the brain's activity, our understanding of brain functions has dramatically changed. As research continues, our understanding of this wondrously complex organ is continuing to evolve.

Models have been developed to try to categorize how different individuals think. Early models focused on the specialization of the right and left hemispheres of the brain. Further research has unlocked other complexities and required the development of more complex models. A common current model was developed by Ned Herrmann and shows the brain as four quadrants. Originally, these were four physiological components of the brain. As research continued, it was determined that brain function was too complex to be precisely represented in this format. However, the model has proven effective as a representation of our preferred thinking style and the four quadrants are currently used as a metaphorical model of how we prefer to process information.

Figure 6.1 shows the evolution of the quadrant model from the hemispherical models. Quadrants A and B represent the left brain, and quadrants C and D represent the right brain. Herrmann's complete model is detailed in his book *The Creative Brain*. These quadrants illustrate personal preferences and a dominant thinking style. However, most people exhibit characteristics of each quadrant to some degree. Successful people are aware of their strong areas and can capitalize on them while compensating for their weaker areas, and possibly building them up.

Tables 6.1 through 6.4 highlight examples of the various thinking skills, preferred learning activities, and practice activities for each quadrant. Examine the lists and try to place yourself in one or more quadrants.

The most successful students pick and choose a variety of strategies that work best for their learning style and specific classes. Observe what other successful students are doing and learn from their example. Remember, your one-of-a-kind brain is very capable of learning new material. Spend some time learning how you learn best and you've paved the way for successful learning.

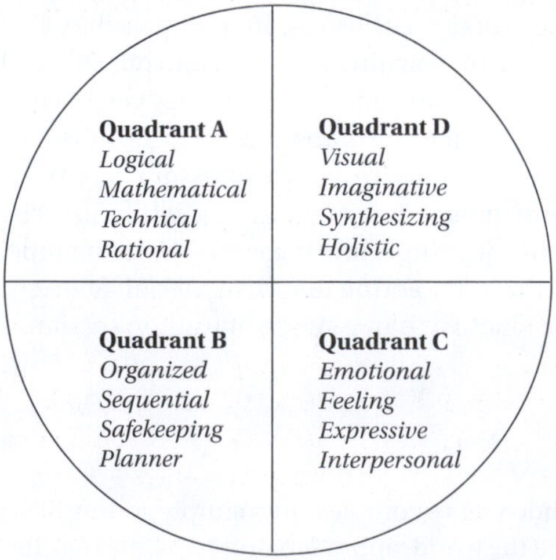

Quadrant A
Logical
Mathematical
Technical
Rational

Quadrant D
Visual
Imaginative
Synthesizing
Holistic

Quadrant B
Organized
Sequential
Safekeeping
Planner

Quadrant C
Emotional
Feeling
Expressive
Interpersonal

Figure 6.1 Herrmann's whole-brain model describing four quadrants of the brain
Source: Copyright Herrmann International.

Table 6.1 Quadrant A

Thinking	Preferred Learning Activities	Practice Activities
Factual Analytical	Collecting data and information	Develop graphs, flowcharts, and outlines from info and data
Quantitative	Listening to lectures	Do a library search on a topic of interest
Technical	Reading textbooks	Join an investment club
Logical	Studying example problems and solutions	Do logic puzzles or games
Rational Critical	Using the scientific method to do research	Play devil's advocate in a group decision process
	Doing case studies	
	Dealing with things rather than people	Break down a machine and identify the parts and their functions

Table 6.2 Quadrant B

Thinking	Preferred Learning Activities	Practice Activities
Organized	Following directions	Cook a new dish following a complicated recipe
Sequential	Doing repetitive, detailed homework problems	Organize a desk drawer or closet
Controlled	Doing step-by-step lab work	Prepare a family tree
Planned	Using programmed learning and tutoring	Be exactly on time all day
Conservative	Listening to detailed lectures	Assemble a model kit by instructions
Structured	Making up a detailed budget	Develop a personal budget
Detailed	Practicing new skills through repetition	Learn a new habit through self-discipline
Disciplined	Making time management schedules	Set up a filing system for your paperwork
Persistent	Writing a "how to" manual	

Table 6.3 Quadrant C

Thinking	Preferred Learning Activities	Practice Activities
Sensory	Sharing ideas with others	Study in a group
Kinesthetic	Learning by teaching others	Share your feelings with a friend
Emotional	Using group study opportunities	Get involved in a play or musical
Interpersonal	Enjoying sensory experiences	Volunteer in your community
Symbolic	Taking people-oriented field trips	Explore your spirituality
	Respecting other's rights and views	Become a penpal to someone from another culture
	Study with background music	

Table 6.4 Quadrant D

Thinking	Preferred Learning Activities	Practice Activities
Visual	Getting actively involved	Daydream
Holistic	Using visual aids in lectures	Play with clay, Legos, Skill Sticks
Innovative	Leading a brainstorming session	Focus on the big picture, not the details of the problem
Metaphorical	Experimenting	Create a logo
Creative	Doing problems with many answers	Invent a new recipe and prepare it
Imaginative	Relying on intuition, not facts	Imagine yourself 10, 20, 50 years in the future
Conceptual	Thinking about the future	Use analogies and metaphors in writing
Spatial		
Flexible		
Intuitive		

The model can also be used to help understand professors who have a different thinking style. Most people teach like they would like to be taught. If you are in a class where your professor has a very different style than you do, use the model to problem-solve ways to get past this hurdle. It is great experience for later in life. In your professional career, you will have to deal with all different styles effectively to be successful. Being aware of the differences and being able to work with them will be a tremendous asset.

If you would like to know which quadrants apply to you, you can take an assessment profile called the Herrmann Brain Dominance Instrument. An academic advisor or testing center on campus should be able to provide you with information about the HBDI. You may also contact Herrmann International at 794 Buffalo Creek Road, Lake Lure, NC 28746, or at their website at www.hbdi.com.

6.8 **Well-Rounded Equals Effective**

Being an effective person goes beyond being a good student. Developing the habits that make you truly effective during your college years will set you up for success in life. Part of being effective is functioning at full capacity. To do this, you must have the various dimensions of your life in order. One analogy that is frequently used is to look at a person as a wheel. A wheel will not roll if it is flat on one side. Similarly, people cannot function optimally if there is a problem area in their life. Five key areas are graphically represented in Figure 6.2. It is important to maintain each area to become as effective as you can be.

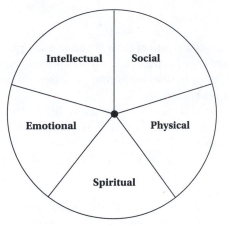

Figure 6.2 Areas of wellness for students

Intellectual

One of the main purposes of going to college is to expand the intellectual dimension of your life. Take full advantage of the opportunities at your institution. Besides engineering courses, schools will require general education courses (languages, humanities, or social science classes). These non-technical classes are a great way to broaden yourself. Many students find these classes a refreshing break from their engineering classes.

In addition to course work, colleges have a wide range of activities to enhance your intellectual development. These range from student organizations to seminars to special events. Planning some non-engineering intellectual activities will help to keep you motivated and feeling fresh for your engineering classes. Some students find that reading a novel or two during the semester also provides a refreshing, intellectual break from engineering studies.

Social

The social aspect of college is one that most students seem to master. Planning appropriate social activities will make your college experience more fun. There is a strong correlation between enjoying your college experience and being successful. Also, establishing a group of friends to socialize with will provide a support structure to help you when times get tough. College friends also provide a network that can be very beneficial later in your professional life.

If you have trouble identifying appropriate social activities, there are campus resources to help you. The first place to go is your residence hall staff. Also, most campuses have a central office for student activities. This office can provide you with a list of student organizations and activities.

Physical

All the studying and preparation is a waste of time if you are not physically able to perform when the tests or quizzes come. Also, you can waste time if you are not able to study efficiently. There are three physical areas that are essential to becoming an effective student: fitness, sleep, and nutrition.

People are more productive when they are physically fit, and students are no exception. Exercise is an activity that will not only help you study better but will allow you to live a longer, more productive life. College is a great time to develop a fitness habit. Fitness activities can include aerobics, jogging, biking, walking, or participating in intramural sports. On every campus there are people who can help you develop a fitness plan that is right for you.

Sleep is an area that many students abuse. Finally they are away from their parents and can go to bed any time they want. The problem is that classes still come at the same time regardless of when you fall asleep. Sleep deprivation reduces a person's productivity, which includes studying. Each person needs a different amount of sleep. Typically, most adults need six to eight hours of sleep. It is important to schedule enough time for sleep or you won't be able to be effective in the classroom. Studies have shown that student performance is actually reduced by staying up very late the night before a test to study. You cover more material the night before the test but are so tired at test time that you can't retrieve the material. Honestly evaluate how much sleep you need. Do you fall asleep in class? If so, you probably aren't getting enough rest. Another test for sleep deprivation is to not set an alarm one morning and see when you wake up. If you are on a schedule, most people will wake up at the same time without the alarm. If you sleep several hours past when the alarm would have gone off, you very possibly are sleep-deprived and need more rest.

The third physical area is diet. Again, Mom and Dad are not there to make you eat that broccoli, so you don't. Our bodies are designed to work properly with the right input. While you don't have to become a diet fanatic, eating a sensible diet will enhance your ability to succeed in your academics.

A final note on the physical dimension pertains to non-dietary substances, such as alcohol. The number-one problem on most college campuses is alcohol abuse. Every year, thousands of very capable students fail due to alcohol. If you choose to indulge, make honest assessments of the impact on your studying. Don't let alcohol become the barrier to your success.

Spiritual

Few would disagree that humans are spiritual beings and function much better and more efficiently with a balanced spiritual component in their lives. Stephen Covey, in his book *The 7 Habits of Highly Effective People,* puts it this way:

> *The spiritual dimension is your core, your center, your commitment to your value system. It's a very private area of life and a supremely important one. It draws*

upon the sources that inspire and uplift you and tie you to the timeless truths of all humanity. And people do it very, very differently.

I find renewal in daily prayerful meditation on the Scriptures because they represent my value system. As I read and meditate, I feel renewed, strengthened, centered and recommitted to serve.

Search for the spiritual activities that keep you renewed. This may be through a small-group study, attending campus worship services, meditating, or immersing yourself in good music for a time of personal reflection. If you aren't sure where to start, try something that is close to what you were used to before college. This will probably be the most comfortable for you. If it doesn't seem to be the right thing for you, branch out from there.

Emotional

Bill Hybels, pastor of a suburban Chicago church, explained the need to attend to this dimension when he addressed a gathering in 1996 in Detroit. He is an avid runner, eats well, and is physically sound. He has many casual friends and is socially active. He has written a number of books and articles and is sound intellectually. Yet he reached an emotional crisis that prevented him from being effective in his job and with his family.

Hybels describes emotional reserves like a gas tank. We need to put reserves into our tank so we have the ability to take it out when we need it. There will be times in your college career, and beyond, when you need those reserves. The lesson Hybels shared is the need to watch the gas gauge and not to let it run out. Students are no different.

A positive step to take to keep the emotional dimension in check is to set up a support network. This network is what you use to make deposits into your emotional tank. Friends are an integral part of the network. Do you have a person or people with whom you can talk honestly about important issues in your life? This might be a close friend, a parent, a relative, an advisor, a professor, or a counselor. Cultivate at least one friendship where you feel the freedom to share honestly.

The second aspect to maintaining emotional reserves is to schedule time that is emotionally neutral or energizing. Many students get themselves into a lot of activities that are emotionally draining and don't schedule time for themselves for recovery. Eventually this catches up with them and adversely impacts their effectiveness.

A final note on this area is that on each campus there are advisors and counselors who are available to talk to you. Don't feel that you have to have a severe psychological problem to simply talk with them. These people are trained to help you and will just listen if that is what you need. They are a great resource for maintaining the emotional dimension and thereby helping to keep your academic effectiveness high.

6.9   **Your Effective Use of Time**

You may be wondering how to find time to maintain all these areas every day (studying two hours for each hour spent in class, keeping a balanced lifestyle, etc.). And what happened to all that fun time you've heard about? Aren't the college years supposed to be the best years of your life?

The answer is effective time management. Time management is a skill that will help you succeed as a student and will serve you well throughout your entire career. Mastering time management will set you ahead of your peers. It is interesting that when engineers and managers are surveyed and asked why they don't take the time to manage their time, the most common response is inevitably that they don't have time. Students have the same feelings.

If we consider the academic requirements with a 15-credit load, we will have 45 hours per week of class and study time. That could be done Monday through Friday from 8 a.m. until 6 p.m., assuming you take an hour for lunch every day. Think about it. That would give you every evening and every weekend free to have fun or to do other activities.

Many students don't believe they can do this, but the hours add up. The key to making such a schedule work is to make every minute count. Treat school like a full-time job, working similar hours. This will not only help you to effectively manage your life during your academic years, but it also will provide excellent preparation for your future work life.

Let's assume that you have a schedule like the one in Table 6.5. This student has Chemistry, Math, English, Computer Science, and an Introduction to Engineering course. On Monday, the day starts at 8 a.m. with two classes back to back. Then there is an hour break before Computer Science. If you are on the job, you need to be

Table 6.5 Sample Class Schedule

	Mon	*Tues*	*Wed*	*Thurs*	*Fri*
8:00	English		English		English
9:00	Chem 101		Chem 101		Chem 101
10:00		Chem Lab		CS Lab	
11:00	CS 101	Chem Lab	CS 101	CS Lab	
12:00		Chem Lab			
1:00				Chem Rec	
2:00	Math	Math	Math		Math
3:00		Engr 101		Engr 101	
4:00					
5:00					
6:00					
7:00					
8:00					
9:00					

working, and we'll define working as attending class or studying. You need to find a place to study during this hour break. This is where many students go astray. They will waste an hour like this between classes. Similarly, there is a two-hour time block before Math class. This is time to grab lunch and get some more studying done.

It will take some planning to make your day effective because you will need to find suitable places to study. You can use these times to do your individual studying or meet with a study partner or group. It is much more efficient to decide in advance what you will study and when. It can be a waste of precious time to try decide which subject you will study during breaks between classes. Plan enough time for each subject so you can study from class to class and not fall behind in any way. For example, you can't do all your studying for Chemistry on Monday, Math on Tuesday, and Computers on Wednesday or you will fall behind the lectures.

Another thing that scares many students away from effective time management is that it sounds too rigid. One of the reasons for going to college is to have fun and meet people. How can you go with the flow if everything is scheduled? The answer is responsible flexibility. If you have your schedule planned out, you can decide when it's a good time to interrupt your work time to have some fun. A criterion for making this decision is recovery time. In your schedule, is there enough non-work time to convert to work time before deadlines arrive in your classes? For instance, taking a road trip the day before an exam would be irresponsible. However, right after you have midterms, there might be a window of time to take such a break and still recover.

Scheduling two hours of study per class period is usually enough to keep up during your first year. However, there are classes or times in a semester when this won't be enough. Once you fall behind, it is hard to catch up. To avoid this, a weekly makeup time is an excellent idea. This is like an overtime period at work. At an aerospace company where I worked, Saturday morning from 6 to 10 was my favorite time to schedule this overtime work. This didn't interfere with anything else in my normal schedule and I could get the work done and still have a full Saturday to enjoy. For some reason, 6 a.m. on Saturday isn't a very popular time for students. However, an evening or two or Saturday afternoon will work, too. These extra time blocks will allow you to keep up and give you the flexibility to take time off with a friend for something fun but still stay on your weekly schedule.

Scheduling Work and Other Activities

No matter how good you get at squeezing in study time between classes, there are still only 24 hours in each day. As a result, there is a limit to what anyone can do in one week. Planning a schedule complete with classes, study time (including makeup time), eating, sleeping, and other activities is essential to deciding how much is too much. If you need to work while you're in school, schedule it into your week. If you join a student organization, plan those hours into your schedule. You need to see where it all fits. A common pitfall with students is that they over-commit themselves with classes, work, and activities. Something has to give. It is easier to drop an activity than to have to repeat a class.

Ray Landis, from California State University at Los Angeles, describes a 60-hour rule to help students decide how much to work. Fifteen credit-hours should equate to 45 hours of effort. That leaves 15 hours for work, totaling 60 hours. Students may be able to work more than this, but each individual has a limit. You need to learn for yourself what your limit is.

If you need to work to make it through college financially, you may need to look at the number of credits you can handle effectively each semester. A schedule can help guide you in determining what is manageable. Before deciding to take fewer credits so you can work more hours, examine the value of the money you can make at your job compared with the starting salary you would expect by graduating a semester earlier. In some cases, it will make sense to work more while taking fewer classes. In others, it will make financial sense to take out a loan, graduate sooner, and pay the loan back with your higher salary.

Calendars

It is important to have a daily and weekly calendar. Another critical item in effective time management is a long-range schedule. For students, this is a semester calendar. This will help you organize for the crunch times, like when all five of your classes have midterms within two days, or you have three semester projects due on the same day at the end of the semester. It is a fact of student life that tests and projects tend to be bunched together. That is because there are educationally sound times to schedule these things, and most professors follow the same model. Given this fact, you can plan your work schedule so that you spread out your work in a manageable way. This is the part you have control over.

Once, while I was in school, we complained to a professor about this on the first day of class. He offered to give the final exam in the next class period and have the semester project due the second week of the semester to free us up for our other classes. We declined his offer and decided it was best to learn the material first.

Organizers: Paper or Electronic?

Calendars and task lists can be kept in numerous forms, and you will need to determine which is the most effective for you. There has been a proliferation of electronic tools and devices that are designed to help people communicate and be organized. Smartphones can store tasks and calendars and can be used to replace paper calendars. Laptops and tablets can be equipped with software such as Microsoft Outlook, Franklin Covey software or a host of freeware apps and programs to manage your tasks and keep appointments. Even with all the technology available, many people still find a role for paper lists or traditional calendars. Just like with the learning styles and study strategies, however, the method you use to organize your tasks needs to work for you. It does not have to impress your friends with how tech-savvy you are.

Do not feel like you have to adopt the latest technology to be effective. The media you use to organize your schedule is a personal choice and needs to work for you. It can be electronic, paper, or a combination, a little of each. The important thing is that it works for you. Our recommendation is to explore different approaches and test how they work and fit for you, your style, and needs. Stay flexible as your needs change and as new approaches and devices come online.

Organizing Tasks

A common method for organizing one's day is to make a to-do list each morning (or the evening before). This is a good way to ensure that everything needing attention gets it. Things not done on one day's list get transferred over to the next.

The one drawback to making a list is determining what to do first. You can approach the tasks in the order you write them down, but this ignores each task's priority. For instance, if you have a project due the following day for a class and it gets listed last, will it get done? A popular planning tool is the Franklin Planner produced by the Franklin Covey Co. (Material here is used courtesy of Franklin Covey.) They stress the importance of prioritizing your life' activities according to your values, personal mission statement, and goals. They suggest using a two-tier approach to prioritizing activities. The first is to use A, B, and C to categorize items as follows:

A = Vital (needs attention today)
B = Important (should be taken care of soon)
C = Optional (no one will notice if it is not done today)

Keep your long-term and intermediate goals in mind when assigning priority. And be sure to write simple statements for each task on your list, beginning each item with an action verb.

Once you have categorized the items, look at all the A priorities. Select the item to do first and give it a 1. This item now becomes A1. The next becomes A2, etc. Now you can attack the items on your list starting with A1 and working through the A's. Once you have finished with the A's, you move on to B1 and work through the B's. Often, you will not get to the C priorities, but you have already decided that it is okay if they don't get done that day.

Table 6.6 shows a sample list for a typical student's day. If you are starting a time management routine, include your list. This keeps you reminded of it, and it gives you something to check off your list right off the bat!

Some lessons can be learned by looking at how this student has scheduled activities. One is that paying the phone bill is not necessarily a top priority, but it requires little time to address. A helpful approach is to look at the quickly done A items and get them out of the way first. Then move on to the more time-consuming items. Also note that a workout was planned into the schedule, right after the chemistry lab. The workout is a priority that comes after some extensive studying. This can help

Table 6.6 Sample Prioritized List of Tasks

Priority	Task
A3	Finish chemistry lab (due tomorrow)
A5	Read math assignment
B3	Do math homework problems
B4	Outline English paper
A4	Work out at Intramural Building
B2	Debug computer program
A6	Annotate math notes
B1	Annotate chemistry notes
A7	Spiritual meditation
C1	Facetime Stacy
C2	Text parents that I am still alive and in school
A2	Pay phone bill
C3	Update status on social media
A1	Plan day

motivate the student to finish a difficult task and then move on to something enjoyable.

Looking at this schedule, you may think that poor Stacy is never going to get called or the e-mail to the parents written. It often happens that the more fun things are, the more likely they end up a C priority. But it is okay to put relationship-building or fun activities high in priority. People need to stay balanced. Maybe today Stacy is a C, but tomorrow she might have that call as an A priority.

The Franklin Covey people provide additional help for identifying priorities. They rate tasks by two criteria, urgency and importance, as shown in Table 6.7. Franklin Covey stresses that most people spend their time in quadrants I and III. To achieve your long-term goals, quadrant II is the most important. By spending time on activities that are in II, you increase your capacity to accomplish what matters most to you. The urgency is something that is usually imposed by external forces, whereas importance is something you assign.

You may well want to use a schedule that shows a whole week at a time and has your goals and priorities listed. An general example is shown in Table 6.8.

A weekly schedule allows you to see where each day fits. It also keeps your priorities in full view as you plan. Franklin Covey offers many versions of their Planner, including the Collegiate Edition. Contact them at (800) 654–1776 or visit their website at www.franklincovey.com.

Stay Effective

There are two final notes on time management that can help you stay effective. The first is to plan to tackle important activities when you are at your best. We are all born with

Table 6.7 Franklin Covey's "Time Matrix" for Daily Tasks

		Urgent	*Not Urgent*	
Important	I	▪ Crises ▪ Pressing deadlines ▪ Deadline-driven projects, meetings, preparations	▪ Preparation ▪ Prevention ▪ Values clarification ▪ Planning ▪ Relationship building ▪ Recreation ▪ Empowerment	II
Not Important	III	▪ Interruptions, some phone calls, texts and social media ▪ Some email, some reports ▪ Some meetings ▪ Many pressing matters ▪ Many popular activities	▪ Trivia, busywork ▪ Excessive social media acivity ▪ Time-wasters ▪ "Escape" activities ▪ Irrelevant email and texts ▪ Excessive TV, movies, or online surfing	IV

Adapted from "Time Matrix", Copyright 1999 Franklin Covey Co.

different biological clocks. Some people wake up first thing in the morning, bright-eyed and ready to work. These people are often not very effective in the late evening. Others would be dangerous if they ever had to operate heavy equipment first thing in the morning, especially before that first cup of coffee. However, once the sun goes down, they can be on a tear and get a lot done. Know yourself well enough to plan most of your studying when you are at your best. It is a common tendency among students to leave studying until late in the evening. This is often because they haven't done it earlier in the day. There are always stories about how the best papers are written at 3 a.m. Some are, but most are not. Know yourself well enough to plan effectively.

And finally, pace yourself. You are entering a marathon, not a sprint. Plan your study time in manageable time blocks and plan built-in breaks. A general suggestion is to take a 5- to 10-minute break for every hour spent studying. This will help keep you fresh and effective.

6.10 Accountability

The single most important concept to being successful is to establish accountability for your goals and intentions. The most effective way to do this is with an accountability partner. This is someone whom you trust and from whom you feel comfortable receiving honest feedback. Ideally, you will meet regularly with this person and share how things are going. Regularly means about once per week. Throughout a semester you can fall behind quickly, so weekly checks are a good idea. The person you choose

Table 6.8 Sample Generic Weekly Schedule and Planner

Weekly Schedule			Sunday	Monday	Tuesday	Wednesday	Thursday	Friday	Saturday
Musts	**Goals**				Day's	Priorities			
<u>Academics:</u>				Run laps		Run laps		Finish CS	Run laps
Math	Finish Prob							Finish Math	
Chem lab	Anal. data							Movie	
Comp Sci	Debug Prog								
English	Essay								
					Appointments/Commitments				
		8:00		English		English		English	
Hall Dance		9:00	Chapel Synagogue	Chem		Chem		Chem	
		10:00			Chem Lab		CS Lab		
		11:00		CS		CS			
		12:00							
		1:00					Chem Rec		
		2:00		Calculus	Calculus	Calculus		Calculus	
		3:00		Meet prof		Engr101		Engr101	
		4:00							
		5:00							
Physical	Run and Ride	Evening							
Mental	Read a novel								
Spiritual	Devotions								
Social	Hall dance								
Emotional	Lunch w/Dave								

can be a peer, a mentor, a parent, a relative, or anyone you feel comfortable with, respect, and trust.

Your accountability could also be to a group of people. Bill Hybels describes the change in his life when he entered into an accountability group with three other men. These men held each other accountable in all areas of their lives—professional, familial, and spiritual. You can do the same as a student.

The first step is to give your accountability partners the standard to which you want to be held. An easy way to do this is to share your goals with them. Let them see and hear from you where you want to be in five years, in 10 years, and beyond that. Let them also see the short-term goals you have set to achieve those longer-term objectives. Then, weekly, share with them how you are doing in meeting your goals.

You will find that it is much easier to stay on course with the help of at least one close friend. Success will come much more easily.

6.11   Overcoming Challenges

We all at some point in our lives encounter adversity. As a student, adversity may take the form of a bad grade on a test or for an entire course, a professor you feel is treating you unfairly, not getting into the academic program of your choice, or any of a myriad other college-related obstacles. Since everyone encounters adversity and failure at some point in his or her life, your personal success is not always dependent on how well you avoid failing, but rather on how you learn from it, move on, and improve. This is especially true as a student.

Almost all students will run into a course or a test where they struggle. The key is to avoid focusing on the negative and continuing to be productive. Every academic advisor has stories about good students who tripped up on a test or in one class and let that affect their performance in other classes.

Focus on Controllables

If something bad happens, figure out what can be done. This may mean writing down a list of actions you can take. Don't focus on things you can't change. If you receive a low grade on a test, you can't change the grade. You may, however, be able to talk to the professor about the test. You can study the material you missed and be ready for the next one.

Keep It in Perspective

Examine the real impact of the situation. What are the consequences of the situation? One class can affect your graduation GPA by about 0.05. This will not produce life-altering consequences, unless you would have graduated with a 4.0! If you are unsure about the consequences, see your professor or an academic advisor and have him or her help you put it in perspective.

Move to the Positive

If you hit a bump in the road, it is very important that you move beyond it quickly and remain effective in your studies. Examine the problem to learn from it. Perhaps you failed a test. This is a great opportunity to examine how you are studying and to take corrective action to improve your effectiveness. You can affect your present and future, but not your past.

Remember, Everyone Has Setbacks

Again, keep things in perspective. If you run into difficulty, you are not alone. There are numerous examples of setbacks and even failures.

Henry Ford neglected to put a reverse gear in his first automobile, but he recovered and went on to be successful. Abraham Lincoln, likely our greatest president, provided a tremendous example of overcoming adversity. Here is a brief list of some of his accomplishments and setbacks:

1831—failed in business
1832—defeated for legislature
1833—failed in business again
1834—elected to legislature
1835—fiancée died
1838—defeated for Speaker
1840—defeated for Elector
1843—defeated for Congress
1846—elected to Congress
1848—defeated for Congress
1855—defeated for Senate
1856—defeated for Vice-President
1858—defeated for Senate
1860—elected President

Lincoln never stopped achieving just because he ran into a setback. You don't have to either.

Another example comes from a grandfather who relayed a story about someone in his family failing:

We were visiting your grandmother's brother and his family. It was the time of year when college was in session, so I was surprised when my nephew, Hughey, opened the door and greeted us. He was in his first year of college at the state university. I jokingly said, "What happened to you, Hughey? Flunk out already?" I sure felt stupid when he replied that in fact he had. After that I felt obligated to try to help him out. That evening, he, his father, and I talked about what he was going to do. I encouraged him to enroll in the other big state university. I had heard a lot of good things about it even though their science and engineering programs were not as prestigious. He enrolled and did very well. Not only did he graduate, he even got his master's degree before taking a job with a top company.

As a footnote to the story, Hughey continued to do well. The job he took was with a Fortune 100 company. He married and had four children and five grandchildren, at last count. He stayed with the same company for his entire career and retired as a senior vice president, one step below CEO. Failing in his first attempt at college didn't stop him. He learned from it, moved on, and did very, very well. You can do the same with the right attitude and effort.

REFERENCES

Covey, S. R., *The 7 Habits of Highly Effective People*, New York, Simon and Schuster, 1989.

Douglass, S., and A. Janssen, *How to Get Better Grades and Have More Fun*, Bloomington, IN, Success Factors, 1997.

Herrmann, N., *The Creative Brain*, Lake Lure, NC, Brain Books, 1988.

Landis, R., *Studying Engineering, A Road Map to a Rewarding Career*, Burbank, CA, Discovery Press, 1995.

Lumsdaine, E., and M. Lumsdaine, *Creative Problem Solving, Thinking Skills for a Changing World*, 2nd ed., New York, McGraw-Hill, 1993.

Maxwell, J. C., *The Winning Attitude, Your Key to Personal Success*, Nashville, TN, Thomas Nelson Publishers, 1993.

Tobias, C. U., *The Way They Learn, How to Discover and Teach to Your Child's Strengths*, Colorado Springs, CO, Focus on the Family, 1994.

EXERCISES AND ACTIVITIES

6.1　Identify three things you can do to keep a positive attitude toward your studies.

6.2　What could you have done if you were in the physics class described in Section 6.2?

6.3　Actually writing goals as described in this chapter is an important element in defining your goals.

    a) Write five goals to achieve by graduation.
    b) Write five goals to reach within 10 years.
    c) Write 10 goals to reach during your lifetime.

6.4　Write some specific goals:

    a) Set a grade goal for each class this semester.
    b) Calculate your expected semester GPA.
    c) Define a stretch goal for your semester GPA.

6.5　Find out what the following GPAs are:

    a) Minimum GPA to stay off academic probation
    b) Minimum GPA to be admitted to Tau Beta Pi
    c) Minimum GPA to be admitted to the engineering honor society in your engineering discipline (e.g., Pi Tau Sigma for mechanical engineering, Eta Kappa Nu for electrical engineering)
    d) Minimum GPA to be on the Dean's List or Honors List
    e) Minimum GPA to graduate with honors

6.6   Identify a particular employer and an entry-level position with that employer that you would like to obtain following graduation. Find out what is required to be considered for that position.

6.7   List the barriers you see to achieving your graduation goals.

a) Categorize them as those within your control and those beyond your control.
b) Develop a plan to address those within your control.
c) Develop strategies to cope with those beyond your control.

6.8   For each class you are currently taking:

a) Map out where you sit in class.
b) Rate your learning effectiveness (A, B, C, D, or F) for each class.
c) For each class not rated an A, develop a plan to increase your learning effectiveness.

6.9   For one week, make a log of everything you do. Break it up into categories of attending class, studying, working, playing, eating, sleeping, physical activity, intellectual pursuits, emotional health, spiritual well-being, and other.

a) Total your time spent on each area. Are you spending the time needed on your studies (two hours for each hour in class)?
b) Is this the kind of schedule that will make you successful? If not, what will you change?

6.10   Review your current class schedule and identify times and places you can study between classes.

6.11   Develop a work schedule. Identify the times you are on the job (that is, studying and taking classes). Identify a makeup time for weekly catchup.

6.12   List all your current courses. Put a "+" next to each one in which you are caught up. Put a "−" next to each one in which you are not caught up right now. ("Caught up" is defined as prepared for the next lecture with all assignments completed and all assigned reading done.)

6.13   Ask one of your professors the main objective of his or her course and report the response. Does this information help you in organizing your studying?

6.14   Annotate your notes from a lecture in one of your courses, then take those notes in to your professor to get his or her comments on your effectiveness. Report the professor's comments.

6.15   Make a list of the students you already know who are pursuing the same major as you. Then make a list of the students you know in your science and math classes. Rate each one according to his or her suitability as a study partner.

6.16   Identify one or more people with whom you might form a study group. Try studying together on a regular basis until the next test. Report on your experience.

6.17   List the study group rules you used in your group studying. Report on the things that went well and those that did not.

6.18   Make a list of problems you can work, in addition to the assigned homework problems, in one of your math, science, or engineering courses. Report on the method you used to identify these problems.

6.19   Visit one of your professors and report three personal things about that professor. (For example: Where did she go to school? Why did he choose the field he is in? What does she like best about her job? What one piece of advice does he have for freshmen that he didn't share in class?)

6.20   After completing Exercise 6.19 above, did this change your feelings about your professor? Does he or she seem more approachable now? Did you learn anything else during your visit that could help you succeed in that class?

6.21   Identify at least one opportunity for you as an undergraduate to become involved in engineering research. What advantages could you gain from this experience?

6.22   Identify the type of environment in which you study the best. Describe it and include a sketch of the location.

6.23   Test your conclusion in Exercise 6.22 with an experiment. Do one assignment in one environment (perhaps in your room, alone and totally quiet) and another in a different environment (perhaps in a more public place, with background noise). Report your findings.

6.24   At the halfway point in the semester, calculate your grades in each class. Compare these to the goals set in Exercise 6.4. Are you on track? If not, identify the changes you will make to reach these goals.

6.25   Define at least one long-term goal in each of the five wellness areas discussed in this chapter. Next, define short-term goals to achieve these long-term goals.

6.26   List at least two needs you have in each of the five wellness areas. List at least two activities you can plan to meet these needs.

6.27   Rate your roundness in the wellness wheel (Fig. 6.3), using 5 for very well, 4 for well, 3 for okay, 2 for some improvement needed, and 1 for much improvement needed. Do you have a well-functioning wheel? If not, identify steps to correct the problem areas. List at least two new activities for each problem area.

6.28   Write a paragraph on how each wellness area can help you succeed:

a) as an engineering student;
b) as a practicing engineer.

6.29   Make a weekly schedule with two hours of studying for every hour spent in class. Identify activities that will help develop the different areas in your life. Is this a sustainable schedule? If the answer is no, identify things that can be dropped.

6.30   Identify on your weekly schedule the times of each day when you study most efficiently. Are these the times you are actually scheduled to study? If not, explain why you've scheduled other items instead of studying at these times.

6.31   Adopt one of the weekly schedule instruments discussed in this chapter. Try using it for two weeks and report the results. Was this the right instrument for you?

6.32   Make a to-do list. Prioritize the list using A, B, and C categories, and then number the items in each category.

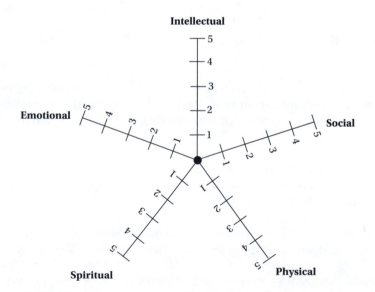

Figure 6.3 Wellness assessment (for Exercise 6.27)

6.33 Place the items of Exercise 6.32 into the four quadrants of Covey's matrix shown in Table 6.1. How many A priorities are in quadrant II? Based on this ranking of your to-do list, would you change any of your priorities?

6.34 List the top three things you are failing to do that would most help you succeed in your career. Briefly explain why you are currently failing to do each item.

6.35 List three people you would consider for an accountability partner.

6.36 Make a list of at least five questions you want your accountability partner to ask you at each checkup.

6.37 Meet with an accountability partner for one month. Report back on your experience. Did it help? Is this something you will continue?

6.38 Call home and tell your parents that you are developing a success strategy for college. Explain the strategy to them.

6.39 Assume that your grades will be lower than expected at the end of the semester, and write out a strategy for recovering from these low grades.

6.40 Make a list of at least three university people who can help you deal with a problem that is preventing you from succeeding. List each by name and position.

6.41 Make a list of your activities over the last 24 hours. Compare the list and the hours spent on each with your planned schedule.

6.42 List your long-term goals, including every area of your life. Fill out a schedule of what you actually did during the last seven days. Put each activity into categories corresponding to one of your long-term goals, or an extra category titled "not related." Grade yourself (using A, B, C, D, or F), indicating how well you did the past week working toward your long-term goals.

6.43 Take the inventory test discussed in Section 6.7 on memory languages. Which is your dominant mode? List three actions you can take to improve your learning based on this dominant mode.

6.44 Which of Herrmann's quadrants best describes your dominant thinking style? Which one is the weakest?

6.45 Which of Herrmann's four quadrants is best suited for engineering? Are there any dominant modes of thinking that are not ideal for engineering? Explain your answers.

6.46 List all your current professors and identify their probable Herrmann quadrant. Compare this with your own style. For the professors with styles that differ from yours, develop an action plan to deal with these differences.

6.47 Explore the availability of electronic organizers and prepare a brief report on the options for students.

6.48 Compare and contrast electronic organizers versus paper ones. What are the benefits of each and which would you recommend to your classmates?

6.49 Interview two professors and report what they use to organize and prioritize their time.

CHAPTER 7

Problem Solving

7.1 Introduction

Seven-year-old Dan could build all kinds of things: toy cars, ramps, and simple mechanical gadgets. Even more amazing was that he could do all this with four simple tools—scissors, hammer, screwdriver, and duct tape. One Christmas, his father asked if he'd like more tools. "No thanks, Dad," was his response. "If I can't make it with these, it's not worth making." It wasn't long, however, before Dan realized that the cool stuff he could create or the problems he could solve were dependent on the tools he had. So he began gathering more tools and learning how to use them. Today, he still solves problems and makes a lot of cool stuff—like real cars, furniture, playgrounds, and even houses. His tools fill a van, a workshop, and a garage. Having the right tools and knowing how to use them are critical components of his success.

Likewise, most people, when they are children, learn a few reliable ways to solve problems. Most fifth graders in math class know common problem-solving methods like "draw a picture," "work backwards," or the ever-popular "guess and check" (check in the back of the textbook for the correct answer, that is). Unfortunately, many people choose to stop adding new problem-solving tools to their mental toolbox. "If I don't know how to solve it, it's probably not worth solving anyway," they may think. Or, "I'll pass this problem over for now and get back to my usual tasks." This severely limits them. Engineers, by definition, need to be good at solving problems and making things. Therefore, filling your toolbox with the right tools and knowing how to use them are critical components to becoming a successful engineer.

The goal of this chapter is to increase the number of problem-solving strategies and techniques you commonly use, and to enhance your ability to apply them creatively in various problem-solving settings. Hopefully, this will be only the beginning of a lifelong process of gaining new mental tools and learning how to use them.

7.2 Analytic and Creative Problem Solving

The toolbox analogy is very appropriate for engineering students. Practicing engineers are employed to solve a very wide variety of problems. In daily practice, engineers can end up filling a number of different roles. These roles can require very different problem-solving abilities. Problem solving can be broken into analytic and creative. Most students are more familiar with analytic problem solving where there is one correct answer. In creative problem solving there is no single right answer. Your analytic tools represent what's in your toolbox. Your creative skills represent how you handle your tools.

To better understand the difference between these two kinds of problem solving, let's examine the one function that is common to all engineers: design. In the Engineering Design chapter (Chapter 12), we detail a design process with 10 steps:

1. Identify the problem.
2. Define the working criteria/goal(s).
3. Research and gather data.
4. Brainstorm for creative ideas.
5. Analyze.
6. Develop models and test.
7. Make the decision.
8. Communicate and specify.
9. Implement and commercialize.
10. Prepare post-implementation review and assessment.

By definition, design is open-ended and has many different solutions. The automobile is a good example. Sometime when you are riding in a car, count the number of different designs you see while on the road. All of these designs were the result of engineers solving a series of problems involved in producing an automobile. The design process as a whole is a **creative problem-solving process.**

The other method, **analytic problem solving,** is also part of the design process. Step 5, Analyze, requires analytic problem solving. When analyzing a design concept, there is only one acceptable answer to the question "Will it fail?" A civil engineer designing a bridge might employ a host of design concepts, though at some point he or she must accurately assess the loads the bridge can support.

Most engineering curricula provide many opportunities to develop analytic problem-solving abilities, especially in the first few years. This is an important skill. Improper engineering decisions can put the public at risk. For this reason, it is imperative that proper analytic skills be developed.

As an engineering student you will be learning math, science, and computer skills that will allow you to tackle very complex problems later in your career. These are critical, and are part of the problem-solving skill set for analysis. However, other tools are equally necessary.

Engineers also can find that technical solutions applied correctly can become outmoded, or can prove dangerous if they aren't applied with the right judgment. Our creative skills can help greatly in the big-picture evaluation, which can determine the long-term success of a solution. The solution of one problem can cause others if foresight is neglected, and this is not easily avoided in all cases. For instance, the guilt that the developers of dynamite and atomic fission have publicly expressed relates to a regret for lack of foresight. Early Greek engineers invented many devices and processes that they refused to disclose because they knew they would be misused. The foresight to see how technically correct solutions can malfunction or be misapplied takes creative skill to develop. What about fertilizers that boost crops but poison water supplies? Modern engineers need to learn how to apply the right solution at the right time in the right way for the right reason with the right customer.

Most people rely on two or three methods to solve problems. If these methods don't yield a successful answer, they become stuck. Truly exceptional problem solvers learn to use multiple problem-solving techniques to find the optimal solution. The following is a list of possible tools or strategies that can help solve simple problems:

1. Look for a pattern.
2. Construct a table.
3. Consider possibilities systematically.
4. Act it out.
5. Make a model.
6. Make a figure, graph, or drawing.
7. Work backwards.
8. Select appropriate notation.
9. Restate the problem in your own words.
10. Identify necessary, desired, and given information.
11. Write an open-ended sentence.
12. Identify a sub-goal.
13. First solve a simpler problem.
14. Change your point of view.
15. Check for hidden assumptions.
16. Use a resource.
17. Generalize.
18. Check the solution; validate it.
19. Find another way to solve the problem.
20. Find another solution.
21. Study the solution process.
22. Discuss limitations.
23. Get a bigger hammer.
24. Sleep on it.
25. Brainstorm.
26. Involve others.

In both analytic and creative problem solving, there are different methods for tackling problems. Have you ever experienced being stuck on a difficult math or science problem? Developing additional tools or methods will allow you to tackle more of these problems effectively, making you a better engineer.

The rest of this chapter is a presentation of different ways of solving problems. These are some of the techniques that have been shown to be effective. Each person has a unique set of talents and will be drawn naturally to certain problem-solving tools. Others will find a different set useful. The important thing as an engineering student is to experiment and find as many useful tools as you can. This will equip you to tackle the wide range of challenges that tomorrow's engineers will face.

7.3   Analytic Problem Solving

Given the importance of proper analysis in engineering and the design process, it is important to develop a disciplined way of approaching engineering problems. Solving analytic problems has been the subject of a great deal of research, which has resulted in several models. One of the most important analytic problem-solving methods that students are exposed to is the Scientific Method. The steps in the Scientific Method are as follows:

1. Define the problem.
2. Gather the facts.
3. Develop a hypothesis.
4. Perform a test.
5. Evaluate the results.

In the Scientific Method, the steps can be repeated if the desired results are not achieved. The process ends when an acceptable understanding of the phenomenon under study is achieved.

In the analysis of engineering applications, a similar process can be developed to answer problems. The advantage of developing a set method for solving analytic problems is that it provides a discipline to help young engineers when they are presented with larger and more complex problems. Just like a musician practices basic scales in order to set the foundation for complex pieces to be played later, an engineer should develop a sound fundamental way to approach problems. Fortunately, early in your engineering studies you will be taking many science and math courses that will be suited to this methodology.

The Analytic Method we will discuss has six steps:

1. Define the problem and make a problem statement.
2. Diagram and describe.
3. Apply theory and equations.
4. Simplify the assumptions.
5. Solve the necessary problems.
6. Verify accuracy to required level.

Following these steps will help you better understand the problem you are solving and allow you to identify areas where inaccuracies might occur.

Step 1: Problem Statement

It is important to restate in your own words the problem you are solving. In textbook problems, this helps you understand what you need to solve. In real-life situations, this helps to ensure that you are solving the correct problem. Write down your summary, and then double-check that your impression of the problem is the one that actually matches the original problem. Putting the problem in your own words is also an excellent way to focus on the part of the problem you need to solve. Often, engineering challenges are large and complex, and the critical task is to understand what part of the problem you need to solve.

Step 2: Description

The next step is to describe the problem and list all that is known. In addition to restating the problem, list the information given and what needs to be found. This is shown in Example 7.2, which follows this section. Typically, in textbook problems, all the information given is actually needed for the problem. In real problems, more information is typically available than is needed to do the calculations. In other cases, information may be missing. Formally writing out what you need and what is required helps you to clearly sort this out.

It is also helpful to draw a diagram or sketch of the problem you are solving to be able to understand the problem. Pictures help many people to clarify the problem and what is needed. They are also a great aid when explaining the problem to someone else. The old saying could be restated as, "a picture is worth a thousand calculations."

Step 3: Theory

State explicitly the theory or equations needed to solve the problem. It is important that you write this out completely at this step. You will find that most real problems and those you are asked to solve as an undergraduate student will not require exact solutions to complete equations. Understanding the parts of the equations that ought to be neglected is vital to your success.

It is not uncommon to get into a routine of solving a simplified version of equations. These may be fine under certain conditions. An example can be seen in the flow of air over a body, like a car or airplane wing. At low speeds, the density of the air is considered a constant. At higher speeds this is not the case, and if the density were considered a constant, errors would result. Starting with full equations and then simplifying reduces the chance you will overlook important factors.

Step 4: Simplifying Assumptions

As mentioned above, engineering and scientific applications often cannot be solved precisely. Even if they are solvable, determining the solution might be cost-prohibitive. For instance, an exact solution might require a high-speed computer to calculate for a year to get an answer, and this would not be an effective use of resources. Weather prediction at locations of interest is such an example.

To solve the problem presented in a timely and cost-effective manner, simplifying assumptions are required. Simplifying assumptions can make the problem easier to solve and still provide an accurate result. It is important to write the assumptions down along with how they simplify the problem. This documents them and allows you to interpret the final result in terms of these assumptions.

While estimation and approximation are useful tools, engineers also are concerned with the accuracy and reliability of their results. Approximations are often possible if assumptions are made to simplify the problem at hand. An important concept for engineers to understand in such situations is the conservative assumption.

In engineering problem solving, "conservative" has a non-political meaning. A conservative assumption is one that introduces errors on the safe side. We mentioned that estimations can be used to determine the bounds of a solution. An engineer should be able to look at those bounds and determine which end of the spectrum yields the safer solution. By selecting the safer condition, you are ensuring that your calculation will result in a responsible conclusion.

Consider the example of a materials engineer selecting a metal alloy for an engine. Stress level and temperature are two of the parameters the engineer must consider. At the early stages of a design, these may not be known. What he or she could do is estimate the parameters. This could be done with a simplified analysis. It could also be done using prior experience. Often, products evolve from earlier designs for which there are data. If such data exist, ask how the new design will affect the parameters you are interested in. Will the stress increase? Will it double? Perhaps a conservative assumption would be to double the stress level of a previous design.

After taking the conservative case, a material can be selected. A question the engineer should ask is: "If a more precise answer were known, could I use a cheaper or lighter material?" If the answer is "No," then the simple analysis might be sufficient. If "Yes," the engineer could build a case for why a more detailed analysis is justified.

Another example of conservatism can be seen in the design of a swing set (see Fig. 7.1). In most sets, there is a horizontal piece from which the swings hang, and an inverted V-shaped support on each end. If you were the design engineer, you might have to size these supports.

One of the first questions to ask is, "For what weight do we size the swing set?" One way to answer this question is to do a research project on the weight distribution of children, followed by doing research into the use of swings by children at different ages, and then analyzing those data to determine typical weights of children who would use such swings. You might even need to observe playground activity to see if several children pile on a swing or load the set in different ways—say, by climbing on it. This might take weeks to do, but you would have excellent data.

On the other hand, you could just assume a heavy load that would exceed anything the children would produce—for this case, say, 500 pounds per swing. That would assume the equivalent of two large adults on each swing. Make the calculation to determine the needed supports, and then ask if it makes sense to be more detailed.

To do this, you must answer the question, "What problem am I solving?" Are you after the most precise answer you can get? Are you after a safe and reliable answer regardless of other concerns?

In engineering, the "other concerns" include such things as cost and availability. In the current example, the supports would most likely be made of common materials.

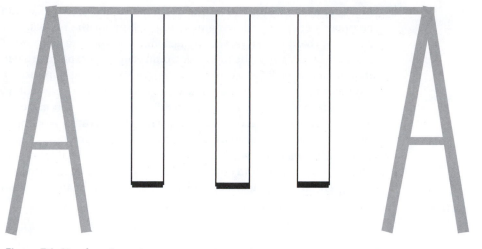

Figure 7.1 Simple swing set

Swing sets are typically made of wood or steel tubing. So, one way to answer the above questions is to quickly check and see if making a more detailed analysis would be justified. Check the prices of the materials needed for your conservative assumption against potential materials based on a more detailed analysis. Does the price difference justify a more detailed analysis? Another way to ask this is "Does the price difference justify spending time to do the analysis?" Remember, as an engineer, your time costs money.

The answer may be, "It depends." It would depend on how much the difference is and how many you are going to produce. It wouldn't make sense to spend a week analyzing or researching the answer if it saves you or your company only $500. If your company were producing a million sets and you could save $5 per set, it would justify that week's work.

Approximation can improve decision making indirectly as well. By using it to quickly resolve minor aspects of a problem, you can focus on the more pressing aspects.

Engineers are faced with these types of decisions every day in design and analysis. They must accurately determine what problem is to be solved, and how. Safety of the public is of paramount concern in the engineering profession. You might be faced with a short-term safety emergency where you need to quickly prioritize solutions. Or you might be designing a consumer product where you have to take the long-term view. Swing sets are often used for generations, after all.

As an engineer, you will need to develop the ability to answer, "What problem am I really solving?" and "How do I get the solution I need most efficiently?"

EXAMPLE
7.1

A manager for an aerospace firm had a method for breaking in young engineers. Shortly after the new engineer came to work in his group, the manager would look for an appropriate analysis for the young engineer. When one arose, he would ask the young engineer to perform the analysis. The engineer would typically embark on a path to construct a huge computer model requiring lengthy input and output files that would take a few weeks to complete. The manager would return to the engineer's desk the following morning, asking for the results. After hearing the response that all that was really accomplished was to map out the next three weeks' worth of work, the manager would erupt and explain at a very high volume that he or she had been hired as an engineer and was supposed to be doing an engineering analysis, not a science project. He would show the young engineer how doing a simplified analysis would take an afternoon to perform and would answer the question given the appropriate conservative assumptions. He would then leave the engineer to work on the new analysis method.

While his methods were questionable, his point is well taken. There are always concerns about time and cost as well as accuracy. If conservative assumptions are made, safe and reliable design and analysis decisions can often also be made.

Step 5: Problem Solution

Now the problem is set up for you actually to perform the calculations. This might be done by hand or by using the computer. It is important to learn how to perform simple hand calculations; however, computer applications make complex and repetitive calculations much easier. When using computer simulations, develop a means to document what you have done in deriving the solution. This will allow you to find errors faster, as well as to show others what you have done.

Step 6: Accuracy Verification

Engineers work on solutions that can affect the livelihood and safety of people, so it is important that the solutions you develop are accurate.

I had a student once whom I challenged on this very topic. He seemed to take a cavalier attitude about the accuracy of his work. Getting all the problems mostly correct would get him a B or a C in the class. I tried to explain the importance of accuracy and how people's safety might depend on his work. He told me that in class I caught his mistakes, and on the job someone would check his work. This is not necessarily the case! Companies do not hire some engineers to check other engineers' work. In most systems, there are checks in place. However, engineers are responsible for verifying that their own solutions are accurate. Therefore, it is important for you while you are a student to develop the skills to meet any required standard.

A fascinating aspect of engineering is that the degree of accuracy is a variable under constant consideration. A problem or project may be properly solved within a tenth of an inch, but accuracy to a hundredth might be unnecessary and difficult to control, rendering that kind of accuracy incorrect. But the next step of the solution might even require accuracy to the thousandth. Be sure of standards!

There are many different ways to verify a result. Some of these ways are as follows:

- Estimate the answer.
- Simplify the problem and solve the simpler problem. Are the answers consistent?
- Compare with similar solutions. In many cases, other problems were solved similar to the one currently in question.
- Compare to previous work.
- Ask a more experienced engineer to review the result.
- Compare to published literature on similar problems.
- Ask yourself if it makes sense.
- Compare to your own experience.
- Repeat the calculation.
- Run a computer simulation or model.
- Redo the calculation backwards.

It may be difficult to tell if your answer is exactly correct, but you should be able to tell if it is close or within a factor of 10. This will often flag any systematic error. Being able to back up your confidence in your answer will help you make better decisions on how your results will be used.

The following example will illustrate the analytical method.

EXAMPLE 7.2   Problem: A ball is projected from the top of a 35 m tower at an angle 25° from horizontal with an initial velocity v_o = 80 m/s. What is the time it takes to reach the ground, and what is the horizontal distance from the tower to the point of impact?

Problem Statement: Given: Initial velocity v_0 = 80 m/s

Initial trajectory = 25° from horizontal

Ball is launched from a height of 35 m

Find: 1) Time to impact

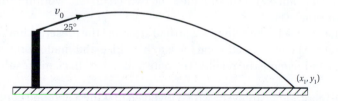

2) Distance of impact from the tower

Equations: Found in a physics text:

(a) $$x_i = x_o + v_o t \cos\theta + \frac{1}{2}a_x t^2$$

(b) $$y_1 = y_o + v_o t \sin\theta + \frac{1}{2}a_y t^2$$

Assumptions: Neglect air resistance.

Acceleration = −9.81 is constant in the vertical direction.

Acceleration is zero in the horizontal direction.

Ground is level and impact occurs at $y = -35$ m.

Solution: Begin with equation (b):

$$-35 = 080t \sin 25° - \frac{1}{2} 9.8t^2 \text{ or } t^2 - 6.90t - 7.14 = 0$$

$$\therefore t = \frac{b \pm \sqrt{b^2 - 4ac}}{2a} = \sqrt{\frac{6.9^2 + 4 \times 1 \times 7.14}{2}} = 7.81 \sec$$

The solution must be 7.81 sec, since time must be positive.
The distance is found using equation (a):

$$x_1 = 0 + 80 \times 7.81 \times \cos 25° = 566 m$$

Estimation

Estimation is a problem-solving tool that engineers need to develop. It can provide quick answers to problems and can help to verify complicated analyses. Young engineers are invariably amazed when they begin their careers at how the older, more experienced engineers can estimate so closely the answers to problems before the analysis is even done.

I was amazed at one such case that I experienced first hand. I was responsible for doing a detailed temperature analysis on a gas turbine engine that was to go on a new commercial aircraft. We presented our plan to some managers and people from the Chief Engineers Office. The plan called for two engineers working full time for a month on the analysis.

One of the more senior engineers who had attended the presentation sent me some numbers the next day. He had gone back to his office and made some approximate calculations and sent me the results. His note told me to check my results against his numbers.

After the month-long analysis, which involved making a detailed computer model of the components, I checked our results with his back-of-the-envelope calculations. He was very close to our result! His answers were not exactly the same as ours, but they were very close. The detailed analysis did provide a higher degree of precision. However, the senior engineer was very close.

What he had done was simply use suggestion #13 from the list of problem-solving methods. He solved a simpler problem. He used simple shapes with known solutions to approximate the behavior of the complex shapes of the jet engine's components.

While estimation or approximation may not yield the precision you require for an engineering analysis, it is a very useful tool. One thing that it can do is help you check your analysis. Estimation can provide bounds for potential answers. This is especially critical in today's dependence on computer solutions. Your method may be correct, but one character mistyped will throw your results off. It is important to be able to have confidence in your results by developing tools to verify accuracy.

Estimation can also be used to help decide if a detailed analysis is needed. Estimating using a best-case and worst-case scenario can yield an upper and lower bound to the problem. If the entire range of potential solutions is acceptable, why do the detailed analysis? What if, for our case, the whole range of temperatures met our design constraints? It would not have justified a month of our work. In this case, the added precision was needed and did lead to a design benefit. As engineers, you will be asked to decide when a detailed analysis is required and when it is not.

Estimation can be a very powerful tool to check accuracy and make analytic decisions. Senior engineers in industry often comment that current graduates lack the ability to do an approximation. This has become more important with the dependence on computer tools and solutions. Computer analyses lend themselves to typos on inputs or arithmetic errors. Being able to predict the ballpark of the expected results can head off potentially disastrous results.

EXAMPLE 7.3   With all of today's sophisticated analytic tools, it may be hard to imagine that accurate measurements could be made without them. Take the size of a molecule. Today's technology makes this measurement possible. Could you have made this measurement in the 18th century? Benjamin Franklin did.

He found the size of an oil molecule. He took a known volume of oil and poured it slowly into a still pool of water. The oil spread out in a circular pattern. He measured the circle and calculated the surface area. Since the volume of the oil remained the same, he calculated the height of the oil slick and assumed that was the molecular thickness, which turned out to be true. His result was close to the correct value.

Do we have more accurate measurements now? Certainly we do. However, Benjamin Franklin's measurement was certainly better than no value. As engineers, there will be times when no data exist and you will need to make a judgment. Having an approximate result is much better than having none at all.

7.4 Creative Problem Solving

Many engineering problems are open-ended and complex. Such problems require creative problem solving. To maximize the creative problem-solving process, a systematic approach is recommended. Just as with analytic problem solving, developing a systematic approach to using creativity will pay dividends in better solutions.

Revisiting our toolbox analogy, the creative method provides the solid judgment needed to use your analytic tools. Individual thinking skills are your basics here. The method we'll outline now will help you to apply those skills and choose solution strategies effectively.

There are numerous ways to look at the creative problem-solving process. We will present a method that focuses on answering these five questions:

1. What is wrong?
2. What do we know?
3. What is the real problem?
4. What is the best solution?
5. How do we implement the solution?

By dividing the process into steps, you are more likely to follow a complete and careful problem-solving procedure, and more effective solutions are going to result. This also allows you to break a large, complex problem into simpler problems where your various skills can be used. Using such a strategy, the sky is the limit on the complexity of projects which an engineer can complete. It's astonishing and satisfying to see what can be done.

Divergence and Convergence

At each phase of the process, there is both a divergent and a convergent part of the process, as shown in Figure 7.2. In the divergent process, you start at one point and reach for as many ideas as possible. Quantity is important. Identifying possibilities is the goal of the divergent phase of each step. In this portion of the process (indicated by the outward-pointing arrows, $\leftarrow \rightarrow$) use the brainstorming and idea-generating techniques described in subsequent sections. Look for as many possibilities as you can. Often, the best solutions come from different ideas.

In the convergence phase (indicated by the inward-pointing arrows, $\rightarrow \leftarrow$) use analytical and evaluative tools to narrow the possibilities to the one(s) most likely to yield results. Quality is most important, and finding the best possibility to move the process to the next phase is the goal. (One common method is to use a matrix to rate ideas based on defined criteria.) If one choice fails to produce satisfactory results, go back to the idea lists for another method to tackle the problem.

Sample Problem: Let's Raise Those Low Grades!

To illustrate the problem-solving process, let's consider an example of a problem and work through the process. The problem we will use for illustration is a student who is getting low grades. This certainly is a problem that needs a solution. Let's see how the process works.

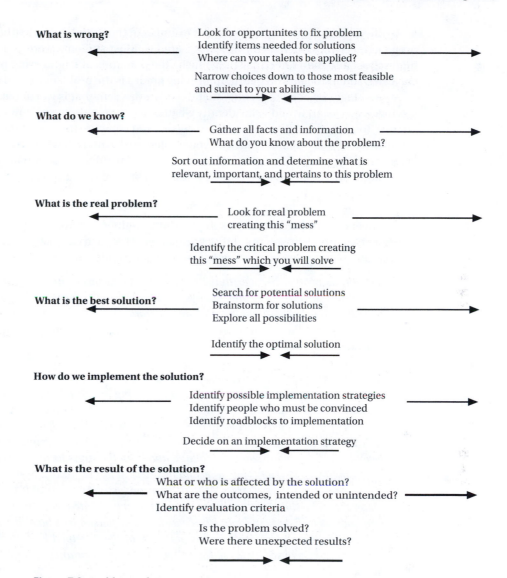

What is wrong?

Look for opportunites to fix problem
Identify items needed for solutions
Where can your talents be applied?

Narrow choices down to those most feasible
and suited to your abilities

What do we know?

Gather all facts and information
What do you know about the problem?

Sort out information and determine what is
relevant, important, and pertains to this problem

What is the real problem?

Look for real problem
creating this "mess"

Identify the critical problem creating
this "mess" which you will solve

What is the best solution?

Search for potential solutions
Brainstorm for solutions
Explore all possibilities

Identify the optimal solution

How do we implement the solution?

Identify possible implementation strategies
Identify people who must be convinced
Identify roadblocks to implementation

Decide on an implementation strategy

What is the result of the solution?

What or who is affected by the solution?
What are the outcomes, intended or unintended?
Identify evaluation criteria

Is the problem solved?
Were there unexpected results?

Figure 7.2 Problem-solving process

What's Wrong?

In the first step of the problem-solving process, an issue is identified. This can be something stated for you by a supervisor or a professor, or something you determine on your own. This is the stage where entrepreneurs thrive—looking for an opportunity to meet a need. Similarly, engineers look to find solutions to meet a need. This may involve optimizing a process, improving customer satisfaction, or addressing reliability issues.

To illustrate the process, let's take our example of the initial difficulty that beginning engineering students can have with grades. Most students were proficient in high school and didn't need to study much. The demands of engineering programs can catch some students off guard. Improving grades is the problem we will tackle.

At this stage, we need to identify that there is a problem that is worth our effort in solving. A good start would be to identify whether it truly is a problem. How could you do this? Total your scores thus far in your classes and compare them with the possible scores listed in your syllabus. Talk with your professor(s) to find out where you are in the classes and if grades will be given on a curve. Talk to an advisor about realistic grade expectations. At the end of this process, you should know if you have a grade problem.

EXAMPLE 7.4   Engineers are trained problem solvers. Many engineers move into other careers where their problem-solving abilities are valued. One such case is Greg Smith, who was called into the ministry after becoming a mechanical engineer.

> *As soon as people discover that I have a degree in mechanical engineering and am in the ministry, they always ask me the same question, "So, do you still plan on using your engineering degree some day?" It is difficult for people to understand that I use my engineering training every day as a full-time minister.*

> *One of the greatest skills I learned at Purdue was problem solving. Professors in mechanical engineering were notorious for giving us more information than we would need to solve the problem. Our job was to sift through all the facts and only use what was needed to reach a solution. I find myself using this very skill whenever I am involved in personal counseling. For example, just this week I was listening to someone unload about all the stress he was experiencing from life. He listed his job, a second business at home, family, and finances as the primary sources of stress. After further conversation, I was able to "solve" his problem by showing that most of his stress was related to only one of the four sources he listed. The root issue was that this person was basing much of his self-image on his success in this one area of life. I helped him regain his sense of balance in seeing his self-worth not in job performance, but in who he was as a whole person.*

What Do We Know?

The second step in problem solving is to gather facts. All facts and information related to the problem identified in the first step are gathered. In this initial information-gathering stage, do not try to evaluate whether the data are central to the problem. As will be explained in the brainstorming and idea-generating sections to follow, premature evaluation can be a hindrance to generating adequate information.

With our example of the student questioning his grades, the information we generate might include the following:

1. Current test grades in each course
2. Current homework and quiz grades in each course
3. Percent of the semester's grade already determined in each course
4. Class average in each course
5. Professors' grading policies
6. Homework assignments behind
7. Current study times
8. Current study places
9. Effective time spent on each course
10. Time spent doing homework in each course
11. Performance of your friends
12. Your high school grades
13. Your high school study routine

Part of this process is to list the facts that you know. To be thorough, you should request assistance from someone with a different point of view. This could be a friend, an advisor, a parent, or a professor. This person might come up with a critical factor you have overlooked.

What Is the Real Problem?

This stage is one that is often skipped, but it is critical to effective solutions. The difference between this stage and the first is that this step of the process answers the question *why*. Identifying the initial problem answers the question of *what*, or what is wrong. To fix the problem, a problem solver needs to understand *why* the problem exists. The danger is that only symptoms of the problem will be addressed, rather than root causes.

In our example, we are problem solving low class grades. That is the *what* question. To really understand a problem, the *why* must be answered. Why are the grades low? Answering this question will identify the cause of the problem. The cause is what must be dealt with for a problem to be effectively addressed.

In our case, we must understand why the class grades are low. Let us assume that poor test scores result in the low grades. But this doesn't tell us why the test scores are low. This is a great opportunity for brainstorming potential causes. In this divergent phase of problem definition, don't worry about evaluating the potential causes. Wait until the list is generated.

Possible causes may include the following:

- Poor test-taking skills
- Incomplete or insufficient notes
- Poor class attendance

- Not understanding the required reading
- Not spending enough time studying
- Studying the wrong material
- Attending the wrong class
- Studying ineffectively (e.g., trying to cram the night before tests)
- Failing to understand the material (might reveal need for tutor or study group)
- Using solution manuals or friends as a crutch to do homework
- Not at physical peak at test time (i.e., up all night before the test)
- Didn't work enough problems from the book

After you have created the list of potential causes, evaluate the validity of each. In our example, there may be more than one cause contributing to the low grades. If so, rank the causes in order of their impact on class performance.

Assume the original list is reduced to the following:

- Poor class attendance
- Studying ineffectively
- Not spending enough time studying
- Not understanding the material

Order these things in terms of their impact on grades. You may be able to do this yourself or you may need help. An effective problem solver will seek input from others when it is appropriate. In this case, a professor or academic advisor might be able to help you rank the causes and help determine which ones will provide the greatest impact. For this example, say the rank order was as follows:

1. Not understanding the material
2. Studying ineffectively
3. Not spending enough time studying
4. Poor class attendance

We determined that while class attendance was not perfect, it was not the key factor in the poor performance. The key item was not understanding the material. This goes with ineffective studying and insufficient time.

Identifying key causes of problems is vital, so let's look at another example.

EXAMPLE 7.5   "What was the confusion?" Defining the real problem is often the most critical stage of the problem-solving process and the one most often skipped. In engineering, this can be the difference between a successful design and a failure.

An example of this occurred with an engine maker. A new model engine was introduced and began selling quickly. A problem developed, however, and failures started to happen. Cracks initiating from a series of bolt holes were causing the engines to fail. A team of engineers was assembled and the part was redesigned. They determined that the cause was high stress in the area of the bolt hole. Embossments were added to strengthen the area. It was a success.

Until, that is, the redesigned engines accumulated enough time in the field to crack again. A new team of engineers looked at the problem and quickly determined that the cause was a three-dimensional coupling of stress concentrations created by the embossments, resulting in too high a stress near the hole where the crack started. The part was redesigned and introduced into the field. It was a success.

Well, it was a success until the redesigned engines accumulated enough time in the field to crack yet again. A third team of engineers was assembled. This time, they stepped back and asked what the real cause was. Part of the team ran sophisticated stress models of the part, just like the other two teams had done. The new team members looked for other causes. What was found was that the machine creating the hole introduced microcracks into the surface of the bolt holes. The stress models predicted that further stresses would start a crack. However, if a crack were intentionally introduced, the allowable stresses were much lower and the part would not fail. The cause wasn't with cracking but with the wrong kind of cracking. The third team was a success because it had looked for the real problem. Identifying the real cause initially, which in this case involved the machining process, would have saved the company millions of dollars and spared customers the grief of the engine failures.

What Is the Best Solution?

Once the problem has been defined, potential solutions need to be generated. This can be done by yourself or with the help of friends. In an engineering application, it is wise to confer with experienced experts about the problem's solution. This may be most productive after you have begun a list of causes. The expert can comment on your list and offer his or her own as well. This is a great way to get your ideas critiqued. After you gain more experience, you will find that technical experts help you narrow down the choices rather than providing more. Also, go to more than one source. This may provide more ideas as well as help with the next step.

In our example, the technical expert may be a professor or advisor who is knowledgeable about studying. Let's assume that the following list of potential solutions has been generated:

- Get a tutor.
- Visit the professors during office hours.
- Make outside appointments with professors.
- Visit help rooms.
- Form a study group.
- Outline books.
- Get old exams.

- Get old sets of notes.
- Outline lecture notes.
- Review notes with professors.
- Do extra problems.
- Get additional references.
- Make a time schedule.
- Drop classes.
- Retake the classes in the summer.
- Go to review sessions.

You must now evaluate the list and decide on the best solution. This convergent phase of the problem-solving process is best done with the input of others. It is especially helpful to get the opinions of those who have experience and/or expertise in the area you are investigating. In an engineering application, this may be a lead engineer. In our case, it could be an academic advisor, a teaching assistant, or a professor. These experts can be consulted individually or asked to participate as a group.

When refining solutions, an effective tool is to ask yourself which of them will make the biggest impact and which will require the most effort. Rank solutions high or low in terms of impact and effort. Ideal solutions are those that have a high impact yet require a low level of effort. A visual tool that helps make this process easier is the problem-solving matrix shown in Figure 7.3. Avoid solutions that fall in quadrant B. These solutions require a great deal of effort for marginal payoff. Solutions that are the most effective are those in quadrant C. These are the ones that are easy to implement and will produce significant results. You may need to include solutions in quadrant D, those requiring high effort and yielding high impact, but evaluate these carefully to ensure that the investment is worth the reward.

Implementing the Solution

Implementing the solution is a step that may seem trivial and not worth discussion, but it is a critical phase of the problem-solving process. In our scenario, the solution selected may have been "do extra problems."

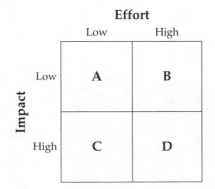

Figure 7.3 A problem-solving matrix

To accomplish this solution, appropriate additional problems must be selected, done, and corrected. Implementation probably requires the assistance of the instructor to help select appropriate problems. Randomly picking additional practice problems could move the effort from quadrant C to D and even B. To have the greatest chances of success, you would want to gain the assistance of your instructor. Other sources of assistance could come from a study group, which poses the challenges of getting them to do extra problems with you.

As with all the other steps, a divergent phase begins the step including such activities as brainstorming. The convergent phase of the step culminates with the selection of the implementation plan.

In engineering applications, implementation can be a very critical phase of the process. Most of the solutions to problems require either additional resources, including money, or the cooperation of other groups that are either not directly affected by the problem or not under your control. Especially early in your career, you will not have control over the people who need to implement solutions, so you will need to sell your ideas to the managers who do. An effective implementation plan will be critical to getting your solutions accomplished.

President Jimmy Carter was educated as an engineer and was one of the most respected presidents as a strong moral person. During his presidency, the country was in an energy crisis and action was needed to address the problems. The story goes that he summoned Tip O'Neill, the Speaker of the House of Representatives, to the White House to explain his plan. Mr. O'Neill responded by giving the president a list of people who would need to be sold on the plan before it went public. President Carter, who was new to Washington, D.C., was going to skip the development of an implementation plan and go ahead and enact the plan. He believed that with the country in such an energy crisis that it was obvious what was needed and cooperation was a given. Tip O'Neill was very experienced in the Washington political system, however, and understood that they needed to spend time gaining the acceptance to their plan before it could be implemented.

Engineering situations can be very similar. While most are not as political as Washington D.C., most require gaining the acceptance of others to implement a solution. In all of these cases, gaining the recognition of those involved is a critical part of the problem-solving process.

Evaluating the Solution

Implementation does not necessarily end the problem-solving process. Just as the design process is a circular process with each design leading to possible new designs, problem solving can also be cyclic. Once a solution has been found and implemented, an evaluation should be performed. As with the other steps in the problem-solving process, this step begins with a divergence phase as the problem solver asks what makes a successful solution and how it will be evaluated. To effectively evaluate a solution, a criterion for success must be established. Sometimes there are objective criteria for evaluation, but in many circumstances the criterion for success is more subjective.

Once the criterion has been established, the evaluation process should be defined including who will evaluate the solution. Often it is desirable to get a neutral person who was not involved in the solution process to be the evaluator.

If the solution is a success, the process may be done. Often in engineering applications, solutions are intermediate and lead to other opportunities. The software industry is a prime example. Software is written to utilize the current computer technologies to address issues. Almost as soon as it is completed, new technologies are available which open up new opportunities or require new solutions.

Sometimes a success does not address the true problem as was the case of the turbine disk failures described earlier. In these cases, the evaluation process may need time before it can be deemed a true success. In other cases, solutions have unintended outcomes that require a new solution even if the initial solution was deemed a success.

Whether a solution is effective or not, there is value in taking time to reflect on the solution and its implications. This allows you as a problem solver to learn both from the process and the solution. Critically evaluating solutions and opportunities is a skill that is extremely valuable as an engineer and is discussed further in the critical thinking section of this chapter.

7.5   Personal Problem-Solving Styles

The creative problem-solving model presented previously is one of the models that can be used to solve the kind of open-ended problems engineers face in everyday situations. You may find yourself in an organization that uses another model. There are many models for problem solving and each breaks the process up into slightly different steps. Isaksen and Treffinger, for example, break the creative problem-solving process into six linear steps:

1. Mess finding
2. Data finding
3. Problem finding
4. Idea finding
5. Solution finding
6. Acceptance finding

Dr. Min Basadur of McMaster University developed another model, a circular one, which he calls Simplex, based on his experience as a product development engineer with the Procter and Gamble Company. This model separates the problem-solving process into eight different steps:

1. Problem finding
2. Fact finding
3. Problem defining
4. Idea finding

5. Evaluating and selecting
6. Action planning
7. Gaining acceptance
8. Taking action

All these problem-solving processes provide a systematic approach to problem solving. Each process has divergent phases where options need to be generated along with convergent phases when best options need to be selected.

Basadur has created and patented a unique method of helping individuals participating in the creative problem-solving process to identify which parts of the process they are more comfortable in than others. This method, called the Basadur Simplex Creative Problem Solving Profile, reflects your personal creative problem-solving style. Everyone has a different creative problem-solving style. Your particular style reflects your relative preferences for the different parts of the problem-solving process.

Basadur's Creative Problem Solving Profile method identifies four styles, each correlating with two of the eight problem-solving steps in the eight-step circular model of creative problem solving. Thus the eight problem-solving steps are grouped into four quadrants or stages of the complete problem-solving process (Fig. 7.4). The Basadur Creative Problem Solving Profile begins with quadrant one, the generation of new problems and opportunities. It cycles through quadrant two, the conceptualization of the problem or opportunity and of new, potentially useful ideas, and quadrant three, the optimization of new solutions. It ends with quadrant four, the implementation of new solutions. Each quadrant requires different kinds of thinking and problem-solving skills.

Basadur describes the four different quadrants or stages as follows.

Generating

Generating involves getting the problem-solving process rolling. Generative thinking involves gathering information through direct experience, questioning,

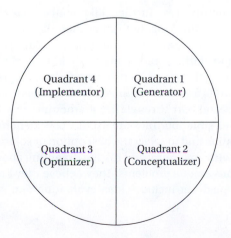

Figure 7.4 The four quadrants of Basadur's Simplex Creative Problem Solving Profile.

imagining possibilities, sensing new problems and opportunities, and viewing situations from different perspectives. People and organizations strong in generating skills prefer to come up with options or diverge rather than evaluate and select or converge. They see relevance in almost everything and think of good and bad sides to almost any fact, idea, or issue. They dislike becoming too organized or delegating the complete problem, but are willing to let others take care of the details. They enjoy ambiguity and are hard to pin down. They delight in juggling many new projects simultaneously. Every solution they explore suggests several new problems to be solved. Thinking in this quadrant includes problem finding and fact finding.

Conceptualizing

Conceptualizing keeps the innovation process going. Like generating, it involves divergence, but rather than gaining understanding by direct experience, it favors gaining understanding by abstract thinking. It results in putting new ideas together, discovering insights that help define problems, and creating theoretical models to explain things. People and organizations strong in conceptualizing skills enjoy taking information scattered all over the map from the generator phase and making sense of it. Conceptualizers need to understand: to them, a theory must be logically sound and precise. They prefer to proceed only when they have a clear grasp of a situation and when the problem or main idea is well defined. They dislike having to prioritize, implement, or agonize over poorly understood alternatives. They like to play with ideas and are not overly concerned with moving to action. Thinking in this quadrant includes problem defining and idea finding.

Optimizing

Optimizing moves the innovation process further. Like conceptualizing, it favors gaining understanding by abstract thinking. But rather than diverge, an individual with this thinking style prefers to converge. This results in converting abstract ideas and alternatives into practical solutions and plans. Individuals rely on mentally testing ideas rather than on trying things out. People who favor the optimizing style prefer to create optimal solutions to a few well-defined problems or issues. They prefer to focus on specific problems and sort through large amounts of information to pinpoint what's wrong in a given situation. They are usually confident in their ability to make a sound logical evaluation and to select the best option or solution to a problem. They often lack patience with ambiguity and dislike dreaming about additional ideas, points of view, or relations among problems. They believe they know what the problem is. Thinking in this quadrant includes idea evaluation and selection and action planning.

Implementing

Implementing completes the innovation process. Like optimizing, it favors converging. However, it favors learning by direct experience rather than by abstract thinking. This results in getting things done. Individuals rely on trying things out rather than mentally testing them. People and organizations strong in implementing prefer situations in which they must somehow make things work. They do not need complete understanding in order to proceed and adapt quickly to immediate changing circumstances. When a theory does not appear to fit the facts they will readily discard it. Others perceive them as enthusiastic about getting the job done but also as impatient or even pushy as they try to turn plans and ideas into action. They will try as many different approaches as necessary and follow up or bird-dog as needed to ensure that the new procedure will stick. Thinking in this quadrant includes gaining acceptance and implementing.

Your Creative Problem-Solving Style

Basadur's research with thousands of engineers, managers, and others has shown that everyone has a different creative problem-solving style. Your particular style reflects your relative preferences for each of the four quadrants of the creative problem-solving process: generating/initiating, conceptualizing, optimizing, and implementing. Your behavior and thinking processes cannot be pigeonholed in any single quadrant. Rather, they're a combination or blend of quadrants: you prefer one quadrant in particular, but you also have secondary preferences for one or two adjacent quadrants. Your blend of styles is called your creative problem-solving style. Stated another way, your creative problem-solving style shows which particular steps of the process you gravitate toward.

Typically as an engineer, you will be working as part of a team or larger organization. Basadur's research has shown that entire organizations also have their own problem-solving process profiles. An organization's profile reflects such things as the kinds of people it hires, its culture and its values. For example, if an organization focuses almost entirely on short-term results, it may be overloaded with implementers but have no conceptualizers or generators. The organization will show strengths in processes that deliver its current products and services efficiently, but it will show weaknesses in processes of long-term planning and product development that would help it stay ahead of change. Rushing to solve problems, this organization will continually find itself reworking failed solutions without pausing to conduct adequate fact finding and problem definition. By contrast, an organization with too many generators or conceptualizers and no implementers will continually find good problems to solve and great ideas for products and processes to develop, but it will never carry them to their conclusion. You can likely think of many examples of companies showing this imbalance in innovation process profiles.

Basadur suggests that in order to succeed in creative problem solving, a team requires strengths in all four quadrants. Teams must appreciate the importance of all four quadrants and find ways to fit together their members' styles. Team members must learn to use their differing styles in complementary ways. For example, generating ideas for new products and methods must start somewhere with some individuals scanning the environment, picking up data and cues from customers, and suggesting possible opportunities for changes and improvement. Thus, the generator raises new information and possibilities—usually not fully developed, but in the form of starting points for new projects. Then the conceptualizer pulls together the facts and idea fragments from the generator phase into well-defined problems and challenges and more clearly developed ideas worth further evaluation. Good conceptualizers give sound structure to fledgling ideas and opportunities. The optimizer then takes these well-defined ideas and finds a practical best solution and well-detailed efficient plans for proceeding. Finally, implementers must carry forward the practical solutions and plans to make them fit real-life situations and conditions.

Skills in all four quadrants are equally valuable. As an engineering student, you will find yourself working on group projects. Often, these projects are designs that are open-ended and are well suited for creative problem solving. Use these opportunities to learn about your problem-solving skills and preferences as well as those of your team members.

EXAMPLE 7.6   Find out what your preferred problem-solving style is by taking the Basadur Simplex Creative Problem Solving Profile. This profile and additional information about the Simplex model can be obtained from the Center for Research in Applied Creativity, www.basadur.com.

Why Are We Different?

What causes the main differences in people's approaches to problem solving? Basadur's research suggests that they usually stem from inevitable differences in how knowledge and understanding—"learning"—are gained and used. No two individuals, teams, or organizations learn in the same way. Nor do they use what they learn in the same why. As shown in Figure 7.5, some individuals and organizations prefer to learn through direct, concrete experiencing (doing). They gain understanding by physical processing. Others prefer to learn through more detached abstract thinking (analyzing). They gain understanding by mental processing. All individuals and organizations gain knowledge and understanding in both ways, but to differing degrees. Similarly, while some individuals and organizations prefer to use their knowledge for ideation, others prefer to use their knowledge for evaluation.

How individuals or organizations combine these different ways of gaining and using learning determines their problem-solving process profile. When you understand these differences, you can shift your own orientation in order to complement the problem-solving process preferences of others. Equally important, you can take various approaches to working with people. You can decide on the optimal strategy

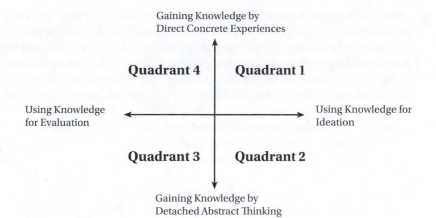

Figure 7.5 The two dimensions of the Basadur Simplex Creative Problem Solving Profile.

Source: The Basadur Problem Solving Model and Profile are copyrighted and are used with the permission of Dr. Min Basadur

for helping someone else to learn something. And you can decide whom to turn to for help in ideation or evaluation. Understanding these differences also helps you interact with other people to help them make the best use of the creative problem-solving process. For example, you can help strong optimizers discover new problems and facts or present new problems and facts to them. You can help strong implementers better define challenges or present well-defined challenges to them. You can help strong generators/initiators evaluate and select from among solutions and make plans, or present to them evaluated solutions and ready-made plans. You can help strong conceptualizers to convince others of the values of their ideas and push them to act on them, or push their ideas through to acceptance and implementation for them (Basadur 1994).

7.6 Brainstorming Strategies

Alex Osborn, an advertising executive in the 1930s, devised the technique of stimulating ideas known as "brainstorming." Since that time, countless business and engineering solutions have come about as a result of his method. Basically, brainstorming is a technique used to stimulate as many innovative solutions as possible. These then can be used in both analytic and creative problem-solving practices.

The goal of brainstorming is to stimulate your mind to trigger concepts or ideas that normal problem solving might miss. This has a physiological basis. The mind stores information in a network. Memories are accessed in the brain by remembering the address of the memory. When asked about ideas for potential solutions, some would come just by thinking about the problem. A portion of your brain will be stimulated, and ideas will be brought to the surface. However, there will be areas of your brain that will not be searched because the pathways to get there are not stimulated. In brainstorming, we get the brain to search as many regions of the brain as possible.

At times it may seem difficult to generate ideas. Isaksen and Treffinger suggest using the method of idea lists. The concept is simple. Write down the ideas you have currently. Then, use these ideas to branch out into new ideas. To help stimulate your thinking in new directions, Alex Osborn developed a set of key words, such as the following:

Adapt—What else is like this? What other ideas does this suggest? Are there any ideas from the past that could be copied or adapted? Whom could I emulate?

Put to other uses—Are there new ways to use the object as is? Are there other uses if modified?

Modify—Could you change the meaning, color, motion, sound, odor, taste, form, or shape? What other changes could you make? Can you give it a new twist?

Magnify—What can you add to it? Can you give it a greater frequency? Make it stronger, larger, higher, longer, or thicker? Can you add extra value? Could you add another ingredient? Could you multiply or exaggerate it?

Minify—What could you subtract or eliminate? Could you make it smaller, lighter, or slower? Could you split it up? Could you condense it? Could you reduce the frequency, miniaturize it, or streamline it?

Substitute—What if I used a different ingredient, material, person, process, power source, place, approach, or tone of voice? Who else instead? What else instead? What if you put it in another place or in another time?

Rearrange—Could you interchange the pieces? Could you use an alternate layout, sequence, or pattern? Could you change the pace or schedule? What if you transpose the cause and effect?

Reverse—What if you tried opposite roles? Could you turn it backwards, upside down, or inside out? Could you reverse it or see it through a mirror? What if you transpose the positive and negative?

Combine—Could you use a blend, an assortment, an alloy, or an ensemble? Could you combine purposes, units, ideas, functions, or appeals?

Creativity expert Bob Eberle took Osborn's list and created an easy-to-remember word by adding "eliminate" to the list. Eberle's word is SCAMPER. The words used to form SCAMPER are as follows:

Substitute?
Combine?
Adapt?
Modify? Minify? Magnify?
Put to other uses?
Eliminate?
Reverse? Rearrange?

Using the SCAMPER list may help to generate new ideas.

Individual Brainstorming

Brainstorming can be done either individually or in a group. Brainstorming individually has the advantage of privacy. There may be times when you deal with a professional or personal problem that you don't want to share with a group. Some research even shows that individuals may generate more ideas by working by themselves. Edward de Bono attributes this to the concept that "individuals on their own can pursue many different directions. There is no need to talk and no need to listen. An individual on his or her own can pursue an idea that seems 'mad' at first and can stay with the idea until it makes sense. This is almost impossible with a group."

When brainstorming individually, select a place that is free from distraction and interruption. Begin by writing down your initial idea. Now you will want to generate as many other ideas as possible. Let your mind wander and write down any ideas that come into your head. Don't evaluate your ideas; just keep writing them down. If you start to lose momentum, refer to the SCAMPER questions and see if these lead to any other ideas.

There are different formats for recording your ideas when brainstorming. Some find generating an ordered lists works best for them. Others will write all over a sheet of paper, or in patterns. Kinesthetic learners often find that putting ideas on small pieces of paper, which allows them to move them around, helps the creative process. Auditory learners might say the ideas out loud as they are generated.

Group Brainstorming

The goal of group brainstorming is the same as with individual brainstorming—to generate as many potential solutions as possible without judging any of them. The power of the group comes when each member of the group is involved in a way that uses his or her creativity to good advantage.

There are a few advantages to group brainstorming. The first is simply that the additional people will look at a problem differently, and bring fresh perspectives. When brainstorming is done correctly, these different views trigger ideas that you would not have come up with on your own.

Another advantage to group brainstorming is that it gets others involved in the problem-solving process early. As you will see in the following section, a key component in problem solving is implementing your solution. If people whose cooperation you need are in the brainstorming group, they will be much more agreeable to your final solutions. After all, it was partly their idea.

To run an effective group brainstorming session, it helps to use basic guidelines. Those outlined in this section have been proven effective in engineering applications. They are also very helpful for use in student organizations and the committees that engineering students are encouraged to join. This process not only covers the generation of ideas, but also provides a quick way to evaluate and converge on a solution. In some situations, you may want to stop after the ideas are generated and evaluate them later.

The guidelines for this process include the following:

1. Pick a facilitator.
2. Define the problem.
3. Select a small group.
4. Explain the process.
5. Record ideas.
6. Involve everyone.
7. Don't evaluate.
8. Eliminate duplicates.
9. Pick three.

Pick a Facilitator

The first step is to select a facilitator who will record the ideas and keep the group focused. The facilitator is also responsible for making sure the group obeys the ground rules of brainstorming.

Define the Problem

It is important that all the participants understand the problem you are looking to solve before generating solutions. Once you start generating ideas, distractions can bring the definition process to a grinding halt. Idea generation would be hampered if the group tried to define the problem they are trying to solve at the wrong time. The definition discussion should happen before the solution idea-generating phase.

Small Group

The group size should be kept manageable. Professor Goldschmidt from Purdue University recommends groups of six to 12 members. Groups smaller than that can work, but with less benefit from the group process. Groups larger than 12 become unwieldy, and people won't feel as involved. If you need to brainstorm with a larger group, break it into smaller subgroups and reconvene after the groups have come up with separate ideas.

Explain the Process

Providing the details of the process the group will follow is important. It gives the participants a feeling of comfort knowing what they are getting into. Once the process begins, it is counterproductive to go back and discuss ground rules. This can stymie idea generation just like a mistimed definition of the problem.

Record Ideas

Record ideas in a way that is visible to the whole group. Preferably, arrange the group in a semicircle around the person recording. A chalkboard, flip chart, whiteboard, or newsprint taped to a wall all work well. It is important to record all ideas, even if they seem silly. Often the best ideas come from a silly suggestion that triggers a good idea. By writing the suggestions in a way that is visible to all, the participants can use their sight as well as hearing to absorb ideas which may trigger additional ideas. The use of multiple senses helps to stimulate more ideas. Also, having all the ideas in front of the participants guarantees recollection and allows for new ideas, triggered by earlier ideas. Using the SCAMPER questions is a great way to keep ideas flowing.

Involve Everyone

Start with one idea from the facilitator or another volunteer and write it down. It is easier to get going once the paper or writing surface is no longer blank. Go around the group, allowing each person to add one idea per round. A person who doesn't have an idea on that round can say "pass." It is more important to keep moving quickly than to have the participants feel like they must provide an idea every time.

By taking turns, all the participants are ensured an equal opportunity to participate. If the suggestions were taken in a free-flowing way, some participants might monopolize the discussion. The power of group brainstorming lies in taking advantage of the creative minds of every member of the group, not just the ones who could dominate the discussion. Often, the best ideas come from the people who are quiet and would get pushed out of the discussion in a free-flowing setting. It matters not when or how the ideas come, as long as they are allowed to surface.

Continue the process until everyone has "passed" at least once. Even if only one person is still generating ideas, keep everyone involved and keep asking everyone in turn. The one person who is generating ideas might stimulate an idea in someone else at any point. Give everyone the opportunity to come up with one final idea. The last idea may very well be the one that you use.

Don't Evaluate

This may be hard for the facilitator or some participants. Telling someone that his or her suggestion is dumb, ridiculous, or out in left field, or making any other negative judgment makes that person (and the rest of the group) less likely to speak up. The genius of brainstorming is the free generation of ideas without deciding if they are good until the correct stage. There are countless times when the wackiest ideas are brought up right before the best ideas. Wacky suggestions can trigger ideas that end up being the final solution. Write each idea down, as crazy as some may be. This makes participants feel at ease. They will feel more comfortable making way-out suggestions and will not fear being made fun of or censored.

Avoiding negative comments is one thing, but if you are facilitating, you need to watch for more subtle signals. You might be communicating indirectly to someone that his or her ideas are not good. If you are writing the ideas down, it is natural to say "good idea" or "great idea." What happens, however, if one person gets "great idea" comments and the others don't? Or what if it is just one person who doesn't get the "great idea" comments?

The participants need to feel completely at ease during brainstorming. Providing negative feedback in the form of laughter, sarcastic remarks, or even the absence of praise will dampen the comfort level and, therefore, the creativity. It is safest not to make any value judgments about any ideas at this point. If you are facilitating, try to respond to each idea in the same way and to keep order in your court, so to speak. If you are a participant, try not to make value statements or jokes about other ideas.

Eliminate Duplicates

After generating ideas, you may want to study the suggestions. In some cases, it is to your advantage to sort out a solution or the top several solutions. Doing this at the end of group brainstorming can be very easy, and also serves to involve the group members in the decision process.

The first step is to examine the list of ideas and eliminate duplicates. There is a fine line between eliminating duplicates and creating limiting categories. This step is designed to eliminate repeated ideas. Ask if they really are the same. If they are similar, leave both. Creating categories only reduces the possible solutions and defeats the purpose of brainstorming.

The next step is to allow the group to ask clarifying questions about suggestions. This is still not the time for evaluation—only clarification. Don't let this become a discussion of merits. Clarification may help in eliminating and identifying duplicates.

Pick Three

Once everyone understands all ideas as best they can, have them evaluate the suggestions. This can be done quickly and simply. First, ask everyone to pick his or her three top choices. (Each person can select more if you need more possibilities or solutions.) After the members have determined these in their own minds, let each member vote publicly by marking his or her choices. Do this with a dot or other mark, but not a 1, 2, and 3. The top ideas are selected based on the total number of marked votes they receive. The winners then can be forwarded to the problem-solving phases to see if any survive as the final solution.

Note that with this method of voting, the group's valuable time is optimized. Time is not wasted determining whether an idea deserves a "1" or a "3." Also, no time is spent on the ideas that no one liked. If all ideas are discussed throughout the process, the ideas that are not chosen by anyone or that are least preferred often occupy discussion for quite a while. It's best to eliminate these likely time sinkholes as soon as possible.

Another benefit of not ranking choices at first is that each member could think that his or her ideas were the fourth on everyone's list, which might be frustrating. Also, how might it affect someone if his or her ideas were ranked low by the other members? Such a discouraged person could be one who would have very valuable input the next time it is needed, or it could be that this person will be someone you need as an ally to implement the group's decision. Either way, there is no need to cast any aspersions on ideas that weren't adopted.

7.7 **Critical Thinking**

Since engineering is so dominated by systematic problem solving, it is easy to get caught up in the methodologies and look at everything as a problem needing a solution or goal. We have presented problem-solving methods that give a systematic approach for analyzing and solving problems. However, many of life's issues are more complex and have no single solution; in fact, some cannot be solved. I once assigned a project to design a program using MATLAB to predict students' financial future and life plans, including a potential spouse. I jokingly suggested that the class use the program when dating to evaluate the financial impact of a potential spouse or life partner ("Sorry, MATLAB says we wouldn't be compatible"). The class laughed at the suggestion. Clearly, whom you marry, or if you marry, is a complex question and not one that can be solved by some stepwise methodology or computational algorithm. Life is more complex than that.

Issues and challenges you will face have a similar complexity and require a critical perspective so that you can effectively address engineering issues with human, environmental, or ethical impacts. For example, many programs have sought to address poverty, and many programs and solutions involved engineering. Poverty, however, has remained a pressing social issue in this country and abroad. Solutions to these larger social issues are complex and often do not have a single or simple solution.

Many of the issues you will be exposed to in your career are problems that cannot be easily or quickly solved. It is easy to get caught up in the processes and methodologies and move to solutions and implementation plans. Sometimes this is appropriate, but other times we need to evaluate our actions and examine the larger issues of our actions. Being able to evaluate why and how you are doing something by taking a step back to gain perspective is an essential skill. At each step in your problem solving, one of the questions you can ask is "why?"

There are times when you need to evaluate your actions, to discover deeper meanings and larger issues. One example involves the design of toys for children who are physically challenged. In the EPICS (Engineering Projects in Community Service, http://epics.ecn.purdue.edu) Program, students work on engineering solutions to local community issues. One of the partnerships in the program is with a clinic that works with children with disabilities. As the students work with the children, families, and therapists, they gain an appreciation for the myriad complex issues facing these children. When the engineering students design toys, they must design them so they can be used with the given physical abilities of the children. This is fairly straightforward.

What the students have learned is that they can address many other issues, such as socialization dimensions, in their designs. Some of the toys are designed so that children with disabilities can more easily interact with their peers in their integrated classroom, when using the toys. The students have also learned to design so that the children with disabilities can control aspects of play when using the toys with their peers. The engineering students have learned about the socialization of children with and without disabilities, and its impact on self-esteem. They have also come to understand the role of play in child development and the barriers children with disabilities may encounter. Armed with this knowledge, the engineering students can design toys that meet a wider spectrum of needs.

When working with issues such as children or adults with disabilities, there are not simple problems to be solved, but complex issues that need to be addressed and pondered. By working through these issues and their multiple dimensions, students can become aware of the social and emotional issues of their work. They can become better designers and better educated about relevant issues.

As engineers, we have grave responsibilities because our work has great potential for good and for harm. While having the highest of ethical standards is essential, keeping a critical eye on the context and direction of your work is equally vital. Creating opportunities to gain perspective and evaluating your actions, experiences, and perceptions is as important as the problem-solving methodology presented in this text. At the beginning of the creative problem-solving process, as well as the design process, we discussed how crucial it was to get the question right. Similarly, when evaluating an issue, asking the right question will enable you to generate the best solution.

We use our personal background and experience to evaluate problems, and this perspective can limit our thinking. When evaluating a solution, getting an outside opinion has value. You must become aware of the perspective you bring to issues based on your background so it does not limit you. Thomas Edison wanted people around him who would not limit their thinking by jumping to conclusions. He would take every prospective engineer to lunch at his favorite restaurant, which was known for its soup, which he would recommend. Invariably, the candidate would order soup. Edison wanted to see if he or she would salt the soup before tasting it. Salting without tasting, he reasoned, was an indication of someone who jumps to conclusions. He wanted people who were open to questioning. Periodically, ask yourself if there are things that you are salting without tasting. Are there issues for which you jump to conclusions based on your background, personality, or thinking style?

Critical thinkers ask the "why?" question about themselves as well as about the things around them. Periodically take a step back and ask "why?" or "what?" about yourself. Why are you studying to become an engineer? What will you do as an engineer to find personal fulfillment? Becoming a critical thinker will make you a better problem solver and engineer, and one who is more engaged in society and more fulfilled.

REFERENCES

Basadur, M., *Simplex: A Flight to Creativity*, Creative Education Foundation, Inc., 1994.

Beakley, G. C., H. W. Leach, and J. K. Hedrick, *Engineering: An Introduction to a Creative Profession*, 3rd ed., New York, Macmillan Publishing Co., 1977.

Burghardt, M. D., *Introduction to Engineering Design and Problem Solving*, New York, McGraw-Hill Co., 1999.

Eide, A. R., R. D. Jenison, L. H., Mashaw, and L. L. Northup, *Introduction to Engineering Problem Solving*, New York, McGraw-Hill Co., 1998.

Fogler, H. S., and S. E. LeBlanc, *Strategies for Creative Problem Solving*, Englewood Cliffs, NJ, Prentice Hall, Inc., 1995.

Gibney, Kate, "Awakening Creativity," *Prism*, March 1998, pp. 18–23.

Isaksen, Scott G., and Donald J. Treffinger, *Creative Problem Solving: The Basic Course*, Buffalo, NY, Bearly Limited, 1985.

Lumsdaine, Edward, and Monika Lumsdaine, *Creative Problem Solving—Thinking Skills for a Changing World*, 2nd ed., New York, McGraw-Hill Co., 1990.

Osborn, A. F., *Applied Imagination: Principles and Procedures of Creative Problem-Solving*, New York, Charles Scribner's Sons, 1963.

Panitx, Beth, "Brain Storms," *Prism*, March 1998, pp. 24–29.

EXERCISES AND ACTIVITIES

7.1   A new school has exactly 1,000 lockers and exactly 1,000 students. On the first day of school, the students meet outside the building and agree on the following plan: the first student will enter the school and open all the lockers. The second student will then enter the school and close every locker with an even number (2, 4, 6, 8, etc.). The third student will then reverse every third locker (3, 6, 9, 12, etc.). That is if the locker is closed, he or she will open it; if it is open, he or she will close it. The fourth student will then reverse every fourth locker, and so on until all 1000 students in turn have entered the building and reversed the proper lockers. Which lockers will finally remain open?

7.2   You are stalled in a long line of snarled traffic that hasn't moved at all in 20 minutes. You're idly drumming your fingers on the steering wheel and accidentally start tapping the horn. Several sharp blasts escape before you realize it. The driver of the pickup truck in front of you opens his door, gets out, and starts to walk menacingly toward your car. He looks big, mean, and unhappy. Your car is a convertible and the top is down. What do you do?

7.3   Your school's team has reached the national championship. Tickets are very difficult to get but you and some friends have managed to get some. You all travel a great distance to the site of the game. The excitement builds on the day of the game until you discover that your tickets are still back in your room on your desk. What do you do? What is the problem that you need to solve?

7.4   You are a manufacturing engineer working for an aircraft company. The production lines you are responsible for are running at full capacity to keep up with the high demand. One day, you are called into a meeting where you learn that one of your suppliers of bolts has been forging testing results. The tests were never done, so no data exist to tell if the bolts in question meet your standards. You do not know how long the forging of the test results has been going on; you are only told that it has been "a while." This means that your whole production line including the planes ready to be delivered may be affected. You are asked to develop a plan to manage this crisis. What do you do? Some background information:

a) Bolts are tested as batches or lots. A specified number are taken out of every batch and tested. The testing typically involves loading the bolts until they fail.
b) The bolts do not have serial numbers so it is impossible to identify the lot number from which they came.
c) The supplier who forged the tests supplies 30 percent of your inventory of all sizes of bolts.
d) Your entire inventory is stored in bins by size. Each bin has a mixture of manufacturers. (To reduce dependence on any one company, you buy bolts from three companies and mix the bolts in the bins sorted by size.)
e) Bolts have their manufacturer's symbol stamped on the head of the bolt.
f) It takes weeks to assemble or disassemble an aircraft.
g) Stopping your assembly line completely to disassemble all aircraft could put your company in financial risk.
h) Your customers are waiting on new planes and any delays could cause orders to be lost.
i) The FBI has arrested those who forged the test and their business has been closed, at least temporarily.

7.5   How much of an automobile's tire wears off in one tire rotation?

7.6   A farmer often makes a trip with a squirrel, acorns, and a fox. During each trip he must cross a river in a boat in which he can't take more than one of them with him each time he crosses. Since he often has to leave two of them together on one side of the river or the other, how can he plan the crossings so that nothing gets eaten, and they all get across the river safely?

7.7   Measure the height of your class building using two different methods. Which is more accurate?

7.8   Pick a grassy area near your class building. Estimate how many blades of grass are in that area. Discuss your methodology.

7.9   How high do the letters on an expressway sign need to be to be readable?

7.10   A company has contacted you to design a new kind of amusement park ride. Generate ideas for the new ride. Which idea would you recommend?

7.11   Estimate the speed of a horse.

7.12   Estimate the maximum speed of a dog.

7.13   An explosion occurs in your building. Your room is intact but sustained damage and is in immediate danger of collapsing on you and your classmates. All exits but one (as specified by your instructor) are blocked. The one remaining is nearly blocked. Quickly develop a plan to get your classmates out. Who goes first and who goes last? Why?

7.14   A child's pool is eight feet in diameter and two feet high. It is filled by a garden hose up to a level of one foot. The children complain that it is too cold. Can you heat it up to an acceptable temperature using hot water from the hose? How much hot water would you need to add?

7.15   How else could you heat the water for the children in problem 7.14?

7.16   Without referring to a table or book, estimate the melting temperature of aluminum. What are the bounds of potential melting temperatures? Why? What is the actual melting temperature?

7.17   Select a problem from your calculus class and use the analytic problem-solving steps to solve it.

7.17   Select a problem from your chemistry class and use the analytic problem-solving steps to solve it.

7.19   Select a problem from your physics class and use the analytic problem-solving steps to solve it.

7.20   Write a one- or two-page paper on a historical figure who had to solve a difficult problem. Provide a description of the problem, solution, and the person.

7.21   Aliens have landed on Earth! They come to your class first but are very frustrated because they cannot communicate with you. Develop a plan to break this communication barrier and show them that we are friendly people.

7.22   You are working on a project that has had many difficulties that are not your fault. However, you are the project leader and your manager has become increasingly frustrated with you and even made a comment that you are the kind of irresponsible person who locks your keys in your car. Your manager is coming into town to see first hand why the project is having problems, and you are going to pick her up. You leave early to take time to relax. With the extra time, you decide to take the scenic route and stop the car next to a bubbling stream. Leaving your car, you go over to the stream to just sit and listen. The quiet of the surroundings is just what you needed to calm down before leaving for the airport. Glancing at your watch, you realize that it is time to leave to pick up your manager. Unfortunately, you discover that you indeed have locked your keys in the car. A car passes on the road about every 15 minutes. The nearest house is probably a mile from your location and the nearest town is 15 miles away. The airport is 20 miles away and the flight is due to arrive in 30 minutes. What will you do?

7.23 You are sitting at your desk in the morning reading your e-mail when your boss bursts into your office. It seems that a local farmer was not happy with your company and has dropped a truckload of potatoes in your plant's entrance. The potatoes are blocking the entrance and must be moved. You have been selected as the lucky engineer to fix this problem. What would you do to move the potatoes and how would you get rid of them? Brainstorm ideas and select the best solution. (Note: You work at a manufacturing plant of some kind that does not use potatoes in its process. The potatoes also appear to be perfectly fine, just in the way.)

7.24 You work in a rural area that is known for chickens. An epidemic has swept through the area, killing thousands of chickens. The Environmental Protection Agency has mandated that the chickens may not be put into the landfill nor can they be buried. Develop a plan to get rid of the chickens.

7.25 A fire has destroyed thousands of acres of forest. Seedlings have been planted in the area but are being destroyed by the local wildlife. Develop a way to reforest the region by preventing the wildlife from eating the seedlings.

7.26 A survey shows that none of the current high school students in your state are planning to major in engineering when they get to college. Your dean has asked your class to fix this dilemma (with no students, she would be out of a job!). Develop a plan to convince the high school students that engineering is worthwhile.

7.27 Select a problem from your math book. How many ways can you solve the problem? Demonstrate each method.

7.28 You and seven of your friends have ordered a round pizza. You need to cut it into eight pieces and are only allowed to make three straight cuts. How can you do this?

7.29 Select an innovative product and research the story of its inventors. Prepare a presentation on how the ideas were formed and the concept was generated.

7.30 With which part of the Basadur problem-solving process are you most comfortable? Write a one-page paper on how your preference affects your problem solving.

7.31 Take the Basadur problem-solving profile test. Report on your scores. How will this information affect your ability to solve problems.

7.32 Organizations have problem-solving styles. Select an organization and prepare a brief report on its problem-solving styles.

7.33   You and your classmates have discovered that your funding for college will be withdrawn after this semester. Develop ideas on how you could obtain the money for the rest of your college expenses.

7.34   Your class has been hired by a new automotive company that wants to produce a brand-new car. Their first entry into the market must be truly innovative. Do a preliminary design of this car. What features would it have to set it apart from current vehicles?

7.35   An alumnus of your engineering program has donated money to build a jogging track that would encircle your campus. The track will be made of a recycled rubberized material. How much of this rubberized material will be needed?

7.36   Every Christmas season it seems that one toy becomes a "I have to have one" craze. You and some classmates have been hired to come up with next year's toy of the season. What is it?

7.37   A professional sports team has decided to locate in your town. A local businessperson is going to build a stadium for the team. Unfortunately, the site is contaminated. The project will fall through unless you can devise a way to get rid of the contaminated soil under the new arena. What do you do with the soil?

7.38   NASA has decided to send astronauts to the other planets in the solar system. Psychologists, however, have determined that the astronauts will need something to keep them occupied during the months of travel or they will develop severe mental problems that would threaten the mission. Your job is to devise ways to keep the astronauts occupied for the long journey. Remember that you are limited to what will fit in a space capsule.

7.39   Landfill space is rapidly running out. Develop a plan to eliminate your city's dependence on the local landfill. The city population is 100,000.

7.40   An appliance manufacturer has hired you to expand its market. Your job is to develop a new household appliance. What is it and how will it work?

7.41   College students often have trouble waking up for early classes. Develop a system that will guarantee that the students wake up and attend classes.

7.42   Select one of your classes. How could that class be more effective in helping you learn?

7.43   Prepare a one-page report on a significant engineering solution developed in the past five years. Evaluate its effectiveness and report on the outcomes (both intended and unintended) of this solution.

7.44   Habitat for Humanity International spearheaded an initiative to eliminate substandard housing within the Georgia county in which it is headquartered. One of the first steps was to develop a set of standards to determine if housing was substandard or not. Develop criteria for your own community for determining whether a housing unit is substandard.

7.45 If your own community were to undertake an initiative like the one presented in the previous question, what would need to be done (both from an engineering and community perspective)?

7.46 Identify one current topic related to engineering or technology that has no single simple solution and write a brief paper discussing issues related to the topic.

7.47 Brainstorm ideas for toys for physically disabled children. Narrow your choices to the top three and identify the benefits of each design. What are the important issues that you would need to consider in your designs?

7.48 This chapter mentioned Edison's "salt test." Identify areas in your own thinking where you might tend to "salt before tasting" (in other words, where you might jump to conclusions or use your own biases to overlook potential options).

7.49 Critically evaluate why you are majoring in engineering. What do you hope to gain by studying engineering?

7.50 Write a one-page paper on your responsibilities to society as a citizen in your community and as an engineer.

7.51 Early one morning it starts to snow at a constant rate. Later, at 6:00 a.m., a snow plow sets out to clear a straight street. The plow can remove a fixed volume of snow per unit time. In other words, its speed is inversely proportional to the depth of the snow. If the plow covered twice as much distance in the first hour as the second hour, what time did it start snowing?

7.52 While three wise men are asleep under a tree a mischievous boy paints their foreheads red. Later they all wake up at the same time and all three start laughing. After several minutes suddenly one stops. Why did he stop?

7.53 Using just a five-gallon bucket and a three-gallon bucket, can you put four gallons of water in the five-gallon bucket? (Assume that you have an unlimited supply of water and that there are no measurement markings of any kind on the buckets.)

7.54 A bartender has a three-pint glass and a five-pint glass. A customer walks in and orders four pints of beer. Without a measuring cup but with an unlimited supply of beer how does he get a single pint in either glass?

7.55 You are traveling down a path and come to a fork in the road. A sign lays fallen at the fork indicating that one path leads to a village where everyone tells the truth and the other to a village where everyone tells lies. The sign has been knocked down so you do not know which path leads to which village. Then someone from one of the villages (you don't know which one) comes down the path from which you came. You may ask him one question to determine which path goes to which village. What question do you ask?

7.56   Four mathematicians have the following conversation:

Alice: I am insane.
Bob: I am pure.
Charlie: I am applied.
Dorothy: I am sane.
Alice: Charlie is pure.
Bob: Dorothy is insane.
Charlie: Bob is applied.
Dorothy: Charlie is sane.

You are also given that:

a) Pure mathematicians tell the truth about their beliefs.
b) Applied mathematicians lie about their beliefs.
c) Sane mathematicians' beliefs are correct.
d) Insane mathematicians' beliefs are incorrect.

Describe the four mathematicians.

7.57   Of three men, one man always tells the truth, one always tells lies, and one answers yes or no randomly. Each man knows which man is which. You may ask three yes/no questions to determine who is who. If you ask the same question to more than one person you must count it as a question used for each person asked. What three questions should you ask?

7.58   Tom is from the U.S. Census Bureau and greets Mary at her door. They have the following conversation:

Tom: I need to know how old your three kids are.
Mary: The product of their ages is 36.
 Tom: I still don't know their ages.
Mary: The sum of their ages is the same as my house number.
Tom: I still don't know their ages.
Mary: The younger two are twins.
Tom: Now I know their ages! Thanks!

How old are Mary's kids and what is Mary's house number?

7.59   Reverse the direction of the fish image below by moving only three sticks:

CHAPTER 8

Graphics and Orthographic Projection

8.1 Introduction

Part of learning the concepts in engineering requires the use of visual information and visual thinking. This requires that engineers have graphical communication skills, not only as students but also as practicing engineers. Communicating by way of graphics requires the ability to visualize information and translate it into visual products such as sketches and drawings. This allows representation of a three-dimensional object in a two-dimensional medium. Whether working with plain pencil and paper or sophisticated computer graphics, this is an important skill for all engineering disciplines. This chapter presents the fundamentals of visualization and graphics and provides readers with basic skills to understand and create simple sketches.

8.2 Orthographic Projection

Three-dimensional objects are represented in two dimensions by using a concept known as orthographic projection. The basic idea is that a three-dimensional object can be accurately represented using multiple two-dimensional drawings taken from several views. Each view shows what the object would look like if its image were projected onto a flat surface. Generally, three views are required for a complete representation, with each of the projections being perpendicular (orthogonal) to one another.

One way to visualize orthographic projection is by using the "glass box" theory. Figure 8.1 shows a step block surrounded by a glass box. Each panel of the glass box represents a projection plane or picture plane. Note that three of the picture planes are marked "Front," "Top," and "Side." For each view we imagine what an observer might see while viewing each picture plane in a perpendicular manner. This yields a two-dimensional image or view of the object on each panel of the glass box. In the front picture plane, the width and height of the step block are displayed. In the top

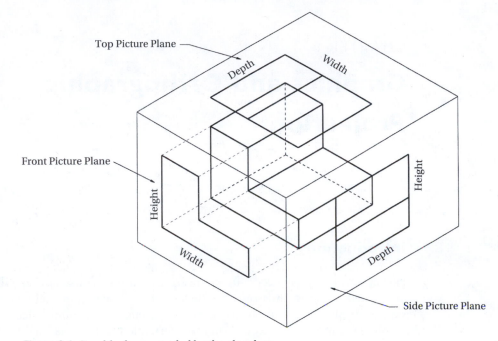

Figure 8.1 Step block surrounded by the glass box

picture plane, the width and depth of the step block are displayed. And in the side picture plane, the depth and height of the step block are displayed. To see the orthographic views, we must imagine the panels of the glass box hinged between the front and top views and between the front and side views. When we open the panels of the glass box (Fig. 8.2), we see the front, top, and side views of the step block. Each view shows only two dimensions of the object, but when we look at all three views, we have a complete picture of the three-dimensional shape. The front view shows the width and height of the object and the depth dimension is suppressed. The top view shows the width and depth of the object and the height is suppressed. The side view shows the depth and height of the object and the width is suppressed. The location of each view relative to the other views is extremely important in describing the object. The front view is the pivotal view. Once the front view is displayed, the top view must appear above the front view, and the side view is shown to the right. The views are never displayed out of position such as you see in Figure 8.3. This would be in violation of the "glass box" theory and thus in violation of basic orthographic projection. It would also lead to confusion when viewed by others, as orthographic projection convention is universally accepted and is the way that engineers and technicians communicate with each other. There are drawing conventions that allow views other than the front view to appear in the lower

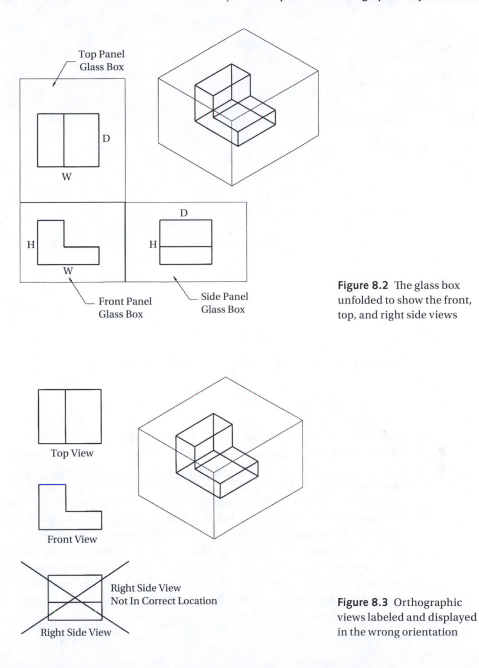

Top Panel
Glass Box

D

W

H

W

H

D

Front Panel
Glass Box

Side Panel
Glass Box

Figure 8.2 The glass box unfolded to show the front, top, and right side views

Top View

Front View

Right Side View
Not In Correct Location

Right Side View

Figure 8.3 Orthographic views labeled and displayed in the wrong orientation

left corner, but all drawing conventions use the "glass box" theory to place the correct views in the right positions.

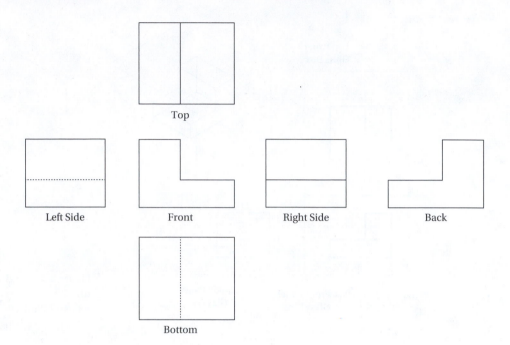

Figure 8.4 The six principal orthographic views of the step block.

Look again at Figure 8.1, the step block inside the glass box. The glass box has six sides and therefore six picture planes can be shown. We already know about the front, top, and side picture planes, which are the most commonly used. The side picture plane is really the *right-side* picture plane and captures the right-side view of the object. The other picture planes of the glass box are the left side, the bottom, and the back picture planes. Figure 8.4 shows all six views of the step block in their proper orthographic positions. These views are called the six "principal orthographic views." As you can see, the left side, bottom, and back views do not yield any additional information regarding the shape of the object beyond what can be seen in the front, top, and right-side views. Only the principal views necessary to visualize and describe the part are used, so typically only three are used. Additional views are only used if necessary for clarity. Occasionally a section view is drawn showing what the object might look like if a plane were to be passed through the object, but again only if such a view is necessary for clarity.

8.3 The Meaning of Lines

Solid lines drawn to represent three-dimensional objects can have various meanings with regard to the object. One of these is to represent the edge view of a surface. Figure 8.5

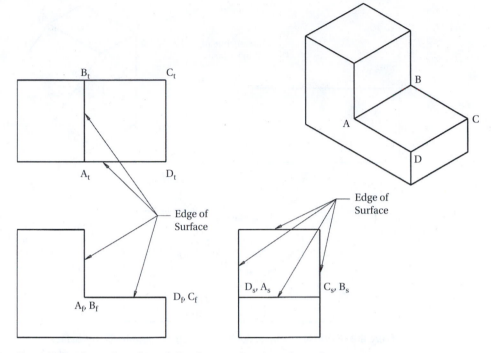

Figure 8.5 A line of an object defined as an edge view of a surface.

shows the step block. Each line in each view represents the edge view of a specific surface. For example, surface ABCD is represented by a single line in the front view. It is also represented by a single line in the right-side view. All the lines of the step block in each of the orthographic views represent the edge of a surface.

A line can also represent the intersection of two surfaces. This is very different from a line that represents an edge of a surface. In Figure 8.6 we see the orthographic views of a block that has an inclined surface (surface ABCD). Surface ABCD is represented in the front view by a single line that is the edge view of the surface. In the top view, line AB is not the edge view of a surface, but it is the intersection of two surfaces. In a similar manner, line CD in the right-side view is the intersection of two surfaces.

Finally, a line can represent the limiting element of a curved surface. Figure 8.7 shows a cylinder. We can draw an infinite number of lines on the outside of the cylinder from the top to the bottom. These lines are parallel to the axis of the cylinder and are called elements. Element lines are often depicted as finer lines, as seen in Figure 8.7. In the front view of the cylinder, note the heavier lines AB and CD. Those lines are neither the edge of a surface nor the intersection of two surfaces. Instead they represent the limiting element of a curved surface. The limiting element is the last element that one can see before the curve begins to turn back on itself. Thus a line is generated to show the limit of the cylinder.

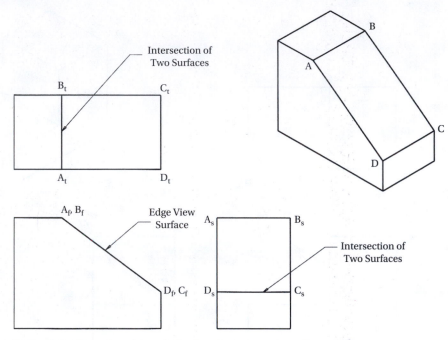

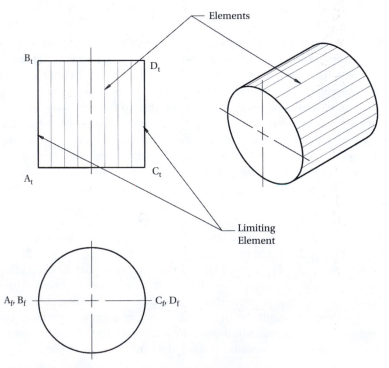

Figure 8.6 A line of an object defined as the intersection of two surfaces.

Figure 8.7 A line of an object defined as the limiting element of a curved surface.

When dealing with real objects and representing them with orthographic views, a line can be defined as the edge view of a surface, the intersection of two surfaces, or the limiting (extreme) element of a curved surface. There are no other definitions of solid lines on an orthographic view of a three-dimensional object.

8.4 Hidden Lines

When representing three-dimensional objects orthographically, we not only must show all of the lines that are visible to the observer, but we must represent the lines that are hidden or not visible as well. Generally, the object itself blocks our view of the details that appear hidden, so we need a way to represent these features. In order to distinguish these hidden lines from visible lines, we display them as dashed lines, sometimes drawn thinner than the solid lines. Figure 8.8 shows a step block with a notch cut out of the back. Two lines in the front view that represent the notch are shown as hidden (dashed) lines. Also, the right-side view has a hidden line that shows the notch.

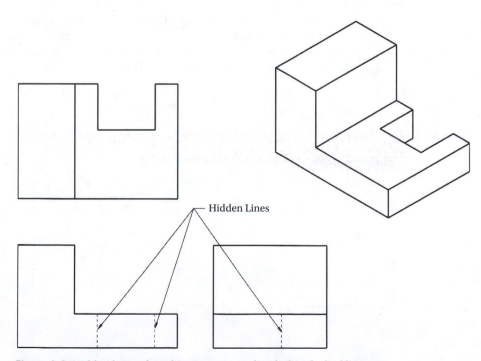

Figure 8.8 Hidden lines of an object represented with thin dashed lines.

8.5 **Cylindrical Features and Radii**

Cylindrical features are considered either positive or negative. Positive cylinders are features that look like axles or soda cans. Negative cylinders are simply circular holes. Objects such as a hollow pipe involve both positive and negative cylinders. When dealing with cylinders, whether positive or negative, we have to include with the shape description a special treatment called a centerline. A centerline is a signal to the interpreter of the drawing that a circular feature is shown. Centerlines are also used to depict the center of a symmetrical part. Whenever you show positive or negative cylinders, the centerlines should be shown. Centerlines are sometimes drawn thinner than either solid lines or hidden lines to further distinguish between them.

Figure 8.9 shows the orthographic representation of a hollow pipe. Note that only a front view and a right-side view are shown. The top view is not shown because it would be identical to the front view and would not show any additional details that are not shown in the front view or right-side view. In the right-side view, note the centerline. In this view where the cylinder appears as a circle, the centerline looks like a crosshair. The center point of the cylinder is represented by a small +. A small gap (about 1/16 inch) is placed on each side of the + with a longer line following that extends about 1/8 inch beyond the curve that is governed by the centerline. This gives the appearance of a crosshair marking the center of the circular object.

In the longitudinal view of the cylinder (front view) we see the centerline as a long line that is broken in the middle. This line represents an axis of symmetry. A common error is to install a crosshair type of centerline on the longitudinal view, but this is wrong. The crosshair type of centerline is used to depict the center of the circular

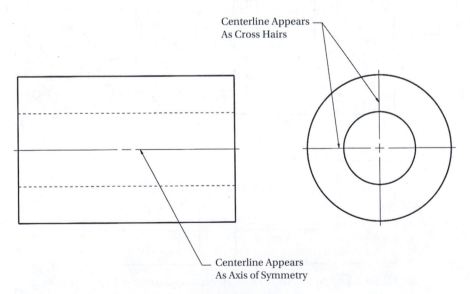

Centerline Appears
As Cross Hairs

Centerline Appears
As Axis of Symmetry

Figure 8.9 Treatment of centerlines of cylindrical objects

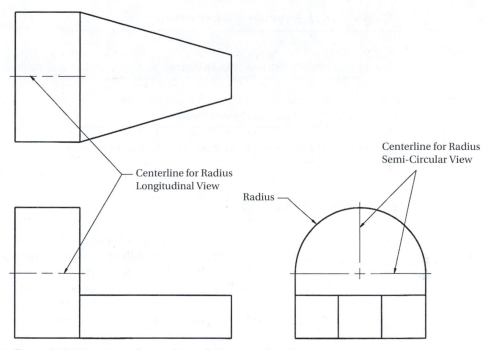

Figure 8.10 Treatment of centerlines of objects with radii

object, while the broken line is used to depict the symmetry of the part. Figure 8.10 shows both types of centerlines in use.

Radii are treated in a similar manner as cylinders, as seen in Figure 8.10. They too have centerlines associated with them. The only difference between radii and cylinders is that radii do not complete 360°, but are something less. In the view where there is a partial circle or arc, the centerline is represented as a crosshair. In the longitudinal view(s), the centerline is a long, broken line and is an axis of symmetry.

8.6 Line Precedence

Sometimes, especially in complex shapes, various line types will get superimposed on each other in a specific view. For example, a visible line and a hidden line may occupy the exact same space in a view. Line precedence tells us what should be shown.

Various line types (visible, hidden, centerline) have a priority with regard to the orthographic drawing. The highest priority belongs to the visible line, the next priority to the hidden line, and the lowest priority to the centerline. Figure 8.11 shows the three types of lines from the highest priority to the lowest priority.

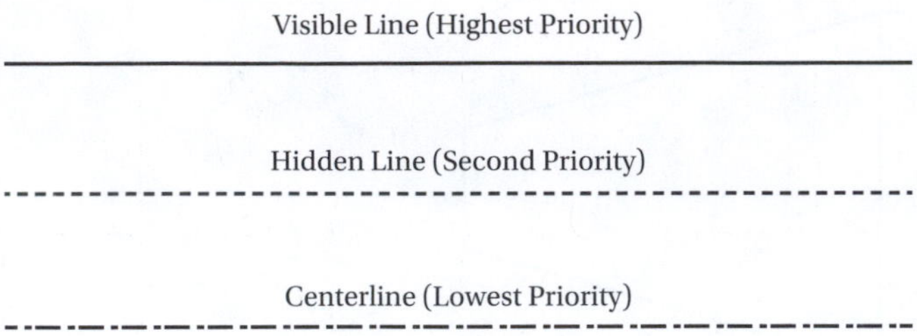

Visible Line (Highest Priority)

Hidden Line (Second Priority)

Centerline (Lowest Priority)

Figure 8.11 Three common standard line types used in engineering drawing and their priorities.

8.7 Freehand Sketching

Freehand sketching is very important to the engineer in that it helps to make a quick graphical representation of an idea that can be easily communicated to others. Sketches are frequently used in meetings with colleagues, technicians, and clients, and in cost quotations. A product often begins with concept sketches during brainstorming sessions, so developing skills in sketching can be very beneficial to the engineer. To develop sketching skills takes practice. In this section we will explore the basics of freehand sketching so that you can practice and develop your skills in this area.

Obviously, lines used in sketching are either straight or curved. An effective way to draw straight lines is to put the pencil at the starting point, look at the finish point of the line, and draw to it. Most right-handed people find it easier to draw lines from right to left, while the opposite is true for left-handed people. To draw sketch lines at various angles, rotate the paper to a comfortable angle.

Sketching curved lines such as circles or radii can be a bit tricky. To sketch circles most accurately, first sketch the centerline. Then mark the radius of the circle on the centerline to use as a guide (Fig. 8.12A). Then, using these marks, construct a light construction box that will contain the circle (Fig. 8.12B). It should not surprise you that this process is called "boxing in." Draw light construction diagonals from each corner of the construction box. Mark the radius of the circle on these diagonals (Fig. 8.12C). You have now defined eight points that will be on the circle. Carefully sketch each quadrant of the circle by connecting the points you have defined. Darken the circle so that it can be seen easily. By making your construction lines extremely light, there is no need to erase them. The finished circle is shown in Figure 8.12D. Try drawing this sketch yourself.

When creating a sketch, keep size and proportions in mind. We do not measure any distances on the sketch, but we want the width, depth, and height to remain in their proper proportions. When sketching, it may be rather difficult to keep the line thicknesses precise; however, your visible lines should be thick and your hidden lines and centerlines should be thin, as discussed earlier in the chapter. A common error when learning to make sketches is to draw them very small. To be effective, a sketch should be at least the size of an index card, and should take up at least half a sheet of paper.

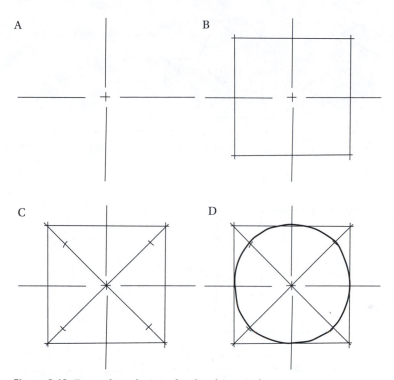

Figure 8.12 Example technique for sketching circles

8.8 **Pictorial Sketching**

In previous sections of this chapter we learned that orthographic drawings were composed of multiple views where each view represented two dimensions. Pictorial sketching involves creating a view of the object in which all three dimensions are shown. Pictorials are used as a method to help us with visualization, which enables us to formulate in our mind's eye what the object looks like. Also, pictorials can be used to present your case for a design recommendation among various non-technical professionals. Pictorials are relatively easy to understand and can be extremely important in selling a design.

Although several types of pictorial sketches can be used, we will limit the discussion here to the two most commonly used by engineers: isometric and oblique. Examples of both types are shown in Figure 8.13.

We will create both an isometric pictorial and an oblique pictorial of the slotted block shown in Figure 8.14. We will create each pictorial in steps so that you can follow the process and be able to create other pictorials after you learn the steps. Isometric sketches are more commonly used, although oblique sketches are slightly easier to draw.

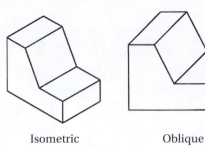

Isometric   Oblique

Figure 8.13 Examples of isometric and oblique pictorial sketches.

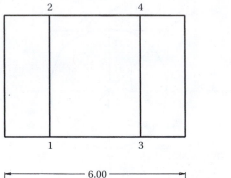

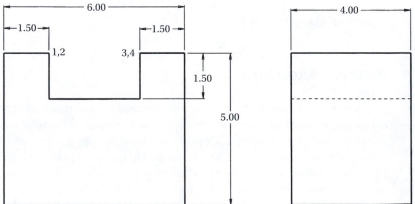

Figure 8.14 Orthographic views of the slotted block.

Isometric Pictorials

The most common orientation for the isometric sketch is described when the observer would be looking down on the object (bird's-eye view). In this orientation, the three isometric axes are shown in Figure 8.15A. The axes are composed of a vertical and two receding axes that measure 30° from the horizontal. Using these axes, we would

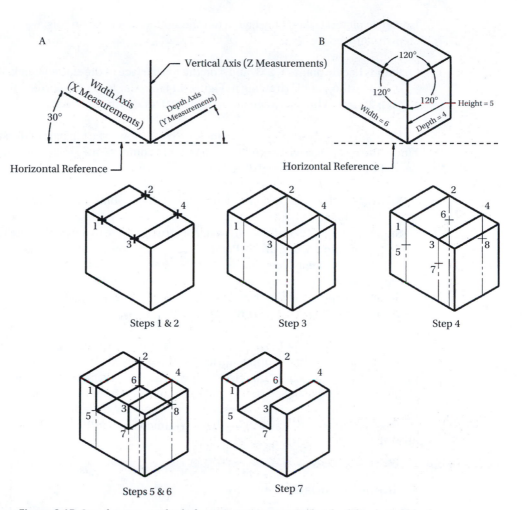

Figure 8.15 Step-by-step method of creating an isometric sketch of the slotted block.

consider the width (*x* measurements) along the receding axis that goes to the left, the depth (*y* measurements) along the receding axis that goes to the right, and the height (*z* measurements) on the vertical axis. We can box in our slotted block by plotting its height (5), width (6), and depth (4) along each appropriate axis (Fig. 8.15B). Note that we have created an isometric prism that encompasses the entire object. This would be similar to having a block of wood (5 × 6 × 4) that we can cut in order to get the desired shape. Also note that the angles at the corner of the isometric prism are each 120°. These angles are equal, which is how the isometric pictorial got its name. The prefix "iso" means equal and the term "metric" means measure, so the word "isometric" means equal measure, and we have equal angles of 120°. We now can develop our isometric pictorial sketch by laying out the slotted block. To do this, we will follow a stepwise procedure.

Step 1. Locate points 1, 2, 3, and 4 on the top surface of the slotted block. Read the orthographic drawing in Figure 8.14 in order to get the correct distances.

Step 2. Draw a line between points 1 and 2. Draw a line between points 3 and 4.

By connecting these points with lines, we have shown the top part of the slot in the block. The next steps will be followed in order to complete the slot by measuring how deep the slot is in the vertical direction.

Step 3. Sketch a vertical construction line through point 1, point 2, point 3, and point 4.

Step 4. Measure 1.5 units down each construction line to determine the locations of points 5, 6, 7, and 8. Point 5 is directly below point 1, point 6 is directly below point 2, etc.

Step 5. As in step 2, draw a line to connect points 5 and 6.

Step 6. Complete the slot by connecting 1–5, 5–6, 6–2, and 6–8.

Step 7. Erase lines 1–3, 2–4, and part of 6–8.

Note that you do not show line 7–8. This line would be hidden because it is blocked from view by the upper surface of the block. With pictorials, we are interested in what we can see. The hidden line in this case is not that important in the shape description of the slotted block, so it is omitted from the sketch. Hidden lines are omitted from pictorials unless they are necessary to define the exact shape of the object. As a result, line 6–8 has been trimmed so that the hidden portion is not shown.

Oblique Pictorials

While isometric pictorials have two receding axes 30° from the horizontal reference line, oblique pictorials have only one receding axis that shows depth. To draw an oblique pictorial, we can simply show the width and height as it appears in the front view of the orthographic drawing. Then the receding axis can be sketched in at any angle desired; however, 45° is the most common. Again, we will show the oblique pictorial construction in a stepwise manner (Fig. 8.16).

Step 1. Create the front view of the slotted block. Draw construction lines that recede at 45°. See Step 1 in Figure 8.16.

Step 2. Along each receding depth line, lay off the depth of the slotted block. See Step 2 in Figure 8.16.

Step 3. Connect the appropriate points to show the completed slotted block. See Step 3 in Figure 8.16.

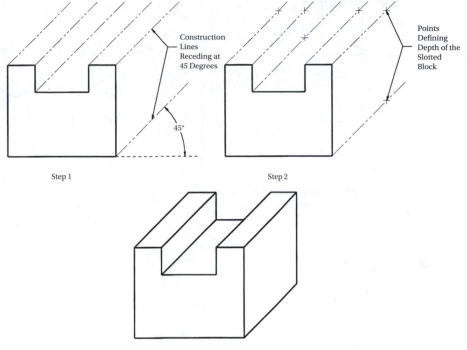

Figure 8.16 Step-by-step method of creating an oblique sketch of the slotted block.

Once again, hidden lines are not shown unless they are necessary to describe the shape of the object. In this case, hidden lines are not necessary, so they are omitted.

Adding Complexity

Three-dimensional pictorials can be constructed rather easily if you plot the three-dimensional coordinates and then connect the points. No matter how complex the object may be, if you can locate each point and connect the points to define the surfaces, you will be successful in creating a pictorial of the object.

Suppose we have an object that is a little more complex than an ordinary slotted block, like what is shown as an orthographic drawing in Figure 8.17. In this object, we not only have a slot, but we also have an inclined surface. To make a pictorial sketch of this object, plot points using three-dimensional coordinates and then connect the appropriate points. As before, start by developing the isometric prism that encompasses the object (see Fig. 8.17). Essentially, follow the same steps as before to create the slotted block. The major difference in this case is that you will have to account for the inclined surface. To do this, locate point A on the bottom part of the slot (see Fig. 8.17).

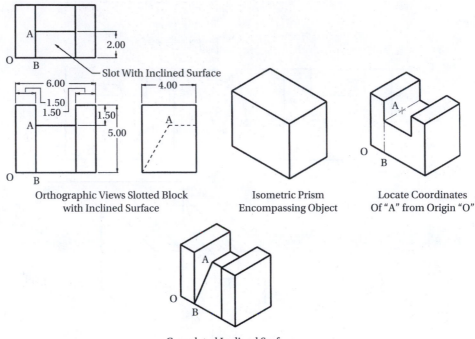

Orthographic Views Slotted Block
with Inclined Surface

Isometric Prism
Encompassing Object

Locate Coordinates
Of "A" from Origin "O"

Completed Inclined Surface

Figure 8.17 Creating a sketch of a slotted block with an inclined surface.

This is accomplished by plotting the three-dimensional coordinates of point A. If we assume point O is the origin of our three-dimensional space, then point A is located 1.5 units down the isometric axis from O, and point B is 3.5 units up a vertical construction line and 2 units to the right along the depth construction line. From point A draw a line to point B, and you have installed your inclined surface. Erase the lines that are not needed to define the shape, and you've completed the isometric pictorial of the object.

Creating orthographic and isometric sketches is made easier by using paper printed specifically for that purpose (Fig. 8.18). Such paper can be purchased at office supply stores, and free printable downloads are also readily available from many online sources.

Cylindrical features shown in isometric sketches have a level of complexity that needs to be addressed. Essentially, we are talking about positive cylinders and negative cylinders or holes. In orthographic views, cylinders appear as circles and are relatively easy to draw using the boxing-in technique. In isometric pictorials, circles will always appear as ellipses (Fig. 8.19). The orientation of the ellipse depends on which surface (front, top, or side) the cylinder appears.

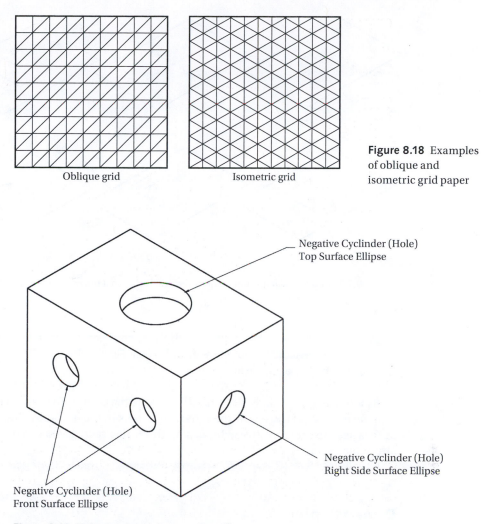

Oblique grid   Isometric grid

Figure 8.18 Examples of oblique and isometric grid paper

Negative Cyclinder (Hole)
Top Surface Ellipse

Negative Cyclinder (Hole)
Right Side Surface Ellipse

Negative Cyclinder (Hole)
Front Surface Ellipse

Figure 8.19 Holes in pictorials appear as ellipses.

To construct an ellipse on the top isometric surface, follow these steps:

Step 1. Sketch in the center lines on the top surface plane (see Fig. 8.20).
Step 2. Plot points on the centerlines that represent the radius of the circle (see Fig. 8.20).
Step 3. Box in the circle by drawing an isometric box through the points created in Step 2 (see Fig. 8.20).
Step 4. Sketch the long radii of the ellipse by completing arcs connecting points A and B and points C and D (see Fig. 8.20).

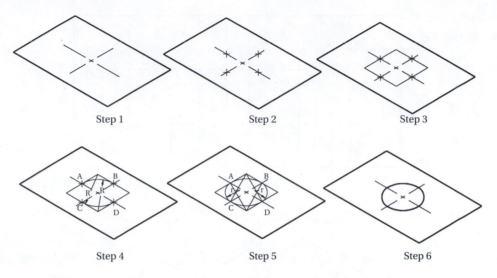

Figure 8.20 Step-by-step method of creating an ellipse on a horizontal surface.

Step 5. Sketch the short radii of the ellipse by completing arcs connecting
points A and C and points B and D (see Fig. 8.20).
Step 6. Finish the top surface ellipse (see Fig. 8.20).

By following these six steps, you will complete the ellipse. To draw an ellipse in the right face or front face, follow the same set of steps in those planes. The same procedure is used to draw positive cylinders. With a little practice you can draw ellipses in any plane.

8.9 Dimensioning

Drawings and sketches are of little use without dimensions. It is beyond the scope of this book to cover all of the dimensioning and tolerancing used by engineers, but we can give you some basic tools. Figure 8.21 shows an orthographic projection of a mounting bracket with dimensions in inches.

To indicate a dimension on a drawing, light lines called **extension** lines are used to extend the feature. They are spaced slightly away from the feature so as not to be confused with an edge line. Arrows pointing to the extension lines called leaders point to the extension lines, and the numerical value of the dimension is placed between them, usually in the center.

When dimensioning a drawing, the outer dimensions must be indicated so that the overall size of the part is specified. Other features must also be indicated so that the size and placement of each is understandable. However, only sufficient

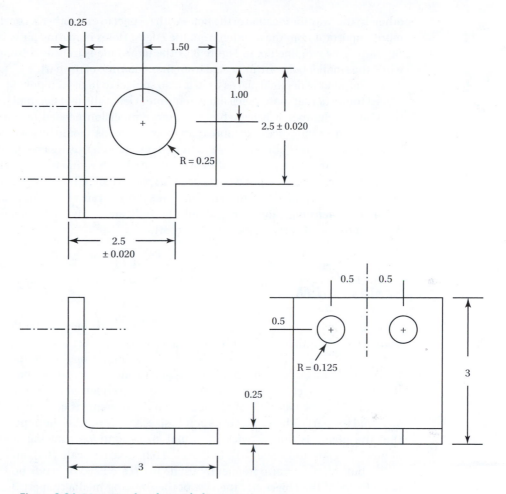

Figure 8.21 Mounting bracket with dimensions

information to specify the feature should be used. If more dimensions than necessary are used, the drawing is said to be over-dimensioned, which can be confusing. As seen in Figure 8.21, the overall size of the part is shown to be 3 inches and a notch is indicated with dimensions in the top view, with only one leg on each side dimensioned. If both the 2.5- and 0.5-inch dimensions were indicated, the part would be over-dimensioned.

Features can be indicated from either edges or centerlines. How they are indicated on the drawing shows which dimensions are to be interpreted as the most important. If a feature is dimensioned from the edge, that dimension should be considered to be important. Holes are always indicated to their centers, not their edges. A small + is usually placed in the center of the hole, with extension lines indicating its location. Holes such as those that are to fit a pattern should be indicated from centerlines and/or each other. In this way the location of the holes with respect to each other is considered to be more important than the location from the edges. Holes in mating parts indicated in the same way will line up. In Figure 8.21 the large hole is indicated from the edges, while the smaller holes are indicated from the centerline of the part.

The number of decimal places used is also subject to interpretation. More decimal places indicate that more precision is required. Generally one decimal place means the widest tolerance is acceptable, with more precision needed for an increasing number of decimal places. No more than three decimals should be used on drawings using imperial units, and no more than two on drawings using metric units. When specific tolerances are required, they are indicated with ± dimensions. In Figure 8.21, dimensions with one, two, and three decimal places can be found, indicating the relative importance of those dimensions. Specific tolerances are indicated on the 2.5-inch dimensions. Tolerances should be indicated only where needed to keep manufacturing of parts simpler and less costly.

8.10   Scales and Measuring

Most technical drawings are drawn to some scale. This means that there is a ratio between the length of the line on the drawing representing the object and the length of the same line on the real-world object. A 1:1 (one-to-one) scale means that one linear unit of measure on the drawing is equal to one linear unit on the real object. In that case the object is drawn to its actual size, often referred to as "full scale." For a scale of 1:2, known as half scale, one unit on the drawing is equal to half of the full size, and the part is shown half of its actual size. A scale of 2:1 is the opposite, meaning that the object is drawn twice its actual size. Common drawing scales used in engineering are full scale (1:1), half scale (1:2), quarter scale (1:4), and tenth scale (1:10), but others are also frequently used. The scale used depends on the size and complexity of the object and the size of the drawing media (paper). The larger the object being shown on the drawing, the smaller the drawing scale has to be on the sheet. Larger paper allows for more detail in more complex objects. Drawing media come in various sizes that are standardized. Other sizes are also used, particularly in architectural drawing, but for engineering drawings, the drawing sizes summarized in Table 8.1 are the most commonly used. Today's drawings are rarely made on large-size paper; most drafting is done today in CAD, where the drawing is made in 1:1 scale regardless of the size of the object, and then plotted on a paper medium to a scale that makes the drawing easy to interpret.

Drawing to scale is best done using special measurement tools that are made for that purpose. While a standard imperial unit ruler is divided into 1/16-inch increments, an engineer's scale is divided into 1/10-inch increments. This makes scaling a drawing simply a matter of using the right increments. A comparison can be seen in Figure 8.22.

To measure a distance using the 10 scale, start by placing the number 0 at one of the endpoints of the line you wish to measure. Read each of the major graduations (1, 2, 3 . . .)

Table 8.1 Standard Paper Sizes in ANSI and Metric Dimensions

ANSI Standard Paper		ISO Standard Paper	
ANSI Size	*Dimensions (in)*	*ISO Size*	*Dimensions (mm)*
A	8½ × 11	A4	210 × 297
B	11 × 17	A3	297 × 420
C	17 × 22	A2	420 × 594
D	22 × 34	A1	594 × 841
E	34 × 44	B1	707 × 1000
		A0	841 × 1189

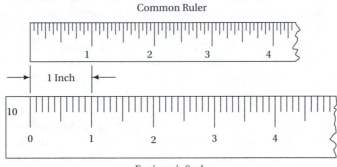

Figure 8.22 Comparison of a common ruler and an engineer's scale

across the scale as full inches and add the number of minor graduations as tenths of an inch. For example, the line in Figure 8.23 measures 3.4 inches. Using the 10 scale of the engineer's scale makes measuring in multiples of 10 very simple. The measurement on the scale can be multiplied by 10, 100, or 1,000. For example, if your drawing were drawn to a scale of 1:100, then the numbers along the scale would read 100, 200, 300, etc., with the minor graduations reading as 10, 20, 30, . . . 90. The line shown in Figure 8.23 would measure 340 inches using a drawing scale of 1:100.

The engineer's scale also has scales other than 10, often including 20, 30, and 50, each of which can be used for scaling drawings up or down to conveniently fit the desired medium size and detail level. The engineer's scale makes measurements in decimal inches. The American National Standards Institute (ANSI) has designated that mechanical parts shall be documented using decimal inches; however, fractional inches are frequently used in actual practice, so engineers should be familiar with them.

In the Système International (SI) measurement system, millimeters are the base unit of measure. ANSI has designated that mechanical parts using metric measurements

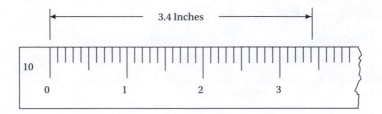

Figure 8.23 Measuring the distance between two points using the 10 scale on an engineer's scale.

must be designated in millimeters. There is good reason for this: if a dimension on a part happens to be 50 millimeters and it is written as just 50 (millimeters or inches is not specified with the dimension on a drawing), then it cannot be interpreted as 50 inches. On the other hand, if the metric unit of 5 centimeters were used instead of 50 millimeters, then the number 5 could easily be misinterpreted as 5 inches. To avoid this confusion, ANSI has designated that mechanical parts defined in SI units shall be drawn using millimeters. For drawings that deal with large areas such as topographic maps, it is permissible to use meters or kilometers.

Since a number of drawings throughout the world are drawn using a metric measure, you should have some practice in measuring using a metric scale that has millimeters as its smallest unit. The 1:1 metric scale is the starting point for learning to measure using a metric scale. The 1:1 scale is divided into 1-mm graduations and is calibrated at 10-mm intervals, as shown in Figure 8.24. This is an example of a full-scale measurement using the metric scale.

Measuring distances on the metric scale is very similar to the method used to measure distances using the engineer's scale. Start by lining up the 0 at the left end of the 1:1 scale to one endpoint of the distance you want to measure. Read the numbers of the major graduations along the scale (10, 20, 30, . . .) as millimeters and add the number of minor graduations directly to the first number. The line in Figure 8.24 measures 37 mm. Like the engineer's scale, the 1:1 scale can be used to measure other multiples of 10, such as 1:10, 1:100, 1:1,000, etc. Simply multiply each of the graduations by 10, 100, 1,000, etc., as you measure them. This means that on a drawing scaled to 1:100, you would read 1,000, 2,000, 3,000, etc., along the scale.

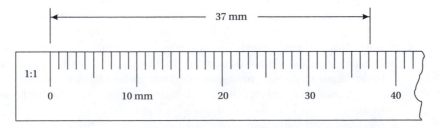

Figure 8.24 Measuring the distance between two points using the 1:1 scale on a metric scale.

8.11 **Coordinate Systems and Three-Dimensional Space**

CAD systems make heavy use of coordinate systems, usually Cartesian, for two- and three-dimensional drawing. This makes sense, because computers can use coordinate positions down to 16 decimal places to plot points, making CAD drawings highly accurate. A typical three-dimensional coordinate system used in mathematics is shown in Figure 8.25. Such a system of coordinates follows the right-hand rule, in which fingers can curl from the x-axis to the y-axis, with the thumb pointing in the direction of the z-axis. In this configuration the x-y and the y-z planes are vertical, with the x-z plane horizontal.

While this coordinate system is standard in mathematics, in computer graphics it is done a little differently. In computer graphics, we want the vertical direction to indicate the z-axis instead of the y-axis, so the axis system is set up differently. In the three-dimensional space defined for a CAD system, the x-y plane is horizontal and the x-z plane is vertical. It still follows the **right-hand rule**, but the axes have been rotated so that the z-axis is vertical.

There is good reason why CAD systems use this type of rectangular coordinate system. Industry uses lots of different computer numerically controlled machines (CNC machines) to manufacture its products. Long ago, it was established that the vertical direction with regard to computer manufacturing would be standardized as the z direction. Therefore, as CAD began to develop, it was natural to build in the same coordinate system, since often the geometry designed in a CAD system can be directly loaded into a CNC machine for manufacturing. Therefore, this type of coordinate system is commonly used when defining points in three-dimensional space for graphical solutions of problems.

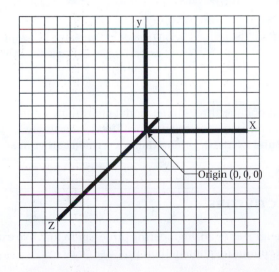

Figure 8.25 Axes for a three-dimensional rectangular coordinate system

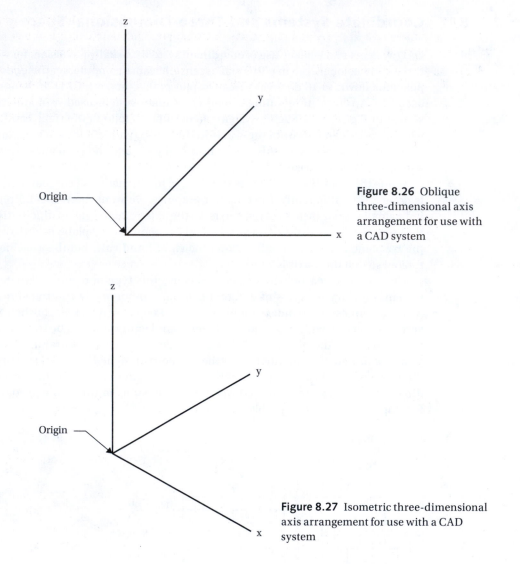

Figure 8.26 Oblique three-dimensional axis arrangement for use with a CAD system

Figure 8.27 Isometric three-dimensional axis arrangement for use with a CAD system

Figure 8.26 shows an oblique three-dimensional coordinate system orientation, while Figure 8.27 shows an isometric orientation. These coordinate systems correspond with the oblique and isometric orientations described earlier in Section 8.8.

EXERCISES AND ACTIVITIES

8.1   Interview a practicing engineer or an engineering professor to find five examples of how they use graphical communication in their professional lives.

8.2 Keep a sketching journal to practice your sketching technique. This idea is similar to writers who keep written journals to document their ideas in words. It is a good habit for an engineer to document ideas using sketches in order to maintain a record of when and how the idea was developed—this is especially critical in the patent application process. Suggestions for your journal include the following: use a variety of projection techniques (orthographic, oblique, isometric); use either a bound notebook or loose sheets of rectangular grid paper contained in a binder; sketch familiar objects that are in your room, apartment, or house for practice.

8.3 Practice sketching straight lines between points given by your instructor on a separate sheet of rectangular grid paper.

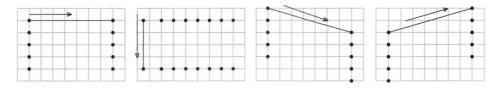

8.4 Practice sketching circles of variable radius on a separate sheet of rectangular grid paper.

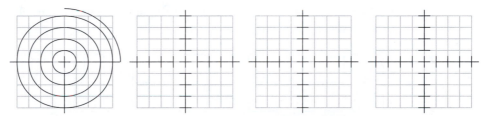

8.5 Practice sketching arcs of variable radius on a separate sheet of rectangular grid paper.

8.6 Practice sketching ellipses of variable radius on a separate sheet of rectangular grid paper.

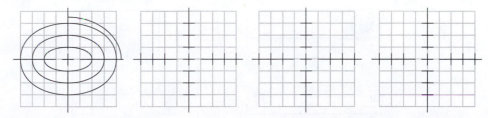

8.7   Using the scales given below, draw lines of the specified length on a separate sheet of paper.

Drawing Scale	Line Length	Drawing Scale	Line Length
1″ = 10′	8.0′	1:1	15 cm
1″ = 10′	6.3′	1:100	5 m
1″ = 2.0′	8.4′	1:10	85 mm
1″ = 20′	54′	1:2	50 mm
1″ = .01′	.035′	1:20	750 mm
1″ = 300′	750′	1:200	14 m
1″ = 1000′	3250′	1:2000	.500 km
1″ = 50′	245′	1:50	4350 mm

8.8   Sketch oblique and isometric drawings of the objects shown in the orthographic drawings below. Use the appropriate type of paper.

8.9   Sketch orthographic drawings of the objects shown in the isometric drawings below. Use an appropriate drawing scale as specified by your instructor.

8.10   Practice sketching isometric cylinders on the isometric grid given below or on a separate sheet of isometric grid paper.

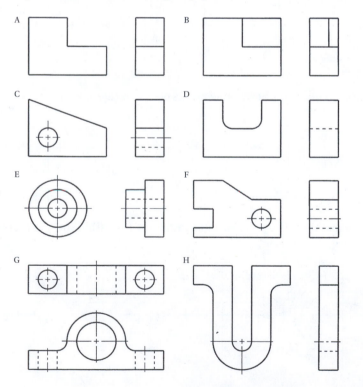

8.11 List the three things that a line can represent on a drawing, and sketch an example of each.

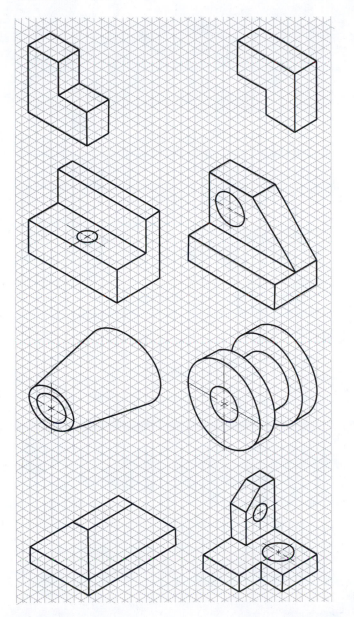

8.12   Select an object from your daily life and draw orthographic and 3D views.

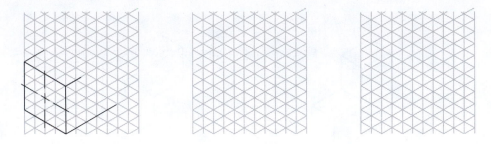

8.13   Draw orthographic and 3D views of an object assigned by your instructor.

CHAPTER 9

Computer Tools for Engineers

9.1 Introduction

Assume that you are a food process engineer, and a dairy company has hired you to design a process to manufacture high-quality powdered milk for a profit. This is no small undertaking! Table 9.1 shows a list of the steps that you must complete.

With this huge project ahead of you, you realize that you need help, so you turn to the computer. Computer tools can facilitate nearly every task you need to perform. These tools include operating systems, programming languages, and computer applications. For example, word-processing applications enable computer users to write papers. To be able to use the computer to solve an engineering problem, you need strong computer skills and a working familiarity with the types of computer tools and applications that are available. Table 9.1 indicates the types of computer tools you would use to complete this project.

The following sections of this chapter will highlight some of the computer tools that an engineer might use to solve problems. Many of the tools will be discussed in the context of solving the dairy design problem.

Table 9.1 Steps Required to Complete the Powdered-Milk Process Design

Step	Computer Tool
Research the dairy industry	Internet
Design the process	Mathematical software
	Design software
	Programming languages
Select equipment	Internet
Perform economic analyses	Spreadsheet
	Mathematical software
Analyze experimental data	Spreadsheet
	Mathematical software
Document the design process	Word processor
Present findings to management	Presentation software

9.2   The Internet

Introduction

The first step required to solve the dairy design problem is to fill gaps in your knowledge base. That is, you need to do research so that you have a better understanding of the design problem. Some things you should learn are:

- How is powdered milk manufactured?
- What are the consumer safety and health issues involved?
- Who manufactures equipment for dairy processing?
- What other dairy companies manufacture powdered milk?
- Who does dairy processing research?

Even under the best conditions, research can take a significant amount of time. So your aim is to avoid excessive trips to the library and the use of bound indexes. It is preferable to access as much information as possible via your computer. The Internet provides such capabilities.

History of the Internet

The Internet is a worldwide network of computers that allows electronic communication with each other by sending data packets back and forth. The Internet has been under development since the 1960s. At that time, the Department of Defense was interested in testing out a concept that would connect government computers. It was hoped that such a network would improve military and defense capabilities (Glass and Ables 2003). Before the end of 1969, machines at the University of Southern California, the Stanford Research Institute, the University of California at Santa Barbara, and the University of Utah were connected.

The linking of those four original sites generated much interest in the Internet. However, extensive customization was required to get those original sites linked, because each site used its own operating system. It was apparent that the addition of more sites would require a set of standard protocols that would allow data sharing between different computers. And so the 1970s saw the birth of such protocols.

Protocols are used to move data from one location to another across the network and to verify that the data transfer was successful. Data are sent along the Internet in **packets**. Each packet may take a different route to get to its final destination. The **Internet Protocol** (IP) guarantees that if a packet arrives at its destination, the packet is identical to the one originally sent. A common companion protocol to the IP is the **Transmission Control Protocol** (TCP). A sending TCP and a receiving TCP ensure that all the packets that make up a data transfer arrive at the receiving location.

Internet applications developed in the 1970s are still used today. The Telnet program connects one computer to another through the network. Telnet allows a user on a local computer to remotely log into and use a second computer at another location.

The **ftp** (file transfer protocol) program allows users to transfer files on the network. Both applications were originally non-graphical; they require that text commands be entered at a prompt.

Internet IP addresses were established in the 1970s. All organizations that become part of the Internet are assigned a unique 32-bit address by the Network Information Center. IP addresses are a series of four eight-bit numbers. For instance, IP addresses at Purdue University take the form 128.46.XXX.YYY, where XXX and YYY are numbers between zero and 255.

In the 1980s, corporations began connecting to the Internet. Figure 9.1 shows the growing number of sites connected to the Internet. With so many sites joining, a new naming system was needed. In came the Domain Naming Service. This hierarchical naming system allowed easier naming of individual hosts (servers) at a particular site. Each organization connected to the Internet belongs to one of the top-level domains (Table 9.2). Purdue University falls under the U.S. educational category; therefore, it has the domain name purdue.edu. Once an organization has both an IP address and a domain name, it can assign addresses to individual hosts. This author's host name is pasture.ecn.purdue.edu. Pasture is a large computer (server) on the Purdue University Engineering Computer Network (ecn), a network of engineering department servers. With the explosion of websites, the number of allowable domain names has increased with new top-level names and extensions.

Year	Hosts
1981	213
1985	1,961
1990	313,000
1995	6,642,000
2000	93,047,000
2005	353,284,000
2010	768,913,036

Figure 9.1 Number of Internet hosts by year
Source: Internet Systems Consortium.

Table 9.2 Example Top-Level Domain Names

Name	Category
edu	U.S. educational
gov	U.S. government
com	U.S. commercial
mil	U.S. military
org	Nonprofit organization
net	Network
XX	Two-letter country code
(e.g., fr, uk)	(e.g., France, United Kingdom)

As Internet Service Providers (ISP), such as America On-Line and CompuServe, came into being in the 1990s, the Internet became more mainstream. The development of the browser named Mosaic made accessing information on the Internet easier for the general public. A *browser* is software that allows you to access and navigate the Internet. Mosaic displays information (data) using both text and graphics. Each page of information is written using the **HyperText Markup Language** (HTML). The most useful aspect of HTML is the hyperlink. A **hyperlink** connects two pages of information. The beauty of the hyperlink is that it allows an Internet user to move through information in a non-linear fashion. That is, the user can surf the World Wide Web (WWW). The Web, not to be confused with the Internet, refers to the interconnection of HTML pages. Some of the most common browsers are Google's Chrome, Microsoft Internet Explorer, and Mozilla Firefox. Oother browsers have been introduced for mobile devices such as Apple's Safari. Updates to these browsers have allowed for the construction of more complicated web pages, such as multiple-frame pages. Sound and video can now be played through the use of helper applications. Greater user interaction with the Web has been made possible through HTML coded references to Java applets (defined later).

Searching the Web

Let's return to the powdered-milk problem. You'd like to use the Internet to answer some of your research questions. First focus on searching for equipment vendors. Assume that you already know about one company that supplies food processing equipment, called "APV." Check to see what kinds of equipment they have available for dairy processing. Using your Web browser, you access APV's website by providing the **Universal Resource Locator** (URL) for APV's home page. In other words, you tell the Web browser what Web address to use to access APV's information. Assume that the URL is "http://www.apv .com." This URL consists of the two primary components of a Web address: the **protocol** (means for transferring information) and the Internet address. The protocol being used here is http (**HyperText Transport Protocol**), which allows access to HTML. The Internet address consists of the **host name** (apv) and the **domain** (com).

APV's URL takes you to APV's introductory (home) page. Once there, you can browse their website for information that pertains to your specific research problem. At one point, you find yourself on a page that describes a one-stage spray dryer designed specifically for temperature-sensitive materials such as milk. A spray dryer is a piece of equipment that converts liquid into a powder, or as in your application, milk to powdered milk. The URL for this page is http://www.apv.com/anhydro/spray/onestage .htm. This Web address is longer than that for APV's home page. The address contains a **path name** (/anhydro/spray/) and a file name (onestage.htm). The path name tells you the location within APV's directory structure of the file that you are accessing with your Web browser, just as you store files in directories on your hard drive.

APV is just one equipment vendor. You'd like to investigate others, but you do not know any other Web addresses. To further your research, you use a Web search engine. A **search engine** matches user-specified keywords with entries in the search engine's database. The engine then provides a list of links to the search results. There are a number of search engines such as Yahoo! (www.yahoo.com) and Google (www. google.com). A keyword search performed using one search engine will yield different results than the same keyword search on another search engine. The reason for the different results lies in the type of information that a particular search engine tends to have in its database.

You decide to use the search engine Google to do a keyword search for "spray dryer." When your search results come up, you immediately notice that your query was not specific enough. You get back links to sites on laundry dryers! There are so many sites on the Internet that well-constructed queries are a must. Each search engine provides links to searching tips. In your case, you find that a keyword search for "spray dryer and food" improves your search results.

Home Pages

Home pages are web pages developed by users that contain information about topics of interest to the user, and often include links to the user's favorite websites. The language of all web pages is HTML. An **HTML document** is a text file containing information the user wants posted to the Web, and instructions on how that information should look when the web page is accessed by a browser. HTML code can be written from scratch using a text editor, or a composer can be used to generate the code. Many word processors and desktop publishers now have the capability of saving files in HTML format so that the document is displayed by the web browser exactly as it was in the word-processing program. The HTML code consists of instructional tags as seen in Figure 9.2. This code generates the Web page shown in Figure 9.3.

Your employer may ask you to design a web page discussing the process you developed through the completion of the powdered-milk process design project. A personal reason for creating your own home page is to make available your résumé and samples of your work to all Web users. This sort of visibility may result in employment opportunities such as internships and post-graduation positions.

More information on creating home pages can be found on the Web or in any book on HTML or the Internet. You may find *HTML: The Complete Reference* by T. A. Powell to be helpful when writing HTML code.

Libraries and Databases on the Internet

Not all information is available through the Internet, but an ever-increasing amount is. New search features are coming online each year and databases are continually emerging and growing. The Internet can be used to locate many sources for research through search features such as Google Scholar (scholar.google.com). Information that is not available through the Web can be located using the Internet, such as through a website of a university's library. Most have sophisticated search tools that can be used to access articles in electronic form or to locate hard copies.

```
<HTML>
<HEAD>
<TITLE>FOODS BLOCK LIBRARY – BASIC ABSTRACT</TITLE>
</HEAD>

<BODY BGCOLOR="#FFFFFF">
<H1>A Computer-Aided Food Process Design Tool for
Industry, Research, & Education</H1>
<H2>Heidi A. Diefes, Ph.D.</H2>
<H2><I>Presented at the Food Science Graduate Seminar<BR>
Cornell University, Fall 1998</I></H2>
<HR ALIGN=LEFT NOSHADE SIZE=5 WIDTH=50%>
<P>
```
There is a great need in the food industry to link food science research with food engineering and process technologies in such a way as to facilitate an increase in product development efficiency and product quality. Chemical engineers achieve this through industry standard computer-aided flowsheeting and design packages. In contrast, existing programs for food processing applications are limited in their ability to handle the wide variety of processes common to the food industry. This research entails the development of a generalized flowsheeting and design program for steady-state food processes which utilizes the design strategies employed by food engineers.
```
</P>
</BODY>
</HTML>
```

Figure 9.2 HTML code for web page shown in Figure 9.3

Contacting People Through the Internet via E-mail

The Internet can also be used to send electronic mail (e-mail) to individuals, groups, or mailing lists. E-mail is currently exchanged via the Internet using the **Simple Mail Transport Protocol** (SMTP) and **Multipurpose Internet Mail Extension** (MIME). To send e-mail, two things are required: an e-mail package and e-mail addresses. There are many e-mail packages from which to choose.

Regardless of the brand, e-mail tools have certain common functions including an ability to send mail, receive mail, reply to mail, forward mail, save mail to a file, and delete mail.

To communicate via the Internet, an e-mail address is required. All e-mail users are assigned a unique e-mail address that takes the form username@domain.name. However, it can be difficult to locate a person's e-mail address. One method of locating an e-mail address is to use one of the many search engines In addition, most colleges and universities have methods for searching for staff and students on the school's website.

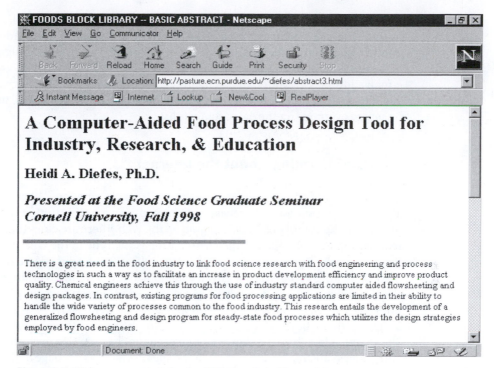

Figure 9.3 Web page generated using HTML code in Figure 9.2

Once you have selected an e-mail address and e-mail software, and after you have signed up with an ISP, you can send and receive e-mail. As part of your dairy design problem, you may keep in touch with the dairy company contacts and other engineering consultants via e-mail. A typical e-mail message might read like the example that follows.

To: *dairyconsult@purdue.edu*
cc: *dairycontacts@purdue.edu*
Attachments: *spdry.bmp*
Subject: *Spray dryer design model*
Message:
I have been working with the spray dryer vendor on a model of the system we intend to use. Attached is a current picture of the model. Please advise.
H. A. Diefes

The above e-mail message contains both the typical required and optional components. An e-mail is sent:

To: username@domain.name or an alias (explained in the next paragraph). A copy (cc:) may also be sent to a username@domain.name or an alias.

Attachments: are files to be appended to the main e-mail message.

Subject: is a brief description of the message contents.

Message: is the actual text of the message.

An **alias** is a nickname for an individual or group of individuals, typically selected by the owner. An e-mail to an alias gets directed to the actual address. Address books, another common feature of e-mail packages, allow the user to assign and store addresses and aliases. For example, in your address book you may group all the e-mail addresses for your dairy company contacts under the alias "dairy contacts." If you put "dairy contacts" in the To or cc line of your mail tool, the e-mail message will be sent to everyone on the dairy contacts list.

Words of Warning About the Internet

With millions of people having access to the Internet, one must be very discerning about the information one accesses. Common sense and intuition are needed to determine the validity of material found on the Web. There are clues that you can use to ascertain the reliability of the website. One is the type of the URL domain name. Pages that have the extensions such as .edu, indicating an educational institution, or .gov, indicating some kind of governmental agency, should have more credibility than a random .com or .net site. Be aware that an extension such as .edu in the URL does not guarantee that the institution sponsors that webpage. A "~" symbol is used to designate that a particular user is in control of the page rather than the institution. The example in Figure 9.3 showed a web page that has the URL pasture.ecn.purdue. edu\~diefes\ abstract3.html. The "~diefes" means that this is a page under the control of the user "diefes" and not the host institution.

Other clues to the trustworthiness of a website include the information on the web page. Is there an author cited on the page? Is that author credible? What is the creation date of the site? Are there references on the site? The Internet has made information readily available but it has also made it easy to post information that is not credible, which places a responsibility on you as the researcher to verify information as credible.

In addition, care must be taken not to plagiarize materials found on the Web. As with hard-copy materials, always cite your sources.

Further, computer viruses can be spread by executable files (.exe) that you have copied. The virus is unleashed when you run the program. Text files, e-mail, and other non-executable files cannot spread such viruses. Guard against viruses attached to executable files by downloading executable files only from sites that you trust. Antivirus programs such as those from Norton or McAfee can be installed to help detect and protect against viruses.

Social media, such as Facebook and MySpace, and other websites such as You-Tube, have rapidly expanded how information is transferred by text, pictures, and video. Many students use these tools to share information and update others on what they are doing or have done. We recommend caution when using the Internet to share information. Once released, you lose control over that content. Once something is on the Internet, it may be there for a very long time. Employers, scholarship committees, and graduate schools can and will do Internet searches on candidates, and there are many stories of students who posted something they thought was funny to their friends but ended up being viewed as evidence of their professionalism—rather, their lack thereof. The Internet is a powerful tool that has fundamentally changed the way we live and communicate, but like any tool it needs to be used with caution.

9.3 Word-Processing Programs

Throughout the powdered-milk process design project, you will need to do various types of writing. The largest document will be the technical report for the design. You also may have to complete a non-technical report for management. In addition, you may write memorandums and official letters, and you will need to document experimental results. The computer tool that will enable you to write all these papers is a **word processor**, such as Microsoft Word.

The most basic function of a word processor is to create and edit text. When word processors first became available in the early 1980s, that was about the extent of their ability. The user could enter, delete, copy, and move text. Text style was limited to bold, italic, and underline, and text formats were few in number.

"Word processing," as understood today, is really a misnomer, since the capabilities of word processors have become indistinguishable from what was once considered desktop publishing. Word processors such as Microsoft Word, have features that are particularly useful to engineers. There are also a growing number of open source or free software packages that are compatible with other word processors including Google Docs. Many of these also have apps for mobile devices. Equation, drawing, and table editors are features that make the mechanics of technical writing relatively easy. Most programs include spelling and grammar checkers and an online thesaurus to help eliminate common writing errors.

Figure 9.4 shows the **graphical user interface** (GUI) for Microsoft Word. All word processors display essentially the same components. The title bar indicates the application name and the name of the file that is currently open. The menu bar provides access to approximately 10 categories of commands such as file, edit, and help commands. Some word processors have a toolbar that gives direct access to commands that are frequently used such as "save," "bold," and "italic." A typical word processor GUI also has scroll bars and a status bar. The scroll bars allow the user to move quickly through the document. The status bar provides information on current location and mode of operation. The workspace is where the document is actually typed. The workspace in Figure 9.4 shows a snippet of the technical information related to the spray dryer of our example.

This textbook you are reading was written using word processors. Throughout these pages you can find examples of tables, equations, and figures created using a word processor. A wide variety of typefaces, paragraph styles, and formats are also in evidence. You also will see images and pictures that have been imported into this document. A user may import materials that are created using other computer applications as well.

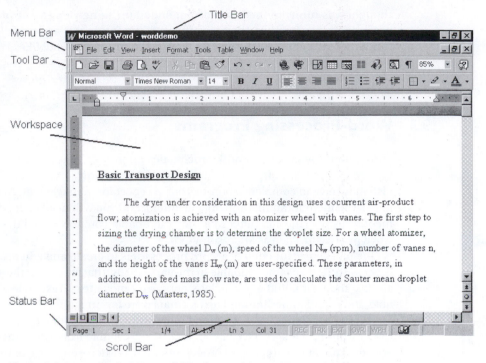

Figure 9.4 Example of a Microsoft Word paper in progress

Documenting is an ongoing process. You will need to return time and again to your word processor to record all subsequent steps of the dryer design process. Periodically, you will use the word processor to write memos and official letters and to prepare experimental reports, technical documents, and non-technical papers.

9.4 Spreadsheets

Introduction

Much data collection is performed during the research and design phases of an engineering project. Data come from a variety of sources including experiments, design calculations, product specifications, and product statistics. While designing the powdered-milk process, you have gathered cost information from your equipment vendors and collected experimental data on powder solubility. You may want to use the cost information in an economic analysis of your process (recall that your process is supposed to make a profit). You may also attempt to analyze your laboratory data. Due to the tabular nature of these data sets, this sort of information is best recorded with a spreadsheet.

Spreadsheet packages have their origins in finance. The idea for the first spreadsheet application, called VisiCalc, came in 1978 from a Harvard Business School student who was frustrated by the repetitive nature of solving financial-planning problems (Etter 1995). VisiCalc was originally written for Apple II computers. IBM came out with its own version of VisiCalc in 1981. In 1983, a second spreadsheet package was released by Lotus Development Corporation, called Lotus 1-2-3. Now there are a number of other spreadsheet packages from which to choose, including Microsoft Excel, and free software and mobile apps including Google sheets. One thing to note about the evolution of the spreadsheet is the increase in functionality. VisiCalc centered on accounting math: adding, subtracting, multiplying, dividing, and percentages. Spreadsheets today have many, many functions.

Common Features

A spreadsheet or worksheet is a rectangular grid that may be composed of thousands of columns and rows. Typically, columns are labeled alphabetically (A, B, C, . . ., AA, AB, . . .) while rows are labeled numerically (1, 2, 3, . . .). At the intersection of a given column and row is a cell. Each cell has a column/row address. For example, the cell at the intersection of the first column and first row has the address A1. Each cell can behave like a word processor or calculator in that it can contain formattable text, numbers, formulas, and macros (programs). Cells can perform simple tasks independently of each other, or they can reference each other and work as a group to perform more difficult tasks.

Figure 9.5 shows the GUI for Microsoft Excel. Notice that many of the interface features are similar to those of a word processor, such as the title bar, menu bar, toolbar, scroll bars, and status bar. Many menu selections are similar as well, particularly the "File" and "Edit" menus. The standard features of all spreadsheets are the edit line and workspace. The edit line includes the name box and formula bar. The name box contains the active cell's address or user-specified name. Within the worksheet in Figure 9.5 is the beginning of the equipment costing information and economic analysis of our example.

Today's spreadsheets have many features in common, including formatting features, editing features, built-in functions, data manipulation capabilities, and graphing features. Let's look at each one of these areas individually.

As with word processing, the user has control over the format of the entire worksheet. Text written in a cell can be formatted. The user has control over the text's font, style, color, and justification. The user can also manipulate the cell width and height as well as border style and background color.

Editing features are also similar to a word processor. Individual cell contents can be cleared or edited. The contents of one or more cells can be cleared, copied, or moved. Excel even has a spell checker.

Spreadsheets have categories of built-in functions. At a minimum, spreadsheets have arithmetic functions, trig functions, and logic functions. Arithmetic functions include a function to sum the values in a set (range) of cells (SUM) and a function to

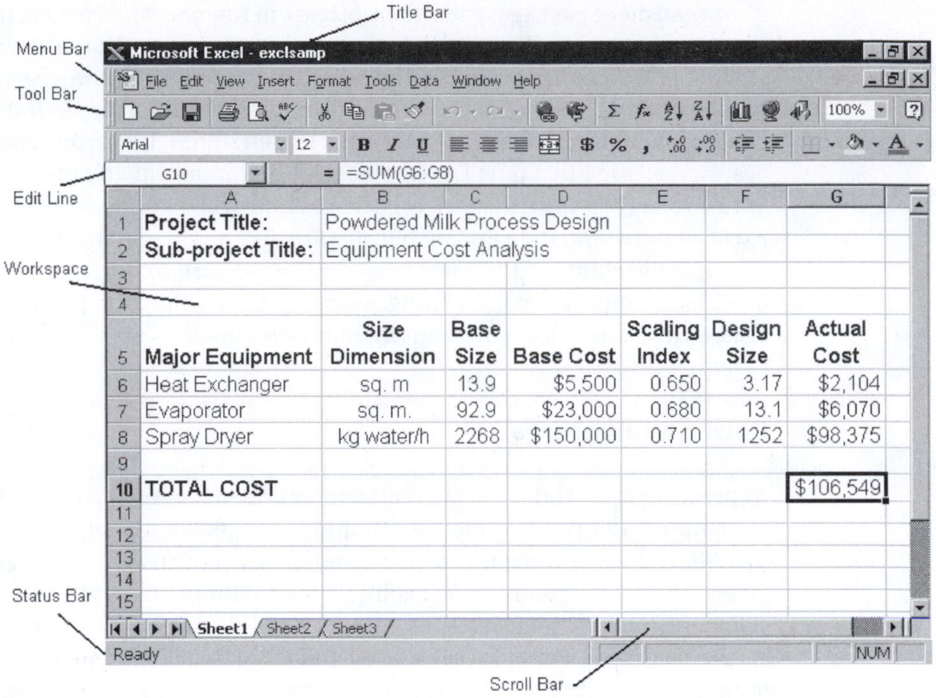

Figure 9.5 Example of a Microsoft Excel spreadsheet

take an average of the values in a range of cells (AVERAGE). Trig functions include SIN, COS, and TAN, as well as inverse and hyperbolic trig functions. Logic (Boolean) functions are ones that return either a true (1) or false (0) based on prescribed conditions. These functions include IF, AND, OR, and NOT. You might use a logic function to help you mark all equipment with a cost greater than $100,000 for further design consideration. The formula to check the heat exchanger cost would be IF(G6>100000, "re-design," "okay"). This formula would be placed in cell H6. If the value in cell G6 is greater than 100,000, the word *re-design* is printed in cell H6. Otherwise, the word *okay* is placed in cell H6.

Spreadsheet developers continually add built-in, special-purpose functions. For instance, financial functions are a part of most spreadsheets. An example of a financial function is one that computes the future value of an investment. In addition, most spreadsheets have a selection of built-in statistical functions.

A user who wants to perform repeatedly a task for which there is not a built-in function can create a new function called a macro. A **macro** is essentially a series of statements that perform a desired task. Each macro is assigned a unique name that can be called whenever the user has need of it.

Spreadsheets also provide a sort function to arrange tabular data. For instance, vendor names can be sorted alphabetically, while prices can be sorted in order of increasing or decreasing numeric value.

Graphing is another common spreadsheet feature. Typically, the user can select from a number of graph or chart styles. Suppose that you are interested in studying ways to reduce the energy it takes to operate the spray dryer. Figure 9.6 is a Microsoft Excel-generated *x-y* plot of such an analysis. Here you are investigating the impact that recycling hot air back into the dryer has on steam demand.

Advanced Spreadsheet Features

Database construction, in a limited fashion, is possible in some spreadsheet applications such as Excel. A database is a collection of data that is stored in a structured manner that makes retrieval and manipulation of information easy. A database consists of records, each having a specified number of data fields. Typically, the user constructs a data entry form to facilitate entry of records. Once a database has been constructed, the user can sort and extract information. You could use the database feature to catalog all expenses during implementation of your powdered-milk process. This would require a record for each piece of equipment purchased. Data fields would include size, quantity, list price, discounts, and actual cost of each item.

Real-time data collection is another advanced feature of spreadsheets. This feature allows data and commands to be received from, and sent to, other applications or computers. During the design phase of the powdered-milk process, you might use this feature to log data during pilot plant experiments. Later, you might use this feature to create a spreadsheet that organizes and displays data coming from online sensors in your powdered-milk process. This would enable line workers to detect processing problems quickly.

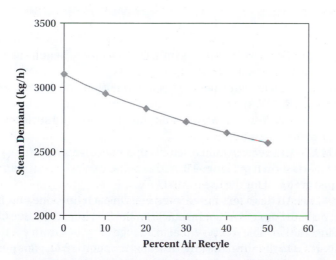

Figure 9.6 An *x-y* plot generated using Microsoft Excel's Chart Wizard

9.5 Mathematics Software

Mathematical modeling is a large component of the design process. For the powdered-milk process, you may need to develop a mathematical model of parts of the process. Problems of this nature are best solved using mathematical or computational software.

MATLAB

MATLAB is a scientific and technical computing environment. It was developed in the 1970s by the University of New Mexico and Stanford University to be used in courses on matrix theory, linear algebra, and numerical analysis (MathWorks 1995). While matrix math is the basis for MATLAB ("MATrix LABoratory"), the functionality of MATLAB extends much farther. MATLAB is an excellent tool for technical problem solving. It contains built-in functions and a command window, used to perform computational operations. (See *Introduction to Technical Problem Solving with MATLAB* by Jon Sticklen.)

MATLAB is best explained through an example. Consider again the powdered-milk design. As with any food process, you must be concerned about consumer safety. The best way to ensure that the dairy company's customers do not get food poisoning from this product is to pasteurize the milk before drying. The first step in pasteurization is to heat milk to a specified temperature. Milk is typically heated in a plate heat exchanger where a heating medium (e.g., steam) is passed on one side of a metal plate and milk is passed on the other side.

The required total surface area of the plates depends on how fast the heat will be transferred from the heating medium to the milk and how much resistance there is to that heat transfer. Mathematically, this is represented by

$$A = Q \times R$$

where A is the surface area of the plates (m^2), Q is the rate of heat transfer (W), and R is the resistance to heat transfer (m^2/W).

From your previous design calculations, assume that you know that the rate of heat transfer is 1.5×10^6 W. The resistance to heat transfer range can be any value from 4×10^{-6} to 7×10^{-6} m^2/W. You may wish to study how the required surface area changes over the range of the resistance.

A typical MATLAB session would resemble that in Figure 9.7. The plot generated by these commands is shown in Figure 9.8. At this point, don't concern yourself with the details, but just get a feel for the use of MATLAB.

Note that MATLAB has a text-based interface. On each line, one task is completed before going on to the next task. For instance, on the first line, the value of Q is set. On the second line, the variable R is set equal to a range of values from 4×10^{-6} to 7×10^{-6} in steps of 1×10^{-7}. On the third line, the arithmetic operation $Q \times R$ is performed. On command lines 4 through 6, MATLAB's built-in plotting commands are used to generate a plot of the results.

```
>       Q = 1.5e6;
>       R = 4e-6:1e-7:7e-6;
>       A = Q.*R;
>       plot(R,A,'s-');
>       xlabel('Area m^2');
>       ylabel('Resistance m^2/W');
```

Figure 9.7 MATLAB code for determining heat transfer surface area

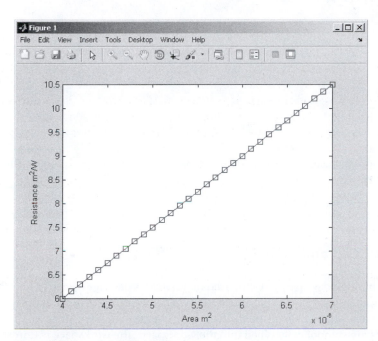

Figure 9.8 Sample MATLAB plot

The commands shown in Figure 9.7 could be saved as a file with a ".m" extension (MATLAB file) and run again and again, or edited at a future date. MATLAB has a built-in editor/debugger interface that resembles a very simple word processor (Fig. 9.9). The debugger helps the user by detecting syntax errors that would prevent the proper execution of the code. The editor shown in Figure 9.9 is displaying the code for the heat transfer area calculation.

MATLAB maintains, and periodically adds, toolboxes, which are groups of specialized built-in functions. For example, one toolbox contains functions that enable image and signal processing. There is also a Symbolic Math Toolbox that enables symbolic computations based on Maple (see next paragraph). SIMULINK is another toolbox, a graphical environment for simulation and dynamic modeling. The user can assemble a personal toolbox using MATLAB script and function files that the user creates.

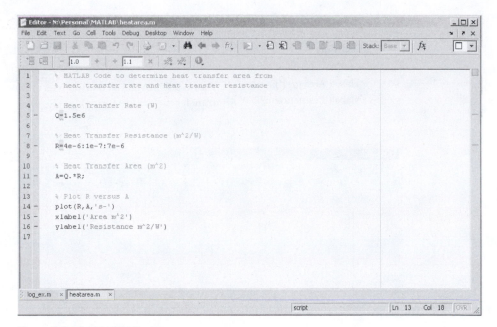

Figure 9.9 The MATLAB editor

MathCad and Maple

MathCad and Maple are GUI-based mathematics packages. Each is 100% WYSIWYG—computer jargon for "What You See Is What You Get." That is, what the user sees on the screen is what will be printed. There is no language to learn since mathematical equations are symbolically typed into the workspace. The workspace is unrestrictive—text, mathematics, and graphs can be placed throughout a document. These packages are similar to a spreadsheet in that updates are made immediately after an edit occurs. MathCad and Maple feature built-in math functions, built-in operators, symbolic capabilities, and graphing capabilities. Perhaps the greatest advantage of MathCad is its ability to handle units. In MATLAB, the user is charged with keeping track of the units associated with variables, but in MathCad, units are carried as part of the assigned value.

You could work the same heat transfer problem in MathCad that you did using MAT-LAB. Figure 9.10 shows the GUI for MathCad and the text used to solve the problem.

EXAMPLE 9.1 The key to being able to use MATLAB to solve engineering problems effectively is to know how to get help with MATLAB's many commands, features, and toolboxes. There are three easy ways to access help.

Method 1: Access the hypertext documentation for MATLAB. At the MATLAB prompt type:

```
>> helpdesk
```

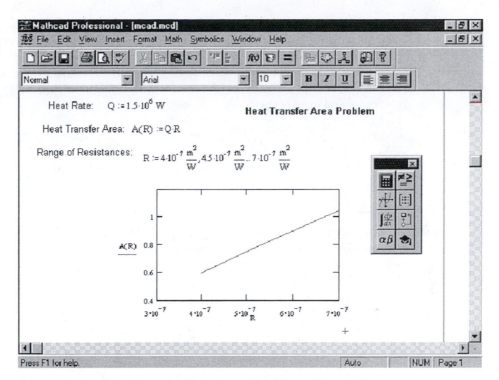

Figure 9.10 Example of a MathCad window and problem-solving session

Method 2: Use the help command. The help command can be used in three ways. Typing help at the MATLAB prompt will generate a list of all available help topics. Try it out.

```
>> help
```

Near the top of this list is a topic called matlab\specfun. To see the commands included in a given topic, the general format is to type >> help topic. To see the commands included in specfun, type:

```
>> help specfun
```

This will generate a list of commands under that topic heading. To get help on a specific command, type >> help command. Let's see how the MATLAB command factor is used:

```
>> help factor
```

The following statement should appear:
FACTOR Prime factors.
FACTOR(N) returns a vector containing the prime factors ofN.
See also PRIMES, ISPRIME.

Even though MATLAB displays the description in capital letters, MATLAB is case-sensitive and will only recognize MATLAB commands in lower case.

Method 3: MATLAB has a means of looking for key words in the command descriptions. This enables you to discover command names when you only know what type of operation you wish to perform. For instance, I may wish to learn how to plot. So I would type:

```
>> lookfor plot
```

A list of all the available commands that have the word *plot* in their descriptions will be generated. How many different commands have the word *plot* in their description?

EXAMPLE 9.2 MATLAB is designed to work with matrices. Therefore, by default, matrix math is performed. Given the following two matrices:

$$A = \begin{bmatrix} 1 & 0 & 4 \\ 7 & 9 & 3 \\ 3 & 6 & 9 \end{bmatrix} \quad B = \begin{bmatrix} 9 & 0 & 1 \\ 3 & 4 & 5 \\ 5 & 9 & 7 \end{bmatrix}$$

Problem 1. Find the product of A and B (Section 9.5):
% Assign each matrix to a variable name
A_matrix = [1 0 4; 7 9 3; 3 6 9]
B_matrix = [9 0 1; 3 4 5; 5 9 7]
% Perform matrix multiplication to find C
C_matrix = A_matrix *B_matrix
The result in MATLAB will be:
C_matrix =

29	36	29
105	63	73
90	105	96

Problem 2. Find the element-by-element product of A and B and set it equal to D (e.g., a11 * b11 = d11):
% Perform element-by-element multiplication to find D
D_matrix = A_matrix. *B_matrix
Note that the notation for element-by-element multiplication is .*. This same notation is used to do any element-by-element operations (e.g., .*, ./, .\, .^). The resulting matrix D is:
D_matrix =

9	0	4
21	36	5
15	54	63

EXAMPLE
9.3

MATLAB can be used to plot data points or functions. We will use both methods to plot a circle of unit radius.

Method 1: Plotting Data Points

First generate a vector of 100 equally spaced points between 0 and 2* π using lin-space. Note that the semicolon is used after each command line. This suppresses the echo feature on MATLAB so the contents of the vectors are not displayed on the screen:

```
>> angle=linspace(0,2*pi,100);
```

Define vector of *x* and *y* to plot later:

```
>> x=sin(angle);
```

```
>> y=cos(angle);
```

Use the plot function to generate the plot:

```
>> plot(x,y);
```

Use the axis command to set the scales equal on the *x*- and *y*-axes. Otherwise you might produce an oval.

```
>> axis('equal');
```

Label the *x*-axis and *y*-axis and title, respectively.

```
>> xlabel('x-axis');
```

```
>> ylabel('y-axis');
```

```
>> title('Circle of Radius One');
```

Method 2: Plotting with a Function

First define the function. The name for this function is func_circle. You may choose any name, but be sure to avoid names that MATLAB already uses. Also note that the .^ is used for the exponent. Remember that MATLAB assumes everything is a matrix. The .^ function indicates that each element in the data should be raised to an exponent individually rather than using a matrix operation:

```
>> func _ circle=('sqrt(1-x.^2)')
```

Plot the function using the command fplot instead of plot over the range –1 to 1:

```
>> fplot(func _ circle,[-1,1])
```

This function is only positive and gives a semicircle. To get the bottom half of the circle, define a second function that is negative:

```
>> func _ neg=('-sqrt(1-x.^2)')
```

Activate the hold feature to plot the second function on the same plot:

```
>> hold on
```

Plot the negative function over the same range:

```
>> fplot(func _ neg,[-1,1])
```

Make the axes equal in length to get a circle:

```
>> axis square
```

Label the *x*-axis and *y*-axis and title, respectively:

```
>> xlabel('x-axis');

>> ylabel('y-axis');

>> title('Circle of Radius One');
```

EXAMPLE 9.4 MATLAB can be used to perform functional analysis. For example, find the two local minimums of the function $y = x\cos(x + 0.3)$ between $x = -4$ and $x = 4$.
Method: First, you need to create a function assignment. Then it is best to plot the function over the specified range because MATLAB's command for finding minimums (fmin) requires that you provide a range over which to look for the minimums:

```
% Assign the function to a variable name as a string
func _ _ y = ëx.* cos (x + 0.3)°

% Plot the function over the entire range

fplot(func _ y,[-4 4])

% Find each local minimum over the specified
sub-ranges

x _ min1 = fmin(func _ y,-2,0)

x _ min2 = fmin(func _ y,2,4)

% Find the y values at the minimum locations. Note
that MATLAB can only evaluate functions in terms % of x.

x = [x _ min1 x _ min2]

y _ mins = eval(func _ y)
```

The MATLAB solution will yield:

```
x _ min1 = −1.0575

x _ min2 = 3.1491

y _ mins = −0.7683 −3.0014
```

EXAMPLE 9.5

MATLAB can also be used to find the roots or zeroes of a function. As with locating the minimums in the previous example, it is best to plot the function first. Define the function of interest. In this case, we will find the zero of sin(x) near 3:

```
>> func _ zero = ('sin(x)')
```

Plot the function over the interval –5 to 5:

```
>> fplot(func _ zero,[-5,5])
```

Use the fzero command to find the exact location of the zero. An initial guess must be provided (in this case it is 3.0):

```
>> fzero(func _ zero,3)
```

MATLAB returns the answer:

```
ans = 3.1416
```

Note that fzero finds the location where the function crosses the x-axis. If you have a function that has a minimum or maximum at zero but does not cross the axis, it will not find it.

EXAMPLE 9.6

MATLAB can be used to write scripts to be used later. On a PC or Mac, select "New M-File" from the File menu. A new edit window should appear. On a Unix workstation, open an editor. In either case, type the following commands:

```
disp('This program calculates the square root of a number')

x=input('Please enter a number: ');

y=sqrt(x);

disp('The square root of your number is')

disp(y)
```

Save this file under the name "calc_sqrt." Note that on a PC or Mac, a ".m" extension will be added to your file name. Using Unix, you will need to add a ".m" to the end of your file name so MATLAB will recognize it as a script file. In MATLAB, type:

```
>> calc _ sqrt
```

MATLAB should respond with the first two lines below. Assuming that "3" is entered and the Return key is pressed, MATLAB will display the last two lines:

```
This program calculates the square root of a number

Please enter a number: 3

The square root of your number is

    1.7321
```

If MATLAB did not recognize "calc_sqrt," check to see that it has a ".m" extension attached to the end of the file name. Alternatively, the file may not be in a directory or folder in which MATLAB is searching.

9.6 Presentation Software

At regular intervals during the design process, presentations must be made to research groups and to management. Presentation programs allow a user to communicate ideas and facts through visuals. Presentation programs provide layout and template designs that help the user create slides.

PowerPoint

Microsoft PowerPoint is a popular electronic presentation package. Figure 9.11 shows a slide view window of PowerPoint. As you have seen in other Microsoft programs, the GUI for this package has a title bar, a menu bar, a toolbar, and a status bar. The unique feature of this tool is the drawing toolbar.

PowerPoint has five different viewing windows. The slide view window displays one slide at a time. Typically, this is the window through which slide editing is done. The outline view window lists the text for each slide in a presentation. The slide sorter view window graphically displays all the slides. As the name implies, slide sorting is done through this window. The notes page view allows the user to attach text with each slide. Since bullets are used on most slides, this feature comes in very handy. It allows the user to come back to a presentation file at a future date and still remember what the presentation covered. The slides can be viewed sequentially using the slide show view. This is the view used during a presentation.

Presentation packages have a number of features that help the user put together very professional presentations. PowerPoint has a gallery of templates. Templates specify color scheme and arrangement of elements (e.g., graphics and text) on the slide. Figure 9.11 shows a slide view with clip art. PowerPoint also has animation capabilities. With the proper computer hardware, sound and video clips also can be incorporated into a presentation.

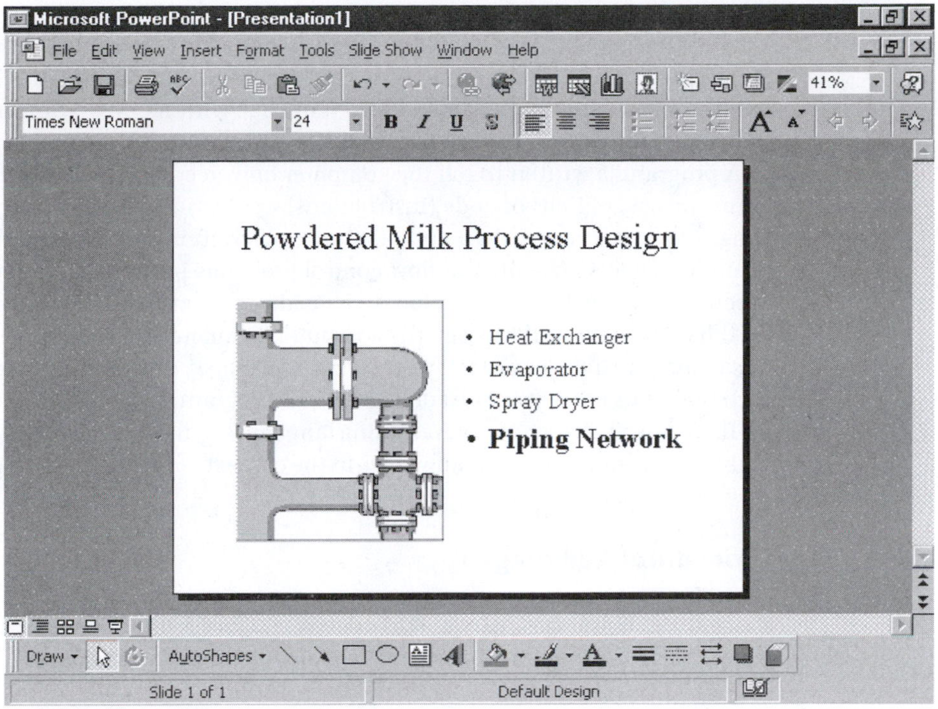

Figure 9.11 Example of a PowerPoint slide view window

9.7 Operating Systems

An **operating system** is a group of programs that manage the operation of a computer. It is the interface between the computer hardware, the software, and the user. Printing, memory management, and file access are examples of processes that operating systems control. The most recognized operating systems are UNIX, originally from AT&T Bell Laboratories; MS-DOS, from Microsoft Corporation; Microsoft's Windows family (Windows 7, Vista, xp, etc.); and the Apple OS for Macintosh computers.

9.8 Programming Languages

Introduction

Programming languages enable users to communicate with the computer. A computer operates using machine language, which is binary code—a series of 0's and 1's (e.g., 01110101 01010101 01101001). Machine language was used primarily before 1953. Assembly language was developed to overcome the difficulties of writing binary code. It is a very cryptic symbolic code that corresponds one to one with machine language. FORTRAN, a higher-level language that more closely resembles English, emerged in 1957. Other higher-level languages quickly followed. However, regardless of the language the programmer uses, the computer only understands binary code. A programming language is an artificial language with its own grammar and syntax that allows the user to talk to the computer.

A program is written to tell the computer how to complete a specific task. A program consists of lines of code (instructions) written in a specific programming language. The program is then compiled or converted into machine language. If the compiler detects no syntax or flow control problems in the code, the user may then execute the code. Upon execution of the code, the computer performs the specified task. If the code is well written, the computer completes the task correctly. If you are unsatisfied with the computer's execution of the task, or the computer is unsatisfied with the user's code, you must debug (fix) the program and try again.

There are a number of programming languages in use. The following sections will highlight the most popular languages in the context of their classification.

Procedural Languages

Procedural languages use a fetch-decode-execute paradigm (Impagliazzo and Nagin 1995). Consider the following two lines of programming code:

```
value = 5

new _ value = log(value)
```

In the second line of code, the computer fetches the variable named "value" from memory, the log function is learned (i.e., decoded), and then the function log(value) is executed. FORTRAN, BASIC, Pascal, and C are all procedural languages.

The development of **FORTRAN** (**For**mula **Tran**slating System) was led by IBM employee John Backus. It quickly became popular with engineers, scientists, and mathematicians due to its execution efficiency. Three versions are in use today: FORTRAN 77, FORTRAN 90, and FORTRAN 95. Each new version has added functionality. The American National Standards Institute (ANSI), which facilitates the establishment of standards for programming languages, recognizes FORTRAN 77 and FORTRAN 90. Thousands of FORTRAN programs and routines written by experts are gathered in standard libraries such as the International Mathematics and Statistics Library (IMSL) and the Numerical Algorithm Group (NAG).

BASIC (Beginner's All-Purpose Symbolic Instruction Code) was created in 1964 at Dartmouth College by John Kemeny and Thomas Kurtz. Its purpose was to encourage more students to use computers by offering a user-friendly, easy-to-learn computing language.

Pascal is a general-purpose language developed for educational use in 1971. The developer was Professor Nicklaus Wirth (Deitel and Deitel 1998). Pascal, named for 17th-century mathematician and philosopher Blaise Pascal, was designed for teaching structured programming. This language is not widely used in commercial, industrial, or government applications because it lacks features that would make it useful in those sectors.

Dennis Ritchie of AT&T Bell Laboratories developed the C programming language in 1972. It was an outgrowth of the BCPL (Basic Combined Programming Language) and B programming languages. Martin Richards developed BCPL in 1967 as a language for writing operating system software and compilers (Deitel and Deitel 1998). Ken Thomas of AT&T Bell Laboratories modeled the B programming language after BCPL. B was used to create the early versions of UNIX. ANSI C programming language is now a common industrial programming language and is the developmental language used in creating operating systems and software such as word processors, database management tools, and graphics packages.

Object-Oriented Languages

Object-oriented languages treat data, and routines that act on data, as single objects or data structures. Referring back to the programming code example, value and log are treated as one single object that, when called, automatically takes the log of value. Objects are essentially reusable software components. Objects allow developers to take a modular approach to software development. This means software can be built quickly, correctly, and economically (Deitel and Deitel 1998).

C++ (C plus plus) is an object-oriented programming language developed by B. Stroustup at AT&T Bell Laboratories in 1985. It is an evolution of the C programming language. C++ added additional operators and functions that support the creation and use of data abstractions as well as object-oriented design and programming (Etter 1997).

Java's principal designer was James Gosling of Sun Microsystems. The creators of Java were assigned the task of developing a language that would enable the creation of very small, fast, reliable, and transportable programs. The need for such a language arose from a need for software for consumer electronics such as TVs, VCRs, telephones, and pagers (Bell and Parr 1999). Such a language also makes it possible to write programs for the Internet and the Web. Two types of programs can be written in Java—applications and applets. Examples of application programs are word processors and spreadsheets (Deitel and Deitel 1998). An **applet** is a program that is typically stored on a remote computer that users can connect to through the Web. When a browser sees that a Java applet is referred to in the HTML code for a web page, the applet is loaded, interpreted, and executed.

9.9 Advanced Engineering Packages

The computer tools introduced so far are useful to all engineers. At one time or another, all engineers need to write papers, work with tabular data, and develop mathematical models. An engineer also needs to be able to solve particular classes of problems. For instance, a food process engineer may need to analyze and optimize processes. A mechanical engineer may need to perform finite element analysis. There are many software tools designed to handle specific types of engineering problems. Described here are just a few key packages that an engineer might encounter.

Aspen Plus

Aspen Plus is software for the study of processes with continuous flow material and energy. These types of processes are encountered in the chemical, petroleum, food, pulp and paper, pharmaceutical, and biotechnology industries. Chemical engineers, food process engineers, and bioprocess engineers are typical users of Aspen Plus. This tool enables modeling of a process system through the development of a conceptual flowsheet (process diagram). The usual sequence of steps to design and analyze a process in Aspen Plus is as follows:

- Define the problem.
- Select a units system (SI or English).
- Specify the stream components (nitrogen, methane, steam, etc.).
- Specify the physical property models (Aspen Plus has an extensive property models library that is user-extendible).
- Select unit operation blocks (heat exchanger, distillation column, etc.).
- Define the composition and flow rates of feed streams.
- Specify operating conditions (e.g., temperatures and pressures).
- Impose design specifications.
- Perform a case study or sensitivity analysis on the entire process (i.e., see how varying the operating conditions affects the design).

The output of Aspen Plus is a report that describes the entire process flowsheet. Overall material and energy balances, steam consumption, and unit operation design specifications are just a sampling of what might be included in the final output report.

GIS Products

Civil engineers, agricultural engineers, and environmental and natural resource engineers work with geographical-referenced data. The Environmental Systems Research Institute, Inc. (ESRI) is the leader in software tool development for **geographical information systems** (GIS). The purpose of these tools is to link geographical information with spatial data. For example, GIS products can be used to study water and wastewater management of a particular landmass (watershed). Specifically, ARC/INFO or ArcView (GIS products) could be used to link data sets concerning drainage and weather with a geographical location as specified by latitude and longitude. GIS products also enable visualization of the data through mapping capabilities.

Virtual Instrumentation

Engineers in all disciplines of engineering have occasions that require them to acquire and analyze data. Modern software tools make it possible to transform a computer into almost any type of instrument required to perform these tasks. One such software tool is LabVIEW, created by National Instruments Corporation. This software allows users to create a virtual instrument on their personal computer. LabVIEW is a graphical programming development environment based on the C programming language for data acquisition and control, data analysis, and data presentation. The engineer creates programs using a graphical interface. LabVIEW then assembles the computer instructions (program) needed to accomplish the tasks. This gives the user the flexibility of a powerful programming language without the associated difficulty and complexity.

The software can be used to control data acquisition boards installed in a computer. These boards are used to collect the data the engineer would need. Other types of boards are used to send signals from the computer to other systems. An example of this would be in a control systems application. LabVIEW allows the engineer to quickly configure the data acquisition and control boards to perform the desired tasks.

The tools also can be used to make the computer function as a piece of instrumentation. The signals from the data acquisition boards can be displayed and analyzed like an oscilloscope or a filter. A sample window of a LabVIEW window performing this function is shown in Figure 9.12. In this example, data were collected and processed through a filter to yield a sine wave.

Software tools such as LabVIEW allow a single computer to take on the function of dozens of pieces of instrumentation with relative ease. This reduces cost and improves the speed of data acquisition and analysis.

Finite Element Tools

Engineers who work with mechanical or structural components include aeronautical, agricultural, biomedical, civil, and mechanical engineers. All must verify the integrity of the systems they design and maintain. Testing these systems on a computer is the most cost-effective and efficient method, and modern finite element tools allow engineers to do just that. One such commonly used tool is ANSYS, created by ANSYS, Inc.

Finite element tools take the geometry of a system and break it into small elements or pieces. Each element is given properties and allowed to interact with the other elements in the model. When the elements are made small enough, very accurate simulations of the actual systems can be made.

The first step in using such a tool is to separate the physical component into a mesh. A sample mesh of centrifugal compressor vanes is shown in Figure 9.13. This mesh defines the physical boundaries of the elements. ANSYS contains aids to allow the user to create the mesh from a CAD drawing of the system. Other tools are also available to create the mesh to be used by ANSYS.

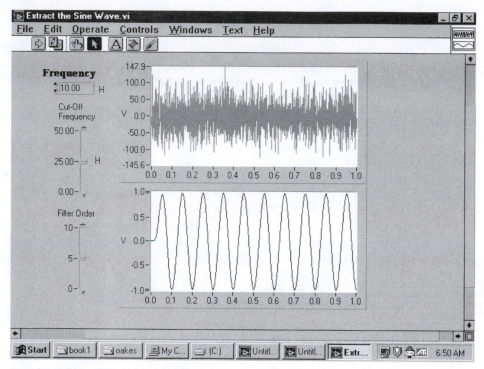

Figure 9.12 Sample LabVIEW window

After the physical mesh is generated, the properties of the elements must be defined. The properties of the elements represent the material that the engineer is modeling. The elements can be solids or fluids and can change throughout the model. Part of the system may be one material, and another part may be different. Initial conditions such as temperatures and loads also must be defined. The engineer can use tools such as ANSYS to perform steady-state as well as transient analyses.

A final useful aspect of these tools is the post-processing capability. Tools such as ANSYS allow the engineer to examine the results graphically. The finite element model can be colored to show temperature, stress, or displacements. The engineer can then visually interpret the results, rather than searching through lists of numbers, which can highlight problems with a design.

Other Tools

We would need an entire book, or more likely a series of books, to touch on all the computer tools engineers use today. There are different sets of computer tools within each discipline and each subspecialty of each discipline. Every day more tools are produced and existing tools improved. New versions are constantly being developed to add features to help engineers perform their jobs more efficiently and more accurately.

Figure 9.13 A sample mesh of a centrifugal compressor and diffuser vanes.

To a large degree, the development of new software tools is capitalizing on rapidly changing computer hardware capabilities. Microprocessors increase in speed according to Moore's Law, which states that microprocessor speed doubles every 18 months. That means that in the four and a half years from your senior year in high school to your first engineering job, processors will go through three doublings—they will be eight times faster. This may not sound impressive, but consider what this has meant since the first microprocessor was developed in 1972. This rate has held until recently—the physics of transistors is now bumping up against atomic forces, but innovations in nanotechnology and biology are seeking to address these issues. Whether Moore's law can continue is controversial in the industry, but do not underestimate the innovative power of engineers. Processor speed has already increased over a million times. As an analogy: compare a snail inching along the ground with a commercial jet flying overhead. The 1972 computer would be the snail and the computer of 2007 would be the jet.

Considering that you will be entering a career that will last over 30 years, the computers you will be using as a senior engineer could be a million times faster than the machines you use now. Will microprocessors continue to advance at this pace? We don't know. We do know, however, that the computers of tomorrow will be much faster than the snails you are using now.

What this means in terms of computer tools is that they will continue to change. Experiment with as many tools as you can. Just as mechanics learn how to use an assortment of tools, so should engineers learn to use an assortment of computer tools. They also should always be ready to learn how to use new tools, because there always will be new tools to learn.

REFERENCES

Andersen, P. K., et al., *Essential C: An Introduction for Scientists & Engineers*, Fort Worth, TX, Saunders College Publishing, 1995.

Bell, D., and M. Parr, *Java for Students*, 2nd ed., London, Prentice-Hall Europe, 1999.

Chapman, S. J., *Introduction to FORTRAN 90/95*, Boston, WCB/McGraw-Hill, 1998.

Dean, C. L., *Teaching Materials to Accompany PowerPoint 4.0 for Windows*, New York, McGraw-Hill, 1995.

Deitel, H. M., and P. J. Deitel, *Java: How to Program*, 2nd ed., Upper Saddle River, NJ, Prentice-Hall, 1998.

Etter, D. M., *Introduction to C++ for Engineers and Scientists*, Upper Saddle River, NJ, Prentice-Hall, 1997.

Etter, D. M., *Microsoft Excel for Engineers*, Menlo Park, NJ, Addison-Wesley, 1995.

Glass, G., and K. Ables, *UNIX for Programmers and Users*, 2nd ed., Upper Saddle River, NJ, Prentice-Hall, 1999.

Gottfried, B. S., *Spreadsheet Tools for Engineers: Excel 97 Version*, Boston, WCB/McGraw-Hill, 1998.

Grauer, R. T., and M. Barber, *Exploring Microsoft PowerPoint 97*, Upper Saddle River, NJ, Prentice-Hall, 1998.

Greenlaw, R., and E. Hepp, *Introduction to the Internet for Engineers*, Boston, WCB McGraw-Hill, 1999.

Hanselman, D., and B. Littlefield, *Mastering MATLAB 5: A Comprehensive Tutorial and Reference*, Upper Saddle River, NJ, Prentice-Hall, 1998.

Heal, K. M., et al., *Maple V Learning Guide*, New York, Springer-Verlag, 1998.

Impagliazzo, J., and P. Nagin, *Computer Science: A Breadth-First Approach with C*, New York, John Wiley & Sons, 1995.

James, S. D., *Introduction to the Internet*, 2nd ed., Upper Saddle River, NJ, Prentice-Hall, 1999.

Kincicky, D. C., *Introduction to Excel*, Upper Saddle River, NJ, Prentice-Hall, 1999.

Kincicky, D. C., *Introduction to Word*, Upper Saddle River, NJ, Prentice-Hall, 1999.

King, J., *MathCad for Engineers*, Menlo Park, NJ, Addison-Wesley, 1995.

MathWorks, Inc., *The Student Edition of MATLAB, Version 4*, Upper Saddle River, NJ, Prentice-Hall, 1995.

Nyhoff, L. R., and S. C. Leestma, *Introduction to FORTRAN 90 for Engineers and Scientists*, Upper Saddle River, NJ, Prentice-Hall, 1997.

Nyhoff, L. R., and S. C. Leestma, *FORTRAN 90 for Engineers and Scientists*, Upper Saddle River, NJ, Prentice-Hall, 1997.

Palm, William J., *MATLAB for Engineering Applications*, Boston, WCB/McGraw-Hill, 1999.

Powell, T. A., *HTML: The Complete Reference*, Berkeley, CA, Osbourne/McGraw-Hill, 1998.

Pritchard, P. J., *Mathcad: A Tool for Engineering Problem Solving*, Boston, WCB/McGraw-Hill, 1998.

Rathswohl, E. J., et. al., *Microcomputer Applications—Selected Edition—WordPerfect 6.1*, The Benjamin/Cummings Publishing Company, Redwood City, 1996.

Sorby, S. A., *Microsoft Word for Engineers*, Menlo Park, NJ, Addison-Wesley, 1996.

Sticklen, Jon, *Introduction to Technical Problem Solving with MATLAB*, Great Lakes Press, www.glpbooks.com.

Wright, P. H., *Introduction to Engineering*, 2nd ed., New York, John Wiley, 1989.

EXERCISES AND ACTIVITIES

9.1   Access your school's home page; surf the site and locate information about your department.

9.2   Perform a keyword search on your engineering major using two different Web search engines.

9.3   Determine if your school has online library services. If so, explore the service options.

9.4   Send an e-mail to a fellow student and your mom or dad.

9.5   Search for the e-mail address of an engineering professional society.

9.6   Find the website of an engineering society. Prepare a one-page report on that organization based on the information found at its website.

9.7   Select the website of the publisher of this text (www.oup.com). Prepare a critique of the website itself. What could be done to improve it? How does it compare with other websites you have accessed? If you wish, you may send your suggestions to the publisher via e-mail.

9.8   Prepare an *oral* presentation on the reliability or unreliability of information obtained from the Internet. Include techniques to give yourself confidence in the information.

9.9   Prepare a *written* report on the reliability or unreliability of information obtained from the Internet. Include techniques to give yourself confidence in the information.

9.10   Do a case study on a situation where inaccurate information from the Internet caused a problem.

9.11   Write a paper for one of your classes (e.g., Chemistry) using the following word processor features: text and paragraph styles and formats.

9.12   Create a table in a word processor.

9.13   Prepare a résumé using a word processor.

9.14   Prepare two customized résumés using a word processor for two different companies you might be interested in working for someday.

9.15   Prepare a two-page report on why you selected engineering, what discipline you wish to enter, the possible career paths you might follow, and why. Create the first page using a word processor. Create the second page using a different word processor.

9.16   Prepare a brief presentation on your successes or frustrations using the two different word processors. Which would you recommend?

9.17   Use a spreadsheet to record your expenses for one month.

9.18   Most spreadsheets have financial functions built in. Use these to answer the following question: If you now started saving $200 per month until you were 65, how much money would you have saved? Assume an annual interest rate of a) 4 percent, b) 6 percent, and c) 8 percent.

9.19   Assume that you wish to accumulate the same amount of money on your 65th birthday as you had in the previous problem, but you don't start saving until your 40th birthday. How much would you need to save each month?

9.20   A car you wish to purchase costs $25,000; assume that you make a 10 percent down payment. Prepare a graph of monthly payments versus interest rate for a 48-month loan. Use a) 6 percent, b) 8 percent, c) 10 percent, and d) 12 percent.

9.21   Repeat the previous problem, but include plots for 36- and 60-month loans.

9.22   Prepare a brief report on the software tools used in an engineering discipline.

9.23   Interview a practicing engineer and prepare an *oral* presentation on the computer tools he or she uses to solve problems.

9.24   Interview a practicing engineer and prepare a *written* report on the computer tools he or she uses to solve problems on the job.

9.25   Interview a senior or graduate student in engineering. Prepare an *oral* presentation on the computer tools he or she uses to solve problems.

9.26   Interview a senior or graduate student in engineering. Prepare a *written* report on the computer tools he or she uses to solve problems.

9.27   Select an engineering problem and prepare a brief report on which computer tools could be used in solving the problem. Which are the optimal tools, and why? What other tools could be used?

9.28   Prepare a brief report on the kinds of situations for which engineers find the need to write original programs to solve problems in an engineering discipline. Include the languages most commonly used.

9.29   Prepare a presentation using PowerPoint that discusses an engineering major.

9.30   Prepare a presentation that speculates on what computer tools you will have available to you in five, 10, and 20 years.

9.31   Use MATLAB Help to learn about the command "linspace."

9.32   What MATLAB command will put grid lines on a plot?

9.33   Perform the following operations using the matrices in Example 9.2.

a) A/B and A./B (right division)
b) A\B and A.\B (left division)
c) A^3 and A.^3
d) A^B and A.^B

Can all of these operations be performed? If not, why not?

CHAPTER 10

Teamwork

10.1 Introduction

More than ever, engineering schools are requiring their students to work in teams. One major reason is that employers want students to have experience with teams as part of their classwork. These teams take the form of collaborative study groups, laboratory groups, design groups as part of individual classes (including most senior design classes), and teams participating in extracurricular competitions such as the solar car, human-powered vehicle, SAE formula car, concrete canoe, bridge design, and many others. This team emphasis in engineering schools mirrors what engineering work is like in the corporate world. Employers seek engineering graduates who have team skills as well as technical skills. The purpose of this chapter is to introduce you to working in teams, to show why teams and collaboration are so important to organizational and technical success, and to give practical advice for how to organize and function as a team. If you study this chapter, work through the exercises, and view some of the references, you should have most of the tools you need to be a successful team member and leader while in engineering school. Apply the principles in this chapter to your team experiences while in college and learn from each experience. Team experiences are not always easy, but a good team can be enjoyable, productive, and memorable. Teams can also be a valuable part of your résumé. Above all, you will have the satisfaction that comes from being a part of great teams.

10.2 Engineers Often Work in Teams

The challenging nature of today's technology dictates that people work in teams. Engineering work has become complex, and no one person can make a meaningful impact by himself or herself. More and more, companies are using interdisciplinary teams as the standard method of carrying out work. Whether it is product design, process design, manufacturing, construction, or many other functions, engineers work in teams. A recent job posting from Tesla Motors included the statement:

> *Our world-class teams operate with a non-conventional automotive product development philosophy of high inter-disciplinary collaboration, flat organization structure, and technical contribution at all levels.*

The greater the problem, the greater the need for the use of teams. Why is the use of teams so popular today?

- There is greater understanding of the power of collaboration.
- Engineers are asked to solve extremely complex problems.
- More factors must be considered in design than ever before.
- Many corporations are international in scope, with design and manufacturing operations spread across the globe.
- Because time to market is extremely important to competitive advantage, "concurrent engineering," inherently a team activity, is widely employed.
- Corporations are increasingly using project management principles.

Power of Collaboration

There is a greater understanding today than ever before of the importance and power of creative collaboration. One recent book on the subject is *Group Genius: The Creative Power of Collaboration*, by Keith Sawyer. The first chapter, "The Power of Collaboration," points out that collaboration has a long history of success in invention and innovation. The book starts with the Wright Brothers, who lived together, ate together, and discussed their airplane project every day. A modern example is the design firm IDEO, which has contributed design ideas to more than 3,000 products in at least 40 industries. IDEO focuses on group improvisation, starting with brainstorming and rapid prototyping to test designs. They create several teams to work on the same project independently, then blend the best ideas together.

In *Organizing Genius: The Secrets of Creative Collaboration*, authors Warren Bennis and Patricia Ward Biederman argue that great groups are able to accomplish more than talented individuals acting alone, stating that "None of us is as smart as all of us." A chapter on the Manhattan Project discusses the development of the atomic bomb during the Second World War by a team of great scientists. One chapter discusses Lockheed's "Skunk Works," developers of the U-2 spy plane, the SR-71 Blackbird, and the F117-A Stealth fighter-bomber used in the Persian Gulf War. Another example is the development of the graphical user interface at Xerox's PARC (Palo Alto Research Center) and its adoption into the Macintosh computer by Steve Jobs and Steve Wozniak. Steve Jobs' goal was to lead his team to create something not just great, but "insanely great," to "put a dent in the universe."

There have been many famous business collaborations in the past. Besides those already mentioned, the list includes Bill Hewlett and David Packard (Hewlett-Packard),

Bill Gates and Paul Allen (Microsoft), Ben Cohen and Jerry Greenfield (Ben & Jerry's Ice Cream), and Larry Page and Sergey Brin (Google).

James Watson and Stanley Crick, winners of the Nobel Prize for discovering the double helix structure of DNA, along with Rosalind Franklin, credit their collaboration for their success: "Both of us admit we couldn't have done it without the other—we were interested in what the answer is rather than doing it ourselves." Crick wrote that "our . . . advantage was that we had evolved . . . fruitful methods of collaboration." Modern physics also found its genesis in collaboration. Heisenberg, Bohr, Fermi, Pauli, and Schrodinger, all Nobel laureates for their roles in founding quantum physics, worked, played, and vacationed together as they developed together their revolutionary new ideas.

Today's Complex Problems

Engineers are asked to solve increasingly complex problems. The complexity of mechanical devices has grown rapidly over the past 200 years. David Ullman, in his text *The Mechanical Design Process*, gives the following examples. In the early 1800s, a musket had 51 parts. The Civil War–era Springfield rifle had 140 parts. The bicycle, first developed in the late 1800s, has over 200 parts. An automobile has tens of thousands of parts.

The Boeing 787 Dreamliner aircraft, with over 2 million components, was designed with modern composite materials, and a focus on fewer parts and lower maintenance costs. Most modern design problems involve not only many individual parts but also many subsystems (mechanical, electrical, controls, thermal, and others), each requiring specialists working in teams.

Many Design Factors

Engineering designers must consider more factors than ever before, including initial price, life cycle costs, performance, aesthetics, overall quality, ergonomics, reliability, maintainability, manufacturability, environmental factors, safety, liability, and acceptance in world markets. Satisfying these criteria requires the collective teamwork of design and manufacturing engineers, marketing, procurement, business, and other personnel. The typical new engineering student, being interested primarily in technical aspects, tends to believe that engineers work in isolation from other factors, focusing on the best technical solution. *This is not true.* The very act of engineering involves solving difficult problems and finding technical solutions while considering numerous constraints, many of which are not technical in nature. Engineers make things happen; they are doers, and they solve problems while interacting with many other functions within the culture of complex organizations. Often acting as members of interdisciplinary teams, engineers are technical experts who, to be truly effective, must understand economics, management, marketing, and the dynamics of the business enterprise.

International Scope

Many corporations are international in scope, with research & development (R&D), design, and manufacturing operations spread across the globe. These operations require teams to work together who may never physically meet. Instead, they meet and share data via electronic means. For example, it is not unusual for design engineers in the United States to collaborate on a design with Japanese engineers for a product that will be manufactured in China. Manufacturing personnel from the Chinese plant also contribute to the project. As another example, software engineers in the U.S., the U.K., and India may collaborate on a software project literally around the clock and around the world. At any time during a 24-hour period, programmers in some part of the world may be working on the project in what is called a virtual team. The world has become a global marketplace, and a firm's supply chain is often just as global. Have your passport ready!

"Time to Market" and Concurrent Engineering

Concurrent engineering, inherently a team activity, is widely employed in order to achieve better designs and to bring products to market more quickly. Time to market is the total time needed to conceive, plan, prototype, create manufacturing capacity, procure materials, create marketing strategies, devise tooling, put into production, and bring to market a new product. In the traditional company, each of these steps is done sequentially, one step at a time. The design engineer tosses a design over the wall to a manufacturing engineer who may not have seen it yet. The manufacturing engineer then (metaphorically) tosses it over the wall to procurement personnel, and so on, usually with disastrous results. Concurrent engineering, on the other hand, is a parallel operation. Like rugby, all of the players run together, side by side, tossing the ball back and forth. R&D, marketing, manufacturing, and procurement personnel are involved from the beginning of the design phase. The use of teams and the development of new technologies such as CAD/CAM, rapid prototyping, shared data, and advanced communications have radically changed the process of engineering. By the use of teams and CAD/CAM technologies, in the face of strong international competition, U.S. automakers have cut the time needed to bring a new car to market from approximately five years to less than three years.

Speed—that is, timely delivery of products to the marketplace—is critical for companies to be profitable. Mark Zuckerberg, CEO of Facebook, commented:

> *Moving fast enables us to build more things and learn faster. However, as most companies grow, they slow down too much because they're more afraid of making mistakes than they are of losing opportunities by moving too slowly. We have a saying: "Move fast and break things." The idea is that if you never break anything, you're probably not moving fast enough.*

Adrian Cockcroft, Netflix's Cloud architect, has commented that one of the things he learned at Netflix is that speed wins in the marketplace. Companies that take too long to commercialize their product may miss a narrow window of opportunity before the competition gets there first. Being first does not guarantee success, but the speed to market of innovative products and technologies is important to economic success. Many products have a limited time on the market. For example, new prescription drugs have an average life of 10 years before their patent expires; then, other companies can make what becomes a generic drug. Delays in getting a new drug to market limits the time of exclusive sales and the revenue that it creates. In many product areas, early product introduction leads to higher profits because early products bring higher profit margins, are more apt to meet customer needs, and are more likely to capture a larger market share.

Developing new products is a key to growth and profits. In a recent study, new products launched in the past three years accounted for 27.5 percent of corporate sales and 28.4 percent of profits. Apple, Inc. has had a string of new product introductions in the past few years, including new models of the iPhone, iMac, and iPad, while introducing the Apple Watch and Apple Music. 3M Corporation CEO L. D. DiSimone, facing flat revenues and stiff competition, adopted a business focus of generating 30 percent of revenues from products less than four years old. The creative thought processes, the innovation needed to create a steady stream of new ideas, and the expertise to manufacture and bring these to market quickly are all functions that teams perform well.

It is important for engineering students to understand the importance of speed, without compromising quality. Delays have many negative consequences. Engineering professors who supervise teams will often provide penalties for failure to meet deadlines. Student teams should recognize the importance of speed and push themselves and their teams to always complete projects on time.

Project Management Is a Team Approach

Project management is widely practiced in industries and government labs. Many graduate engineers discover on their first job that engineers frequently work in project-oriented teams. Unfortunately, most engineering students never take a project management class. Project management principles were developed by the U.S. Navy in the late 1950s as a way to productively manage the Polaris missile project, and by DuPont at about the same time as a way to manage scheduling. The typical corporation is organized by divisions and functional groups. For example, a corporation may have separate research, manufacturing, engineering, human resources, product development, marketing, and procurement departments. Project management is a way of organizing individuals not by function but by products or projects. Selected individuals from research, engineering, manufacturing, human resources, procurement, and other departments are gathered together as a team for a particular purpose, such as solving a corporate problem, developing a new product, or meeting a crucial deadline. A project management approach is inherently a cross-functional team approach and an excellent way to solve problems. Successful project management, however, is not an easy task.

The project manager's job is to complete a project, on time, within budget, with the personnel that are provided, while meeting the customer's requirements. Unfortunately, the project manager is rarely able to hand pick the team's personnel, and supervisors and customers often set the project completion time and budget. In other words, project managers are virtually never given all the people, time, or money they wish for. These tough constraints are often mirrored in student design teams. The instructor picks the team members, the student team doesn't have much money or resources to work with, and the instructor determines the timeframe (usually within the time constraints of a semester). These conditions are uncomfortable for students, but they will prepare you for the real engineering world.

Project managers use tools that can be successfully applied to nearly any team leader function. Project managers plan work requirements and schedules and direct the use of resources (people, money, materials, equipment). One popular project management tool that student teams should learn to use is the Gantt chart (a simple Gantt chart is pictured in Fig. 10.1). A Gantt chart organizes and clearly presents information on task division, sequencing, and timing. Tasks are divided into sub-tasks, and the shading conveys the sequence in which the tasks should be completed and how long a task is expected to take. Gantt charts can also show who is responsible for each task.

Project managers monitor projects by tracking and comparing progress to predicted outcomes. They use project management software such as Microsoft Project (which was used for Fig. 10.1) for planning and monitoring purposes. Project managers must carefully guard the scope of the project from excessive **scope creep**, which is the slow increase in the amount of work that a project team is required to do. A project is successful if it is completed on time, within budget, and at the desired quality. Unfortunately, most projects fail to deliver on all of these requirements. For more information on project management, see Chapter 11, "Project Management," or consult a textbook on project management. The Project Management Institute (PMI) publishes the book *PMBOK Guide: A Guide to the Project Management Body of Knowledge*. Resources can be found online at www.pmi.org.

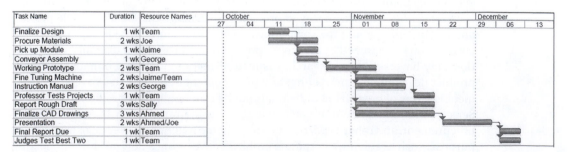

Figure 10.1 Example Gantt chart

10.3 Team Organizational Structures

Some engineering design teams are asked to create an organizational chart. Preparing this chart gives the team the opportunity to define roles and to give thought to the relationship between the team and the team leader. Most student teams, in part because they don't know any better, default to a traditional team structure such as that shown in the first block in Figure 10.2. This structure is easy to draw, but it may or may not accurately represent the best way the team can operate. It implies a strong leader who largely directs the actions of the group, possibly with little participation or discussion from the other team members. It suggests separation between the leader and the other team members.

The second structure portrays a participative leadership/team model in which the leader is positioned in close proximity, with short, direct communication paths to all of the members. The figure implies direct accountability of the leader to all of the members and dependence of the leader on the participation of all the members.

The third structure is similar to the second except it emphasizes the leader's role as a working member of the team. It suggests a flat structure where the leader is an equal, rather than a hierarchical structure where the leader is above the team. For this structure it is easy to visualize the leadership function shifting around the ring from member to member as situations arise needing the special expertise of individual members.

1. Traditional Team Structure: Top-down direction from leader.

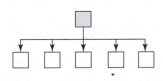

2. Participative Team Structure: The leader is accountable to team members, receives input from members, gives direction to members.

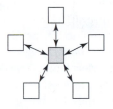

3. Shared Leadership Team Structure: The leader is also a participating team member; the leadership function may be rotated among team members.

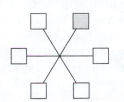

4. The relationship between a student-directed team and their instructor.

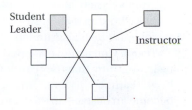

Figure 10.2 Possible team structures

The fourth structure shows the relationship between a student team and its instructor. The student team may be a design group, a research team, or a collaborative study unit. The instructor, though not part of the team, will be nearby and will serve as an important resource to the team. The instructor may be asked to advise the team on administrative issues, to act as a technical consultant, or to assist in intervention or disciplinary action for a nonperforming team member. All teams—in industry, in the university, and in private organizations—ultimately operate under the authority of a leader or administrator.

There is no one *right* structure. None of these structures is inherently good or bad. Examining these example structures gives the team the opportunity to define roles and to give thought to the relationship between the team and the team leader. A team needs to choose a structure that effectively models how the members want to work as a team. The ultimate success of the team then hinges upon the team and its leader(s) working together to accomplish the team's purpose. As a team works, members may find that their operational structure may shift periodically in response to different situations and challenges that they face.

10.4   Team Growth Stages

Simply forming teams to solve problems may seem like a good idea, given the evidence and cases from the previous sections, but teams are not a cure-all. Teams require nurturing. Successful team development is a process, and can be uncomfortable at times. Whether in a corporate, academic, or other setting, all teams must pass through several developmental phases before they truly become productive. These stages are titled "forming," "storming," "norming," "performing," and (in the case of project teams) "adjourning." Unfortunately, successfully completing the first four stages and becoming a productive, performing team is not guaranteed. *Many teams in practice never reach the performing stage*; they remain mired in the forming, storming, or norming stages. Most teams pass through the early forming period fairly quickly. Unfortunately, some teams get stuck in the storming phase, a very difficult period for a team. Every team's challenge is to grow through these stages and achieve performance.

Work teams are widely used in industry. Many modern manufacturing facilities use teams of workers to operate the factory, with the shift supervisor being the team leader. These teams are often stable for long periods of time and do not experience the adjourning phase. Project teams exist at various levels, often made up of professionals from a variety of disciplines. All teams need help to progress through the stages. Teams who stop their growth at the storming phase will not be productive, and their members will be disgruntled.

Teams who achieve the performing stage are extraordinarily productive; there is nearly nothing that they cannot accomplish, and the quality of work life can be quite high. Teams at this level can become self managing, and traditional managers tend to become an obstacle to the team's success.

Stage 1: Forming

The first stage of team development is forming. In this stage, typical dialog may consist of the following: "Nice to meet you. Yeah, I'm not sure either why we are here. I'm afraid this might be a lot of work." In the forming stage, team members usually do not know each other and tend to be formal. Initial conversations tend to avoid anything controversial. Team members get acquainted with one another, with the leader, with the team's purpose, and with the overall level of commitment (workload) required. Team members begin to learn about one another's personalities, abilities, and talents, but also about one another's weaknesses and idiosyncrasies. Individual team members are initially shy, reserved, self-conscious, and uncertain. A key role of the leader in this stage is to lead team ice-breaking activities and facilitate discussion, encouraging all to speak and quieting some who might tend to dominate the conversation. Another role of the leader is to help the team begin to focus on the task at hand.

Stage 2: Storming

The second stage of team development is called storming or the "bid for power" stage, which describes what happens. In this stage, typical dialogue (or more likely private thoughts) may consist of the following: "Do I have to work with this team? What did I do to deserve this? There clearly aren't any superheroes on this team, including that ditzy leader. How can I exert my influence on the team? How are we supposed to solve this mess of a problem?" During the storming stage, the enormity and complexity of the task begins to sink in, sobering and discouraging the participants. "We are supposed to do what? By when?" Teams are rarely formed to solve easy problems, only very difficult and complex ones. Typically, time schedules are unrealistically short, and budgets are inadequate. Further complicating the issue, teammates have learned enough about their fellow team members to know that there are no superheroes, no saviors whom they can count on to do it all. (One person doing all the work is actually a team failure.) Some team members may not initially hit it off well with the others. Cliques or factions may emerge within the team, pitted against others. Since the leader's weaknesses (all leaders have weaknesses) are by now apparent, some individuals or factions will often vie for leadership of the team. Because more than one person will usually try to be the leader, different factions will often fight with each other for dominance. Though possibly under siege, the role of the leader is critical during the storming stage. The leader must help the team members to focus on their collective strengths, not weaknesses, and to direct their energies toward the task, not petty disagreements. To be a successful team, team members do not need to like one another or to be friends. A professional knows how to work productively with individuals with widely differing backgrounds and personalities. Everyone must learn the art of constructive dialog and compromise.

Stage 3: Norming

The third stage of team development is norming. In this stage, typical dialog (or private thoughts) may consist of the following: "You know, I think we can do it. The team knows that I'm good at certain things, but other people are good at other things. True, there are no superheroes, not by a long shot, but once we stopped fighting and started listening to one another, we discovered that these folks have some good ideas. Now if we can just pull these together . . ." Norms are shared expectations or rules of conduct. All groups have some kinds of norms, though many times they are unstated. Do you recall a time you joined a group or team and felt subtle influences to act, dress, look, speak, or work in a particular way? The more team members work together, the more they tend to converge to some common perspectives and behaviors. During the norming stage, team members begin to accept one another instead of complaining and competing. Rather than focusing on weaknesses, personality differences, and turf battles, they acknowledge and use one another's strengths. Individual team members find their place in the group and do their part. Instead of directing energies toward fighting itself, the team directs its collective energy toward the task.

The key to this shift in focus is a collective decision to behave in a professional way, to agree upon and adhere to norms. Possible norms include working cooperatively as a team rather than individually, agreeing on the level of effort expected of everyone, conducting effective discussions and meetings, making effective team decisions, and learning to critique one another's ideas without criticizing the person. One suggested norm is that all team members are expected to be at all meetings, or to communicate clearly in the event that they cannot attend. During the norming stage, feelings of closeness, interdependence, unity, and cooperation develop among the team. The primary role of the leader during the norming phase is to facilitate the cohesion process. Some team members will lag behind the majority of people in embracing norms. The leader and others on the team must artfully nudge individuals along toward group accountability and focus on the task. At this stage, leadership is beginning to be shared among members of the team, and leadership may shift to a different person, depending on the issue.

Stage 4: Performing

The fourth stage of team development is performing. In this stage, typical dialog (or private thoughts) may consist of the following: "This is a fun team. We still have a long way to go, but we have a great plan. Everyone is pulling together and working hard. No superheroes, but we're a super team." In this stage, teams accomplish much. They have a shared, clear vision. Responsibilities are distributed. Individual team members accept and execute their specific tasks in accordance with the planned schedule. They are individually committed and hold one another accountable. On the other hand, there is also a blurring of roles. Team members pitch in to help one another, doing whatever it takes for the team to be successful. In a performing team, so many team members have taken such significant responsibility for the team's success that the spotlight is rarely on a single leader anymore.

Typically, whoever initially led the team is almost indistinguishable from the rest of the team. If the original leader is at a different level (such as a manager responsible for a team of hourly paid employees), that person must adjust his or her management style to be compatible with the team. The team leader is now a coordinator, and perhaps the liaison for the team, not a traditional boss. Autocratic management styles will inhibit the team and get in the way of the team obtaining results. Stage 4 teams can be incredibly productive, but people outside of the team may not understand how the team is operating. If you are ever responsible for such a team, support them, but be careful not to get in the way. Remember that a team of people working together can perform much faster than a single traditional manager.

Stage 5: Adjourning

The fifth and final stage of team development is adjourning. Project teams are typically assembled for a specific purpose or project, and the team is disbanded when that goal is accomplished. If the team was successful, there is a definite feeling of accomplishment, even euphoria, by the team members. On the other hand, an underperforming team will typically feel anger or disappointment upon adjournment.

These stages, originally developed by Bruce W. Tuckman in the mid-1960s, appear in some form in most books that discuss teams. Some leave off the final step, adjourning, because not all teams have a limited life. Others switch the order and give the steps as forming, conforming (instead of norming), storming, and performing.

10.5 What Makes a Successful Team?

Have you ever been a member of a great team? It's a great experience, isn't it? On the other hand, how many of you have ever been part of an unsuccessful team? It was a disappointing experience, wasn't it? How do you measure or evaluate a team's success or lack thereof? What specific factors make a team successful? What factors cause a team to fail?

First it is important to define what a team is and is not. A team is not the same as a group. The term "group" implies little more than several individuals in some proximity to one another. The term "team," on the other hand, implies much more. *A team is two or more persons who work together to achieve a common purpose.* The two main elements of this definition are "purpose" and "working together." All teams are groups, but not all groups are teams.

All teams have a purpose and a personality. A team's purpose is its task, the reason the team was formed. Its personality is its people and how the team members work together. Different teams have their own style, approach, dynamic, and ways of communicating that are different than other teams. For example, some teams are serious, formal, and businesslike, whereas others are more informal, casual, and fun-loving. Some teams are composed of friends; others are not. Friendship, though desirable, is not a requirement for a team to be successful. Commitment to a common purpose and to working together is required. Regardless of style or personality, a team must have professionalism. A team's professionalism ensures that its personality promotes productive progress toward its purpose.

Attributes of a Successful Team

The successful team should have the following attributes:

1. *A common goal or purpose:* Team members are individually committed to that purpose.
2. *Leadership*: Though one member may be appointed or voted as the team leader, ideally every team member should contribute to the leadership of the team.
3. *Each member makes unique contributions to the team's project:* A climate exists in the team that recognizes and appropriately uses the individual talents and abilities of the team members.
4. *Effective team communication:* There are regular, effective meetings, with honest, open discussion. The team is able to make decisions.
5. *Creative spark:* There's excitement and creative energy. Team members inspire, energize, and bring out the best in one another. There is a "can do" attitude. This creative spark fuels collaborative efforts and enables a team to rise above the sum of its individual members.
6. *Harmonious relationships among team members*: Team members are respectful, encouraging, and positive about one another and the work. If conflicts arise, there are peacemakers on the team. The team's work is both productive and fun.
7. *Effective planning and use of resources*: This involves appropriate breakdown of the tasks and effective use of resources (people, time, and money).

Team Member Attributes

Engineers spend more time in team member roles than in team leader roles, so team member attributes are extremely important for a successful team. Individual team members of a successful team should have these attributes:

1. *Respectful*: Has mutual respect for other members regardless of gender, ethnicity, background, or other issues that are not relevant to the work. This carries over into behaviors, such as being on time for meetings and meeting team deadlines.
2. *Attendance*: Attends all team meetings, arriving on time or early. Dependable, like clockwork. Faithful, reliable. Communicates in advance if cannot attend a meeting.

3. *Responsible*: Accepts responsibility for tasks and completes them on time, needing no reminders or cajoling. Has a spirit of excellence, yet is not a perfectionist.
4. *Abilities*: Possesses abilities the team needs, and contributes these abilities fully to the team's purpose. Does not withhold self or draw back. Actively communicates at team meetings.
5. *Creative, energetic:* Acts as an energy source, not an energy sink. Conveys a sense of excitement about being part of the team. Has a "can do" attitude about the team's task. Has creative energy and helps spark the creative efforts of everyone else.
6. *Personality*: Contributes positively to the team environment and personality. Has positive attitudes and encourages others. Acts as a peacemaker if conflict arises. Helps the team reach consensus and make good decisions. Helps create a team environment that is both productive and fun. Brings out the best in the other team members.

10.6 Team Leadership

The Need for Leadership

Every successful human endeavor involving collective action requires leadership. Great teams need great leadership. Without leadership, humans tend to drift apart, act alone, and lose purpose. They may work on the same project, but their efforts, without synchronization and coordination, interfere with, rather than build on, one another. Without coordinated direction, people become discouraged, frustrations build, and conflicts ensue. Money is wasted, time schedules deteriorate, and the greatest loss is a loss of human potential. A failed team effort leaves a bitter taste not easily forgotten. The thrill of team success, on the other hand, is sweet.

The Leader's Role

The leader ensures that the team remains focused on its purpose and that it develops and maintains a positive team personality. The leader challenges and guides the team to high performance and professionalism. The role involves both building the team and building the project. In support of these objectives, the leader must do the following:

1. *Focus on the purpose:* Help the team remain focused on its purpose and be tenacious.
2. *Be a team builder:* The leader may actively work on some project tasks, but the most important task is the team, not the project. The leader builds, equips, and coordinates the team so it can accomplish its purpose and succeed as a team.

3. *Plan well and effectively use resources (people, time, and money):* Assess and use the team members' abilities effectively.
4. *Run effective meetings:* Ensure that the team meets regularly and that the meetings are productive.
5. *Communicate effectively:* Effectively communicate the team vision and purpose. Praise good work and provide guidance for improvement of substandard performance.
6. *Promote team harmony by fostering a positive environment:* If team members focus on one another's strengths instead of weaknesses, conflicts are less likely to arise. Conflict is not necessarily evil, however. The effective leader must not be afraid of conflict. Conflicts are an opportunity to improve team performance and personality, and to refocus it on its purpose.
7. *Foster high levels of performance, creativity, and professionalism:* Challenge team members to do the impossible, to think creatively, and to stimulate one another to high performance.
8. *Eliminate barriers that stand in the way of team members getting their work done:* Teams are often faced with organizational barriers that inhibit productive work. An effective leader will view eliminating these barriers as a one of his or her responsibilities.

Team Membership

Every team needs all the leadership it can get. Most teams have a single individual who is either voted or appointed as its leader. In the best teams, however, many members contribute leadership in ways that support and complement the appointed leader. We spend most of our time as a project or team member, even if we are in a leadership position. Like the appointed leader, every member should focus on the purpose, help build the team, plan, recognize the gifts of others, contribute to effective meetings, communicate well, promote a harmonious team environment, and be creative. Ideally, every member, together with the leader, should help build the team while being a working member at the same time. Author Gene Dixon proposed an organization where everyone is a "leader-follower." Such a team would be highly productive while being a joy to work in.

Author John Maxwell defines leadership in this way: "Leadership is influence." Every team member has the opportunity to considerably influence the team's performance. How important is the leadership from the entire team? It is possible to have a truly outstanding leader, but if the team chooses to use its influence to not follow the leader, to undermine the leader, and to promote team disunity, the team will fail. Conversely, a team with a relatively weak leader can be successful if the team members pitch in and use their influence to build the team. Ultimately, the entire team—both the members and the appointed leader—must work together to build a great team. It takes a team to make a great team.

Leadership Styles

According to Schermerhorn, "leadership style is a recurring pattern of behaviors exhibited by a leader." It is the tendency of a leader to act or to relate to people in a particular way. There are two general categories of leadership style—task-oriented leaders and people-oriented leaders. A **task-oriented leader** is highly concerned about the team's purpose and the task at hand. The task-oriented leader likes to plan, carefully define the work, assign specific task responsibilities, set clear work standards, urge task completion, and carefully monitor results. A **people-oriented leader** is warm and supportive toward team members, develops rapport with the team, respects the feelings of followers, is sensitive to team members' needs, and shows trust in followers.

For a team, is one style of leadership better than another? A successful team needs both styles of leadership in some proportion. Earlier it was stated that a team has a purpose and a personality. The team must be task-oriented, but it also must cultivate a team personality and a positive environment in which team performance can flourish. This is further evidence for the need for individual team members to provide leadership that complements the leadership of the appointed leader. If the appointed leader tends to be more task-oriented, the team can balance this with a people orientation. If the appointed leader is people-oriented, the team may need to assist in details of task management. The best leaders welcome complementary leadership from the teams.

Every leader exhibits some of both tendencies, and this has been organized into the managerial grid shown in Figure 10.3. A manager who is low on concern for results and low on concern for people (1,1) will not survive in the job long. A person who has a high concern for people and a low concern for results (1,9) is considered a "Country Club" manager, and will also not survive long in the job. Someone who has a low concern for people and a high concern for results (9,1), the Authoritarian, may be very difficult to work for but is often successful in corporate America. Middle-of-the-Road managers are often successful because they deliver results without causing major organization problems. A true (9,9) Team Leader tends to be more of a theoretical goal than a reality.

10.7 **Effective Decision Making**

Every team makes decisions in order to accomplish its work. The ability to make high-quality decisions in a timely manner is a mark of a great team. Not surprisingly, the *process* of how decisions are made greatly affects the *quality* of these decisions. Unfortunately, most teams arrive at decisions without even knowing how they did it. In this section, we present a variety of ways in which decisions are typically made and discuss their advantages and disadvantages. Though it is recommended that a team decide in advance what decision-making process it will use, the best teams use a variety of different means of making decisions depending on the decision circumstances. The

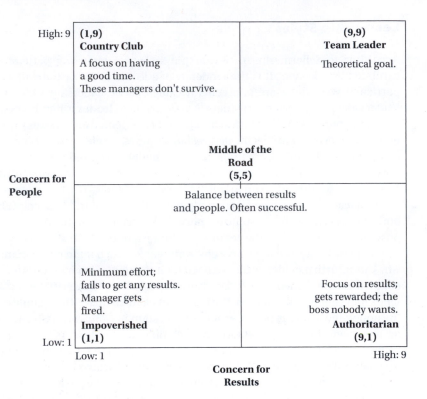

Figure 10.3 Managerial grid

following classification of ways teams make decisions is drawn from the books *Why Teams Don't Work* by Harvey Robbins and Michael Finley and *Leadership* by Andrew DuBrin.

- **Consensus.** A decision by consensus is a decision in which all the team members find common ground. It does not necessarily mean a unanimous vote, but it does mean that everyone had an opportunity to express his or her views and to hear the views of others. It also means that everyone can support the decision that has been made. The process of open sharing of ideas often leads to better, more creative solutions. Unfortunately, achieving consensus can take much time and become unwieldy for larger teams (more than about 15 people). It takes more time to make a consensus decision, but it takes less time in executing the decision because everyone supports it. Overall, consensus decision making usually leads to better decisions and in the long run does not take more time.

- **Majority.** Another way to make a decision is by majority vote. Take a vote and whatever receives the most votes wins. The advantage of this approach is that it takes less time than reaching consensus. Its disadvantage is that it provides for less creative dialog than consensus, and there will always be a faction—the minority—who lose the vote and may become alienated. Those in the minority may slow or stop the decision from being implemented.

- **Minority.** Sometimes a small subset of the team—for example, a subcommittee—makes the decision. The advantage of this is that it may expedite the decision. Its disadvantage is that there is less overall team communication, and some team members may be prevented from making contributions to the decision.

- **Averaging.** Averaging is compromise in its worst form; it's the way Congress and some committees arrive at decisions. Averaging is often accomplished with haggling, bargaining, cajoling, and manipulation. Usually no one is happy except the moderates. The advantage of averaging is that the extreme opinions tend to cancel one another out. The disadvantages are that there is often little productive discussion (as in consensus) and that the least informed can cancel the votes of the knowledgeable.

- **Expert.** When facing a difficult decision, there is no substitute for expertise. If there is an expert on the team, that person may be asked to make a decision. If a team lacks an expert on a particular issue, which is often the case, the best teams recognize this and seek the advice of an expert. The advantage of this is that in theory the decision is made with accurate, expert knowledge. The disadvantage is that it is possible to locate two experts who, when given the same information, disagree on the best course of action. How can team members be sure that the expert gives them the best advice? Team members may be divided on which expert to consult, as well as on their assessment of the expert's credentials.

- **Authority Rule: Decision.** This occurs when there is a strong leader who makes decisions without discussing the details with or seeking advice from the team. This works well with small decisions (particularly of an administrative nature), with decisions that must be made quickly, and with decisions that the team is not well qualified to contribute to. There are many disadvantages to this approach, however. The greatest is that the team's trust in its leader will be undermined. Team members will perceive that the leader does not trust them and is trying to circumvent them. If one person is continually making all the decisions, or if the team is giving up its responsibility to act together and by default, forcing the decision onto an individual, then this is a group, not a team.

- **Authority Rule: Consultation.** This can be a very effective way of making decisions. It is understood from the outset that the final decision will be made by one person—the leader or a delegated decision maker—but the leader first seeks input. There are two widely practiced ways of doing this. First, the leader may talk individually with a select few people. These might be thought of as leaders on the team, but more likely they are people who the leader thinks will agree. The second method is where the entire team is consulted. The team meets and discusses the issue. Many, but perhaps not all, viewpoints are heard. Now more fully informed, the leader makes the decision. The advantage of this method is that the team members, being made part of the process, feel valued. They are more likely to be committed to the result. This type of decision-making process requires a leader and a team with excellent communication skills, and a leader who is willing to make decisions. In a modification of this, the leader may help the team to make a decision, and the leader becomes a member with an equal voice. In another modification, the leader delegates the decision to the team and has no voice. The team can use any of the above methods.

10.8   Attitudes Toward Team Experiences

If you have an opportunity to become part of a team, how should you approach it? What should your attitudes be?

1. Determine to give your best to help the team grow as a team and to accomplish its purpose. Don't just tolerate the team experience.
2. Do not expect to have perfect teammates. You're not perfect; neither are they. Don't fret that your friends are not on your team. Make friends with those on your team.
3. Be careful about first impressions of your team. Some team members may seem to know all the right things to say, but you may never see them again. Others might at first look like losers, but in the long run they are the ones who always attend meetings, make steady contributions, and ensure the success of the team. Be careful about selecting a leader at the very first team meeting. Sometimes the person who seems at first to be the most qualified to lead will never make the commitment necessary to be the leader. Look for commitment in a leader.
4. Be a leader. If you are not the appointed leader, give your support to the leader and lead from your position on the team. Don't be scared off by team experiences in the past where the leader did all the work. Be a leader who encourages and grows other leaders.
5. Help the team achieve its own unique identity and personality. A team is a kind of corporation, a living entity with special chemistry and personality. No team is just like any other team. Every team is formed at a unique time and setting, with different people and a different purpose. Being part of a productive team that sparkles with energy, personality, and enthusiasm is one of the rewards of teamwork. Being part of a great team is a great memory.
6. Be patient. Foster team growth, and give it time to grow. Teammates are at first strangers, unsure of one another and of the team's purpose. Watch for and help the team grow through the stages of forming, storming, norming, and performing.
7. Evaluate and grade yourself and your team's performance. Document its successes and failures. Dedicate yourself to be one who understands teams and who acts as a catalyst, one who helps the team perform at a high level. Note team experiences in your undergraduate portfolio, especially times when you took a leadership role. Prepare to communicate these experiences to prospective employers.

10.9 Documenting Team Performance

Gain the Most from Team Experiences by Critical Evaluation

The primary aim of this chapter has been to give you tools and insights into teams that will significantly increase the probability of you being part of (and being an active leader in) great teams. You will likely be involved in many teams during your college years, in your professional life, and in volunteer capacities. Teams are everywhere, but few people know how to make teams work. For teams, Voltaire's words are applicable: "Common sense is not so common." If you work hard to be uncommonly sensible about teams, you will be of great value to an employer. Approach every team experience as an opportunity to learn and to grow as a team member and leader.

An excellent way to gain the most benefit is to critically evaluate each team experience. Some suggested dimensions for evaluating teams are as follows:

1. Did the team accomplish its purpose? Did it get the job done?
2. Did it do the job well? Were the results of high quality? If not, why?
3. Did the team grow through all of the developmental stages (forming, storming, norming, performing)? Were there any detours?
4. Reflect on the team's personality. Did the team enjoy working together? Did team members inspire one another to greater creativity and energy?
5. Evaluate team members on a report card like the one shown in Table 10.1.

Table 10.1 Team Member Report Card

Criteria	Team Member				
	Ahmed	George	Jaime	Joe	Sally
1. Attendance: Attends all team meetings, arriving on time or early. Dependable, faithful, reliable.					
2. Responsible: Gladly accepts work and gets it done. Spirit of excellence.					
3. Has abilities the team needs. Makes the most of abilities. Gives fully, doesn't hold back. Communicates.					
4. Creative. Energetic. Brings energy, excitement to team. "Can do" attitude. Sparks creativity in others.					
5. Personality: Positive attitudes, encourages others. Seeks consensus. Fun. Brings out best in others. Peacemaker. Water, not gasoline, on fires.					

Average Grade:

Grading Scale: 5, Always; 4, Mostly; 3, Sometimes; 2, Rarely; 1, Never

6. Evaluate the team leader(s). Were they effective? Why or why not?
7. Honestly evaluate your own contribution to the team.

Report Team Evaluations to Supervisor

In addition to learning more about teams, another reason to document team performance is to inform your instructor (when you are in college) or your supervisor (when at work) about the team's work. Virtually all teams have a supervisor to whom the team is accountable. It is important that you report data to them. Some argue that members of student teams should all get the same grade; in general, this is a common practice. But students should know from the start that if they do not participate in a significant way in the team's work, then their instructor may adjust the individual's team grade downward. In order for the instructor to make such determinations, providing quantitative information on team members' contributions can be very helpful. Sometimes instructors ask team members to grade one another, but the reasons for the grades may be unclear. The act of team members evaluating one another on specific team success criteria gives concrete evidence for team member grades. These criteria are even more valuable to the team if it is made clear that they will be the basis for evaluating the team.

There are a variety of methods for teams to assess themselves:

1. A few open-ended questions. For example, on a 100-point scale, grade each team member on participation and on contribution to the team. Provide examples.
2. Provide a detailed questionnaire, including attendance, responsibilities, contributions, creativity, personality, etc. A simple example is shown in Table 10.1.

There is a wide range of other options between these two. The types of assessments done in college courses can vary widely and usually depend on the preferences of the instructors. Such detailed assessments are rarely done in industry.

REFERENCES

Bagley, Rebecca O., "Speed to Market: An Entrepreneur's View," *Forbes*, May 1, 2013.

Bennis, Warren, and Patricia Ward Biederman, *Organizing Genius: The Secrets of Creative Collaboration*, New York, Addison-Wesley, 1997.

Blake, Robert R., Jane S. Mouton, and Alvin C. Bidwell, "Managerial Grid," *Advanced Management—Office Executive*, vol. 1 (9), 1962, pp. 12–15.

Cooper, Robert G., and Scott J. Edgett, *Successful Product Innovation*, Product Development Institute, 2009.

Dixon, Gene, "Can We Lead and Follow?," *Engineering Management Journal* , vol. 21 (1), 2009, pp. 34–41.

DuBrin, Andrew J., *Leadership*, 8th ed., Boston, Cengage Learning, 2016.

Katzenbach, Jon R., and Douglas K. Smith, *The Wisdom of Teams: Creating the High-Performance Organization*, New York, HarperBusiness, 1993.

PMBOK Guide: A Guide to the Project Management Body of Knowledge, 5th ed., Project Management Institute, 2013.

Robbins, Harvey A., and Michael Finley, *Why Teams Don't Work: What Went Wrong and How to Make it Right*, Princeton, NJ, Peterson's/Pacesetter Books, 1995.

Sawyer, Keith, *Group Genius: The Creative Power of Collaboration*, New York, Basic Books, 2007.

Schermerhorn Jr., John R., *Management*, 5th ed., Hoboken, NJ, John Wiley and Sons, 1996.

Scholtes, Peter R., Brian L. Joiner, and Barbara J. Streibel, *The TEAM Handbook*, 2nd ed., Edison, NJ, Oriel Incorporated, 1996.

Schrage, Michael, *Shared Minds: The New Technologies of Collaboration*, New York, Random House, 1990.

Tuckman, Bruce W., "Developmental Sequence in Small Groups," *Psychological Bulletin*, vol. 63, 1965, pp. 384–389.

Ullman, David G., *The Mechanical Design Process*, 2nd ed., New York, McGraw-Hill, 1997.

Whetten, David A., and Kim S. Cameron, *Developing Management Skills*, 3rd ed., New York, Harper Collins, 1995.

Zuckerberg, Mark, "Mark Zuckerberg's Letter to Investors: 'The Hacker Way'," *Wired*, February 1, 2012.

EXERCISES AND ACTIVITIES

10.1   Some individuals consider teams as the answer to all human and organizational problems. Others think that being part of a team is a pain. Comment on these propositions.

10.2   From your perspective, is it easy or hard for humans to work together as teams? Does it come naturally, like falling out of bed, or is a lot of work involved in making a great team?

10.3   Describe in your own words the growth stages of a team. Give at least two examples from your own team development experiences. Have you been a part of a team that developed to a certain point and stalled? A team that progressed quickly to a productive, performing team? A team that experienced prolonged rocky periods but ultimately made it to the performing stage? If possible, note factors that led to the success and failure of these teams.

10.4   With a team of three to five people, role play a sample of conversation(s) and team behaviors for each of the five team growth stages: forming, storming, norming, performing, adjourning.

10.5 With a team of three to five people, role play a sample of conversation(s) and team behaviors for teams making decisions by consensus and three other decision-making methods given in the chapter.

10.6 There are two definitions (source unknown) for the TEAM acronym: "Together Excellence, Apart Mediocrity" and "Together Everyone Achieves Much." These communicate the message that by working together humans can achieve more than what they could achieve working alone. Other quotes related to this are "The whole is greater than the sum of the parts" and "None of us is as smart as all of us." Do you believe that this is true? Why, or why not? Does it depend on the situation? Are some problems more suited to teams? Give examples.

10.7 Look up the word "synergy" and give its basic meaning and its root meaning from the Greek. In an introductory biology textbook look up examples of "synergism." Write and briefly describe three examples of synergism from the biological world. Discuss lessons that humans can learn from these biological examples.

10.8 There are a number of historical examples where individuals have collaborated and accomplished what likely would not have been possible had they acted alone. Have you ever been part of a fruitful collaboration? If so, what? Why was it fruitful? Or have you been part of a failed collaboration? If so, why?

CHAPTER 11

Project Management

11.1 Introduction

Businesses plan how they will maintain and grow their products and services. These business plans are created and implemented through a process called strategic management. Strategy creation is usually carried out by the executives of the firm or organization. Business and government strategies are implemented by carrying out projects, and these involve many people throughout the organization. Some examples of projects include the following:

- New products, such as the latest generation of mobile electronics
- New capital investments, including factories, laboratories, oil wells, mines, and farms
- New processes to make new products or to make existing products more efficiently
- Highways, bridges, electric power distribution, and oil and gas pipelines

Today's projects are complex. The science, engineering, design, and implementation of modern projects are simply too much for one person to be able to do. Firms require teams of people, having many skills, focused on the common goal of implementing their projects.

Today's projects require detailed planning. Many projects fail to meet their goals, and one common problem is a failure to plan in sufficient detail. Quite often, one group does not realize what other parts of the project team are doing. Also, people do not understand how their own work is being used, or why timelines are critical. The failure to understand the broad aspects of a project causes people to work in isolation, and that leads to inefficiencies, lost time, delayed results, higher costs, and poorly delivered projects.

Engineering projects require different tasks to be completed by different people at different times. All these tasks must come together in the end and meet specific deadlines and customer requirements. Projects that involve multiple people and steps must be organized and implemented systematically to ensure success. As with the design process and problem-solving models, following systematic approaches to proper project planning will produce improved results and better solutions.

Employers want students to have experience working on projects. This is because so much industrial, commercial, and government work by engineers involves project work. By having experience in the classroom, students can be introduced to the challenges and the joys of working on projects and can demonstrate to potential employers that they understand the basic project-management tools involved.

In a classroom setting, the impact of missing deadlines or having a project failure is relatively low. Once you enter the workforce, however, missing deadlines can result in lost customers, revenue, and jobs. Taking advantage of your time in class to practice project-management techniques can pay large dividends later in your career.

At the beginning of a simple project, the discipline of developing and managing a plan may seem like a waste of time, but it will lay the foundation so that you understand the tools when the projects are large enough to require them. Spending time to plan the project may seem as if you are losing time when you could be doing something productive, but the planning you do upfront will pay off, allowing your team to work more quickly and efficiently, resulting in a better product. This chapter highlights some of the tools that can be applied to planning and managing a project within an engineering classroom.

11.2   The Triple Constraints

Most projects must deliver three primary items:

- Customer satisfaction (sometimes called project scope)
- Cost
- Time, or schedule

These are known as the triple constraints, and they determine the success or failure of a project.

- **Customer satisfaction.** The project was created with a customer in mind who had certain requirements, or needs. Did you meet the needs of the customer? Did you deliver everything that the customer needed, and is the customer happy with the results? The size of the project (how many items, how many features, etc.) is called the project scope, and customer satisfaction is often referred to as the scope. You want to deliver not only what was promised, but at a quality level that makes the customer elated with your work.
- **Cost.** Projects have budgets, and the project manager is responsible for delivering the project at or under budget. No project manager ever got into trouble for delivering a project under budget, but going over a budget can lead to customer dissatisfaction and other problems (including the loss of your job). An academic project can have costs in terms of both time and money.
- **Time.** Projects have schedules, and these can be as important as cost. For example, one campus construction project promised new dorms at the beginning

of August, ready for the students coming for the fall semester. During the early summer, the old dorms were demolished, because the new dorms were scheduled to be completed in time for fall. By mid-August it was obvious that the new dorms would not be completed for several more months. The university had to put students in motel rooms for most of the fall semester (at a very high extra cost). Time must be managed closely, and schedules often slip.

Project success can often be managed by balancing the triple constraints. You can often make up a deficit in one area by adding or subtracting from one or more of the other areas. For instance, if the scope needs to be increased, it often costs money or time or possibly both. If the project is behind schedule, you can often speed up the timeframe by spending more money on overtime or hiring more people. If costs are too high, decreasing scope often saves money and time, but may irritate the customer. Creative solutions are often available.

The triple constraints are so important to project management that they are sometimes called the "Golden Triangle" (Fig. 11.1). All three are interdependent, and the project must achieve a proper balance between them.

Most projects fail to deliver all three of the triple constraints. Either cost, schedule, or customer satisfaction, or combinations of the three, miss the original target. This failure has prompted increased use of project-management training and tools and has helped fuel the strong growth of the Project Management Institute (PMI) and its certification of project managers (PMP). A classic example of a failed project is the Big Dig, a highway project to take Interstate I-93 underground in downtown Boston. The finished project was billions of dollars over budget (it went over budget by 190 percent), it was finished nine years late, and it leaks water; the ceiling occasionally falls down on moving traffic. The Alaskan Way Viaduct project in Seattle (using a tunnel-boring machine called "Big Bertha") is also having delays and cost overruns.

Figure 11.1 The project management golden triangle

11.3 Student Example Project

Many engineering courses use team projects to give students experience in managing a project while working together in teams. The following student project will be used in this chapter to demonstrate various tools to help plan and execute the project. Microsoft Project is the software that will be used to develop the planning tools.

Your assignment is to develop a plan for designing a low-cost robot to assist police department bomb squads. Many police departments need remote-controlled robots to help remove potentially dangerous packages. You must create a set of tasks and responsibilities to conceptualize, design, build, and demonstrate a prototype robot. Steps in the project include the following:

- Concept development
- Project plan, including a work breakdown structure (WBS), network diagram, and Gantt chart
- Detailed engineering design of the hardware
- Detailed engineering design of the software
- Determine the cost to produce the first 25 units
- Build a prototype
- Demonstrate the prototype by participating in the semester's Robot Race

1. Create a rough conceptual design of the robot in sufficient detail to support the creation of a project plan. The design of the robot is up to you but should be compatible with your ability to build a prototype using the supplied robotics kit.
2. Create a WBS, network diagram, and Gantt chart detailing all tasks, schedule, and responsibilities of the project.
3. Build a prototype of the robot using the supplied materials. The prototype needs to be capable of moving across the room, picking up a ping-pong ball, and bringing it back. Both hardware and software development will be demonstrated with this prototype.
4. Because both speed and agility are important features of the robot, the robots will race each other to see which is best able to perform the tasks in #3.
5. Submit a report, detailing the work involved, including the network diagram and Gantt chart in #2. The report should explain the Gantt chart and should include plans to ensure that the project is completed on time and under budget and meets the customer's requirements.

11.4 Creating a Project Charter

The first step in project planning is establishing a project charter, which is a project summary. This may seem obvious, but you must define what the project will involve and how your team will know it has finished. If you have a customer or someone real who will receive the design, you should share the charter document with them to ensure that the expectations are the same for your team and the customer.

The elements of a charter include the following:

1. Description—Describe and summarize what you or your team will be doing. This summary will allow all those involved or affected by the project to understand what is needed. Include the needs of the community you are addressing.
2. Objectives—List the project objectives. Why are you doing the project? What will you or your team achieve? What issues are your team addressing that would not be addressed otherwise? What are the problems you are solving with this project?

3. Outcomes or Deliverables—What are going to be the project results? When the project is finished, what will be left behind by you and your team? Be specific. When you have achieved these outcomes, everyone should understand you have completed the project.
4. Duration—When will the project be started, and when will it meet the objectives and deliver the outcomes?
5. Customers—Who will benefit from the project? Who will receive the project deliverables? Who are you designing for?
6. Stakeholders—Who will be affected by your project other than your customer? Who has a vital interest in the project's success? Stakeholders are those who will need to be kept informed of the project's progress, outcomes, and results.
7. Team membership and roles—If your plan involves a team, list the team members and their roles. (Some of the roles you may want to assign are outlined later in this chapter.) Listing contact information for all of the team members makes the charter document more valuable and useful for those who will receive it.
8. Planning information—What other information do you need for the project plan? If the project involves a design, what are the steps in the design process? Do other constraints or expectations need to be summarized before planning the project?

The project charter is a useful tool. It gives the project vision and allows all team members, community partners, and stakeholders to comment on the vision before beginning. It helps define the roles and responsibilities of your team members as well as any others involved. It forces your team to think about the project's requirements at the start and serves as a place to summarize these results. A concise project charter becomes an excellent communication tool for your team, community partners, and stakeholders.

A project charter for the Bomb Squad Robot project can be created directly from the project description in Section 11.3.

11.5   Task Definitions

Once you have defined the project charter, the next step is identifying the tasks that need to be completed to achieve the objectives and outcomes. In most engineering projects, *many* tasks will have to be completed to finish the project. Your team might start this process by brainstorming what needs to be done for the project, but do not stop there. Many students will lay out their plans and stop with incomplete tasks. The areas of work might look like this:

- Plan
- Design
- Build
- Deliver

Because these are broad topics, or areas of work, they are often called Level 1 tasks.

For plans to be effective, they need to be detailed enough to manage. A detailed plan should allow you and your team to do the following:

1. Understand when tasks must start and when they must be finished.
2. Hold team members accountable for progress during the project.
3. Identify all needed resources, materials, funding, and people.
4. At any time, determine if your project is on schedule.
5. Manage your people and resources by shifting from one task to another as needed to complete the project.
6. Determine when you have finished.

If a project plan is too general, it defeats the planning purpose. A detailed plan will allow you and your team to identify who needs to be working on what task each week, so each task needs to be as detailed as possible. Break larger tasks into smaller tasks whenever possible.

For plans to be effective, they need to be complete, and this will require your team to take advantage of available resources to identify all the tasks. Chances are you are not the first ones to attempt a project like this. You will have some past experience to draw upon. Make use of your entire team and people outside of the team. Gather as much data as possible to develop a comprehensive list.

Include tasks associated with other tasks. If something needs to be ordered, put in time to research the component so you can place the order and have time for shipping. If you need approvals or signatures for components, incorporate those tasks. The more detailed tasks supporting the Level 1 topics are considered Level 2 tasks. As you build the list of tasks, indent the Level 2 items, much like creating an outline, as shown in Table 11.1. If a further level of detail is needed, add Level 3 actions. Any number of levels can be added as needed to organize the work. Just like an outline from an English class, similar tasks are grouped together. This process will make later steps faster and easier to complete. This is the beginning of the WBS.

Along with tasks, your project plan should have milestones. A *milestone* is a deadline or key event for some deliverable. Deliverables may be the completion of subcomponents for your project, or they may be when all parts have been ordered and have arrived. In a design, each phase completion of the design process could be a milestone. The primary benefit of intermediate milestones is that you can monitor your progress to know if you are on schedule. If part of the project falls behind, you may be able to redirect people or resources from another part of the project to move that part back on schedule. When defining tasks, think of places where you could use milestones as markers or measures of your team's progress.

11.6 **Schedule**

Now that you have compiled the task list, you need to determine the required task times. In business settings, tasks are often measured in working days. For your project, working days are a good unit of time.

Table 11.1 Task List for Bomb Squad Robot Project

Concept Development
    Identify the hardware team
    Identify the software team
    Brainstorming meeting
    Conceptual design
    Agree to concept
Project Planning
    Approval of concept
    Create WBS
    Create network diagram
    Create Gantt chart
    Budget approval
    Schedule approval
    Assign team roles
Hardware Design
    Initial sketches
    Obtain robotics kit
    Hold design meeting
    Detailed design
    Design approved by team
Software Design
    Download Commander app
    Program onboard computer
Build Prototype (to be detailed later)
    Cost estimation for 25 units
Robot Race (to be detailed later)
Write Report
Turn in Report; End of Semester

Be sure to include the full time needed for tasks. If a component needs to be ordered, include the time it will take your institution to place the order, and include shipping time. If parts need to be made, how far in advance will you need to place the order for materials, and how long will it take to start and complete the work? Realize that a task may take only an hour to perform, but it may take a week before someone can start on it. Schedule an appropriate time, looking at start dates and end dates, from request to obtaining what you need. The difference between the start date and the end date is called the *duration of the task*.

Working days do not always work with students' schedules. To plan your tasks, you will need to develop a timescale in place of working days that works for your team. Your class probably does not meet every weekday, and you and your classmates have other commitments, so this project is not a full-time job. For your class, everyone has an expectation of the number of hours each person will spend each week on the project, and you can use that as a model to develop your timescale.

Scheduling can be done in weeks, but if you miss one week, you may be missing a significant amount of the available time. Your team can use any of a number of timescales, but your team members should be held accountable for their progress. They should not be expected to work full time.

When assigning times to tasks, for either your team or outside groups, use prior experience or get input on your estimates. Be conservative with your time estimates. The first time you need to do something, it will take much longer than you originally thought (probably three or four times as long). Simple things like ordering a part through your college's system will take some time because you must learn the system. If you need to make a circuit board, the first time will include learning how to make it. Include time for training, if needed, before you complete a task. We suggest you detail the tasks as much as possible, and the times you assigned can serve as a check for the task detail level. If tasks span several weeks, ask if they can be broken into smaller tasks. If possible, break your tasks into small enough components of not more than one or two weeks. Longer times for tasks make it easier for team members to procrastinate and for your team to get behind. If you wait too long, it may be hard to recover from something falling behind. Short timelines make holding each other accountable, and project management, easier.

With your organized task list, the next question is task order. The first step in this process is to determine the task relationships and sequencing. You have already grouped related tasks in your outline. The next step is to put those tasks in the correct sequence. Identify predecessors, which are tasks that must be completed first. For example, you cannot buy parts until after you have the money approved. You cannot build something until you know what you are going to build. Many tasks have predecessors, and these should be identified as part of the sequencing of tasks. Take the tasks that your team identified and prioritize what must be done.

Some tasks must be done in sequential order: Number 1, then Number 2, etc. Some tasks can be done concurrently: Number 1 and Number 2 at the same time. There are multiple people on the team, and different people can be working on different things at the same time. Not everything must be done sequentially. Where tasks can be done concurrently, do so in order to save time.

11.7 **Work Breakdown Structure (WBS)**

For every task, someone needs to be identified as the person responsible. This can be a single person, a few people, or a group. The idea is to assign people based on their skills; in industry, marketing issues are handled by the marketing group, not the engineering group. Use people's strengths and knowledge, but share the tasks among the project team and even the load. If one person is doing all of the work, then the team has failed in its job.

	❸	Task Name	Duration	Start	Finish	Predecessors	Resource Names
1		⊟ Concept development	10 days	Tue 03/01/16	Mon 03/14/16		
2	▥	Identify the hardware team	3 days	Tue 03/01/16	Thu 03/03/16		team
3	▥	Identify the software team	3 days	Tue 03/01/16	Thu 03/03/16		team
4		Brainstorming meeting	2 days	Fri 03/04/16	Mon 03/07/16	2,3	team
5		Conceptual design	5 days	Tue 03/08/16	Mon 03/14/16	4	Jaime/Fernando
6		Agree to concept	0 days	Mon 03/14/16	Mon 03/14/16	5	team
7		⊟ Project planning	12 days	Tue 03/15/16	Wed 03/30/16		
8		Approval of concept	5 days	Tue 03/15/16	Mon 03/21/16	6	team
9		Create WBS	1 day	Tue 03/22/16	Tue 03/22/16	8	Sara
10		Create network diagram	1 day	Wed 03/23/16	Wed 03/23/16	9	Sara
11		Create Gantt chart	1 day	Wed 03/23/16	Wed 03/23/16	9	Sara
12		Budget approval	5 days	Tue 03/22/16	Mon 03/28/16	8	Heewon
13		Schedule approval	5 days	Thu 03/24/16	Wed 03/30/16	11	Corey
14		Assign team roles	3 days	Wed 03/23/16	Fri 03/25/16	9	team
15		⊟ Hardware Design	16 days	Tue 03/22/16	Tue 04/12/16		
16		Initial sketches	5 days	Tue 03/22/16	Mon 03/28/16	8	Fernando
17		Obtain robotics kit	1 day	Tue 03/29/16	Tue 03/29/16	12	Fernando
18		Hold design meeting	3 days	Wed 03/30/16	Fri 04/01/16	16,17	team
19		Detailed design	5 days	Mon 04/04/16	Fri 04/08/16	18	Fernando
20		Design approved by team	2 days	Mon 04/11/16	Tue 04/12/16	19	team
21		⊟ Software design	4 days	Mon 04/04/16	Thu 04/07/16		
22		Download Commander app	1 day	Mon 04/04/16	Mon 04/04/16	18	Corey
23		Program onboard computer	3 days	Tue 04/05/16	Thu 04/07/16	22	Corey
24		⊟ Build prototype	10 days	Fri 04/08/16	Thu 04/21/16		
25		Hardware build	5 days	Wed 04/13/16	Tue 04/19/16	20	Jaime/Fernando
26		Software integration	2 days	Fri 04/08/16	Mon 04/11/16	23	Corey
27		Functional test	1 day	Wed 04/20/16	Wed 04/20/16	25,26	Jaime/Corey
28		Team demonstration	1 day	Thu 04/21/16	Thu 04/21/16	27	team
29		Cost estimation for 25 units	5 days	Wed 04/13/16	Tue 04/19/16	20	Heewon
30		⊟ Robot Race	6 days	Fri 04/22/16	Mon 05/02/16		
31		get spare batteries	1 day	Fri 04/22/16	Fri 04/22/16	28	Sara
32		test run	2 days	Fri 04/22/16	Mon 04/25/16	28	team
33	▥	Race day	0 days	Mon 05/02/16	Mon 05/02/16	32	class
34		Write report	5 days	Wed 04/20/16	Tue 04/26/16	20,29	team
35	▥	Turn in report	0 days	Mon 05/02/16	Mon 05/02/16	34	team

Figure 11.2 Work breakdown structure for the Bomb Squad Robot project

When all tasks are identified, start and stop times defined, task predecessors listed, and responsibilities assigned, then the initial WBS has been completed.

Computer software is available to help organize this work, and such software is widely used in industry to manage projects. The most popular software is Microsoft Project, which is easy to learn because it works much like Microsoft Office programs. The WBS for the Bomb Squad Robot project is shown in Figure 11.2 and demonstrates the use of Microsoft Project software. Notice that the outline from Table 11.1 makes up the task list. Other items are added in the appropriate cells to provide all of the basic scheduling information.

11.8 **Network Diagrams**

The U.S. Navy developed the Program Evaluation and Review Technique (PERT) in the 1950s to manage the Polaris missile project. At about the same time, the Critical Path Method was developed by DuPont. The two methods are very similar and are used in many industries as a major project tool to track and order tasks. Today these tools are commonly called network diagrams, though the term PERT is still often used.

In a network diagram, each task from the WBS is represented by a box that contains a brief description of and duration for the task. Related tasks (showing predecessors) are connected with lines. A portion of a simple network diagram for the Bomb Squad Robot project, starting with task #8, is shown in Figure 11.3. A close-up view of the network, as created in Microsoft Project, is shown in Figure 11.4, showing the details of each task. These details are taken directly from the WBS.

The network diagram shows pathways of tasks. Some tasks have to be done in series, and others can be done in parallel. These series can be laid out as the overall project plan is laid out. This type of diagram is known as an Activity on Node (AON), where the tasks are shown in boxes, or nodes, and the interrelationships are shown

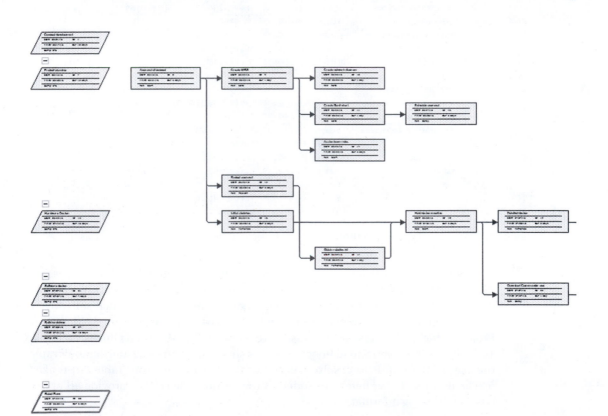

Figure 11.3 Partial network diagram for Bomb Squad Robot project

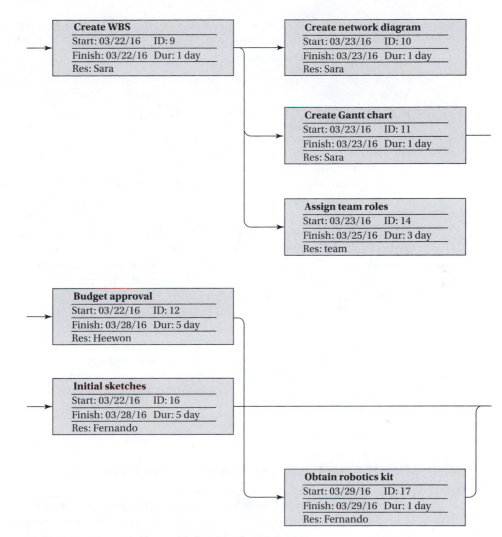

Figure 11.4 Close-up of network diagram, from MS Project

with arrows. Another type of diagram, known as the Activity on Arrow, shows activities on the arrows. Most computer programs use AON diagrams.

Network diagrams are great what-if tools during the initial planning stage. Post-it notes, three-by-five cards, chalkboards or whiteboards, or **project management software** can be used to capture the process flow, to find ways to reduce project completion time, and to use the available resources and people. What needs to be done, when, and by whom? By mapping out the process, a team can apply its resources and allow process tracking. The planning process using this tool can help people think of parallel tasks and what can be going on at the same time.

The sum of the times along each path gives the length or duration for each path. The longest path is the critical path. The critical path will set the project length unless the task times can be reduced.

11.9   Critical Paths

The critical path is the series of tasks that will pace the project; it is the longest string of dependent project tasks. Tasks on the critical path will delay the project if they are not completed on time. An example of a critical path is the mathematics sequence in an engineering curriculum. If a student delays a semester of calculus early in his or her program, it typically delays graduation by one semester. Mathematics is on the critical path for graduation in an engineering program. In industry, there are often delays in obtaining funding or other types of approvals, which can seriously delay the completion of a project. There are domino effects: a delay in one critical item will often affect the timing of many other tasks, and if the delay is on the critical path, the entire project will be delayed.

As you plan for your project, you have to know which tasks can be compressed or rearranged and which cannot. Some tasks can be accelerated by using more people, while others cannot. The classic analogy is having a baby: nine people cannot have the same baby in one month. Some tasks take time and special attention needs to be paid to tasks on the critical path, which cannot be compressed. This information will be valuable during the project if you need to rearrange tasks or change the work of team members.

11.10   Gantt Charts

While the network diagram is for organizing tasks and placing priorities on timing and resources, it may be difficult to interpret for a project with many tasks, especially for people outside your team. A Gantt chart is a popular project management charting method that is easy for people to understand.

A Gantt chart is a horizontal bar chart frequently used in project management where tasks are plotted versus time. Henry L. Gantt, an American engineer and social scientist, developed the Gantt chart in 1917, and it has become one of the most common project planning tools. In a Gantt chart, each row represents a distinct task. Time is represented on the horizontal axis. Typically, dates run across the top of the chart in increments of days, weeks, or months. A bar represents the time when the task is planned for, with the left end representing the expected starting time for each task and at the right end for the completion of each task. Tasks can be shown to run consecutively, with one starting when the previous one ends. They can be shown to run in parallel or overlap. An example of the Gantt

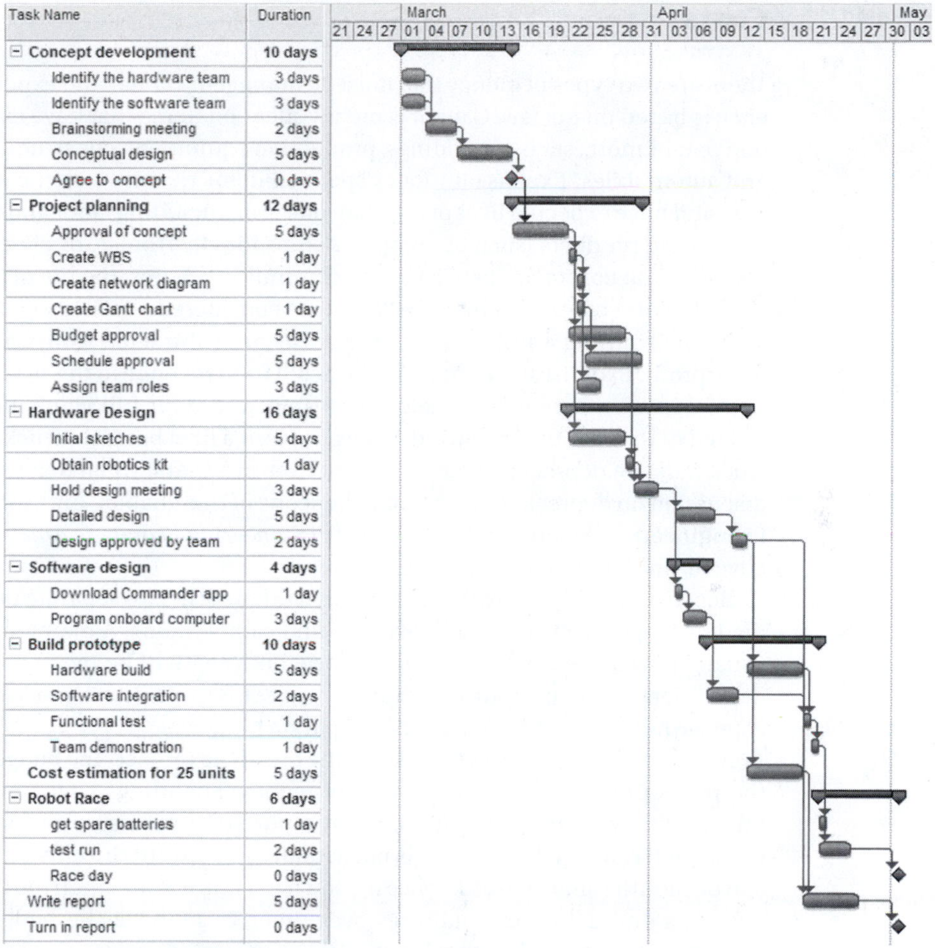

Task Name	Duration
⊟ Concept development	10 days
Identify the hardware team	3 days
Identify the software team	3 days
Brainstorming meeting	2 days
Conceptual design	5 days
Agree to concept	0 days
⊟ Project planning	12 days
Approval of concept	5 days
Create WBS	1 day
Create network diagram	1 day
Create Gantt chart	1 day
Budget approval	5 days
Schedule approval	5 days
Assign team roles	3 days
⊟ Hardware Design	16 days
Initial sketches	5 days
Obtain robotics kit	1 day
Hold design meeting	3 days
Detailed design	5 days
Design approved by team	2 days
⊟ Software design	4 days
Download Commander app	1 day
Program onboard computer	3 days
⊟ Build prototype	10 days
Hardware build	5 days
Software integration	2 days
Functional test	1 day
Team demonstration	1 day
Cost estimation for 25 units	5 days
⊟ Robot Race	6 days
get spare batteries	1 day
test run	2 days
Race day	0 days
Write report	5 days
Turn in report	0 days

Figure 11.5 Gantt chart for Bomb Squad Robot project

chart for the Bomb Squad Robot project, created with Microsoft Project, is shown in Figure 11.5.

Notice that tasks and milestones are included on the Gantt chart, with milestones being shown as a rhombus. Color coding can be used, with the critical path usually shown in red. If a project is partially completed, that bar can be shaded in proportion to the work completed on that task. If the task is half finished, half of that bar will be shaded. A vertical line can be drawn for the current date, and progress can easily be checked based on the line.

To make a presentation of your project easier, fit your Gantt chart onto one page wide. People can often deal with projects that are more than one page long, but keep your chart to one landscape page width.

11.11 Costs

There are two types of money that must be managed: capital and expense. The difference is based on tax law. Capital is money spent on items that have expected lives of one year or more, such as buildings, production equipment, construction equipment, and automobiles. Expense is money spent on items that are a part of everyday business and have expected lives of less than one year, including labor and materials used in creating products (such as computer chips for electronics, or chocolate chips for cookies). The cost of an item is not the determining factor; the life of the item is, but capital items costing less than $5,000 can be considered expense.

In determining a firm's income tax, costs are subtracted from revenue to determine profit. Expense items may be subtracted from revenue in the year they occurred. Capital items must be depreciated over time, and their full value may not be subtracted in the year they occurred. For example, if a firm bought a truck, the cost of the truck must be depreciated, or spread out over five years, when determining taxes. A discussion on depreciation is beyond the scope of this chapter, but more can be found in engineering economics books such as *Engineering Economic Analysis* by Newnan, Lavelle, and Eschenbach.

Because capital and expense are treated differently, two separate budgets must be made. Most projects have both expense and capital items. A maintenance project is usually pure expense. Constructing a new building will be mostly capital, but any demolition or site clearance within the project is considered expense. Capital and expense money must be managed as separate budgets.

A budget is the forecast of costs over the life of the project. The budget is essentially the project plan, expressed in terms of money. Spending money over time is not linear. Money is first committed, when a purchase requisition is written to order parts, material, or labor. Money is not actually spent until the bills are paid, and this can be months later. The difference is shown in Figure 11.6, which illustrates committed costs and actual expenditures over time. To manage the budget, the project manager needs to understand how and when money is spent.

11.12 Personnel Distribution

The tools that have been described can help order tasks, but project management means getting the right people on the right tasks. Some models of project planning will have people identified early in the process based on their expertise and availability. This works when people are separated into departments with other responsibilities. For most student projects, we recommend assigning people after developing a draft of the plan. The people on your team will need to be assigned tasks that allow the project plan to be met. Sketching out who is the most qualified for and/or interested in each task is a great way to start. The team must look at the work distribution and make an equitable assessment. The work needs to be balanced whenever possible and must account for the team members' expertise. When you have assigned people their tasks, add them to the WBS.

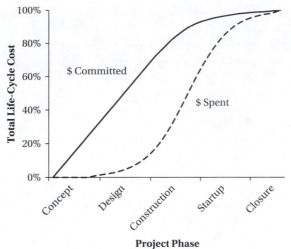

Figure 11.6 Project life cycle costs

11.13 **Documentation**

Do not leave work documentation until the end. As you assemble your project plan, include documentation milestones of the early parts of the project. In many courses, a report is due at the end, and that becomes the motivation for writing things, but look at the project's needs. If you are delivering a product to a customer or the local community, you may need user manuals or maintenance procedures. These should be drafted well before the products are delivered so they can be reviewed and edited. Documenting each step will make it easier if you or someone else readdresses any of the completed tasks.

Another practical reason for documenting as you go along is work reduction. Reports are typically due at the end of a semester or quarter. Leaving all or most of the documentation until the end adds more work during project completion. There will inevitably be details that appear or were overlooked that you will have to deal with near the project's end. Leaving too many details or large tasks, like documentation, for the end will add unnecessary team burdens and will reduce your work quality.

The complete Gantt chart for the example project is shown in Figure 11.7.

11.14 **Team Roles**

The roles on a team vary greatly with the type of project. A new product development project is much different than a construction project, and each requires different skillsets. In project management, individual team members are brought into the project because of their specific skills and given responsibilities accordingly. They

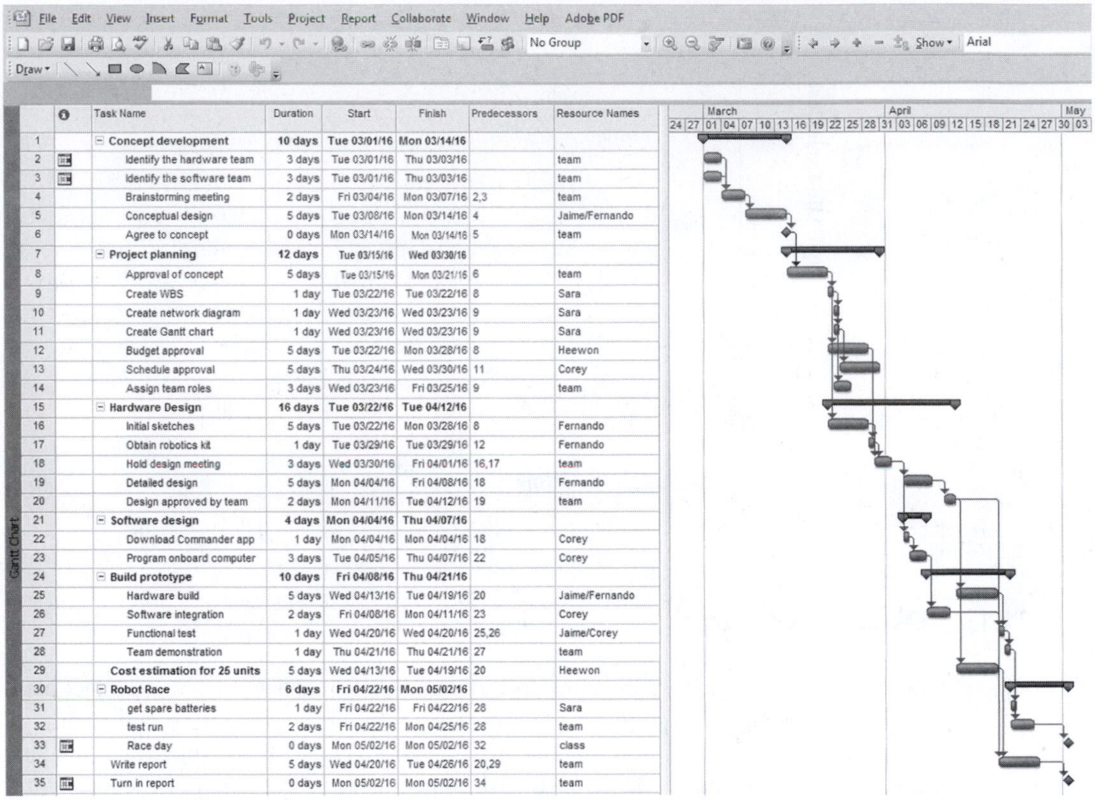

		Task Name	Duration	Start	Finish	Predecessors	Resource Names
1		− Concept development	10 days	Tue 03/01/16	Mon 03/14/16		
2		Identify the hardware team	3 days	Tue 03/01/16	Thu 03/03/16		team
3		Identify the software team	3 days	Tue 03/01/16	Thu 03/03/16		team
4		Brainstorming meeting	2 days	Fri 03/04/16	Mon 03/07/16	2,3	team
5		Conceptual design	5 days	Tue 03/08/16	Mon 03/14/16	4	Jaime/Fernando
6		Agree to concept	0 days	Mon 03/14/16	Mon 03/14/16	5	team
7		− Project planning	12 days	Tue 03/15/16	Wed 03/30/16		
8		Approval of concept	5 days	Tue 03/15/16	Mon 03/21/16	6	team
9		Create WBS	1 day	Tue 03/22/16	Tue 03/22/16	8	Sara
10		Create network diagram	1 day	Wed 03/23/16	Wed 03/23/16	9	Sara
11		Create Gantt chart	1 day	Wed 03/23/16	Wed 03/23/16	9	Sara
12		Budget approval	5 days	Tue 03/22/16	Mon 03/28/16	8	Heewon
13		Schedule approval	5 days	Thu 03/24/16	Wed 03/30/16	11	Corey
14		Assign team roles	3 days	Wed 03/23/16	Fri 03/25/16	9	team
15		− Hardware Design	16 days	Tue 03/22/16	Tue 04/12/16		
16		Initial sketches	5 days	Tue 03/22/16	Mon 03/28/16	8	Fernando
17		Obtain robotics kit	1 day	Tue 03/29/16	Tue 03/29/16	12	Fernando
18		Hold design meeting	3 days	Wed 03/30/16	Fri 04/01/16	16,17	team
19		Detailed design	5 days	Mon 04/04/16	Fri 04/08/16	18	Fernando
20		Design approved by team	2 days	Mon 04/11/16	Tue 04/12/16	19	team
21		− Software design	4 days	Mon 04/04/16	Thu 04/07/16		
22		Download Commander app	1 day	Mon 04/04/16	Mon 04/04/16	18	Corey
23		Program onboard computer	3 days	Tue 04/05/16	Thu 04/07/16	22	Corey
24		− Build prototype	10 days	Fri 04/08/16	Thu 04/21/16		
25		Hardware build	5 days	Wed 04/13/16	Tue 04/19/16	20	Jaime/Fernando
26		Software integration	2 days	Fri 04/08/16	Mon 04/11/16	23	Corey
27		Functional test	1 day	Wed 04/20/16	Wed 04/20/16	25,26	Jaime/Corey
28		Team demonstration	1 day	Thu 04/21/16	Thu 04/21/16	27	team
29		Cost estimation for 25 units	5 days	Wed 04/13/16	Tue 04/19/16	20	Heewon
30		− Robot Race	6 days	Fri 04/22/16	Mon 05/02/16		
31		get spare batteries	1 day	Fri 04/22/16	Fri 04/22/16	28	Sara
32		test run	2 days	Fri 04/22/16	Mon 04/25/16	28	team
33		Race day	0 days	Mon 05/02/16	Mon 05/02/16	32	class
34		Write report	5 days	Wed 04/20/16	Tue 04/26/16	20,29	team
35		Turn in report	0 days	Mon 05/02/16	Mon 05/02/16	34	team

Figure 11.7 WBS and Gantt chart for Bomb Squad Robot project

can be held accountable for their area. Some of the roles that are common on professional teams and more advanced student teams include the following:

- **Project Leader or Coordinator**—The team is organized with a designated leader, or there might be rotating responsibilities. In either case, someone must be designated as a project leader or coordinator. This person's primary function is to ensure that the project is completed on time and at or below budget by monitoring and tracking the progress of the milestones and tasks according to your plan. The coordinator maintains the timelines and will be the person to shift responsibilities if tasks fall behind in part of the team. During regular meetings, the timelines should be checked and reported, and this person will be responsible for bringing this to the meetings. This person may also act as a liaison by interacting with people outside of the team. A diligent project leader will increase the likelihood that the team meets its goals.
- **Procurement**—If the project requires much material to be ordered from outside the university, the team may designate one person to learn the purchasing system and to track the progress of team orders.

- **Financial Officer**—If the project requires purchases that are tied to a budget, designate one person as a financial officer to manage the team's expenses. Going through the estimates at the beginning provides a roadmap, but it will need to be adjusted. Having one person responsible for monitoring progress will make identifying problems faster and easier. One person should be identified early in the process to be the lead person in developing the initial budgetary estimates.
- **Scheduler**—Depending on the size of the project, one person may be identified as the scheduler, the person responsible for maintaining the timeline. This person will develop expertise with the computer programs that produce the WBS, the network diagrams, and the Gantt charts. The team leader often does this in small projects, but larger projects often require someone else to focus on scheduling.

11.15 Agile Project Management

Traditional project management builds a project schedule using a defined scope, estimated task times, a network of task dependencies, and estimated availability of resources. Decades of experience have led to the development of tools to manage risks, control costs, integrate scope changes, etc. The underlying assumption is that *better planning of the complete project will lead to better outcomes*. While this is often the case, the use of these tools does not guarantee success. In the areas of software development and information technology, the track record of success is actually much worse than the average project. Changing scope and changing customer requirements during a project lead to major problems with traditional project management tools. When the scope changes during a project, better planning does not help.

In 2001 the Agile Software Development Alliance (ASDA) was created as a result of an informal gathering of IT experts. They developed a set of 12 principles inspired by a group of core values. The principles and values are known as the Agile Manifesto. The values encompass the following:

- Prioritizing interactions with individuals above following a rigid set of processes and tools
- Meeting the customer's requirements by delivering functioning software on a timely basis, without delays caused by documenting the work
- Cultivating a spirit of collaboration with the customer instead of haggling over contract negotiations
- Having the flexibility to adapt to change requests rather than insisting on the predefined scope

Agile project management approaches have been successfully used in a number of areas outside of software development. Agile approaches can be helpful when any of the following four factors are present:

- Poorly defined scope
- Unknown and perhaps unknowable task times

- Unknown number and set of tasks, implying unknown task dependencies
- Unknown availability of resources

For more information regarding Agile project management, the following are recommended readings (see the reference section for complete information):

1. Beck et al., *Manifesto for Agile Software Development*
2. Highsmith, "Agile Project Management: Principles and Tools"
3. Nicholls, Lewis, and Eschenbach, "Determining When Simplified Agile Project Management is Right for Small Teams"

REFERENCES

Beck, Kent, Mike Beedle, Arie van Bennekum, Alistair Cockburn, Ward Cunningham, Martin Fowler, James Grenning, Jim Highsmith, Andrew Hunt, Ron Jeffries, Jon Kern, Brian Marick, Robert Martin, Steve Mellor, Ken Schwaber, Jeff Sutherland, and Dave Thomas, *Manifesto for Agile Software Development*, 2001 (http://agilemanifesto.org/).

Brooks, Frederick P. Jr., *The Mythical Man-Month*, Essays on Software Engineering, Anniversary Edition, Boston, Addison Wesley, 1995.

Fox, Terry L., and J. Wayne Spence, "Tools of the Trade: A Survey of Project Management Tools," *Project Management*, vol. 29, no. 3, 1998, pp. 20–27.

Highsmith, Jim, "Agile Project Management: Principles and Tools," *Cutter Consortium Agile Project Management*, vol. 4, no. 2, 2003, pp. 1–37.

Meredith, Jack R., and Samuel J. Mantel, Jr., *Project Management, A Managerial Approach*, 6th ed., New York, John Wiley and Sons, 2006.

Modell, Martin E., *A Professional's Guide to Systems Analysis*, 2nd ed., New York, McGraw Hill, 1996.

Newnan, Donald G., Lavelle, Jerome P., and Eschenbach, Ted G., *Engineering Economic Analysis*, 12th ed., New York, Oxford University Press, 2014.

Nicholls, Gillian M., Neal A. Lewis, and Ted G. Eschenbach, "Determining When Simplified Agile Project Management is Right for Small Teams," *Engineering Management Journal*, vol. 27, no. 1, 2015, pp. 3–10.

PMBOK Guide: A Guide to the Project Management Body of Knowledge, 5th ed., Project Management Institute, 2013.

Smith, Karl A., *Project Management and Teamwork*, New York, McGraw-Hill, 2000.

EXERCISES AND ACTIVITIES

11.1 Write a project charter for your project.

11.2 Deliver your project charter to one of your stakeholders, and write a paragraph on his or her reaction.

11.3 Develop an outline for all of the project tasks.

11.4 Select five project tasks that you anticipate doing in the first month and estimate the time required for each task. Be prepared to present your findings, including the resources you used, to the class.

11.5 Develop a network diagram for your project and present it to the class.

11.6 Identify the critical path for your project. List the tasks on the critical path.

11.7 Write a one-page essay on ways to shorten the project's critical path.

11.8 Create a Gantt chart for your project.

11.9 Write a one-page paper that compares and contrasts the benefits and limitations of a network diagram and a Gantt chart.

11.10 Contact a practicing engineer and find out what methods are used for project management. Write a one-page paper on her or his experience.

11.11 Prepare a short oral presentation on a project management software package.

11.12 Write a one-page paper on the benefits and uses of Microsoft Project software.

11.13 Write a summary of details that may appear at the end of your project not included in your plan.

11.14 Write a one-page analysis of our team's progress when you are halfway through your semester or quarter.

CHAPTER 12

Engineering Design

12.1 What Is Engineering Design?

Engineers create our physical environment through design and construction of the infrastructure and design and manufacture of the objects we use in everyday life. Engineers work with their hands, think in numbers, and follow the **engineering design process**, a continuous, iterative procedure in which a feedback loop forces convergence to an acceptable solution for a specified problem. Design is the pinnacle of the engineering profession, the foundation upon which all other engineering activities are built.

Engineering design ranges from the extraordinarily complex, such as the smartphone and associated wireless networks, to the straightforward, such as a woodscrew. It should come as no surprise that few of the building blocks of our current environment have had smooth trajectories to completion or market: true innovation breaks new ground and presents unexpected challenges. You would be forgiven for not knowing that the Brooklyn Bridge in New York City represents an extraordinary feat of civil engineering (beautifully narrated in Ken Burns' PBS documentary *Brooklyn Bridge*). Opening with the death of the lead engineer in 1870, moving through life-threatening conditions inside the caissons and placement of the west tower on aggregate rather than bedrock, the design evolved to overcome each of these challenges and to yield the greatest bridge of its time, first opened to the public in 1883.

Move forward to the 1990s, when Boeing started work on a replacement for the 767 aircraft. When the final design for the 787 Dreamliner was approved in 2003, it was the first commercial passenger aircraft to have been entirely designed and tested with computer aided design (CAD) software. To reduce cost, this approach made possible the distributed manufacture of very large sub-assemblies designed for easy final construction (BBC 2015). However, unanticipated manufacturing problems pushed delivery from late 2008 to fall 2011, showing that, regardless of the level of sophistication of your tool, there will always be new and often unpredictable obstacles that require iterative thinking.

All accredited engineering programs include a major engineering design experience that builds upon the fundamental concepts of mathematics, basic sciences, the humanities and social sciences, engineering, and communication skills. The Accreditation Board for Engineering and Technology (ABET) writes that

Engineering design is the process of devising a system, component, or process to meet desired needs. It is a decision-making process (often iterative), in which the basic sciences, mathematics, and the engineering sciences are applied to convert resources optimally to meet these stated needs.

As described by ABET, the design experience[1] within a program matches the requirements for that discipline:

Students must be prepared for engineering practice through a curriculum culminating in a major design experience based on the knowledge and skills acquired in earlier course work and incorporating appropriate engineering standards and multiple realistic constraints.

In some respects it is more interesting to look at what ABET left out from their new definition in the interests of being concise. Previous descriptions included:

Among the fundamental elements of the design process are the establishment of objectives and criteria, synthesis, analysis, construction and evaluation. . . . it is essential to include a variety of realistic constraints, such as economic factors, safety, reliability, aesthetics, ethics, and social impact.

All these considerations remain within broad-reaching phrases such as "convert resources optimally" and "meet desired needs." Through automatic incorporation of people and the planet into the process, ABET reminds the profession to adopt a holistic approach to design in the interests of the prosperity of the human race.

Learning the principles of engineering design at any level is a significant yet rewarding challenge. There's nothing like the excitement of standing with a group of your peers gazing at a physical object that is the output of your imagination and hands-on labor. En route to this final outcome, hard lessons are learned—compromise, teamwork, and failure—and not always at the same pace.

With appropriate projects, you can easily *learn* the engineering design process. It will probably be a class like no other. On occasion you will crave intervention by your instructor to provide the *right* answer to an open-ended question or to *tell* the class what to do. However, if you are to really learn and experience the engineering design process, you need to be allowed to *fail* during the process, to comprehend that learning through *failure* is one of the most important components of engineering design (Petroski 1992, 2006). One of the more difficult roles for your instructor is stepping back and allowing you the freedom to formulate and implement your ideas, knowing that not all will succeed. There is a fine line to be drawn between useful failure as an integral component of the learning process and lack of progress arising from your limited technical knowledge and hands-on skills. Failure during the design process is not failure in the conventional sense—it is an understanding that the current

---

[1] Not to be confused with drafting or art-related work, ABET adds that "Course work devoted to developing computer drafting skills may not be used to satisfy the engineering design requirement."

solution needs improvement, that it can be, and needs to be, *better*. You quickly appreciate that engineering design is an iterative, time-consuming, and often frustrating process. Engineers have to make things *work*, and not all ideas *work*.

Although closely linked, compromise and teamwork do not necessarily go hand in hand. Technical compromise is required to satisfy the resources available; design compromise is needed to accommodate different ideas. For each, the team has to work together and arrive at a consensus, and this often involves learning to work with people you don't necessarily like or respect, often with very different ideas and opinions from yours. Through this dialog, you learn about the need to be a good listener and to provide critical feedback, however harsh it may appear to be. Put another way, you need to gain the confidence to voice your own opinions and ideas while being receptive to critical feedback; this is the information exchange that feeds iterative design.

Communication, through many different types of media, is another essential component of the engineering design process, as will be covered in Chapter 13, "Technical Communication." Throughout the design process, your activities must be communicated clearly and concisely without ambiguity. At each stage of the design, discussions, decisions, what was done, and why it was done must be documented so that another engineer or team of engineers can replicate what you have accomplished. All engineering and scientific endeavors are built upon the work of others.

Students graduating with an engineering degree from an accredited program will have participated in one or more design experiences as part of their education. Current practice is to assemble teams of individuals with different expertise and backgrounds to find solutions to engineering problems. Most programs include an introductory design experience in the freshman year in which you learn and document your understanding of the process itself. Subsequent classes build upon this foundation with progressively more challenging design coursework in which you apply your prerequisite background in math, science, and related fields to a variety of discipline-specific engineering problems. In most programs, courses that include design constitute between 20 and 25 percent of the total curriculum.

If you are feeling a little intimidated after reading this section, don't be. Jump into engineering design with both feet and you'll soon find out what fun it is. As we have already said, the most exciting part is seeing your ideas come to life in some type of mathematical model or prototype.

12.2 **The Engineering Design Process**

From the outset, it is important to realize that there isn't a specific recipe or formula for design. Indeed, some would argue that design itself cannot be taught—rather, we learn the principles of design through experiences that grow in complexity. The solution of one problem or completion of one design opens up opportunities for subsequent designs or modifications.

Straightforward projects may require only a short, simple process, but more complex products need a multistep approach with several stages. Whether the outcome is

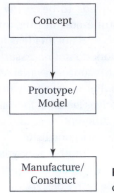

Figure 12.1 The engineering design process (top view)

a product to be manufactured thousands or millions of times or a one-time solution, such as a road bridge, an interplanetary space probe, or a prosthetic limb, the engineering design process is always used.

Discussion of this process usually commences with describing the *problem* to be solved by the problem solvers (the engineers). The third entry in the *Oxford English Dictionary* definition of "problem" is "a difficult or demanding question; (now, more usually) a matter or situation regarded as unwelcome, harmful, or wrong and needing to be overcome; a difficulty." Nowadays, when we use the word "problem" we are invariably describing something that is perceived as wrong. Representing engineers as question answerers hardly seems destined for universal approval, so, bowing to accepted norms, we will use "problem" in the OED sense of "a difficult or demanding question."

Irrespective of the physical implementation of the result, the solution to any problem can generally be found from three separate iterations of the engineering design process (Fig. 12.1). In the first iteration, the overall concept for a solution is created for input to a second in which one or more prototypes or models are used to develop a final design for the construction or manufacture phase.

The exact implementation of the process will depend upon the item and its usage. The prototype and analysis stages required for the design of soda bottles, with a lifespan of months, and for a road bridge, with an expected lifetime of decades, will be radically different. For example, the embodied energy and end-of-life disposal will probably be much more important for the soda bottle than the bridge.

The Five-Stage Design Process

Three representations of the engineering design process are outlined in Table 12.1. The first entry describes a composite path from initial idea to product in the marketplace; this was condensed by NASA to the procedure reproduced in column 2. The more widely adopted core process is listed in column 3 and shown in schematic form in Figure 12.2; note that the *improve* stage represents an iteration of the loop. We will

Table 12.1 Three Representations of the Engineering Design Process

Composite	NASA	Core
Identify the problem/product innovation	Identify the problem/product innovation	Define
Define the working criteria/goals	Define the working criteria/goals	
Research and gather data		
Brainstorm/generate creative ideas	Brainstorm possible solutions	Imagine
	Generate ideas	
Analyze potential solutions	Explore possibilities	Evaluate
	Select an approach	
Develop and test models	Build a model	Create
		Test
	Refine the Design	Improve
Make the decision		
Communicate and specify		
Implement and commercialize		
Post-implementation assessment		

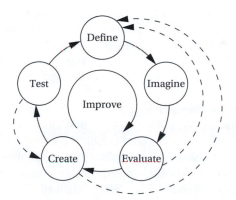

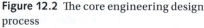

Figure 12.2 The core engineering design process

approach engineering design from this perspective with due acknowledgment to the manufacturing and succeeding stages.

To illustrate how each of the five stages of the engineering design process is applied in the realization of a complex product, we will discuss the design of an automated teller machine (ATM). A working model for classroom implementation of the process comes from the design of an imaginary backpack taken through the same procedure.

All projects begin with the appointment of a project manager/coordinator or team leader who is responsible for driving the design and ensuring that the key elements of each stage are satisfied before the project moves forward. This individual assembles a team of engineers and non-engineers with varying backgrounds and expertise; some will be required throughout the project, others for select stages.

Stage 1: Define

Engineers are problem solvers; the problems they solve relate to the specific or perceived needs of our society. For example, a new prosthesis may be required to overcome a particular handicap, stricter pollution standards might demand higher-efficiency engines, or new computer software might be required to increase the efficiency of a manufacturing process.

The first task for the team, with all groups represented, is the construction of a clear, concise, unambiguous definition of the problem to be solved.[2] Put another way: What is the question to be answered? The problem definition stage is probably *the* most important, for it points the team in a particular direction to achieve a well-defined result. Loosely written, ambiguous problem statements can lead to an enormous waste of resources with little to no chance of success. Concomitant with construction of the problem statement is the establishment of a preliminary list of working criteria and goals that the design must address.

To approach a definition, the team can first ask "What is known about the problem?" There may be a multitude of sources outside as well as within your team that can assist with writing the problem statement. Many organizations have a research and development unit comprising scientists and engineers with the training and expertise to assist with problem evaluation. Critical input is provided by sales and marketing who, through contact with your customers, provide valuable input on problem identification and the desired functionality of the solution. You, as an engineer, are designing something *for* someone else—your project is a response to an actual request or perceived demand from a customer. Too often the real needs of customers are not assigned a fitting priority when defining the challenge to be addressed.

To ensure that the pertinent issues and background of the project are thoroughly explored, access to good information is essential. Your first questions might be the following: "How reliable are the data? Do the data conflict? If so, can the differences be resolved?" Given an open-ended challenge, it is likely that available information may appear inconsistent. However, careful reading of the source often resolves anomalies. For example, different criteria yield different conclusions! The team members active in this phase of the process must determine what types of information they require and where they reside. In-person resources could include marketing departments, technical salespersons, and reference catalogs.

Nowadays, everyone begins an information quest with an Internet search, which in itself initiates the iterative design process. Adjusting search terms to find the information sought contributes to a better comprehension of the challenge and often directs the team toward sources they may not have initially considered. You will quickly learn how to seek, gather, evaluate, and organize information through detailed notes, files, pictures, sketches, and other supporting materials—continuously updated as new data become available. The assistance of librarians, trained in the retrieval and evaluation

---

[2] If you have not already done so, review the section of Chapter 7 which discusses the *real* problem.

of factual, accurate, unbiased data with known provenance, can be enlisted to expand the search to include:

- Books
- Government publications
- University research laboratories
- Peer-reviewed journals and proceedings
- Independent research organizations
- Engineers, professors, and other scientists
- Patent searches and listings (U.S. Patent Office: www.uspto.gov.)
- Professional associations (technical and non-technical)
- Trade journals and publications
- Newspapers and magazines

As soon as is practical, a formal, preliminary statement of the problem should be constructed. For example, the need for a new automobile safety device for infants might evolve into "Develop a better child restraint system that will protect children involved in automobile collisions." From this, the design process may commence with a search for what the competition, if any, has accomplished:

1. What information has been published about the problem?
2. Is a solution to the problem already available?
   a) If the answer is yes:
      i. Who is providing it?
      ii. What are the advantages of their solution?
      iii. What are the disadvantages of their solution?
      iv. How is it made?
      v. What is the cost?
      vi. Was the solution well received?
   b) If the answer is no:
      i. Is there a fundamental reason why no one has worked on this?
      ii. Did someone try and fail?
3. Are there legal issues to consider?
4. Are there environmental concerns to be addressed?
5. What core questions need to be asked and answered before we write our problem statement?
6. How are we going to evaluate our success or failure as we proceed through the design process?

Once the problem has been identified, the team should establish working criteria to validate possible solutions through each stage of the engineering design process. These preliminary goals help the team members focus their efforts on the desired objective as they work through the process and could include the following:

- How much will it cost?
- Will it be difficult to produce?

- What will be its size, weight, strength?
- Does it look good?
- Is it easy to use?
- Will it be safe to use?
- Are there any legal concerns?
- Is it reliable and durable?
- What happens at the end-of-life?
- Is this what the customer truly wanted?
- Will customers want to purchase it?
- Will customers purchase this in preference to a competitor's product?

Having overall objectives for the project provides a means of evaluating, monitoring and, if necessary, adjusting the problem statement as the design process evolves. With each input to the process, the preliminary working criteria and overall goals are reviewed and modified as necessary. For example, one or more criteria may no longer apply, or new issues may surface that necessitate the addition of extra criteria or modification of goals. It cannot be overemphasized that each stage of the engineering design process is an iterative loop in itself.

Using the example of new standards implemented to increase gas mileage and reduce emissions in automobiles, the initial goals for the design might be "To develop an automobile engine that produces 25 percent less emissions while increasing gas mileage by 10 percent."

Stage 2: Imagine

The basic concept for this stage of the engineering design process is to creatively develop as many potential solutions to the problem as possible. A major method of generating multiple ideas is called *Creative Problem Solving* and uses a technique called *brainstorming*.[3] For this, the project leader brings together a group of individuals with varying backgrounds and training to solve a particular problem. The more diverse the group, the broader the range of ideas and the greater the chances of finding an optimal solution to the problem. Such a group may include engineers, scientists, technicians, shop workers, production staff, finance personnel, anthropologists, managers, computer specialists, and potential customers.

In the brainstorming process, every spontaneously contributed idea is recorded. None is deemed too wild or illogical—preliminary judgments and negative comments are not permitted. The goal is to develop a wide-ranging list of possible solutions to the problem at hand. The group leader will encourage participants to suggest random thoughts and ideas using initial triggers such as "Who? What? Why? When? Where? How? How often?"

Anyone who has participated in a brainstorming session knows that it is a very enjoyable, highly stimulating creative process best introduced when the core team dynamics are in place—that is, when each individual is receptive to the ideas of others

[3] For a thorough discussion of creative problem solving and brainstorming strategies, see Chapter 7 sections 7.4 and 7.6.

and comfortable voicing his or her own opinions. If a single session does not provide sufficient input, the process should be repeated for as long as it remains productive.

Stage 3: Evaluate

We now move to evaluate the best ideas imagined in the brainstorming stage[4] with more sophisticated analysis techniques. If many potential solutions have emerged, two or more rounds of elimination may be required to reach a manageable list. Early narrowing of the proposed ideas could include the following:

- Eliminating duplicates—While it is important not to create limited categories, repeated ideas should be eliminated and similar ones retained.
- Allowing clarifying questions—This helps identify duplicates.
- Asking for preliminary evaluation by a vote—The ideas with the most votes go forward for more detailed analysis. Note that a vote in this context represents an informed decision. Limited resources preclude a thorough investigation of all proposed solutions; the team members must therefore use their collective judgment and expertise to choose what they believe to be the most sensible and productive ideas.

Potential solutions are now subjected to thorough technical and financial analysis, the design challenge dictating the method of evaluation. Complex systems may require characterization with mathematical models; in contrast, a verbal discussion could suffice for a simple device. In practice, the evaluation of ideas from a brainstorming session is fraught with uncertainty that depends on access to the qualitative and quantitative design constraints. For example, the mechanical loading of a material with specified dimensions is easily predicted; much less certain, and financially unpredictable, is whether a proposed solution can be built if the design is novel and/or the technology is new. While individuals with varying expertise are involved in evaluation, the engineer is of primary importance. It is here that the engineer's training in mathematics, science, and general engineering principles is applied extensively to critique the potential solutions. Clearly, the certainty of the evaluation of these initial ideas increases with experience and knowledge gained.

As an example, let's consider the types of evaluation that could be applied to the development of an automobile bumper designed to withstand a 20-m.p.h. crash into a fixed object barrier:

Analysis using basic engineering principles and laws: Does each of the proposed solutions satisfy Newton's laws of motion?[5]

---

[4] Note: Chapter 7, Section 3 discusses the analytic method in detail and is directly related to the material in this stage.

[5] Obviously the laws of the universe always have to be obeyed, but this may not have been fully thought through when the idea was proposed.

Table 12.2 A Decision Table

Working Criteria	Points Available	#1	#2	#3
Cost	10	8	9	10
Production difficulty	15	8	12	14
Size, weight, strength	8	7	7	5
Appearance	10	7	9	8
Convenient to use	5	3	4	4
Safety	15	8	11	10
Legal issues	5	4	4	4
Reliability/durability	15	7	9	13
Recyclability	7	4	3	5
Customer appeal	10	7	8	9
Total	**100**	**63**	**76**	**82**

Computer analysis techniques: One frequently used method for structural assessment is *finite element analysis*, in which a component is broken up into segments for numerical analysis. Here, the effects of the impact can be analyzed as a head-on crash and compared to a 45° angle or a side-impact collision. As each section is analyzed, the worst-case scenario can be evaluated.

Estimation: How does the desired performance compare to that predicted? If the early prediction was that some of the solutions would perform better than others, how did they actually perform against the estimate[6]?

Analysis of compatibility: Each of the possible solutions and their related mathematical and scientific principles are compared to the working criteria to determine their degree of compatibility. For example: How did each bumper solution meet the criteria of being cost-effective? What is the size, weight, and strength of each of the proposed solutions? How easy is each to produce?

Common sense: Do the results appear reasonable when evaluated in a simple form? Does the solution make sense compared to the goal?

Economic analysis: Are cost factors consistent with predicted outcomes?

Conservative assumptions: As discussed in Chapter 7, this technique can be most useful in analysis in building in safeguards until more data are generated.

After all possible solutions have been examined and their behavior has been compared to the list of working criteria, a decision table (Table 12.2) eliminates those proposals that have not performed well. The evaluation stage is another critical part of the engineering design process because poor evaluation could eliminate the best solution while promoting inferior or even dangerous ideas. Potential solutions that satisfy all the requirements will likely be authorized to proceed by the project

[6] Review Chapter 7, Section 3 for a more thorough discussion of estimation.

manager. If none of the proposed solutions is deemed satisfactory, it probably means that the problem statement and working criteria need revision.

Stage 4: Create

Here the engineering design process starts to get exciting—the creation of some type of comprehensive physical or mathematical model that enables visualization and exploration of the operational aspects of your design. As for the evaluation process, the actual manifestation of the model will depend upon what you have designed. Is it practical or useful to build a full-scale model or prototype? If it's a large bridge that crosses a river, you will likely generate one or more mathematical models, CAD drawings, and renderings and possibly build a physical model to help you envision how the bridge fits into the existing terrain. If it's a skateboard, full-scale models are built and used as intended. However, with today's composite materials and extensive CAD tools, it is likely that a combination of mathematical and physical models will be used for even the simplest designs. Models commonly used by engineers and others include the following:

- **Full-scale models or prototypes:** If practical, building the product as designed is the best solution. These models, often called prototypes or mock-ups, are very useful in helping engineers visualize the actual product. Such models are useful for comparatively small items such as hinges, furniture, tools, dishwashers, etc.
- **Scale models:** Smaller models are built to simulate the proposed design and are unlikely to include all the features and functions of the actual end product. Such models may be used to depict large structures, such as dams, highways, bridges, or perhaps the entire body of an aircraft and help engineers visualize the actual outcome.
- **Diagrams or graphs:** These tools help designers visualize the basic functions or features of a particular part or product. They could be the output of an electrical circuit of the operating unit of the product, or a visualization of how the electronic components may eventually be assembled.
- **Computer models:** There is a plethora of computer modeling software, generally referred to as CAD, available to transform designs into three-dimensional, on-screen wire-frame drawings, or fully rendered images, for extensive mathematical analysis. Linking the same software to rapid prototyping 3D printers, yields solid individual parts or complete models for use in test procedures.
- **Mathematical models:** Complex electronic systems, such as those whose operation relies on stochastic processes, might require more abstract mathematical analysis to enable fair comparison of their behaviors.

Although the creation stage will involve computer specialists, shop workers, testing technicians, mathematicians and others, engineers guide the overall process using their scientific knowledge and expertise.

Stage 5: Test

In actuality, every idea and model is tested as soon as it appears in the engineering design process. However, it is the results of the tests undertaken on the models from the creation stage that establish the foundation for decisions made about the future of the project. Performing a variety of tests on each of the models allows for comparison and evaluation against the working criteria and the overall goals that have been established. These could include the following:

- **Function:** How well does the design fulfil the role for which it was intended?
- **Form:** Is the overall shape, size, and weight conducive to function?
- **Safety:** Is the product safe for consumer use?
- **Ease of manufacture:** How easily can the product be built? How much labor is required? What ergonomic concerns are involved in manufacture or assembly?
- **Strength:** Under what forces or loads is failure likely, and with what frequency might it occur?
- **Reliability:** How reliable is the product over its simulated life cycle?
- **Durability:** How will the product fare during use? What is its predicted lifespan?
- **End-of-Life:** What happens to the product when it no longer functions? Can the components be recycled?
- **Quality consistency:** Is product quality consistent in the various stages of manufacture? What conditions need to be controlled to ensure uniformity?
- **Consistency of testing:** Are the testing methodologies and their results self-consistent? In the bumper design example, the testing methodology might be different for a head-on impact test to that for a 45° crash test to minimize anomalies.

A means for the consistent and accurate evaluation of data acquired from testing all the models and prototypes must be established. The working criteria developed in the first stage of the design process are used to evaluate the results from testing the output of the creation stage using a **decision table** to visualize and rank their advantages and disadvantages. The table typically lists the working criteria in the first column with a **weighted available point total** for each of the criteria in the second; the order of priority is determined by the team. The third and succeeding columns list the **performance scores** for the possible solutions.

While none of the proposed solutions scored near the ideal, solution #3 outperformed the others. With this information, the project manager and team leaders now make the final decision to "go" or "no-go" with the project. They may decide to pursue solution #3, to begin a new process, or even to scrap the entire project. We will assume they decide to proceed with implementation of the design.

Manufacture

The outcome of the five-stage design process is, in essence, a final design for implementation—input to the manufacturing or construction stage. While the engineering design process is still followed, the path is much more clearly defined. The capabilities of the manufacturing plant or the construction procedure will generally be well understood as will the properties of the materials used and the performance of the end product. Iteration during manufacturing or construction is kept to a minimum to reduce costs although it may be necessary to develop manufacturing procedures and design assembly routes. Changes in the final process should be limited to small adjustments of manufacturing variables such as temperature, pressure, or dimension, for example.

For this to be achieved, there must be complete, thorough, and transparent communication, reporting, and specification of all aspects of the design for input to the manufacturing or construction phase. Team member engineers, skilled craft workers, computer designers, production personnel, and other key individuals work together to write and collate the appropriate documentation. This will include detailed written reports, summaries of technical presentations and memos, relevant e-mails, diagrams, drawings and sketches, computer printouts, charts, graphs, and all other relevant and cataloged material. This information is critical to those involved in determining final approval for the project, as well as the group involved in designing the final implementation of the product. They require complete knowledge of all parts, processes, materials, facilities, components, equipment, machinery, and systems that will be involved in the manufacturing or production of the product. All serious issues must now be resolved since costs now escalate dramatically.

Communication is an essential tool throughout the design process, but especially in this stage. If team members cannot adequately "sell" their ideas to the rest of the organization and cannot appropriately describe the exact details and qualities of the product or process, then possible, good solutions might be ignored. Training materials, operating manuals, computer programs, or other relevant resources for use by the sales team, the legal staff, and prospective clients and customers are now created. In addition to the engineers, project manager, and team leaders, many others, representing a variety of backgrounds and areas of expertise, now become involved:

- **Management and key supervisory personnel** make the ultimate decisions concerning the proposed project. They are concerned with the long-term goals and objectives of the organization, determining future policies and programs that support these goals, and making the economic and personnel decisions that affect the overall health of the organization.
- **Technical representatives** include skilled craft workers, technicians, drafters, computer designers, machine operators, and others involved in manufacturing and production. The members of this group will have primary responsibility for getting the product out of the door.
- **Business representatives** comprise the following:
  - **Human resource personnel** to hire new individuals
  - **Financial personnel** to handle budget details and financial analysis questions
  - **Purchasing personnel** to procure the required materials and supplies
  - **Marketing and advertising personnel** to promote the product
  - **Sales personnel** to sell and distribute the product

- **Attorneys and legal support staff** handle a variety of legal issues including patent applications, insurance, and risk protection analysis.

When all parties are in agreement that all criteria have been satisfied and the overall goal has been achieved, the project moves to implementation and commercialization. With the product in full production, the project team is terminated and the product is considered a regular offering in the company's overall product line.

The project manager, key supervisory personnel, and team members who had significant input to the project gather for a final review and assessment. The product's performance is examined, as are the latest data on production efficiency, quality control reports, sales, revenues, costs, expenditures, and profits. An assessment report is prepared that details the product's strengths and weaknesses, outlines what has been learned from the overall process, and suggests ways that future teams can improve the quality of the process.

12.3   Using the Engineering Design Process—ATM

In this section, to provide you with a better appreciation for the processes involved in getting a product to market and the team effort required for implementation, we speculate on the process used to design an ATM. With the concept of an ATM already well established, the remainder of the process is outlined as a flowchart in Figure 12.3.

Background

The AAA Engineering Company is a small engineering subsidiary of the BCD Research Corporation, a world-renowned research and development facility in the areas of computer vision, real-time image processing and design, and other advanced video applications.

For many years, AAA has supplied products and services to the banking industry and has dominated the field in such areas as security systems, bank cards, and personal identification number (PIN) systems. The parent company, BCD, realizes that the future of face-to-face banking transactions is rapidly changing and that the security of these new types of transactions is a major problem. The challenge for BCD and its subsidiary AAA is to develop a new electronic security system to validate banking transactions. They apply their vision technology to the increasing popularity of the ATM industry.

Stage 1: Define

AAA appoints a project manager to oversee the entire program with a mandate to launch a new product to resolve the issue. The AAA team must first identify the problem

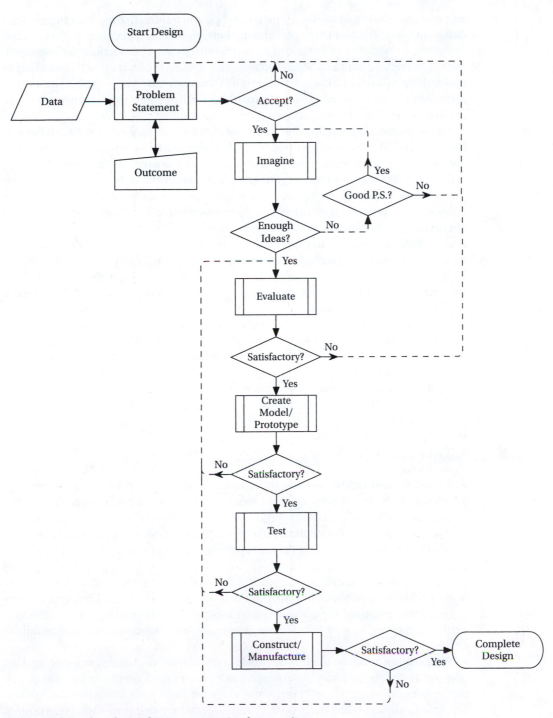

Figure 12.3 Product design from concept to implementation

and generate ideas to better understand its scope and nature. In the initial stages of the design process, the team will probably include engineers, physicists, and computer designers, in addition to personnel from manufacturing, production, management, sales, finance, and banking. The group members discuss the increasing role of ATMs in the banking industry and the current use of a PIN and security access card: an individual approaches an ATM linked to a compatible financial institution, inserts an access card, and enters a PIN to initiate a transaction.

The team members realize that there have been increasing problems with fraudulent use of ATMs. AAA's experience in the banking field and growth of its computer vision products may provide an opportunity for them to expand their operations into this area. Following this premise, the group members work to better understand the overall situation and locate sources to help them narrow their focus on the perceived problem. They begin by consulting their own research and development staff and sales and marketing personnel. In addition, they contact managers and supervisors who may have ideas regarding the expansion of current product offerings through modifications and upgrades. External sources of information include prospective banking industry customers, the Internet, libraries, specialized journals, trade shows, conferences, and the competition. An example of an Internet search for ATM design might resemble that of Figure 12.4.

With these resources, the team can focus on the challenge and develop an initial problem statement, "Eliminate fraudulent access to ATM accounts." From this the team develops working criteria to evaluate possible solutions:

- **Cost:** The machine must be affordable.
- **Reliability:** The accuracy of the solution must be as close to 100 percent as possible.
- **Security:** The security of the cardholder must be fraud-proof.
- **Technical feasibility:** The solution must be easily produced with current technology.
- **Convenience to the customer:** The solution must be easy to use.
- **Acceptance:** The solution must not appear complicated or peculiar, or the public may not use it.
- **Appearance:** The solution must have an aesthetically pleasing appearance.
- **Environmental:** The solution and its components should be recyclable.

After extensive discussions, the team members decide to re-examine their preliminary problem statement. Following additional review of the working criteria, the project goal becomes: "Design and implement an alternate method of identifying a cardholder at an ATM, while focusing on reducing the acceptance of fraudulent entries."

Establishing a goal for the project provides a means of evaluating, monitoring, and possibly changing the focus of the process as it evolves through the design stages. For the project manager, these criteria and the goal become a checkpoint, a means for periodically assessing the project's progress, and help determine when the project is ready to proceed to the next stage.

ATM design   🔍

Images   Videos

ATM Design Considerations

ATM networks are made up of three distinct components: endpoint elements (users), **ATM** switches, and Interfaces, Consider the guidelines discussed in the following...

TN technet.microsoft.com/en-us/library/cc977617.aspx   More results

---

ATM Design Specialists - ICL **ATM Design**

ICL **ATM Design** manufacture and **design ATM** surrounds, window panels and decals. Find out more about our quality **ATM** products here.

🖥 atmdesign.co.uk/Index.html

ATM Design: Make your **ATM** Signage Standout

Expert designers ensure your custom **ATM** Surrounds, Kiosks, Enclosures, Toppers, and other products meet your exact **design** specifications.

💿 companionsystems.com/services/design.html

TMS Design - ATM Solutions

TMS **Design** The Leader in the **ATM** Environment Industry. Since 1987, TMS **Design** Inc. has been the leader in the **ATM** environment industry with it's products and services ...

🔺 tmsdesign.com

The History of **ATM Design** Is Crazy | Co.**Design** | business ...

And even though **ATM** usage is declining, banks are still trying to reinvent the experience. It took some time to endear Americans to the Automated Teller Machine, the ...

◗ fastcodesign.com/3040725/asides/the-history-of-atm-design ...

ATM Designs | LinkedIn

ATM Designs is a self owned company specialized in designing, creating and maintaining tailored websites for individuals and small businesses.

in linkedin.com/company/atm-designs

Figure 12.4 Sample Internet search results for ATM design

The team next moves to collect data on ATM fraud using PINs. In addition to insight from current public opinion and available market research, they investigate available products to explore the possibilities for adaptation of their current product lines, and new developments for the possible introduction of a new product.

They supervise a disassembly of a competitor's product to discover how it is designed and manufactured—a process termed *reverse engineering*. Clearly, the manufacture of a direct copy of a competitor's product may well run into legal problems if it is protected by trademarks or patents, in addition to being potentially unethical. However, the information garnered from such a disassembly exercise may help overcome a design or manufacturing challenge without incurring legal problems. To prevent competitors from discovering closely guarded secrets, manufacturers will often add red herrings to a product or design to derail attempts at reverse engineering.

Throughout the design process, the team members maintain accurate notes, records, and files for the information they collect. Sketches and diagrams of similar products or competitors' models may be useful when they begin to develop their solutions.

Stage 2: Imagine

Prior to the brainstorming session, the team decides to re-evaluate the list of preliminary working criteria and assigns a weighted percentage of importance to each of the factors. After eliminating the criteria of "appearance" and "environmental" from their list, the resulting working criteria and their respective assigned weights are as follows:

Cost = 10 percent
Reliability = 25 percent
Security = 10 percent
Feasibility = 15 percent
Convenience of use = 20 percent
Public acceptance = 20 percent

A team of engineers, sales staff, managers, supervisors, research and development staff members, production workers, computer specialists, and selected clients from the banking industry is assembled. Their assignment is to imagine as many options as possible for uniquely identifying ATM customers in a series of creative problem-solving/brainstorming sessions. Instructed to make no judgments or negative comments on any idea, they propose the following:

1. *Modification of the current use of PIN access codes:* Adjust current access methods to tighten security.
2. *Individualized signature verification:* An ATM attendant matches the cardholder's signature to one written on the spot by the person standing at the machine.
3. *Machine recognition signature verification:* A scanner inside the ATM matches the cardholder's signature to one written on the spot by the person standing at the machine.
4. *Voice recognition:* The person speaks his or her name into an ATM equipped with a speakerphone and an internal computer matches the voice to that of the cardholder recording on file.
5. *Speech pattern recognition:* The person speaks a complete sentence or longer phrase into an ATM equipped with a speakerphone and an internal computer matches the voice characteristics and speech pattern to that of the cardholder recording on file.
6. *Fingerprint match:* The ATM has a fingerprint pad for the person to press and an internal computer compares the print to one on file for the cardholder.

7. *Blood match:* The ATM has a device to prick the person's finger and collect a small blood sample for instant DNA compatibility analysis with the cardholder's profile by an internal computer.

8. *Hair sample match:* The person inserts a hair sample, with root attached, into the machine for instant DNA compatibility analysis with the cardholder's profile by an internal computer.

9. *Eye iris match:* The ATM is equipped with a camera that takes a picture of the person's iris structure and an internal computer performs a comparison to the cardholder's profile on file.

10. *Breath analyzer match:* The person breathes onto a sensitized glass plate, the output of which is compared to the cardholder's profile on file by an internal computer.

11. *Dental identification match:* The ATM is equipped with a sanitized plate attached to a computer sensor that the person bites. The output is compared to the cardholder's profile by an internal computer.

Stage 3: Evaluate

The team now evaluates each of the 11 ideas against the working criteria to decide which should be taken forward to the creation stage.

ANALYSIS OF ALTERNATIVE IDEAS

---

1. *Modification of the current use of PIN access codes:* Adjust current access methods to tighten security.
   a) Similar to current solution
   b) Not reliable or secure
   c) Neutral cost

2. *Individualized signature verification:* Bank employee matches cardholder's signature on file with one written on the spot.
   a) High cost
   b) Not customer-friendly
   c) Difficult to implement

3. *Machine recognition signature verification:* Scanner inside the ATM matches the cardholder's signature on file with one written on the spot.
   a) Technology is currently available.
   b) Public would probably accept this method.
   c) Higher cost to implement
   d) Long-term reliability is untested.

4. *Voice recognition:* ATM is equipped with a microphone to capture cardholder's name for match by an internal computer to voice on file.
   a) Technology is currently available.
   b) Higher cost

    c) Easy to use

    d) Long-term reliability is untested.

5. *Speech pattern recognition:* Similar to above, cardholder speaks a complete sentence or longer phrase for match by an internal computer to pattern on file.

    a) Expensive

    b) Technology not as well developed

6. *Fingerprint match:* ATM has fingerprint pad to capture cardholder's fingerprint for comparison by internal computer to one on file.

    a) Technology is currently available.

    b) Moderate cost

    c) Easy to use

    d) Long-term reliability is good.

7. *Blood match:* ATM collects small blood sample from cardholder's finger for instant analysis by an internal computer for DNA compatibility.

    a) High cost

    b) Good reliability

    c) Completely unacceptable to the public

8. *Hair sample match:* Cardholder submits hair sample, with root attached, into ATM for instant analysis by an internal computer for DNA compatibility.

    a) High cost

    b) Good reliability

    c) Public acceptance would probably be very poor.

9. *Eye iris match:* ATM is equipped with a camera to take a picture of the iris structure and perform a comparison to profile on file.

    a) Technology is currently available.

    b) Moderate cost

    c) Very easy to use

    d) Long-term reliability is excellent.

10. *Breath analyzer match:* Cardholder breathes onto a sensitized glass plate for analysis by internal computer to match pattern on file.

    a) Technology not fully developed

    b) Expensive to implement

    c) Public acceptance would be poor.

11. *Dental identification match:* ATM is equipped with sanitized plate for cardholder to bite for analysis by internal computer to match pattern on file.

    a) Technology not developed

    b) High cost

    c) Public acceptance would be extremely poor.

The proposed solutions are evaluated against the working criteria using a decision table. Those that do not perform well in the various tests are eliminated and we are left with five options. After review by the project manager, three are deemed to satisfy the working criteria and authorized to proceed to the next stage.

Stage 4: Create

Models are now developed and built for the three proposed solutions: #4 (voice recognition), #6 (fingerprint match), and #9 (eye iris match).

A variety of different diagrams and sketches for each concept are developed and reviewed. The team members have many different styles, options, and units to consider. For each idea, they decide to develop a computer-generated model and also have craft workers in the model shop construct the prototypes shown in Figures 12.5 through 12.7.

Idea #1: Fingerprint Scanner with Monitor, Keypad

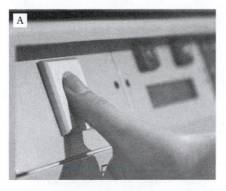

Idea #2: Fingerprint Scanner, Small

Idea #3: Fingerprint Scanner, Larger

Idea #4: Fingerprint Scanner System

Figure 12.5 Idea concepts for fingerprint identification.

Internal Eye Scan Unit

Idea # 1: ATM Wall Unit with
Internal Eye Scan

Idea # 2: ATM Free-Standing
Unit with Internal
Eye Scan

Idea # 3:   Remote
Eye-Scan Unit

Idea # 4: ATM Top-Mounted
Eye Scan Unit

Figure 12.6 Idea concepts for eye iris identification.

Voice Recognition Unit

Idea # 1: ATM Wall Unit with
Voice Recognition

Idea # 2: ATM Free-Standing
Unit with Voice
Recognition

Idea # 3: Ceiling or Wall-
Mounted Voice
Recognition Unit

Figure 12.7 Idea concepts for voice recognition identification.

Stage 5: Test

Based on the working criteria, the team decides to concentrate on specific tests for the prototypes that include technical feasibility, quality performance, error rate, and ease of use. The results of the tests indicate that proposed solution #9 (eye iris identification) performs very well, as do the other two. For solution #9, the following observations were made:

- **Technical feasibility:** Current specialty cameras and computer systems can identify a human being by scanning the iris of the eye and extracting over 400 identifying features from which a digital profile is created for database matching.
- **Quality performance:** Eye scans from 512 individuals were captured, some of whom had their iris features stored in the database. The program correctly identified all those on file, rejecting everyone with no matching profile.
- **Error rate:** This was measured to be 1 in 131,578 cases, by far the lowest rate among any type of biometric testing, including fingerprint and voice recognition. In addition, the iris scan provides long-term reliability since the iris does not significantly change with age as do other body parts.
- **Convenience of use:** The technology available provides a hidden camera able to identify an approaching customer and zoom in on the right iris. As the cardholder draws closer, security access control makes a positive identification in a few seconds.
- **Cost estimation:** While not a test per se, sufficient data are available to estimate the cost of this system at $5,000 per unit. While the initial estimate is high for machines that typically sell for $25,000 to $30,000, it is predicted that the cost will drop as units are mass produced.

The compiled data are evaluated by the project manager who determines that the results consistently meet the working criteria and overall goal. From these data a decision table is constructed (Table 12.3).

The results confirm that the eye scan identification system delivered the best performance, and the project manager decides to move ahead with this solution.

Table 12.3 ATM Security System Decision Table

Working Criteria	Points Available	Voice	Fingerprint	Iris
Cost	10	6	8	8
Reliability	25	20	22	24
Security	10	9	7	9
Technical feasibility	15	15	14	15
Convenience of use	20	15	16	18
Public acceptance	20	16	17	18
Total	**100**	**81**	**84**	**92**

Manufacture and Commercialize

The management of AAA gives approval for the iris scan identification project to move into production, starting with the formation of a new team to undertake all the necessary preparation for manufacture of the ATM units. Questions concerning overall costs and financial and labor commitments have been resolved. The finance department has developed a project budget, and the purchasing department has begun the process of obtaining bids for the needed parts and supplies. The legal staff is resolving final legal issues, including patent applications and copyright materials.

A group of engineers has been selected to oversee the production startup, which includes obtaining the necessary production space, facilities, equipment, personnel, and production timetables. When in place, a pilot production process begins.

The marketing, advertising, and public relations staffs contact prospective clients and customers to promote the eye scan ATM system. Sales personnel involved with selling and distributing the product are trained to work with clients and provided with appropriate promotional literature including training and operations material.

With all parties in agreement that the criteria have been satisfied and the overall goal has been achieved, production begins. Once the eye scan ATM system is in full production and commercialization, the project manager, key supervisory personnel, and team members who had significant input to the project gather for a final project review and assessment. As described earlier, the project team is disbanded and the iris scan ATM system's performance is reviewed, including the latest data on production efficiency, quality control reports, sales, revenues, costs, expenditures, and profits. A comprehensive report containing all inputs to the design and detailed aspects of the product's behavior is compiled for use as a reference for future project managers and teams to consult.[7]

12.4 Using the Engineering Design Process—Backpack

Background

The backpack is an ubiquitous item, especially on school and university campuses. Indeed, it is so omnipresent throughout our society that any attempts to produce a universal solution are probably doomed to failure. However, posed as an open-ended problem, the design of a universal backpack is an almost perfect vehicle to learn the engineering design process. When we start to think about a backpack, we discover that the possible variations are almost limitless (Fig. 12.8). This is principally because the item is used for many different activities by individuals who span a broad range of

---

[7] The iris identification system is currently being piloted in several British and western U.S. locations. The information presented in this case study is a simulated application of the engineering design process.

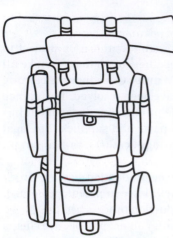

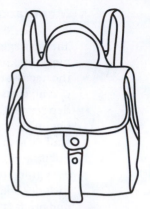

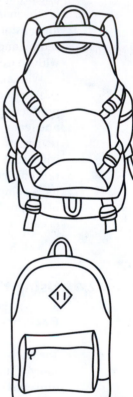

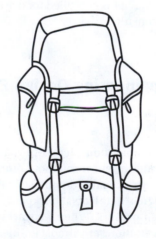

Figure 12.8 Ideas for backpacks

age, physical size and strength, each with very different demands. Notwithstanding our concern that a universal backpack design is probably unattainable, it is still an intriguing design problem if for no other reason than the enormous existing market—annual sales of backpacks run into the millions.

Stage 1: Define

In addition to the mechanical, biomechanical, materials, and production engineers recruited for this task, the team would probably include fashion designers, marketers, anthropologists, college students, social scientists, luggage designers, and travel specialists, to name but a few. In our hypothetical scenario, the team creates a mind map (Fig. 12.9), a useful construct for the visual organization of information or ideas. Broad classification of the arenas for backpack use might include school, college, urban commuting, aircraft carry-on baggage, and backcountry trekking. Since each of these end uses will have different working criteria and goals, the team quickly decides to narrow the focus to optimize an initial design. They eliminate consideration of backpacks for school, backcountry trekking, carry-on baggage, and urban commuting in favor of college use. They conclude that this choice represents a compromise design that, with modification, could be quickly adapted to include a broader section of the market such as school use.

The team includes members with a technical background since they are knowledgeable in the behavior of materials, their mechanical properties, and the constraints these impose. Other members of the team are responsible for answering questions pertaining to the appearance and customer satisfaction of the design. College students use

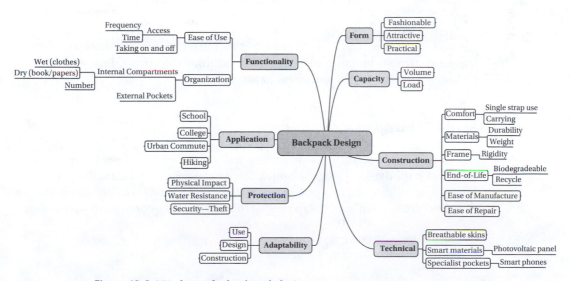

Figure 12.9 Mind map for backpack design

backpacks to carry around books, computers, snacks, sports clothing, and anything else they might require during the day. The group discusses the function and use of different backpack models and collects information related to the problem. Wishing to incorporate new materials and developments into their solution, they use public opinion and market research to understand the lifestyles and expectations of current students.

The team starts with the working problem statement: "Design the ultimate college backpack." They also pose ancillary questions:

- "What does a student use a backpack for?"
- "What are the most important functions of a backpack?"
- "What do students look for when they purchase a backpack?"

Working from their mind map, market research data, and examples of the best-selling backpacks, they develop the following baseline criteria to drive their design:[8]

- **Durability:** The backpack will be heavily used in a plethora of environments.
- **Protection:** It is likely the backpack will contain a laptop computer or tablet, which needs to be protected against physical shock and remain dry.
- **Function:** The backpack has to satisfy a variety of operational requirements, and it needs to do them well.
- **Organization:** The backpack needs to accommodate student requirements for organization of items—for example, separation of expensive textbooks from wet sports clothing.
- **Ease of Use:** The backpack will be frequently accessed and taken on and off.
- **Comfort:** The backpack might be carried for reasonable periods of time.
- **Cost:** The solution must be affordable to the average student.

After discussion, the team re-evaluates its preliminary problem statement and amends the goal statement to: "Design and manufacture a rugged backpack for college student use to accommodate, organize and protect electronic devices, textbooks, notebooks, a water bottle, and extra clothing."

Stage 2: Imagine

Before embarking on their brainstorming session, the newly assembled team members review the data and assign weighted percentages of importance to the factors they retain. After discussion, they conclude that "Ease of Use" and "Comfort" fall within the category of "Function" and assign weights to the remaining criteria as follows:

Durability = 25 percent
Protection = 25 percent

---

[8] These suggestions make no claim to be the right answer—they are here to stimulate your creative thinking.

Function = 25 percent
Organization = 15 percent
Cost = 10 percent

The team, now comprising mechanical and biomechanical engineers, marketers, luggage designers, anthropologists, university students, and others who can offer a different perspective on the problem, start the brainstorming process and generate this list of ways to improve student backpack design:

1. Internal waterproof, padded compartment to protect laptop/tablet
2. Custom exterior pockets for wet clothing, smartphones, etc.
3. Integral, external waterproof cover used to protect against heavy rain
4. Integrated, interior, lightweight frame to provide independent structural support—protect papers, books from getting squashed
5. External straps to accommodate wet clothing
6. Replace compartment zippers with Velcro for rapid access.
7. Make external pockets detachable.
8. Overall access controlled by single zipper with integrated lock

At this juncture, the project manager is comfortable with the results and allows the team to move ahead in the design process.

Stage 3: Evaluate

To reduce the number of options for prototyping, the team evaluates the ideas.

ANALYSIS OF ALTERNATIVE IDEAS

1. *Waterproof, padded compartment.* Protect laptop/tablet.
   a) Easy to use
   b) Similar to current designs
   c) Convenience level not as high
2. *Custom exterior pockets.* Provide specialized storage.
   a) Moderate cost
   b) Specification too open-ended
   c) Technology is available.
3. *Integral waterproof cover.* Use in event of heavy rain.
   a) Very convenient
   b) Moderate cost
   c) Easy to use

4. *Interior frame.* Structural rigidity.
   a) Moderate manufacturing cost
   b) Provides degree of protection
   c) Prevents backpack from being compressed
5. *External straps.* Use to transport wet clothing.
   a) Convenient
   b) Low cost
   c) Items lost if improperly secured
6. *Replace zippers with Velcro straps.* Rapid access to compartments.
   a) Low cost
   b) Not very reliable
   c) Lose effectiveness over time
7. *Detachable external pockets.* Customized design.
   a) Low cost
   b) Choice of size
   c) Easily lost
8. *Single zipper with built-in lock.* Protect contents from theft.
   a) Moderate cost
   b) Easily performs intended function
   c) Increases access time and is generally inconvenient

After comparison to the working criteria, four solutions—option #3 (waterproof cover), #4 (internal frame), #7 (detachable pockets), and #8 (built-in lock)—are retained for further development.

Stage 4: Create

Several prototype backpacks incorporating one or more of the proposed ideas are built and supplemented with CAD models to explore their structural behavior with an eye to the use of composite materials. After reviewing the analysis and criteria, the team decides to continue the process with all the options suggested.

Stage 5: Test

The team of engineers and technicians now initiates the testing process. They develop procedures to measure the performance of the different ideas against the criteria developed in Stage 1 and construct a decision table (Table 12.4).

Tests show that solutions #3, the integral waterproof cover, and #4, the interior frame, are the preferred options so the project manager suggests a combination of these solutions to complete the design process. The team members organize their data and prepare to move to the manufacturing stage.

Table 12.4 Backpack Design Decision Table

Working Criteria	Points Available	Cover	Frame	Pockets	Lock
		#3	#4	#7	#8
Durability	25	15	18	10	15
Protection	25	25	20	10	20
Function	25	16	18	17	14
Organization	15	10	10	12	15
Cost	10	5	5	8	3
Totals	**100**	**71**	**71**	**57**	**67**

Manufacture and Commercialization

The vital role of communication in the final stages of the engineering design process has already been discussed. In the case of a backpack, the investment required to tool up for the manufacture of perhaps hundreds of thousands of backpacks will be considerable, leaving little room for error or uncertainty. Assuming that all the relevant documentation has been collated and checked for accuracy, process and production engineers will be called in to design and install a manufacturing facility. In parallel with this, the necessary managerial, marketing, sales, and legal structure will be put in place for commercialization of the backpacks.

The last phase of the engineering design process is commissioning the production line. Yields from the initial production runs will be used to review thoroughly the reproducibility of the product and the reliability of the manufacturing process. The properties of the backpacks produced are compared against the production specifications to ensure that there are no weak links, such as poor stitching or misalignment of component parts.

Adjustments and fine tuning will no doubt be required before the design team meets one last time to evaluate the data from manufacturing and sales, and the performance of the improved backpacks. The production efficiency, sales, revenues, costs, expenditures, profits, and quality control reports are reviewed and support the role of the backpack as an integral component of the company's products. The group produces a final report for reference by future project teams.

REFERENCES

BBC, *Time-lapse Film Shows How a Boeing Dreamliner is Built* (9/30/15) /http://www.bbc.com/news/business-34406362 (accessed December 30, 2015).

Burns, K., *Brooklyn Bridge*, Walpole, NH, Florentine Films, 1982.

Petroski, H., *To Engineer is Human: The Role of Failure in Successful Design*, New York, Vintage Books, 1992.

Petroski, H., *Success Through Failure: The Paradox of Design*, Princeton, NJ, Princeton University Press, 2006.

EXERCISES AND ACTIVITIES

12.1   Disassemble one of the devices suggested below and put it back together. Sketch all the parts and illustrate how they fit together to make the device operate. List at least three ways you think the design could be improved. Choose one of the following devices: a) flashlight, b) lawn sprinkler, c) sink faucet, d) stapler, e) inkjet printer, f) computer mouse.

12.2   Prepare a list of questions to be resolved in defining each of the following engineering challenges:

a) Develop an improved manual gearshift for a mountain bike.
b) Develop a chair, with back support, for an individual weighing 150 lbs using nothing other than cardboard boxes.
c) Develop a hands-free flashlight.
d) Develop a theft-proof bicycle lock.
e) Develop a container to easily carry 10 gallons of water for five miles.
f) Develop a secure storage area on a bike for a helmet.
g) Develop an efficient open-air woodstove for baking bread.

12.3   Develop a list of working criteria that could be used in deciding whether to:

a) Accept a co-op job offer from Company A or Company B.
b) Study overseas for the fall semester or remain on campus.
c) Buy a new car or repair your old one.
d) Drive to school or use public transport.
e) Change your major or remain in engineering.
f) Purchase a new computer or upgrade your current model.
g) Live in a detached house or an apartment.
h) Eat no meat for a week.

12.4   Identify five product, structure, or system designs you think can be improved. Pick one and write a preliminary problem statement for the engineering design process.

12.5   Using an item from your list in Exercise 12.4, develop a list of reference materials that would be used in developing possible solutions to the problem. Provide specific examples.

12.6   Get together with three other classmates and brainstorm at least 30 ways to use one of the following objects:

a) Two-foot length of string
b) Ping-Pong ball
c) One plastic soda bottle
d) Page of notebook paper and a two-inch piece of tape

  e) 10,000 plastic grocery bags
  f) Obsolete cellphone
  g) Deck of playing cards
  h) Yo-yo
  i) Metal coat hanger
  j) Empty plastic milk container
  k) 15 paper clips
  l) 1 lb of rotting vegetables
  m) Newspaper and 12 inches of masking tape

12.7 Using the result you developed in Exercise 12.4, prepare a decision table for three possible alternative solutions.

12.8 Read a current printed newspaper or magazine article that describes and discusses a groundbreaking product, device, or system. Prepare a four-page report that analyzes and explains, in detail, each stage of the design process that was probably required in the development of this product or device. Apply the engineering design process discussed in this chapter to your product or device, and use specific examples to support your statements. Make sure you list your sources in the body of the paper or in a bibliography.

12.9 Assemble a team of three students from your class and use the engineering design process to develop a portable, garage-like covering for a bicycle that can be stored somewhere on the bike when not in use.

12.10 Assemble a team of four students from your class and use the engineering design process to build the tallest possible tower that can support a 12-oz can of soda/pop. **Materials:** one roll of masking tape, one package of straws, and one can of soda/pop. **Constraints:** Your tower must be freestanding and remain upright for five seconds under load to qualify for measurement. You have 30 minutes to complete this exercise.

12.11 Assemble a team of five students from your class and use the engineering design process to build a bridge that spans at least three feet between supports in contact with the ground. **Materials:** one roll of duct tape, 10 feet of rope, a 40-inch by 75-inch piece of cardboard, and 14 cardboard slats each 12 inches by 2 inches. After the design is complete, 40 minutes will be allowed for bridge construction and testing. **The test:** One of your team members must walk heel-to-toe across the bridge; extra points are awarded for spans exceeding three feet.

12.12 Assemble a team of eight students to design backpacks for a use other than the one proposed. Start with your own mind map, establish a different set of working criteria, and then follow the engineering design process.

CHAPTER 13
Technical Communications

Engineers must possess the technical skills to complete engineering analysis, evaluation, and design. However, these skills are essentially useless if they cannot be communicated to the range of audiences with whom engineers work. These audiences will have a wide variety of technical skills and might come from backgrounds extremely different from yours. To communicate effectively as an engineer, you need to understand your audiences, the impact you wish to have on these audiences, and the conventions of communication these audiences will expect you to follow (Fig. 13.1). In other words, you will need to employ a rhetorical strategy to make your attempts at communication with these disparate audiences successful.

The rhetorical triangle shown in Figure 13.1 illustrates the considerations to be made in any communication scenario. The triangle represents three questions you can ask yourself to help you to make good choices in any rhetorical context:

- *Who is my audience?* Do I know them personally? What are their attitudes? Needs? Values?
- *What is my purpose?* Am I trying to explain something? Share something? Sell something? How do I want the audience to respond?
- *What is expected of me as a communicator?* Are there conventional approaches to this task? How do I make the desired impression?

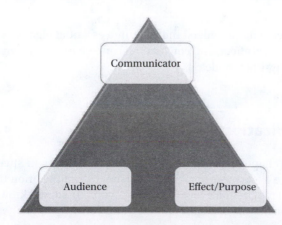

Figure 13.1 The rhetorical triangle

Of course, these questions overlap; they are parts of an overall context for communicating. For example: Is the context professional? Is it technical? What are the time constraints? Still, it is difficult to even begin composing effectively until you have answered the three questions on the rhetorical triangle.

Rhetoric is an often misunderstood word. It can have negative connotations, such as when people dismiss something as "political rhetoric" or when they think of a rhetorical question merely as something that doesn't require an answer. However, rhetoric is all around us, and it is important to all of us. Anything that explains, persuades, or motivates is a form of rhetoric. The lessons of your teachers, the advertisements in various media, and the words and images on your T-shirt are all forms of rhetoric.

You probably make many rhetorical decisions in your daily communication without recognizing them as forms of rhetoric. You would likely wear a suit to a job interview because you are concerned with the impression you will make. You know the conventions for professionalism in your field and dress appropriately. This is a decision about how you will present yourself visually to a specific audience (the interviewer) for a specific purpose (to get a job). This decision implies a strategy involving visual rhetoric, although you would likely never use those terms to describe it.

In engineering, you will be communicating with people who are making decisions concerning your work. These decision makers have minimal time, and you want to have a maximal impact. Therefore, try to keep your communications brief and to the point. Words often have the least impact; instead, information can be communicated with pictures, charts, and graphs. Whereas other types of communication place great emphasis on complex language and nuance (which require more words to explain), technical communication usually requires you to get to the point quickly and simply.

> As we get started, here are three golden rules of communication for engineers:
> 1. KEEP IT BRIEF!
> 2. KEEP IT SIMPLE!
> 3. SHOW IT!

This chapter will help you to understand the three essential elements of effective communication (audience, effect, and conventions) and give tips on communicating in visual, spoken, and written modes.

13.1 Visual Communication

Later sections in this chapter deal with issues involving images in specific media. First, though, there are a few basic principles of visual communication that apply to all media. This section touches on a few basics that everyone should know, including graphics, image composition, color compatibility, and you.

- *Graphics*—A graph or chart that looks acceptable in your technical report might not work as well in a presentation. Posters and Microsoft PowerPoint slides have to be understood more quickly than those in reports and other written documents because the audience usually does not control the pace of a presentation. For this reason, it is a bad idea to simply accept charts and graphs in the form that programs like Microsoft Excel produce by default (Fig. 13.2). Always make sure that labels and numbers on axes of graphs are in fonts that can be easily read. You might also want to remove unnecessary lines from charts or move legends onto the chart itself to enlarge the display of data (Fig. 13.3). The generic chart in Figure 13.2 might be acceptable for a report if it is properly captioned and discussed in the report's text, but changes to the layout will make the same information more readable in a technical presentation.

- *Image composition*—Many people assume that when you compose a picture, you place the subject of the photograph in the center. This might be true in some cases, such as when you are merely showing an item as part of a slideshow. However, artists know that photographic subjects are far more striking when they are slightly off center. The guide generally used for this approach is called the *rule of thirds.* Imagine a tic-tac-toe board laid over top of your image (Fig. 13.4). Major components of that image should fall on those lines with the most powerful aspects of the picture at the intersections.

- *Color compatibility*—The use of color in communication is critically important. A little basic understanding of color theory can help you a lot with color choices. If you look at a color wheel on this book's accompanying website (Fig. 13.5), you'll see that colors that are directly opposite each other are called *complementary colors.* These are colors that work well together, such as the very well-known combinations of red and green, or purple and yellow (gold). However, when using

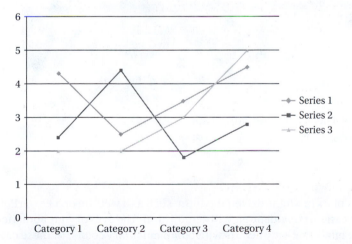

Figure 13.2 A generic chart generated by Microsoft Excel's default settings.

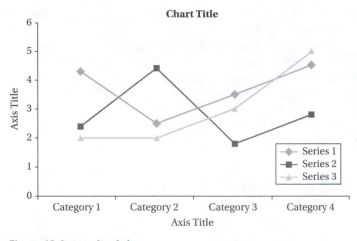

Figure 13.3 An edited chart

Figure 13.4 Photograph illustrating the "rule of thirds."
Source: Teresa Bowles

these colors in a light-driven medium, such as a website or a PowerPoint presentation, the font often will actually hurt the eyes of your audience due to a strobe effect that only occurs with complementary colors. In these cases you can instead choose something called a *split complement*. To do this, look at your color wheel, select the base color you know you want to use, and then draw either an

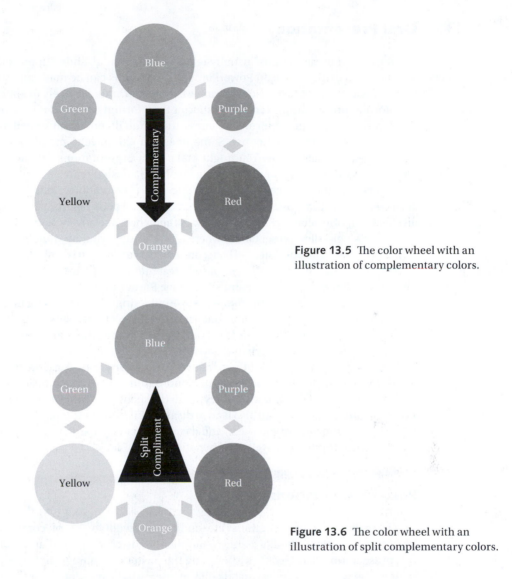

Figure 13.5 The color wheel with an illustration of complementary colors.

Figure 13.6 The color wheel with an illustration of split complementary colors.

equilateral or isosceles triangle with that original color as its topmost vertex (Fig. 13.6). The angles that form the base will point to split complementary colors.

- *You*—You communicate visually through your clothing, your posture, and your expression. Remember that audience members begin taking visual cues about you from the moment they can see you, perhaps even before you begin your presentation. If you run around with a panicked look on your face and papers falling out of your briefcase as you try to load your presentation onto the computer in the last minute before you are scheduled to begin, you will not fill your audience with confidence in your message. Look and act confident and competent, and your audience will be more open to whatever you have to say.

13.2   **Oral Presentations**

The term "oral presentations" usually refers to multimodal, slide-supported presentations, often using Microsoft PowerPoint. Presentations can be made and supported in many ways, and as technology advances, more methods are likely to appear.

Informal presentations tend to be shorter than formal ones and do not require as much preparation time or formal business attire. You do not usually need to prepare a PowerPoint presentation, either. Sometimes you will have to make impromptu or spontaneous presentations. You might make such presentations in a meeting with your community partner or at a meeting involving a design team.

Posters are another means of supporting oral presentations. Many students are surprised to hear that posters exist beyond grade-school book reports, yet posters can be an effective tool for sharing ideas with a diverse audience. A relatively new variation of this sort is the single-slide presentation, where a talk is supported by what is essentially a poster in digital form, but instead of being presented on an easel, it is projected on a screen.

Another example, Prezi, is cloud-based presentation software offering a nonlinear alternative to the basic PowerPoint slide show. Rather than going through a stack of slides, the presenter guides the audience around a single canvas, going into depth in specific areas by zooming in and out on different sections. This format can be a compelling method for single-slide presentations. On the other hand, this format can also invite new problems, such as drawing attention to the new format and therefore away from the content. The nonlinear nature can also be confusing if the presentation is poorly organized. Because Prezi is cloud-based, rather than residing on your computer's hard drive, some also worry about the security of proprietary information. Prezi can be useful, but it can also lead to disaster. Like any new tool, be sure you have a need for the tool before you use it, and if you do decide to use it, be sure you become familiar with its pitfalls as well as its benefits.

Poster Presentations

Traditional poster presentations are audience-driven forms of communication. Poster presentations usually feature a number of posters displayed at the same time in a presentation hall. People walk among the posters, stopping when a title catches their eye, at which time they can talk one on one with the presenter. For this reason, it might be effective to think of the poster as a visual abstract. It should cover the major points of a project but should not go into the sort of detail one would expect from a written document. Often, the most difficult part of poster composition is paring down the details of a complex project to fit the limited space of a poster.

Poster Composition

Posters are not meant to display every detail of a project; rather, they should be a series of conversation starters and supporting visuals that tell the story of your project. If you try to present too much information on a poster, the audience will find the poster intimidating and impenetrable. The following are tips to help you compose a better technical poster:

- *Titles*—Keep them brief and descriptive. Many times, the titles of technical papers are very long and contain all sorts of jargon. Long titles are attention killers on a poster. Remember that a title should grab the audience members' attention as they pass. Once you have their attention, then you can begin to explain the poster's content.
- *Images*—Limit the number of images and make sure they are relevant. Some rhetoricians recommend a limit of five to seven images per poster; however, it is difficult to give a range because pictures and posters both come in a variety of sizes. Four 5×5-inch pictures on a 2×3-foot poster might look cluttered, especially if there is a lot of accompanying text. The same number of 5×5-inch pictures and text on a 3×4-foot poster might not look cluttered at all. Use your best judgment to determine if you have too many images. It is a good idea to caption each image, as well as to number the images, so that they can be referred to in the poster text.
- *Image resolution*—Raster images (e.g., photographs) taken from the web usually have a resolution of 72 pixels per inch (dpi). The standard for printed images is 300 dpi. This is why enlarged pictures sometimes become blurry. If you want to use a 5×5-inch photograph, make sure that it has a high enough resolution to look sharp at that size: $(5 \times 300) \times (5 \times 300)$. Images with a resolution of 1500×1500 or higher will be sufficient for a 5×5-inch photograph.
- *Fonts*—Make sure that your fonts can be easily read from a distance of six to eight feet. The easiest way to do this is to locate the smallest font on your poster, zoom in to 100 percent view on the computer screen, and then step back six feet. If you can read your font, then your font sizes are sufficient. It is also wise to avoid novelty fonts because many of them are difficult to read and some could be perceived as unprofessional.
- *Charts, tables, equations, and graphs*—Like images, these elements should be limited in number. Also as with images, it is difficult to give a specific range of the number of images to use. Use your best judgment to ensure that the poster does not look cluttered. Charts, tables, equations, and graphs should also be thoroughly captioned and numbered.
  - Be careful using equations and tables. Although these are clearly important for understanding concepts and organizing data, they are not always the best way to present information. In the same way that a picture is said to be worth a thousand words, a graph is worth a thousand equations, and a chart is worth a thousand tables. Audiences tend to understand things shown visually more easily than they follow the logic of an equation or table. Whenever you have the option to explain a concept using an image, a chart, or a graph, you probably should!
- *Color*—Color can help organize your poster, or it can be a distraction. Background colors can bring unity to portions of a poster, and they can imply

groupings of similar information. However, colors can also have connotations that are not helpful to you. For example, using your school colors from college might reveal your pride in your school, but it might also make your audience think of you as a student when it would be to your benefit to have them think of you as a mature professional.

— It is also important to be aware that colors have connotations, and sometimes their connotations can shift over time. For example, if you ask members of the younger generation what they think of when they see the color green, they will probably mention environmental issues. In fact, the phrase "going green" is often meant to imply that environmental concerns are being taken into account. Conversely, to older audiences, green might carry very different connotations, implying greed or perhaps envy. They might think of being "green with envy" or think of earning some "greenbacks."

— Usually, when audience members see colors on a chart or a graph, they will try to assign a meaning to that color. Often, that meaning is intentional and is explained through a color key somewhere on the chart itself. However, because poster contents sometimes come from multiple sources and perhaps multiple authors, it is important to be consistent with color denotations throughout your poster. Whatever red might mean on the first graph we see, its meaning should not change as we proceed to the next charts and graphs. Be consistent with color!

- *Layout*—In Western cultures, readers usually begin reading in the upper left corner, moving downward through columns the same way one would read a newspaper.

— Posters are not newspapers, so they aren't always composed in columns. If you deviate from this pattern, you'll need to include numbers or arrows or some similar clue to help readers know how their gaze should move over the poster.

— Composing in columns is an established convention for poster layout. Figure 13.7 illustrates how information is usually presented in three or four vertical columns if the poster is meant to be viewed in a portrait orientation (where the height exceeds the width).

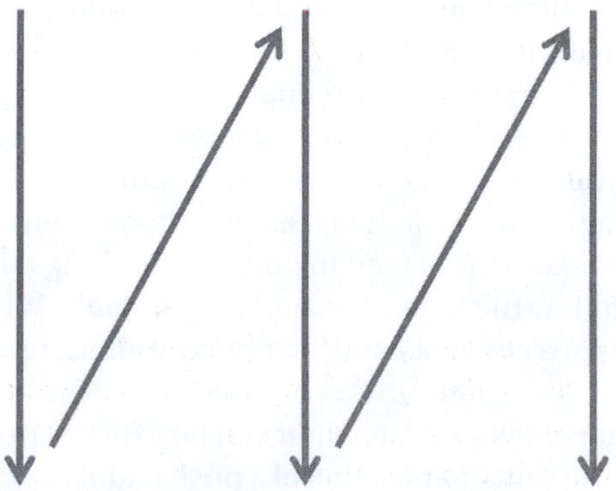

Figure 13.7 The conventional layout of most technical posters

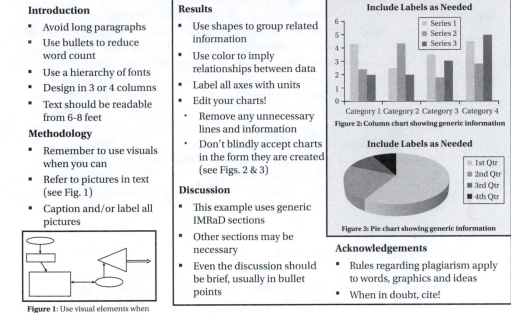

Include a Brief, Descriptive Title

Names, Credentials, and Affiliations Usually Go Here

Introduction

- Avoid long paragraphs
- Use bullets to reduce word count
- Use a hierarchy of fonts
- Design in 3 or 4 columns
- Text should be readable from 6-8 feet

Methodology

- Remember to use visuals when you can
- Refer to pictures in text (see Fig. 1)
- Caption and/or label all pictures

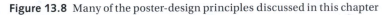

Figure 1: Use visual elements when appropriate (photos, charts, & graphs)

Results

- Use shapes to group related information
- Use color to imply relationships between data
- Label all axes with units
- Edit your charts!
  - Remove any unnecessary lines and information
  - Don't blindly accept charts in the form they are created (see Figs. 2 & 3)

Discussion

- This example uses generic IMRaD sections
- Other sections may be necessary
- Even the discussion should be brief, usually in bullet points

Include Labels as Needed

Figure 2: Column chart showing generic information

Include Labels as Needed

Figure 3: Pie chart showing generic information

Acknowledgements

- Rules regarding plagiarism apply to words, graphics and ideas
- When in doubt, cite!

Figure 13.8 Many of the poster-design principles discussed in this chapter

- *Rhetorical triangle*—As always, ask yourself the three questions about audience, purpose, and conventions implied by the rhetorical triangle shown earlier! Remember the three golden rules of communication: keep it brief, keep it simple, and show it with pictures, charts, and graphs.

Figure 13.8 illustrates many of the poster-design principles discussed in this chapter. This poster is not meant to be a template, but it does illustrate several of the principles discussed in the section on poster composition.

Presenting a Poster

Remember that a poster is not always a standalone presentation. Although it might have a life beyond the event for which it is composed, it will usually have to be presented in person at a conference or other similar event. The following are some tips for making the most of your poster presentation.

- *Poise*—Don't hide from your audience. Inexperienced poster presenters often try to avoid answering questions about their posters by avoiding eye contact with audience members. This defeats the purpose of composing the poster in the first place! Don't be pushy, but be enthusiastic and friendly. A smile will go a long way toward helping you share your project with a large audience.
- *Message*—Have a two-minute talk prepared that covers the basic elements of your project. When conference-goers approach you and your poster, they might have a level of curiosity without having formulated any questions. Your brief talk orients them to your project and introduces them to the elements of your poster.
- *Co-presenters*—If you are presenting a poster as a group, everyone in the group should be conversant on all aspects of the poster, even if the actual work on the project was segmented. Sometimes more than one audience member will engage the group, making it impossible to pass off the presentation to co-presenters.
- *Practice*—Practice presenting the poster to friends and family before you go to the conference. A few practice runs can help you identify the areas where you are weakest in explaining your project, as well as help you find some good phrases to use in your answers.

PowerPoint Presentations

A good presentation renders a service to your audience. A presentation is an opportunity to take audience members from where they are to where you want them to be. The best presentations will do the following:

- Give valuable information the audience probably would not have had otherwise.
- Be in a form that the audience can put into immediate use.
- Motivate and inspire the audience to want to put the presented information into immediate use.

The following tips cover some specific elements of presentations and can help you become a stronger presenter.

The Speaker

Many people would rather do almost anything than speak in front of a group. Being nervous about giving a presentation is normal. It might help you to remember that you aren't expected to be an expert public speaker; you are expected to be a competent engineer. A little nervousness can even endear you to your audience. Just don't let your nervousness take over and ruin your talk. The following techniques will help you combat your nervousness.

- Breathing is important; when you get nervous, your respiration rate tends to speed up and your body can go into adrenaline overdrive (a fight-or-flight mechanism). Controlling your breathing is one way to stay calm.

- — Take three deep breaths before you start.
  - — Practice square breathing, which is inhaling seven counts, holding seven counts, exhaling seven counts, and holding seven counts.
- Visualize yourself successfully presenting your talk.

Practice

If you practice your presentation, you will know how long it takes to complete. If you are nervous, practice will enable you to build in several places in your presentation where you can check yourself and slow down if necessary. The more you practice, the more familiar you become with the material that you are presenting. This familiarity can be important if your information is technical. Showing the audience you are familiar and comfortable with the presented information will help them (and you) gain confidence in your subject. Most likely, you will explain the concepts better as you understand them more fully.

- Have your presentation ready a few days in advance. Being finished usually helps to cut down on your nervousness.
- Rehearse your delivery. When you practice, make it realistic; stand up and actually say aloud the words you intend to say. Many people only practice by scrolling through their slides and going over their lines mentally, perhaps mumbling a few phrases to themselves. This buries your words in the back of your mind. Standing and saying your lines aloud can help you keep the words you wish to say on the tip of your tongue!
- Most people tend to finish their presentations ahead of time because they speak more quickly when nervous.
- Work with your nervousness by tailoring the talk to your strengths. People tend to be most nervous during the initial stages of speaking. One or more of these tips can be used throughout the presentation.
  - — Use humor. This makes the atmosphere less formal. Start your presentation with a joke, especially one that pertains to your presentation. However, be careful making jokes if you haven't done everything you were supposed to accomplish. If you were supposed to present ten deliverables and you only delivered eight, avoid humor. Failure to do what you were supposed to do can make the best joke seriously unfunny!
  - — Use props. It gives you something to do with your hands. A show and tell will interest the audience. Be sure the audience can see your prop. If it is too small for people in the back to see, you might be able to pass it around, but sometimes an audience is too large for this method to work smoothly. In fact, if the crowd is too large, sometimes the props can be lost or broken. Weigh the pros and cons carefully before bringing in a fragile or one-of-a-kind prop.
  - — Use your nervous habit. Work with that habit to your audience's benefit. If you like jingling change (coins) in your pockets, pull something visual out of your pocket to start your presentation.
  - — Use trivia. Focus your audience's attention or teach the audience something interesting that pertains to your talk.

Stage Presence

In the theater, how and where you stand is vital. Because delivering a presentation is essentially a type of theater, these tips can help you make the most of your performance as a speaker:

- Never turn your back to your audience! Not only is it a bit rude, but it also breaks eye contact with your audience and often makes it difficult for your audience to hear you clearly.
- Speak loudly and clearly. If you know that you tend to get nervous and talk too fast when presenting, you can overcompensate a bit by forcing yourself to speak a little slower than your usual pace. It might also help to visualize your words bouncing off the wall at the back of the room. (In smaller rooms, however, be aware that it is possible to be too loud. Nobody likes being shouted at.)
- Plan your choreography if you are presenting in a group. Where will everyone stand? Will you all be on one side of the screen or will you flank it? Will one person advance the slides for all speakers? It's best to answer these questions beforehand. Additionally, your team will look successful if you have a common element in your clothing (for example, everyone wears a white shirt, or has some red in their outfit, such as a tie or scarf).
- Stay out of the projection field. Nobody looks professional or convincing with words on his or her face. If the light from the projector is causing you to squint, then simply step to the side until it doesn't.
- Either stand still or move purposefully. Movements prompted solely by nervousness hurt your stage presence.
  - If you tend to be enthusiastic, use your energy to keep your audience engaged. Make sure, however, that your movements are strong and purposeful. Keep your hands above your elbows if you tend to talk with your hands. Take a few purposeful strides forward as you make strong points. Don't take pointless small steps with stiff arm movements.
  - If you tend to have a more reserved persona, use your poise to project a sense of calm confidence. Stand fairly still, and try to control the pace of your speech. Don't be afraid to move, but if you do move, stay calm, reserved, and controlled in your movements.
- Eye contact is essential. Always avoid talking to the screen instead of to your audience.
  - You can control your audience with your eyes. (If you doubt this, go to a crowded place like a shopping mall, and stare up at the ceiling. It will not take long before you notice everyone around you looking up to see what you are looking at!) Being aware of this control is important because you will keep your audience focused on you, the speaker, by making brief eye contact with audience members throughout the room.
  - In a group presentation, you can do your part when you are not speaking. Any distraction by nonspeakers draws audience eyes away from the speaker. You can help keep audience eyes on the speaker by standing still and looking either

at the speaker or at the slides. When you are speaking, make eye contact with your audience; when you are not speaking, break eye contact with the audience, and look at the speaker or the slides. You will send an additional message of team unity if the non-speakers look in the same place and use the same body language—for example, hands folded in front of you and looking at the speaker.

— If you are explaining data to your audience, it can be difficult to explain specific data on a slide without turning your back to your audience. Using animation to guide your audience's eyes is the best approach. Simply have whatever you want to highlight turn yellow or have a red circle or an arrow appear. However, if you must use a laser pointer or a pointing stick, be sure to point directly to the spot where you want the audience to look, and immediately turn to face your audience again.

Organization

In terms of organizing the information itself, a sample outline of a presentation could be as follows:

- *Introduction*—Who are you, what are your credentials, and why are you presenting?
- *Need*—What is the rationale (need) for your project?
- *Solution*—You have a design or a solution that addresses the need, and here it is. Include major design or solution features and why you did what you did. (Explain how your work specifically addressed the needs.) Be sure to show your final design or explain your solution in its entirety before you begin to address the individual parts.
- *Bottom line*—What are the benefits? How much it will cost?
- *Summary*—Review the main points of the entire presentation. Whatever main concepts you want your audience to remember are the concepts to include on your last slide.

Q & A (Questions and Answers)

Some people like to include a "Questions" slide at the end of the presentation, but rhetorically, this doesn't make sense. Usually, the last point you make will be the most emphatic one. (Think about why your favorite concert performers always play their big hit song last; they want to plant it in your head so you'll go around humming it all week.)

- It is usually better to end on a summary slide or perhaps on an image of your final design than it is to finish on a slide featuring the word "questions," especially because you will usually ask for questions at that time anyway.
- If you have a supervisor who insists on a questions slide, make good use of it. Instead of a blank slide with the word "Questions?" at the top, or one that features

cliché clip art, put something useful on the slide, like a computer-aided design and computer-aided manufacturing (CAD/CAM) image or a schematic. Try to make it something that will prompt thoughtful questions.

- Try to anticipate what questions your audience might ask so that the answers will come to you more easily. Because you are familiar with the contents of your talk, you should know which concepts are difficult to understand or explain. These are areas where you can expect questions. If you can prepare just-in-case slides, slides that are appendices to the main presentation and are only shown if needed during the Q&A, then you will look very well prepared to your audience.
- Remember that questions from your audience are a good sign. If your audience has questions for you, that is a good indication that you have held their interest throughout the presentation.
- Never be defensive in answering questions. If someone asks about information you have already presented, resist the urge to respond in an I-already-told-you-so tone. Presentations relay a lot of complex information in a short amount of time; it is understandable that someone might have missed something. Instead, be grateful that the audience member is interested enough in your project to ask.
- If you get caught unprepared for a question, don't dodge or bluff. Admit it when you're stumped. If you made a mistake, own it. Being honest will always make you look better than trying to hide your mistakes. Another technique is to say that you aren't sure of the answer to the question, but that based on what you do know, here's your best educated guess. In this way, people who may be trying to test your knowledge can understand your train of thought.

Equipment

If possible, try to test your slides on the computer you will use to give your presentation. If that is not possible, try to plan for some common problems:

- Always back up your presentation in more than one place and on more than one medium. For example, you can email yourself a copy of your slides as .pptx and .pdf files, and also save them to a portable drive.
- Ensure the proper video projection system is available, up to date, and compatible. You can often bring your own laptop as a backup.
- For overhead projectors, such as ELMOs, if you can read the document when you place it on the projector, then the audience can read it. You might have to focus, but your document will not be backward or upside down.

Slide Design

Many of the best practices for slide design are the same as the best practices for poster design. The biggest difference between the two media is that posters are very limited spaces. PowerPoint slides are free, and although each slide features a finite amount of space, you can use as many slides as necessary to relay your message.

- If you simply follow the default format that PowerPoint seems to suggest, you will most likely design slides in a topic/subtopic format (Fig. 13.9). This format features a large banner across the top of the slide and a series of bullet points. When your audience members must remember the individual points, this format can make sense. However, slides of this sort are usually most useful as a teleprompter for the speaker. For the audience, they become tedious very quickly!
- More advanced users of PowerPoint could include some sort of visuals or video clips to supplement this format (Fig. 13.10). The key to a good presentation is often the inclusion of high-quality visual elements; however, a small picture shoved into the corner of a slide does little to overcome the shortcomings of this format.
- Because visuals are the key to a strong presentation, it is a good idea to make the visual elements big, sometimes even taking up the full slide. (Remember Rule 3: If you can show it, show it.) The assertion and evidence slide format (Fig. 13.11) makes good use of the entire slide space.
- The assertion/evidence format assumes that most minor points will be made orally. That's fine if a broad understanding on the part of your audience is your goal; however, if the audience needs to understand and retain minor points, animated callout boxes can be added and removed from this format as necessary to give supplemental information (Fig. 13.12). This format allows the audience to see the slide's central idea, the data on which that point is made, and any minor points that the audience might also need to hear.
- No single slide format works for every occasion. To make better choices, remember the three golden rules of communication: keep it brief, keep it simple, and show it! Also remember to consider the three questions from the rhetorical triangle: Who is my audience? What is my goal? What does my audience expect from me?

Getting Started

Now that we have talked about everything you can do before a presentation to maximize its effectiveness, we should think about the presentation itself. The first 15 to 30 seconds of a presentation are critical; establish eye contact and rapport if you are going to turn down the lights for slides or a PowerPoint presentation. You can establish a connection with the audience in several ways:

- You can start by saying, "Good morning (or afternoon or evening), my name is [name], and I am going to discuss [subject.]"
- You can establish the level of formality. For example, you can tell the audience that it is okay to ask questions at any time.
- You can establish your tone with humor (telling a joke), or asking a trivia question with a small reward (e.g., a sticker, pen, or other small item) for the winner.
- Interacting with the audience is an excellent way to establish good rapport. After you introduce yourself and the subject of your talk, you can involve the audience by asking a general question. Those interested might raise their hands with an answer. This information could be useful to you (you can learn something about the audience for a better presentation) and can provide you with a good lead-in to the rest of your talk.

Topic/Sub-topic Format Slides

- Sometimes make sense
- Should be limited to short bullets
- Work best when parts of speech are lined up or "parallel"
- Can be more effective when bullets are animated to appear one at a time
- Are often more useful to the speaker as a teleprompter than they are to the audience
- Quickly become tedious to your audience!

Figure 13.9 Topic and subtopic slide

Topic/Sub-topic Format Slides with Visuals

- Work better than words alone
- Often focus more on words than graphics
- Allow for specific details to be read and heard by the audience
- Should follow the same rules as topic/sub-topic slides without visuals
- Also tend to become tedious

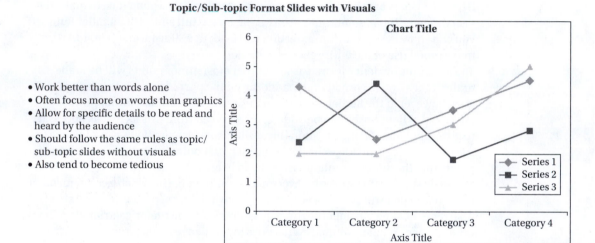

Figure 13.10 Topic and subtopic slide with a visual

Assertion/evidence slides make better use of space and are easily understood by the audience

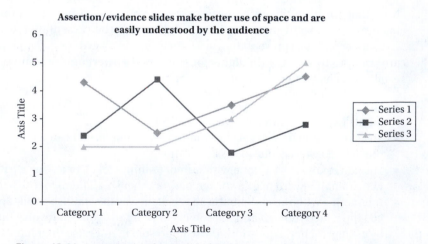

Figure 13.11 Assertion/evidence slide format

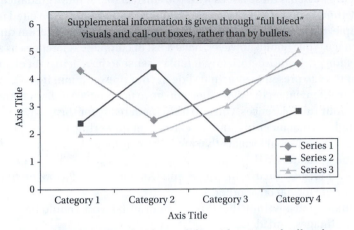

Assertion/evidence slides make better use of space and are easily understood by the audience.

Supplemental information is given through "full bleed" visuals and call-out boxes, rather than by bullets.

Figure 13.12 Assertion/evidence slides with animated callout boxes

- During your presentation, monitor your audience. Do they look engaged, confused, or bored? If the audience members look engaged, don't change anything you're doing. If they look confused, you could ask them a question, or say, "I know that this information is difficult and some of you look confused. Can I clear up anything for you?" If they appear bored, you might re-engage your audience with another trivia question or by asking the audience to become active.
- The conclusion of your talk is important. You need to let your audience know that you have finished—for example, "This concludes my presentation. Thank you. Do you have any questions?"
- Stay within time limits! When your talk runs long, you are usually cutting into someone else's time. If you make a presentation or meeting drag on unnecessarily, very few people will appreciate it.

Citations

Most students have no problem remembering the rules regarding plagiarism when writing a paper; however, these rules are often misunderstood or forgotten when communicating in a different medium, such as PowerPoint. The following tips can help you make ethical choices when composing slides:

- Copyright and plagiarism are different issues. Copyright is a legal issue; plagiarism is an ethical issue. Students can use copyrighted material for educational purposes under Fair Use conventions. However, what and how much you can use are limited, so discuss these issues with your instructor for more guidance.

- Many times, you can place a reference on the same slide as the cited material. For example, if you borrow an image from a source, put a brief citation directly below the image. You should, however, have a list of complete references as a part of your slide package. Rhetorically, it makes sense to have it in a set of appendices, rather than to present it as a final slide in your presentation. If you give the acknowledgment on the slide, then you have not plagiarized, yet if anyone wants to see your full references, you will have them ready for the Q&A.
- No single format for citing sources exists in engineering; however, there are several conventional formats that are fairly common. Again, your instructor can give you guidance on this topic.
- Whatever format you use, be consistent! Don't use MLA format on one slide, then author-date format on another.
- Remember that anything that is not your original work should be cited, including sounds, images, and ideas.
- When in doubt, cite! It is far better to be safe than sorry.

13.3 Written Documents

It is a common misconception among engineering students that once they graduate from college, they won't have to write much. Nothing could be further from the truth. Engineers can expect to write technical reports, proposals, technical memos, résumés, cover letters, and a variety of other documents throughout their careers, and contrary to popular belief, you will not be able to delegate this work to office staff. Understanding what is expected of you in any written rhetorical situation is a vital element to succeeding as an engineer.

Technical Reports

A technical report is a document that explains a technical process, such as an experiment or a design. It is generally broken into three parts:

1. *Front matter*—This section usually includes an abstract, a table of contents, a table of figures, and possibly a list of abbreviations. These elements are like a roadmap letting the audience know where the report will take them, so be careful that the tables and lists are accurate and updated throughout the editing process.
2. *Main body*—This section usually follows a generic format called IMRaD (Introduction, Methodology, Results, and Discussion).
   a) *Introduction*—Details the motivation for the project, scope of the work, and any background information
   b) *Methodology*—Details steps of a procedure or design process, usually making good use of graphics, such as CAD/CAM images, schematics, and photographs

    c) *Results*—Details the findings from the procedures described in the methodology section. Often, this information is shown in tabular form.

    d) *Discussion*—Details the conclusions that can be gleaned from the procedure, as well as any recommendations that follow. The discussion should discuss larger implications of the work and should let the reader know what, if any, action is necessary as a result of the project's findings.

3. *End matter*—This section might contain a glossary, an index, and any necessary appendices.

A variety of other elements might be required for a technical report:

- Qualifications of authors
- Budget estimates
- Limitations
- Research protocols
- Error management
- Literature review
- Regulatory requirements
- Glossary of terms
- Specifications
- Technology overview

> Another—perhaps easier—way to think about the main parts of a report is consider an ABC format:
> 1. **Abstract** gives a summary of the main points.
> 2. **Body** supplies the supporting details.
> 3. **Conclusion** gives readers what they need to act.

Technical reports differ in style from the kind of writing you might have done in English class. Technical reports are usually written in third-person point of view, meaning they don't use the first-person pronouns *I* and *we*. Technical reports do not directly address the reader using the second-person pronoun *you*. They also avoid flowery prose and complex sentence structure that pays strict attention to poetic elements of rhythm, alliteration, metaphor, and simile. Instead, they follow the three golden rules from the beginning of the chapter:

1. *Keep it brief*—Technical reports avoid unnecessarily complex phrasing. For example, instead of saying "as a result of," a better technical writer will say "because." Instead of redundant phrases like "current status" or "past history," the experienced technical writer will simply say "status" or "history."

2. *Keep it simple*—Instead of long paragraphs, technical writers usually use bulleted lists, which can be read quickly and easily. The best technical writers also avoid being repetitive unless they are trying to emphasize a point. Additionally,

headings and subheadings are used to move from point to point, rather than transitional words or phrases. Headings are hierarchical, with the more important headings in larger, bolder fonts.

3. *Show it*—A technical report usually illustrates its points with pictures, charts, and graphs. These figures should be captioned, numbered, and referred to in the text with phrases like "as shown in Figure 1."

Make sure you fully address the scope of the project in a technical report. If you were asked to cover ten items, it is not sufficient to cover only eight or nine. And make sure your information is correct: a well-written but wrong answer is useless.

Abstracts and Executive Summaries

One of the unique features of technical reports is the abstract. Abstracts are written for researchers or other specialized readers who need a brief summary of the contents of the full report. A well-written abstract will define the scope of the problem addressed in the paper, will give major findings, and will comment on how these findings relate to theory. An executive summary is essentially the same sort of document, except that its audience is the executives who will often determine the future of your project. For this reason, it will often contain more information related to cost, staffing requirements, and other issues important to someone who is overseeing your project.

Abstracts and executive summaries should have the following characteristics:

- They should be short, usually a paragraph or two.
- They should be written last, after the rest of the report is completed.
- They should be able to stand alone, containing all the important points of the report, and no points that are not in the report.
- All acronyms and abbreviations should be defined.
- They should not refer to tables, figures, or appendices contained in the main report.

Proposals

Written proposals are documents that describe a need or problem, define a solution, and request funding or other resources to solve the problem. Proposals can be solicited (as when an organization asks for proposals) or unsolicited (as when you send in a proposal and ask for funding or for a project to be implemented). Solicited proposals involve a request for proposal (RFP) or a request for application (RFA), which contain the following guidelines:

- What the proposal should cover
- What sections it should have
- When it should be submitted

- To whom it should be sent
- How it will be evaluated with regard to other proposals

For proposals, you follow the ABC format for report writing, but you concentrate on addressing a need:

- *A:* The *abstract* gives the summary or big picture for those who will make decisions about your proposal. The abstract usually includes some kind of hook or grabber, which will entice the audience to read further. The abstract will include the following:
  — The purpose of the proposal
  — The reader's main need
  — The main features you offer and related benefits
  — An overview of proposal sections to follow

- *B:* The *body* provides the details about your proposal. Your discussion should answer the following questions:
  — What problem do you want to solve or what issue do you want to address, and why?
  — What are the technical details of your approach?
  — Who will do the work, and with what?
  — When will it be done, and how long will it take?
  — How much will it cost?

Commonly Used Proposal Lingo

RFP or rfp: Request for proposal

RFA or rfa: Request for application

PI: Principal investigator; the person responsible for the administration of the project

Co-PI: The person responsible for the administration of the project. If there are two or more people primarily in charge or with significant duties for a project, then they are co-PIs.

Proposal: The document that you prepare to request funds for a project

Grant: The money that you receive if you've written a successful proposal and won the contract

Contract: Most granting agencies require you to sign a contract—a legally binding agreement that you will complete the objectives in the project using the money provided by the agency in a time period specified by the agency.

Program officer: The primary contact person at the agency that is conducting the RFP. Any questions that you might have about your project would be addressed to the program officer. These people can often give you extra information that might not be contained in the RFP.

Proposal reviewer: A person who reads the proposal and critically reviews it. Reviewers check to make sure that the science and engineering are sound, the project is important, the objectives are measurable, the proposed plan of work is excellent, the proposal is well written, and the budget is reasonable.

Resubmission: Grant proposals are hard to get! Usually there are many more potential projects to fund than money to go around. Many federal granting agencies fund only about 10 percent of the projects submitted. This means that some excellent ideas and projects will not receive funding. If your project is not funded, look at the comments from the proposal reviewers (these are usually included with a letter telling you that you didn't receive funding). If you do a good job addressing these comments and resubmit your proposal during the next funding cycle, your chances for funding on this resubmission are much better.

Typical sections in the body portion of the report will be given to you in the RFP, or else you will have to develop them yourself, making sure you address the following:

- A description of the problem or project and its significance
- A proposed solution or approach
- Personnel
- Schedule
- Cost breakdown of funds requested

Give special attention to establishing need in the body. Why should your proposal be chosen over others against whom you are competing? What makes it unique?

- *C:* The *conclusion* makes your proposal's main benefit explicit and makes the next step clear. This section gives you the opportunity to control the reader's final impression. Be sure to do the following:
  — Emphasize a main benefit or feature of your proposal
  — Restate your interest in doing the work
  — Indicate what should happen next

Memoranda

A memorandum, or "memo" for short, is a brief technical communication intended to share information with decision makers. You write a technical memo when you need to document your work in writing. Typical examples include memos to go into a personnel file (congratulations or job improvement details) or those to go into technical files (e.g., engineering projects you are working on). To start, memos contain introductory information that summarizes the memo's purpose. The memo's conclusion restates the main points and contains any recommendations.

Résumés

A résumé is one of the most important documents you will ever create. It sells you and your qualifications. Generally, there are two types of résumés: skills résumés and experience résumés. *Skills résumés* are for people who have not yet completed significant work experience. They highlight the skills and talents that will benefit a potential employer, even if you have little or no technical work experience. *Experience résumés* highlight prior work experience related to the job for which you are applying. In general, most engineering students in the first two years of study and those who graduate without technical work experience will use a skills résumé; engineering students with co-op or internship experience in an engineering context would use an experience résumé.

The following sample and bulleted explanation is intended to assist you with more detailed information on crafting a résumé. A sample evaluation form is included, so you can evaluate your own résumé. Here are a few important things to remember about résumés:

- A résumé should be crafted for a specific position or job. This approach shows the employer that you care about the position. Your résumé's objective section, if you choose to include one, should reflect some specificity as a result.
- You might need to change your résumé for scannable format, if you apply for jobs with larger engineering companies. These companies use a computer to scan each résumé submitted to them for keywords. If so, you should consult your career services office or look at the company's recruitment information to determine these keywords and use them on your résumé. The scanner will target you and your résumé as a potential employee more quickly.
- If possible, keep your résumé to one page. Many employers frown on résumés longer than one page.
- Do not include a section entitled "References." Companies assume you have references who can detail your experience and qualifications, so don't waste valuable space on this section.

The format of a résumé is important. Your information should be arranged to be visually pleasing and readable. Use the following strategies in terms of format:

- List dates in reverse chronological order (starting with your most recent information).
- Use at least 11-point font size (preferably 12-point) and professional-looking fonts, such as Times New Roman, Arial, or Calibri.
- Use one-inch margins around the page.

Cover Letters

You should always send a cover letter with your résumé. This document introduces you to the company in a more conversational manner. A good cover letter is no longer than one page and does the following:

- Identifies the position for which you are applying (by number, if so listed)
- Explains why you should be considered for the position (establish a match between the position description and the skills and experience you possess)
- Highlights specific aspects of your accomplishments, especially those that are noteworthy and will make you stand out with respect to other applicants
- Provides detailed information not in your résumé

Cover letters should contain the following elements:

- Your contact information (address, email, and phone number) and the date
- The full name and contact information of the person at the company or school to whom you are addressing the letter
- A formal greeting (e.g., Dear Dr. Jekyll or Mr. Hyde)

The opening paragraph should contain an explanation of how you heard about the position and should specifically identify the position itself. For example:

- As you suggested during last week's career fair at <Your University>, I have enclosed a copy of my résumé for your consideration.
- I am submitting my résumé in response to the position description for a consulting engineer (position #EN1234) posted on (fill in what website you found it on).
- You can also state your interest in the position, and briefly highlight your experience, education, or both, that make you well suited to the position.

The central section of the letter will contain one or more paragraphs in which you describe and draw attention to relevant aspects of your education or experience.

- Indicate your knowledge of and enthusiasm regarding the company
- Expand on things especially of interest to the company/school and things that are important but that you couldn't detail in your résumé

The concluding paragraph should refer to the résumé for further details and politely but confidently ask for an interview.

- Include your contact information (phone and e-mail).
- Thank the reader for his or her consideration.
- Include a formal closing, and make sure you sign the letter.

The following sample letter contains further information on writing a successful cover letter.

123 Main Street
Citytown, NY 12345
June 6, 2015

Jane Doe
Manager
ACME Company
321 Maple St.
Citytown, NY 12345

Dear Ms. Doe:

The first paragraph should get right to the point. If you are applying to a job that is listed by a number, include that number in your letter. Your opening paragraph might read something like this: "As you suggested during last week's career fair at <Your University>, I have enclosed a copy of my résumé for your consideration. After reading the job description on your company's website, I believe that my skills and my strong work ethic make me a good candidate for the position of field engineer (position #1234)."

The second paragraph in any business letter is usually where you go into detail. After having read your to-the-point opening paragraph, the reader can read the middle paragraphs for more specifics. Here is where you can make the items listed on your résumé come to life. Brief narratives about your past experiences and specific skills make a better impression than generic statements about your qualifications. Instead of broadly referring to your leadership skills or work ethic, tell a story where you had to be leader or recount events that demonstrated your high level of commitment. Remember to include details and avoid clichés.

The final paragraph offers an opportunity for follow-up. It might read something like this: "I would welcome the opportunity to further discuss this position with you. If you have any questions or need further information from me, please contact me by phone at (555) 555-5555 or by e-mail at jsmith@email.net. I look forward to hearing from you."

Sincerely,

John Smith (Your signature belongs in the space between your closing and your name.)

Enclosure (This lets your reader know that there was an additional document in the envelope or as an email attachment: your résumé.)

13.4 Revising and Editing

Many people misunderstand what is meant by revision. Some think that every draft of a written document is essentially a restart, and they resent having to do the same task multiple times. The fact is that the process of composing a document is similar to the process of engineering design. In the same way that a draft of an engineering design is not a final design, the first draft of a document should not be a final draft. It should go through the same sort of iterative process of testing, reviewing, and redesigning that engineers regularly employ. Engineers who bring their engineering skills to bear on their communication projects will find far greater success than those who do not.

Whenever you revise a document, it's best to think of big-picture issues first. Although it's always a good idea to catch grammatical errors as you go, it is possible to focus on them too soon. Worrying about commas and periods when you should be figuring out chapters and paragraphs is akin to worrying about what color you will paint a machine before it has been designed.

Note for Group Work

If you are in a group of four, no fewer than eight eyes should pass over every word in your group report. It is not fair to dump the task of editing onto any one person, especially the poor lead author. All of your names are on the report, so you should all be responsible for its contents. Additionally, you will each have different strengths and weaknesses as writers. This means you will each bring something different to the table when it comes time to revise and edit. This approach will also help to maintain a single voice and tone throughout the paper rather than having a disconnected assemblage of voices, tones, and so on.

Revision

Revision focuses on saying what you want to say and organizing it in the best possible method. Revision is a time for making sure that you have not digressed from your topic, that you are providing the deliverables the task demands, and that you are saying what you need to say in an organized fashion. If you worry about spelling and grammar at this stage, you are likely to spend time correcting sentences that might ultimately need to be cut. You might also be less likely to make changes that need to be made because you consider the sentence fixed in terms of grammar.

Make Use of Active Voice

In sentences written in the passive voice, the subject is acted upon by someone or something. In those written in the active voice, the subject of the sentence performs the action expressed by the verb.

- Passive voice: The treatment system was designed by the engineer.
- Active voice: The engineer designed the treatment system.

Strategy for Finding and Correcting Passive Voice

Passive verbs consist of a "to be" verb (am, is, are, was, were, be, and been) and a main verb.

Examples: was conducted, is provided, have been returned

Step 1: To find verb combinations that might be passive, use the "Edit/Find" feature in Microsoft Word to search for "to be" verbs (am, is, are, was, were, be, and been).

Passive: An experiment **was** conducted by technicians last week.

Step 2: Ask yourself: Is the subject performing the verb, or is the verb happening to the subject? If the subject is performing the verb, your sentence is active and probably doesn't require editing. If the verb is happening to the subject, then the sentence is in passive voice.

Passive: An experiment **was** conducted by technicians last week.

The experiment isn't doing anything; the technicians are. Therefore, this is passive, and it needs rewording.

Step 3: Reword the sentence.

Better: Technicians conducted an experiment last week.
Here, the technicians are doing something; therefore, this sentence is in the active voice.

Editing

Once you have gone through several drafts and are sure you are on message and well organized, it is time to turn your attention to details like mechanics, usage, spelling, and punctuation. There are several methods that work well for finding grammatical errors.

- *Method 1*—Read your paper aloud to someone else. Your mouth will catch errors your eyes will not see. When you do this, be sure to have a pencil handy if you use a physical print out or be ready to type if you are using an electronic version so that when you trip over a phrase you can mark it and see if there was a reason on the page for the mistake.

- *Method 2*—Have someone read your work to you. This will give you an indication of how your words flow through another person's mind. Once again, be sure to mark places where he or she trips over words or has other problems reading aloud. This is often an indication of a problem on the page.
- *Method 3*—Lots of focused editing sessions. Rather than sitting down and trying to catch every error in one sitting, focus on a single issue. For example, read the document once for issues of passive or active voice, and walk away. Come back later to address commas. Another time, hit spellcheck. Lots of short sessions work better than one long grueling one.

13.5 Conclusion

In this chapter, you have learned the basics you need for an effective rhetorical strategy. Remember the three questions illustrated by the rhetorical triangle: Who is my audience? What is my purpose? What is expected of me as a communicator? Let these questions guide you through an iterative process regardless of whether you are working on a visual, spoken, written, or multimodal project. Bring your engineering skills to bear on your communication projects, and you will be a successful communicator.

REFERENCES

Alley, M., *The Craft of Scientific Presentations: Critical Steps to Succeed and Critical Errors to Avoid*, New York, Springer, 2003.

Burnett, R. E., *Technical Communication*, 3rd ed., Belmont, CA, Wadsworth, 1994.

Cox, M. R., *What Every Student Should Know about Preparing Effective Oral Presentations*. Boston, Pearson Allyn and Bacon, 2007.

Markel, M. H., *Technical Communication*, 8th ed., Boston, Bedford/St Martin's, 2007.

O'Hair, D., R. A. Stewart, and H. Rubenstein, *A Speaker's Guidebook: Text and Reference*, 4th ed., Boston, Bedford/St. Martin's, 2001.

Searles, G. J., *Workplace Communications: The Basics*, 5th ed., Boston, Longman, 2011.

Williams, R., *The Non-Designer's Design Book: Design and Typographic Principles for the Visual Novice*, 3rd ed., Berkeley, CA, Peachpit, 2008.

EXERCISES AND ACTIVITIES

13.1 Write a one-page answer to the following questions.

a) What color combinations do you see often in your daily life? How do color combinations affect your mood or your thinking about a subject?

b) Find a photograph from at least 20 years ago. How has society's color palette changed since then?

c) Look at an advertisement. Did the creator of the ad employ a complementary color scheme? A split complement? Something else entirely?

d) Look at the clothing you are wearing right now. What does it say to others about who you are as a person? What clothes do you have that might make a different statement?

e) Think about your most effective teachers. What do they do to make you pay attention to their lessons? Make a list of adjectives to describe the way they present themselves to the class.

f) What causes you to lose faith in someone's message? What sorts of behavior make you roll your eyes in contempt at a speaker? How might you avoid these pitfalls as a speaker?

g) Think about a PowerPoint presentation you either must give soon or recently have given. How would you present that information if you were not able to use any words on your slides?

h) It is easy to see how a technical report or a proposal is different from a story or a poem, but how are they similar?

i) What grammatical rule or principle do you routinely get wrong? How can you learn that rule once and for all?

j) How do you use writing in your everyday life, outside of school and work (grocery lists, texting, diary entries, etc.)? How can you make similar writing acts useful to you as an engineer?

13.2 Interview one of your professors or a practicing engineer to learn more about speaking skills. You might ask them some or all of the following questions:

a) Do you ever get nervous before a public speaking engagement? If so, what do you do to relax? What advice do you have for me to relax if I get nervous?

b) How do you prepare for an oral presentation?

c) Do you employ any methods for engaging the audience at the beginning of a talk?

d) How do you keep your audience interested throughout your talk?

e) Do you have any advice for me as a public speaker?

13.3 Go to a public speaking engagement at your university with a speaker of regional or national prominence. Critique this person's speaking style with respect to the guidelines for oral presentations presented in this chapter. What did the speaker do particularly well? What could he or she have improved on? What things did the speaker do that you would like to integrate into your own speaking repertoire?

13.4 Build a paper airplane or other origami figure and write out the directions for building it as you go. Then give your directions to a friend or colleague and ask him or her to build the same paper airplane or origami figure. Watch your friend as he or she builds it, and compare your friend's final product to yours. What did your friend do correctly? What did you do well to enable his or her success? Where did your friend go wrong? What might you have done differently to help your friend avoid his or her mistakes?

13.5   Share a recent slide or poster presentation with a friend or colleague via the Internet. How does using Internet conferencing software (e.g., Skype, GoToMeeting, Connect, etc.) change the way you present your information? What parts of the presentation convert easily to a virtual environment? What new challenges does the virtual environment raise?

13.6   Choose an organization that you are interested in working with for a specific purpose—for example, a community project, summer internship, co-op, or future employer. Write a résumé and cover letter to this organization.

CHAPTER 14

Ethics and Engineering

14.1 Introduction

In addition to technical expertise and professionalism, engineers are also expected by society and by their profession to maintain high standards of ethical conduct in their professional lives. This claim sounds pretty uncontroversial. It also sounds boring.

But if you take a second to think about it more carefully, it actually raises some challenging—and rather interesting—questions. If the idea is just that engineers should understand the difference between right and wrong, isn't that true for everyone, including non-engineers, anyway? Isn't that already just covered by the law, or by workplace rules? Why do we need a special chapter, or for that matter a whole college course, to explain that? Moreover, where do claims like this come from? Who gets to make rules about what engineers can and can't do, and why should you care?

These are important questions, and any good engineer should be able to answer them. This chapter is designed to help you answer them, or at least to start you thinking more clearly about what you'll need to know in order to answer them, by walking you through the basic concepts, questions, problems, and insights central to engineering ethics.

We'll begin with the most basic and most general features of ethics as a field of study before progressing to an analysis of professional ethics in a broad sense. Then we'll be in a better position to review and discuss the ethical obligations particular to engineering. By the end of the chapter, you'll better understand why it's crucial that all engineers learn to think carefully about ethics and the particular obligations attached to the practice of their profession.

Here is a brief overview of the sections of this chapter:

1. Since engineering ethics is a branch of ethics, we start by explaining exactly what ethics is, and how it is related to other familiar concepts connected to how we live our lives, like morality, law, and even personal preferences.
2. Then, having clarified the nature of ethics, we turn our attention to what ethics has to do with engineering in particular. That is, we begin examining the subfield of ethics we call engineering ethics, which explores and clarifies the ethical obligations of professional engineers. With those clarifications in place, we will

be in a better position to understand where such obligations come from and why they aren't just arbitrary lists of rules. This will also help to establish the role that codes of engineering ethics—as laid out by various professional engineering societies—are supposed to play in addressing these ethical problems.

3. It will then be useful to examine one of these codes in detail so that we can better understand what they contain and even why they contain what they do. So in the final section we'll take a look at one such code in detail, namely the National Council of Examiners for Engineering and Surveying (NCEES) Model Rules of Professional Conduct. The discussion will also be supplemented with some model answers to ethical issues, which provide a useful test for comprehension of the content of ethical codes. In addition, broader discussion and essay questions are included to encourage more principled and independent thinking about ethical issues.

14.2   The Nature of Ethics

Ethics is the philosophical study of morality. It is the study of which actions and ways of living have moral value (i.e., are morally right and wrong, good and bad, permissible and impermissible to do) and why. It is concerned with what justifies such moral claims and how we can apply those justifications to real cases of moral uncertainty in order to live our lives well. But to better understand what all of this means, we need to clarify a few basic concepts first.

To begin, the term "morality" refers to the act of making distinctions between good and bad, right and wrong, and permissible and impermissible conduct. This is something that all people do—we all have the basic capacity for morality because we all learn to make distinctions based on moral values in very basic ways, and to differentiate between better and worse ways to live our lives. "Murder is morally wrong" and "helping people in need is morally right" are examples of the kinds of basic moral distinctions we regularly make. This is what we call morality.

But notice that in making these moral distinctions, or in stating these ideas about how properly to live and act, we have not yet said anything about *why* these things are morally right or wrong, what justifies these claims, how we know whether such claims are justified, or whether those same justifications can be used to determine what else has moral value. Such questions are difficult philosophical questions about morality that belong to the field of ethics. To study such questions, and to attempt to answer them, is to do ethics. That is what it means to say that ethics is the philosophical study of morality.

While most people are capable of making basic moral distinctions, few people spend much time thinking carefully about the source and proper justification behind them. That is what ethics is about, and what the subject matter of ethics demands of us. Now, when you begin asking these questions in the context of engineering—by examining claims about what it is right or wrong for engineers to do—you engage in the study of engineering ethics. While most people don't spend a lot of time doing ethics, it's something we require of professionals, including engineers, because of the very nature of such professions. But before we get to all of that, we need to clarify just a few more things about the nature of ethics generally.

First, ethics is the study of *moral* values like good and bad and right and wrong, but it's important to understand that there are other kinds of values that are not moral values, and so therefore not a part of ethics. For example, there are aesthetic values that concern judgments of beauty (e.g., whether this particular bridge or that particular painting is truly beautiful or ugly), and there are also personal values and preferences (e.g., whether playing sports or making art is important to me, or whether I enjoy the taste of coffee or wearing the color purple). Whether the Empire State Building is beautiful is indeed a question about value, but it is not a question about moral value (i.e., about how it is better or worse to live one's life). Consequently, such value judgments are not a part of ethics.

Distinguishing between the different kinds of values matters because moral values, which we settle and justify by doing ethics (i.e., by careful philosophical reflection on how and why we make moral distinctions), are among the most basic and important to human society. The reason for this is perhaps already obvious: judgments about how it is proper to act or right to live one's life bear on all other features of our lives in virtue of being concerned with how *best* to live, or what it means to live *well*. In the next section we will begin reflecting on ethical questions to determine what kind of progress we can make toward answering such big, important questions. But for now the crucial point is that the deliverances of these ethical reflections—the moral claims we have good reason to accept—are so fundamental to our lives that they constrain and direct our pursuit of other values in significant ways.

For example, building a skyscraper upside down might satisfy our aesthetic preferences for novel designs, but the pursuit of such aesthetic values must always be constrained by such ethical considerations as whether the design jeopardizes the safety of the people who live and work in the building because we have compelling reasons to believe that taking actions that jeopardize the well-being of others are not among the actions conducive to achieving the life well lived.

Second, such ethical considerations are so fundamental that they may guide and constrain our social institutions, which operate by establishing and enforcing *policies* such as rules and laws. The moral values and obligations that we establish by doing ethics are more fundamental than policies that tell people what they can and cannot do. Policies are constrained by ethics, but ethical matters are not established or constrained by policy. That is why it is essential to distinguish doing ethics from making and enforcing policy.

Policies are rules or laws that groups of people or organizations, like governments, engineering societies, and even sports leagues, put in place to keep things running smoothly. While it's true that in some cases those groups may create laws or policies based on moral positions, even in such cases it is nevertheless the moral positions that justify the policies, not the other way around. For example, governments have laws prohibiting murder, and while those laws reflect moral positions on murder (e.g., that it is morally wrong), they do not explain or justify such moral positions and cannot be used for that purpose. Instead, the moral positions may help to justify the laws.

One way to see this is to reflect on what we would think if the government repealed laws prohibiting murder tomorrow. Murder would now be legal, but that wouldn't suffice to show that murder is now morally acceptable, or among the actions conducive to living well. Rather, we would continue to think that murder is morally wrong regardless of whether it is legal because that claim is well established by the study of ethics. And, indeed, when laws violate our moral values or conflict with the moral deliverances of ethics, we work to change them so that they better respect our more basic moral values. Unethical laws like racist Jim Crow laws and sexist laws that denied women the right to vote needed to be changed precisely because they were immoral. That is the sense in which the deliverances of ethics—justifiable, rational positions about what is morally right and wrong and why—are more basic than laws.

We can illustrate the same point on laws that govern engineering practices in particular. Suppose that a law governing motor vehicles did not currently require a vehicle recall in the case when engineering defects were found in its construction. Nevertheless, if the defect were shown to violate ethically established moral standards, such as by causing harm to people, then the law would have to be changed to conform to the ethical standard. So here, too, other kinds of values or standards must be suppressed or adjusted in the case of a conflict with ethical standards, which are always of overriding importance.

In the same way that moral values are more basic than laws, they are also more basic than rules. Sometimes organizations adopt rules that reflect moral values, but in such cases the moral values explain the rules rather than vice versa. As we shall see, this is precisely the case with codes of engineering ethics, which are essentially lists of rules or obligations that govern professional engineers. It's not that these rules are the basis of, or justification for, the moral obligations we expect engineers to respect. Rather, the policies reflect moral values that we already have good reason to demand that professional engineers respect, and we then codify them into rules to help others recognize and reflect on them.

Finally, moral values are also more basic than non-moral personal values and preferences. As we shall see, this is especially true in the case of moral values connected to the practice of a profession—that is, to the ethical standards governing the profession. If one's personal values conflict with ethical standards governing professional engineers, then one must suppress those values to resolve the conflict in favor of the ethical standards. For example, if one personally values finding the highest-paying job, but it turns out that the job in question involves some ethically wrong conduct, then it is one's ethical duty not to accept that job, but to take a lower-paying, more ethical job instead. Later on in the chapter we'll say more about why it's reasonable to expect professionals like engineers to prioritize ethical obligations over their own personal preferences and non-moral values.

The Subject Matter of Ethics

We have seen that ethics is the study of difficult philosophical questions about morality, and further that such questions are of fundamental importance to our lives.

Moral values are more fundamental than our laws, organizational rules, and even our personal tastes and preferences because they are about how we ought to conduct our lives. But how do we establish such values? How can we make progress toward answering such big questions as how best to conduct our lives? To answer these questions we have to better understand the subject matter of ethics.

Ethics is the philosophical study of morality. Consider some natural philosophical questions about morality. What is it that makes a particular action right or wrong? Is lying always wrong, or is it sometimes acceptable? What about polluting? Is it ever acceptable to pollute and, if so, under what circumstances? How do we know? What justifies such claims? These are just some of the many, many questions in ethics. They are all philosophical questions about the nature of morality.

But while they're all about morality, they ask about slightly different features of morality. So to help keep them better organized, we can divide them up based on the kind of information they are seeking. For our purposes there are two broad categories into which we can organize these questions about morality. And in fact philosophers organize the study of ethics into different subfields depending on which category of questions they're addressing, so the two categories of questions correspond to two broad subfields of ethics. The first category, called "normative ethics," deals with broader questions about what makes any given action right or wrong, good or bad. The second category, called "applied ethics," deals with assessing the moral status of particular actions.

Normative Ethics

What, in general, is it right or wrong to do? How should I act? What kind of person should I be? What kinds of actions should people take or avoid taking? What principles or guidelines can we use to determine what we should do under any given circumstance? These questions are questions about moral prescriptions called "norms." Norms are standards—claims about what kinds of actions people should take, or claims about what kind of people they should be. The study of moral norms is, quite naturally, called "normative ethics." These are the basic or central questions in ethics, and so they also figure prominently in engineering ethics.

To answer a normative ethical question is to provide an answer to a general question about what kinds of things people ought to do or what kinds of people they ought to be. For example, suppose I think that in general, people ought always to do whatever makes them happiest. Because this is a general answer to a normative ethical question about what people ought to do in order to be moral—it is a claim about what is actually true of moral behavior—it functions as a kind of **normative principle**. It is a principle that prescribes behavior, or sets moral norms, across a range of cases.

Note that it is not a personal principle. It does not say what one person happens to thinks, nor does it reflect personal values or preferences. Rather, it is a principle that is claimed to apply to everyone, to be *true* of morality. It says that the *right* thing to do—what the good person must in fact do—is to take actions that promote his or her own happiness. So we have a proposal for a moral principle, for what is true of morality, and the question is whether we should accept it. How might we make progress on this question?

One way to test whether this is a good principle to endorse—whether it is actually true of morality—is to see whether it leads to prescriptions we would actually be willing to endorse in a range of difficult cases. If it does, we have a reason to consider endorsing and defending the principle. If not, we have a reason to look for a better principle. When we run this test with the current proposal, it quickly becomes clear that it faces serious problems.

For example, suppose someone we'll call Smith is tempted to murder his neighbor because he thinks he would enjoy doing it. Whether this is permissible is clearly a moral question, and so we should be able to use the proposed principle to sort it out. In this case, because the principle says that the right thing for any moral agent to do is to take actions that increase his or her own happiness, it seems to say that it is morally permissible for Smith to murder his neighbor since doing so would increase Smith's happiness. But that seems patently absurd. Such a principle seems to make a mess out of commonsense morality. At this point, we have to choose between whether to abandon the principle and seek a better one or, alternatively, to try to defend or repair it.

When one develops a normative ethical principle into a coherent ethical framework or viewpoint by providing argumentation for it, defending it against objections, and demonstrating its philosophical merit by showing that it makes sense of moral reasons, one begins to develop a **normative ethical theory**. Normative ethical theories may be built upon just a single principle, they may involve multiple principles, or they may appeal to considerations that don't really involve formulating principles at all. But all such theories involve careful and rigorous philosophical argumentation that demonstrates the merits of their approach to thinking about the nature of right and wrong, answering difficult moral questions, and solving moral dilemmas. A good normative ethical theory offers one competing way of consistently answering difficult questions about what it is right or wrong to do.

Philosophers have been developing, refining, and arguing about normative ethical principles and theories for a long time, and at present there is no clear consensus about which existing normative ethical theory is the right one. Because the matter is a philosophical one, achieving absolute certainty seems unlikely. Still, it would be wrong to say that we have made no progress on this issue. Only a handful of normative ethical theories have withstood rigorous philosophical scrutiny through the years.

As a result, ethicists find it helpful to draw on those durable, longstanding normative theories—to apply the principles or frameworks furnished by those theories—in addressing difficult moral dilemmas. Because ethical theories are such a helpful resource for determining how we ought to conduct ourselves, it will be useful to very briefly sketch a couple of the more durable competing theories of normative ethics so that we can see how ethicists use them to deal with real moral dilemmas. We'll take simplified versions of the two most prominent normative ethical theories as illustrations of how such theories generally work. For the moment, we can hold off on making any judgments about which, or either, of these theories is true, or worth endorsing. We simply want to illustrate the relationship between moral judgments, moral principles, and moral theories.

Normative Ethics: Utilitarianism

One longstanding normative ethical theory says that right actions are those that produce the best consequences for everyone involved in or affected by the situation, where "best consequences" is to be understood as maximizing the well-being or utility of as many of those people as possible. This is (roughly) a version of the ethical theory that philosophers call **utilitarianism**. Utilitarianism can be summed up with the slogan "Right actions are those that produce the greatest good for the greatest number." Because it insists that the moral value of an action is to be determined by assessing its consequences, the view is said to be a **consequentialist** ethical theory.

Now let us consider how utilitarianism works with a simple example that comes originally from the philosopher Peter Singer. Suppose you are walking home from school and you encounter a small child drowning in a very shallow puddle of water. To save the child would require only that you quickly run over to the puddle and pick him up. However, you are wearing a brand-new pair of expensive boots that are likely to be ruined by walking through all of the mud created by the puddle, and you admittedly find that prospect rather upsetting. It seems you face a kind of moral dilemma. You can save the child and ruin your boots, or preserve your boots and allow the child to die (and let us suppose for the sake of the demonstration that you don't have enough time to unlace and remove the boots before the child drowns). What is it morally right to do?

If utilitarianism is true, then the right action is the one that maximizes the well-being of everyone involved in the situation. In this case, that includes the child's well-being, your own well-being, and presumably also the well-being of the various people who care about the child as well as the well-being of the various people who care about you. Once we have taken all of these factors into consideration in our calculation of the consequences, it is probably pretty clear that the potential action that is likely to produce the greater net utility—to maximally enhance the welfare of everyone involved—is the act of sacrificing your boots and pulling the child from the puddle.

This example is rather simple, and as cases grow in complexity we may even need to think more carefully about how to assess the potential consequences, much as we would in a complicated cost/benefit analysis. But the basic idea is always the same. The morally right action is the one that produces the greatest consequences for everyone involved, where those consequences are set in terms of maximizing welfare, or increasing pleasure and decreasing pain.

In the case just given, utilitarianism seems to prescribe actions that accord with our intuitions about what we should do. You may have had the moral intuition, long before we started actively reflecting on the potential consequences, that you are morally obligated to save the child even if it damages your boots. The fact that utilitarianism produces an answer that is compatible with this intuition seems like a consideration in favor of its truth. Of course, there are tons of potential moral dilemmas we could use to test the theory, and in some of those dilemmas it seems to say things that are utterly counterintuitive. If somehow we could manipulate the story about the drowning child and the boots in a way that made it seem clear that saving the child would produce less overall utility than saving the boots—suppose for example the child were a completely isolated orphan and that your boots were a family heirloom passed down through many generations of an important family, or something to this effect—then it seems like utilitarianism, if true, would force us to admit that it may be morally permissible to let the child die. Critics of utilitarianism sometimes try to show that there are such cases in which utilitarianism gets the moral facts wrong, but that dispute remains contentious. So, while utilitarianism offers a useful and consistent way of testing moral actions, it has its limits.

Normative Ethics: Deontology

Let us continue our demonstration of how normative ethical theories work by considering one prominent alternative to utilitarianism, which was developed by the 18th-century German-speaking philosopher Immanuel Kant. Reflecting on an alternative to utilitarianism will help us to demonstrate how normative theories about what it is morally right to do really can come into conflict, and how difficult it can be to determine which is the right way to think about morality.

The theory in question goes by a number of different names, but it will suffice to simply call it "deontology," a fancy word derived from the Greek word for "duty." Roughly, deontology (or duty theory) is a normative ethical theory according to which the moral rightness of actions is to be determined not by the consequences of the action in question but rather by whether the action is done for the right reasons. But what are the right reasons? Kant argues that the right reasons are those that we could sensibly permit *everyone* to act from. Thus before we take an action, we ask what would happen if everyone in the world acted on the same reasons that are about to guide our action. If the result would be an untenable state of affairs—a society we couldn't live in—then the reason is problematic and so cannot be a justifiable basis for taking an action. For example, suppose I'm deciding whether I should lie to the bank about my income in order to get a loan on a brand-new luxury car even though I know I won't be able to make the loan payments. Is this morally permissible?

If I were to decide to tell the lie in order to acquire the car, the reason behind the action would be something like, "It is acceptable to lie in order to acquire things that I want even though I can't pay for them," or "It is acceptable to make promises of repayment that I know I cannot keep." We now simply ask whether it's coherent to will that everyone else act on such guiding reasons. The answer is obviously negative: if everyone were to tell lies in order to acquire expensive goods with no intention of repayment, the entire practice of purchasing items on credit would collapse. Such a situation is simply not tenable.

Using Normative Theories

Now that we have a basic grasp of two different normative ethical theories and how each works, we can better understand how such theories might help us make progress on difficult questions about how properly to live our lives. It might not be obvious that they can offer much help at all, especially because they disagree about how to go about assessing the moral status of our actions.

Notice that, even though utilitarianism and deontology disagree about how to assess the cases, they may nevertheless converge on the same view about whether the action in question is permissible. For example, we have just seen that deontology tells us that it is wrong to lie in order to get the car loan. What does utilitarianism say about this case?

It should be obvious that the utilitarian analysis reveals that the action in question is problematic. While telling a lie in order to acquire a cool car may temporarily enhance your *own* well-being, it is likely to do quite a bit of damage to the well-being of a number of other parties involved in the case. Perhaps most obviously, it seems to leave someone else, like the loan-granting institution, on the hook for the cost of the car. That in turn contributes to that institution's unwillingness to provide expensive loans to others (including you), making loans more expensive and more difficult to acquire for others who actually do intend to repay them, which is sure to decrease overall utility. On top of all that, your lie will eventually catch up with you, damaging your own credit score and your own ability to purchase things on credit, decreasing your utility as well. In adding up all of that damage to the utility or well-being of the many parties involved, it's clear that the action is wrong. So it seems that in this case utilitarianism and deontology agree that the action in question is wrong. But do they agree about why?

Notice that if the utilitarian account is true, then telling lies about your income in order to acquire expensive material goods you cannot repay is wrong, but not because it is somehow wrong *in principle*. It is only wrong because under the particular circumstances it would happen to produce consequences that would fail to maximize overall well-being. Given these particular circumstances, such dishonest promises would result in more unhappiness or dissatisfaction for everyone involved than the opposite. Nevertheless, had the circumstances been different, perhaps the analysis *could* have shown otherwise and the action would have been acceptable. This is not so for deontology.

Notice that if deontology is true, lying in order to acquire a big loan that you know you cannot repay is *never* going to be something that *everyone* could do. So even though it may be possible for *just* you to do it so long as most borrowers do in fact actually repay their loans, you could never will that your reasons be reasons that everyone could act from. And so it seems that, though utilitarianism and deontology agree that the action is wrong, they don't agree on the explanation. For utilitarianism, there may be cases in which such lies are acceptable, so long as it turns out such lies could actually maximize the overall well-being of the parties involved. But according to deontology, the action is always wrong regardless of whether there are any such cases precisely because that action depends upon reasons that could never be made universal.

As it turns out, this is not a minor disagreement. It is rather a substantive disagreement about the nature of moral right and wrong, and one that we have to consider carefully in reflecting on the moral status of any action. Deontology maintains that there are some actions whose reasons will never be permissible because we could never will them to be universal. Utilitarianism is more flexible but less principled than that. According to the utilitarian theory of moral right and wrong, all that matters is that the action produces a net gain in well-being under the particular circumstances. For the utilitarian, it is the actual results rather than the guiding reasons that make the difference. Morality is about actually maximizing well-being rather than broad, principled reasons. Who is right?

As we said at the start of this section, that's a difficult philosophical question, and there is no consensus on it among philosophers. Demonstrating which of these normative ethical theories, if either, is the true account of moral right and wrong is a matter to be settled by careful argumentation. Nevertheless, even without knowing which of the many competing normative ethical theories is the right one, we can try to make progress on particular moral dilemmas by carefully applying a variety of such theories to the dilemma. When they converge on a common answer despite offering different justifications, it's a safe bet that the common conclusion is the right one, or at least a reasonable one. This is why understanding how normative ethical theories work can be incredibly valuable, even for people who don't think they care much for philosophical questions. Moreover, such theories are valuable because they teach us how to think carefully and *consistently* about the decisions we make so that we don't end up contradicting ourselves when we try to justify difficult decisions about how we should conduct ourselves.

Applied Ethics

We turn now to the second category of questions about morality. These are much less abstract and much more specific questions about what it is right or wrong to do in a particular situation, morally speaking, or whether someone should or shouldn't take a particular action in particular circumstances. Should I pretend that I am sick in order to get an extension on my homework? Is it morally acceptable to dispose of chemicals by dumping them into a river? Is it morally wrong for your professor to grade your assignments using a secret random lottery? Such questions have very practical implications for what we do and how we live. So while we might answer them with a simple "yes" or "no," we also have to be able to justify our answers. And, just as importantly, when other people answer these questions or act in ways that presuppose answers to them, we expect them to be able to justify their answers.

Like the questions we face in normative ethics, these questions are philosophical and difficult to answer. Nevertheless, there really are better and worse answers. Formulating answers might require thinking about general moral principles or norms. Suppose a lazy professor decides to grade assignments using a random lottery. Grading assignments by random lottery is wrong because random lotteries have nothing to do with the quality or accuracy of the work and the subject matter of the course. To grade students on irrelevant factors is to treat them unfairly—to give them something other than what they actually deserve. It seems like a reasonable general principle that people should get what they deserve. We can formulate and apply something like this basic moral principle of fairness to help us decide what to do.

Our answers to questions about what it is right or wrong to do thus seem to appeal to general principles that, if reasonable, could also be useful in explaining other cases as well. Random grading is wrong because, at least in part, it is unfair. The principle of fairness also helps to explain why it is wrong for a university to accept students based on their looks rather than their academic achievements, why it is wrong to sentence two thieves guilty of identical crimes to very different jail times, and why it is wrong to reap the benefits of a mutual agreement without shouldering the attendant burdens, such as by driving a new car and skipping the payments. Because we answer these practical questions—questions about what it is right or wrong to do—by applying general principles or norms to particular cases, such questions belong to the branch of ethics called *applied ethics*. Applied ethics is the attempt to apply moral principles or norms (about which we argue in normative ethics) to real cases, or to formulate such principles that could be applied to other cases as well.

So applied ethics and normative ethics address different questions about morality, but they are closely related. Answering questions in applied ethics seems to require formulating or testing principles and theories that we have developed in normative ethics. And, in turn, how we formulate general principles and theories in normative ethics may depend on how we are inclined to respond to particular cases and decision scenarios. They may be different types of questions, but they clearly bear on each other in important ways.

Which subfield of ethics does engineering ethics belong to? Questions in engineering ethics—questions about how engineers ought to behave—seem like rather specific questions about what the engineer should do. If I design a toy that injures children, am I in any way responsible for those injuries? If my boss asks me to design software for cars that could, in theory, be used to circumvent compliance with regulations on automotive emissions, is it really my fault if the product is used for that purpose? The specificity of such questions, and the fact that answering them will require justification by appeal to broader moral principles, ethical theories, and moral norms, reveals that they belong properly within the purview of applied ethics. These are indeed questions in applied ethics.

Nevertheless, because they are united by their own distinctive subject matter, we can carve out an even more specific subfield for them within applied ethics. They belong to the group of questions in applied ethics concerned with how professionals, like engineers, are obligated to behave. Broadly, that branch of applied ethics is called *professional ethics*. It is concerned with understanding the obligations that people have in virtue of their being a part of a profession. And because the profession in question is engineering, we can go one step further and carve out a distinctive subfield of professional ethics called *engineering ethics*. We turn next to the nature of engineering ethics.

EXAMPLE 14.1   What is the difference between morality and ethics?

a) They are synonyms; there is no difference.
b) Morality, unlike ethics, is based on religion.
c) Morality is differentiating between right and wrong; ethics is the philosophical study of such moral distinctions.
d) Ethics is differentiating between right and wrong; morality is the philosophical study of such ethical distinctions.

Solution: C. Morality is the act of differentiating or distinguishing between right and wrong, good and bad, proper and improper. Nearly all people have some basic capacity for making such distinctions, and most people believe they know the difference between right and wrong. But explaining what explains and justifies those distinctions is considerably harder. That involves addressing philosophical questions about morality—that is, doing ethics.

EXAMPLE 14.2

What is the relationship between morality and the law?

a) Laws sometimes try to enforce moral positions but can't be used to justify them.
b) Laws only sometimes try to enforce moral positions but can always be used to justify them.
c) Laws always enforce moral positions but can't be used to justify them.
d) Morality and law are completely separate matters that never overlap.

Solution: A. Sometimes we make laws that encourage moral behavior, but the existence of a law isn't what makes a behavior moral or immoral. Hence (b) is false because laws cannot be used to justify moral positions. (c) is false because it's not true that laws always try to enforce morality. Many laws are simply practical requirements not concerned with morality. And (d) is false because morality and the law do overlap, particularly because moral considerations may be used to justify implementing or changing laws (but not vice versa).

14.3 The Nature of Engineering Ethics

Engineering ethics, as we have just said, is the study of questions within professional ethics concerned with how professional engineers are obligated to behave. But as we asked at the beginning of the chapter, why are there special obligations for engineers? Aren't all people already expected to behave ethically anyway? To answer these important questions, we must first understand some basic facts about the nature of professions like engineering and why professional positions come with special obligations.

The Nature of Professions

There are important differences between jobs and professions. You may already recognize that certain people, like medical doctors, nurses, lawyers, college professors, teachers, government officials, engineers, and so on, are regarded as having professional status. We consider these professions rather than just jobs. Why is this?

The answer is that all professions share some essential features. They are roles or positions that a healthy society needs filled, and because of this they tend to involve significant levels of interaction with, or service to, members of the public. Because the roles involve difficult work, extensive training, and high levels of expertise, and because we naturally care about the well-being of society, we are understandably selective about who gets to fill those roles. We require professionals to have special training, and in many cases special certification, to ensure that they know about, and care about, the kind of work they are doing. In return for their willingness to meet those demands and serve important social needs responsibly, society grants the people who fill those positions some **autonomy**—a right to self-governance—to exercise their own unique skills within it, and rewards faithful practitioners with greater esteem and a better paycheck than typically comes with ordinary jobs. As should now be obvious, engineering shares these features.

The Nature of Professional Ethics

Once we better understand the nature of professions, it becomes a lot easier to see why those positions come with special moral obligations that extend beyond the obligations one has in an ordinary job or as an ordinary citizen. Because the work that professionals do has significant impact on the lives of others in society, we must carefully regulate, and demand self-regulation of, the kinds of things that anyone may do from inside the profession. If we didn't attend carefully to the obligations of professionals, standards would deteriorate, real people would be hurt, society would lose its faith in the profession, and the profession would eventually disintegrate. Imagine how your view of medical doctors would change if you found out that standards on training and certification were being ignored, or that individual doctors were completely free to use any surgical technique without consideration of established professional standards.

It's not especially surprising, then, that so many professions have their own branches of ethics, like medical ethics, legal ethics, research ethics, and, of course, engineering ethics. These special branches of professional ethics are each concerned with analyzing the special obligations that arise from the particular features of a profession, and establishing what must be done in order to maintain the integrity of the profession.

Of course, all of these professions will contain some of the same basic ethical standards. Some of those standards are just general standards that apply to all people, like standards of honesty and fair dealing with other people. And some are standards that apply to any profession, like standards on how to interact with customers or clients, or how to enter faithfully into contracts. Nevertheless, each profession will also have some distinctive ethical concerns and obligations that arise from the nature of that particular profession's work.

Notice, then, that when we talk about the particular ethical obligations of engineers, we are not merely talking about the ordinary moral obligations of people who happen to be engineers (or doctors, lawyers, etc.) but about the particular obligations that arise for *anyone* who occupies the position of professional engineer. As we will see in the next section, it is precisely these obligations—essential to maintaining the integrity of the profession itself—that appear in codes of professional ethics. Those codes reflect, rather than arbitrarily dictate, obligations that one cannot ignore without undermining or doing damage to the profession itself. This explains why, as we said at the beginning of the chapter, ethical values must take priority over personal preferences and values, especially in the context of professions. Personal values or preferences that undermine the integrity of the profession must always be suppressed.

For example, an engineer who is submitting a bid to a potential client on an engineering project might personally decide to factor in a somewhat higher rate of profit on the project for himself or herself than is usual in such projects. This is a kind of personal preference or value. As long as there is an open bidding process (where others are free to submit possibly lower bids), and as long as the personal pricing would not affect the quality of the work that would be done, then the pricing level remains a purely personal or economic decision and there is no conflict between the personal value and the engineer's ethical obligations.

But if an engineer were to decide to maximize his or her profit in another way—by submitting a low bid, but then secretly drastically compromising on the quality of materials used so as to achieve maximum personal profit—then pursuit of that personal value would result in a violation of engineering ethics standards, which require the use of high-quality materials and construction methods acceptable to all parties signing the initial project agreement. This would be a case in which a conflict between personal standards (maximum profit) and engineering ethical standards (honoring the contract and using high-quality materials and methods) must be resolved in favor of supporting the relevant engineering ethics standards.

Importantly, the same ethical constraints that individual engineers must place on personal values and preferences also apply at the level of firms and corporations. The ethical standards on engineering must override the non-moral values, interests, and preferences of the whole engineering firm or corporation. Thus the example just given of legitimate versus ethically illegitimate maximizing of profits would apply just as well if the preference in question were the interest of some corporation or economic group. It would be equally problematic for a corporation or engineering firm to seek profit at the risk of lower engineering quality of products or services, just as it would for an individual engineer. Thus it is the duty of an engineer to uphold engineering ethics standards even at the level of corporate projects. That is why, in extreme cases, engineers may face the difficult obligation to uphold ethical standards when they come into conflict with the non-ethical interests of a boss or corporate structure. In cases in which engineers are encouraged or asked to engage in projects that violate standards of engineering ethics for the sake of corporate profit, it is especially helpful to understand why the professional standards cannot be compromised. As we have now seen, because such standards are essential to the integrity and existence of the profession itself, they must take precedence over such non-ethical values and interests.

As we prepare to wrap up this section, it's important to reflect on the ways in which the values and standards that stem from engineering ethics intersect with other values and standards, including legal values, technical and scientific values, and personal values. Let's begin with the relationship between ethical and legal standards.

Although (as we saw previously) there is an important distinction between policies like laws and moral values, nevertheless it is generally true that professionals have an ethical obligation to obey the prevailing laws in a given jurisdiction. We are now in a better position to understand the nature of this obligation. Professions, like engineering, are special roles that serve the interests of society. Laws that govern society are also concerned with protecting the interests of everyone in society. Naturally, then, in serving particular social interests professionals must also be responsive to the most basic requirements that facilitate the interests of society, which are codified in the law. Of course, complex cases may arise in which the ethical standards of a profession come into conflict with existing laws. Given the importance of upholding ethical standards, in such cases professions may object to those laws and work to change them. Prominent cases tend to arise in professions connected with the practice of medicine. For example, pharmaceutical companies may raise ethical objections to the use of their products in legally sanctioned executions of criminals. Nevertheless, in such cases it's important to recognize that the conflict involves the whole profession and the standards that govern it rather than one individual person.

Next, let's consider the relationship between ethical values and values associated with the technical expertise involved in the practice of a profession like engineering, such as technical and scientific values. While ethical values play a fundamental role in the integrity of the profession, they are not the only values involved in the practice of the profession. The point of this chapter is to clarify the ways in which ethical values interact with the pursuit of other values for professionals. So it's worth briefly reflecting on the harmony between ethical values and other non-ethical values involved in the practice of a profession.

Any scientific community dealing with pure sciences or applied sciences such as engineering will have technical and scientific values concerning the proper and correct ways to construct theories, carry out calculations, make statistically satisfactory estimates where exact results are not possible, and so on. These in themselves are not ethical values (good science is clearly not the same as good ethics), but they intersect with ethical values in important ways. Professionals like engineers have an ethical obligation to use good scientific methods at all times because those values also protect social interests. For example, attention to scientific values may keep a bridge from collapsing and killing people. Scientific and technical values facilitate practices that are valuable and helpful to society rather than harmful, misleading, or dangerous. And so here too (as with the law) there is a close correlation between practicing good science and being an ethical engineer.

Finally, it's important to briefly reflect on the relationship between the values of engineering ethics and personal values. Naturally, most of engineering ethics concerns an engineer's behavior while engaged in professional engineering activities. But one should be aware that engineering ethics codes also generally include prohibitions on unethical behavior while off the job as well, if those activities would affect public perceptions of one's professional integrity or status. This would include activities such as gambling, which might tend to bring the profession of engineering into disrepute; deceptive or other inappropriate forms of advertising of services; or any other activities suggesting a lack of integrity or trustworthiness in an engineer. Given the preceding, it's now easy to see why the ethical demands of a profession must interact with the personal values of those who occupy professional positions. Protecting the integrity of the profession requires placing constraints on the personal values of the people we permit to practice that profession.

EXAMPLE 14.3 Select the best and most relevant answer that matches the following statement. Engineers should follow their professional code of ethics because:

a) It helps them avoid legal problems, such as getting sued.
b) It provides a clear definition of what the public has a right to expect from responsible engineers, one that is crucial to the integrity of the profession.
c) It raises the image of the profession and hence gets engineers more pay.
d) The public will trust engineers more once they know engineers have a code of ethics.

Solution: These choices are a little harder than those in previous questions, because even the wrong choices do have some connection with ethics. But as long as you follow the initial instruction (choose the best and most relevant answer), you shouldn't have any difficulty answering it.

Choice A: Avoiding legal problems is generally a good thing, but it's not the most relevant reason why the code should be followed (and in some cases, strictly following the code might make it more likely that you'd get sued [e.g., by a disgruntled contractor if you refuse to certify shoddy workmanship]).

Choice C: Raising the image of the profession is a good thing, but not for the reason of getting more pay; that would be to act from self-interest rather than ethical motivations.

Choice D: Increased public trust is generally a good thing, but the mere knowledge that there is such a code means little unless engineers actually follow it, and for the right reasons.

Choice B is the best and most relevant answer. The code of ethics is designed to promote the public welfare, and hence the public has a right to expect that responsible engineers will follow each of its provisions, as clearly defined by the code.

EXAMPLE 14.4 Engineers should act ethically because:

a) If they don't, they risk getting demoted or fired.
b) The boss wants them to.
c) It feels good.
d) Society grants an individual professional privileges on the condition that he or she will also accept the obligations that sustain the profession and make those privileges possible.

Solution: As with Example 14.3, some of the wrong answers here do have some connection with ethics, so as before the task is to pick the best and most relevant answer.

Choice A: Even though ethical behavior may usually improve one's job security, it's not the most relevant reason for being ethical.

Choice B: Doing what the boss wants is a self-interested reason for acting, not an ethical reason.

Choice C: Even though we hope that being ethical will make engineers feel good, it's not the main reason why they should be ethical.

Choice D is the right answer. Society permits highly qualified individuals to practice a profession that serves social needs on the condition that they do so in ways that are consistent with the very purpose of the profession. This is a simple but powerful insight that serves as a recurring theme throughout the chapter.

EXAMPLE 14.5

The primary obligation of registered professional engineers is to:

a) The public welfare
b) Their employer
c) The government
d) Their personal values

Solution: As before, the best and most relevant answer should be chosen.

Choice B: Engineers do have ethical obligations to their employer, but it's not their foremost or primary obligation.

Choice C: As with the employer, engineers do have some obligations to the government, but again, it's not the first or primary one.

Choice D: As we have seen, one's personal values are subject to constraint by one's ethical obligations.

Choice A is the correct response. The foremost, primary obligation of engineers is to the public welfare. This is because the practice of the profession, which impacts public welfare, is a privilege granted by society on the condition that practitioners act in the interest of the public.

EXAMPLE 14.6

Registered professional engineers should undertake services for clients only when:

a) They really need the fees.
b) Their own bid is the lowest one.
c) They are fully technically competent to carry out the services.
d) Carrying out the services wouldn't involve excessive time or effort.

Solution: Again, let's review all the options.

Choice A: Personal financial need is a matter of self-interest, not of ethics.

Choice B: The competitive status of a bid is a matter of economics, not ethics.

Choice D: The amount of time involved in carrying out services is also a matter of economics and self-interest, rather than of ethics.

Choice C is the correct response. Ethical engineers will not undertake to provide services for clients unless they are fully technically competent to carry them out.

14.4 Codes of Ethics and the Obligations of Engineers

As we have now seen, engineers have special ethical obligations, beyond the moral obligations of ordinary people, which arise from the very fact that engineering is a profession that serves the public interest. Society permits individuals to benefit from the practice of the profession on the condition that they uphold the ethical standards that preserve the very existence of that profession. This insight is crucial because it explains why the ethical obligations of engineers are not arbitrary lists of rules or something made up by some goody two-shoes.

The constraints on what one may do as an engineer are constraints that all rational people will recognize because they are constraints without which there would *be* no such profession. They are constraints without which one could not *be* an engineer in the first place. That is why all engineers must think carefully about the ethical obligations that come with the practice of the profession. To put it bluntly, people who violate ethical standards governing the practice of their profession reveal that they lack the intellectual capacity to appreciate why such behaviors are contradictory and irrational. But what exactly are the specific ethical obligations?

In this section, we turn to the specific content of those obligations. What specific things are professional engineers obligated to do? To examine these obligations systematically, it will be helpful to look at a real code of engineering ethics, since such codes outline the specific obligations of engineers. In examining such codes, though, it's crucial to remember that any such code merely *reflects* or summarizes genuine ethical obligations that are established by careful philosophical reflection on the nature of engineering and its relationship to moral principles conducive to living good lives and forming good societies. Such codes do not arbitrarily stipulate what engineers may or may not do, nor are they put together secretly by people who are obsessed with making rules. Rather, codes of engineering ethics simply report the deliverances of doing ethics, and with a bit of careful ethical reflection, anyone can understand why codes of professional ethics contain what they do.

For the purposes of exploring one such code, we will use one of the most relevant ones—the National Council of Examiners for Engineering and Surveying (NCEES) Model Rules of Professional Conduct—as the basis for the discussion. The NCEES developed its Model Rules—as the name literally says—to *model* the rules governing professional conduct for state licensing boards. They did this to help those licensing boards, which control registration of professional engineers, to build their own set of rules governing people who practice the profession in the state. Thus the NCEES Model Rules are intended to be applicable to all such state licensing boards. Consequently, studying them will provide a good understanding of the specific ethical requirements that professional engineering societies and licensing boards recognize as necessary demands on professional engineers.

The Content of a Code of Engineering Ethics

There are many different codes of engineering ethics. So important to the profession are ethical obligations that many individual professional societies—including the National Society of Professional Engineers (NSPE), the American Institute of Chemical Engineers (AICE), the Institute of Electrical and Electronics Engineers (IEEE), the American Society of Civil Engineers (ASCE), and so on—each have their own formulation of those obligations. But while each organization may have its own particular code, all such codes reflect the same basic content, though they may be organized into different sections and presented differently depending on their precise purpose.

For example, the NSPE's Code of Ethics for Engineers, which was designed to be a helpful guide for practicing engineers, contains four sections: a preamble (or opening explanation of the purpose of the document), fundamental canons (or governing principles) of the profession, rules of professional practice, and additional professional obligations.

By contrast, the NCEES Model Rules of Professional Conduct (we can refer to the document as the "Model Rules" or just "Rules" for short), which was designed to help state licensing boards establish their own official rules binding the engineers it has licensed, contains two broad sections. It contains a preamble and rules of professional conduct, and then subdivides the rules section into three categories, one for each of the major parties with which professional engineers interact: society, employers and clients, and other engineers (licensees). Despite the different organization and intended audiences of different documents, the ethical content is quite similar. For simplicity, we'll follow the structure of the NCEES Rules. Let's review each of the sections, beginning with the preamble. Note that you can easily obtain a copy of the NCEES Rules online. We strongly encourage you to obtain a copy and use this section of the chapter to guide you through its content.

The Preamble

The preamble to the NCEES Rules contains two crucial statements. First, like all such preambles, it identifies the purpose of the document. That purpose is (a) to help safeguard life, health, and property, (b) to promote the public welfare, and (c) to maintain a high standard of integrity and practice among engineers. Second, it makes explicit an idea we have already encountered several times in this chapter, namely that the practice of professional engineering is a privilege rather than a right. That idea is crucial to all such codes of professional ethics. Let's consider each of these ideas more carefully.

The Purpose of the Document

First, the document identifies its purpose as helping to protect life, health, and property; to promote the public welfare; and to maintain a high standard of integrity and practice among engineers. If we consider what each of these components has in common, it should be clear that the core idea can be summarized as follows: the purpose of the engineering profession is to safeguard and promote the well-being of the public, and that is why its integrity must be continuously maintained.

Notice that there are two ways to attend to public welfare. The first is negative: *avoid* acts that cause harm in the course of one's professional duties. The second is positive: *take* measures that will safeguard or preserve people from possible future harm. For engineers, the more positive measures might include such things as building devices with extra "fail-safe" features included that make harmful consequences of their use as unlikely as possible.

An additional way to actively promote public welfare in the context of practicing one's profession is to take active steps so that one's professional activities will result in definite benefits and improved conditions for the general public. For example, for an engineer planning a new highway, this rule would require not only planning and building it in a safe manner (as previously discussed), but also doing such things as choosing the shortest feasible route between its endpoints, or choosing the route that would permit the most efficient road-construction techniques, hence maximizing the utility of the highway to the general public and minimizing its cost to them as well.

Overall, then, ethically responsible engineers take great care not to cause harm to others, they take whatever extra steps are necessary to minimize risks of potential harm to others, and they take active steps to serve the public interest. In doing these things, they maintain a profession the very purpose of which is to permit highly qualified individuals to carry out technological projects that improve society.

Professional Engineering Is a Privilege Rather than a Right

Second, the preamble makes explicit another idea we have already encountered several times in this chapter, namely that the practice of professional engineering is a privilege rather than a right:

> *Engineering registration is a privilege and not a right. This privilege demands that engineers responsibly represent themselves before the public in a truthful and objective manner.*

One must earn an engineering registration—one does not automatically have a right to that status. And the granting of it requires that person, in return, to adhere to ethical standards that make it possible to have such professions at all.

Codes typically point this out explicitly—not because it packs a rhetorical punch and sounds cool, but rather because it is crucial for those who occupy professional positions to think clearly about the nature of their responsibilities *as* professionals. As we have now seen repeatedly, the professions are positions that society grants some highly qualified individuals the right to occupy and benefit from on the condition that they conduct themselves in ways that protect the integrity of the profession.

The remainder of the NCEES Rules, like all codes of engineering ethics, then follows up the preamble by filling in the details about what constitutes proper professional conduct. We turn now to those details, which constitute the second component of the document: the rules of professional conduct. That section is organized into three subsections, which we'll cover in turn: the engineer's obligations to society, to employers and clients, and to other engineers.

The Engineer's Obligation to Society

The first subsection of the rules of professional conduct addresses **the engineer's obligation to society**. There are eight rules in this section. Here we review them, paraphrasing for brevity and clarity:

I (a). While performing services, the engineer's foremost responsibility is to the public welfare.

This rule is also featured in a related form in the preamble, as a duty to promote the public welfare. The idea here of responsibility to the public welfare includes the idea of safeguarding the public from harm, as more fully spelled out in the next rule:

I (b). Engineers shall approve only those designs that safeguard the life, health, welfare and property of the public while conforming to accepted engineering standards.

We have already commented on these ideas in the preamble. Here we only add that these two rules together imply a much broader context of responsibility for engineers than those arising from any one task or project. Designs and materials that seem perfectly adequate and ethically acceptable within the bounds of a given project may nevertheless be unacceptable because of wider issues about the public interest.

For example, until recently a refrigeration engineer could have specified Freon (a chlorinated fluorocarbon [CFC] product) as the prime refrigerating agent for use in a product, and defended it as an efficient, inexpensive refrigerant with no risks to the purchaser of the appliance. However, we now know that there are significant risks to the public at large from such chemicals because of the long-term damage to the environment they cause when they leak out, perhaps many years after the useful life of the product is over. Rules (a) and (b) tell engineers that they must always keep such wider, possibly longer-term issues in mind on every project they work on.

The third rule takes us into more difficult territory:

I (c). If an engineer's professional judgment is overruled resulting in danger to the life, health, welfare or property of the public, the engineer shall notify his or her employer or client and any authority that may be appropriate.

This is an important rule, which reflects the recurring theme in this chapter that the obligations incumbent on the professional role take precedence over the particular desires of the individual occupying it. The rule essentially requires that engineers refuse to give up on the struggle to protect public welfare. Notice that this may place the engineer in a difficult position if his or her employer or client is among those contributing to the problem that needs reporting. But the engineer's duty in such cases—which arises from his or her accepting the agreement to carry out the duties of the role in return for receiving the privileges—is clear: "any authority that may be appropriate" must be notified, even if the employer or client tries to prevent it.

Sometimes cases of this kind are referred to as "whistleblowing." Whistleblowing is the act of stepping outside the approved chain of command for reporting concerns in an organization in order to report ethical problems and prevent harmful consequences that may otherwise go ignored. There are a number of very difficult and contentious philosophical and ethical questions about whistleblowing that are well worth careful consideration. Notice for example that there may be some ambiguity about what constitutes an appropriate authority. Furthermore, the question of whether one's professional judgment is sound, or genuinely a professional rather than a personal judgment, is often contentious.

For example, during the construction of San Francisco's Bay Area Rapid Transport (BART) rail system in the late 1960s and early 1970s, three engineers blew the whistle on the project over concerns about the safety of the train's automatic control system. To this day, some difficult and contentious questions remain about whether the act of whistleblowing was justified by proper professional judgment. For example, the engineers who continued to voice concerns about the automatic control system were not actually authorized to investigate that system. The point is that it would be a mistake to claim that the rules in codes of ethics can settle all relevant ethical issues and dilemmas for real professional engineers acting in difficult circumstances. What the rules require of engineers is a commitment to activities that maintain the integrity of the profession and serve the public good. But whether that in turn requires blowing the whistle on one's employer—an activity that may come with significant consequences for the whistleblower—remains something that individual engineers must learn to think clearly and consistently about. This is a case in which applying all the tools of engineering ethics, including the analysis of past cases and the application of normative ethical principles and theories to potential actions, becomes crucial. We expect that all engineers will take seriously the obligation to take only those actions that they are prepared to carefully justify.

The next two rules implement the ethical requirement that engineers ought to tell the truth in the specific context of their professional duties.

I (d). Engineers shall be objective and truthful in professional reports, statements, or testimonies and provide all pertinent supporting information relating to such reports, statements, or testimonies.

I (e). Engineers shall not express a professional opinion publicly unless it is based upon knowledge of the facts and a competent evaluation of the subject matter.

Note that the duty as mentioned in Rule I(d) is not simply to be truthful in what one says, but also to be forthcoming about all pertinent or relevant information in reports, etc. Rule I(d) also mentions being "objective," which adds the element of being unbiased and basing one's beliefs and reports only on objective, verifiable matters of fact or theory. Rule I(e) expands the idea of being objective in one's reports—others should be able to rely upon one's professional opinion, and they can do this only if one knows all of the relevant facts and is completely competent to evaluate the matter being dealt with.

The take-home point from Rules I(d) and I(e), then, is that professional engineers have obligations to respect, promote, and proliferate truthful claims. Why is this? Here it's helpful to recall our discussion in Section 14.3 about the intersection of ethical and scientific values. Engineering is a highly specialized profession that requires extensive technical training. The efficacy, and safety, of engineering projects hinges on its having the facts right. That is the same reason that we require all professional engineers to undergo rigorous training and certification. It's sensible, then, that the practice of the profession should require a respect for the scientific virtues that facilitate the effectiveness and safety of the profession's projects.

Rule I(f) picks up on a related point:

I (f). Engineers shall not express a professional opinion on subject matters for which they are motivated or paid, unless they explicitly identify the parties on whose behalf they are expressing the opinion, and reveal the interest the parties have in the matters.

Rule I(f) expresses what is sometimes called the duty of full disclosure. Even if one honestly seeks to be truthful and objective, as in Rules I(d) and I(e), doubts might still be raised about one's motivation or objectivity unless one reveals on whose behalf one is expressing an opinion, and the interests that such persons have in the case. Professional engineers must worry about the various ways in which practitioners' commitment to truth might be compromised because the success of the profession itself depends on a continuous commitment to scientific truth. If we think of science as the enterprise that aims to uncover truths about the world, and engineering as a kind of applied science, then it certainly matters whether our practices guide us toward scientific truth. Telling people what they want to hear and making objective claims under conditions of compromised objectivity are practices disruptive to the pursuit of truth. They are also disruptive to the faith that other engineers, clients, and the public will place in the profession's commitment to the truth. Note that Rule I(f) is also related to the issue of conflicts of interest—see Rules II(f) and II(h) later in the chapter on obligations to employers and clients.

Rule I(g) might be called the "clean hands" rule (shake hands only with those whose hands are as ethically clean as your own):

I (g). Engineers shall not associate in business ventures with nor permit their names or their firms' names to be used by any person or firm that is engaging in dishonest, fraudulent, or illegal business practice.

It isn't sufficient to be completely ethical in one's own (or one's own company's or firm's) practices; one must also ensure that others do not profit from one's own good name if their own activities are unethical in some way. This rule is clearly related to Rule I(a) concerning the public welfare—one must promote this in one's external dealings just as much as in one's own activities.

The next rule is closely related to Rule I(c):

I (h). Engineers who have knowledge of a possible violation of any of the rules listed in this and the following two parts shall provide pertinent information and assist the state board in reaching a final determination of the possible violation.

Rule I(h) generalizes Rule I(c), a duty of disclosure when one's professional judgment is overruled, to a duty of disclosure in the case of any of these rules, when one has knowledge of possible violations of them. In terms of the public welfare, it is very important that each profession regulates itself in this way, so as to minimize or eliminate future infringements of its rules. Strict adherence to this rule also will lead to wider appreciation and respect for the profession of engineering because of its willingness to clean its own house in this way.

The Engineer's Obligation to Employers and Clients

The second subsection of the rules of professional conduct addresses **the engineer's obligation to employers and clients**. This subsection is concerned with ensuring responsible professional behavior with respect to interactions with one's employers and with one's clients. The first two rules address difficulties that arise when professional engineers overstep their expertise or competence.

II (a). Engineers shall not undertake technical assignments for which they are not qualified by education or experience.

II (b). Engineers shall approve or seal only those plans or designs that deal with subjects in which they are competent and which have been prepared under their direct control and supervision.

Rules II(a) and II(b) require an engineer to be professionally competent, both in undertaking technical assignments and in approving plans or designs. Rule II(b) in fact requires a double kind of knowledge, that one has technical competence in the matters to be approved and that one has had direct control and supervision over their preparation. Only thus can one be sure that one's approval is legitimate and warranted. The justification for these rules is by now quite clear. Engineering is a learned profession, and its integrity depends upon a commitment among its practitioners to restrict themselves to exercising their expertise only on those matters to which it actually applies. Overstepping one's expertise, like dishonesty, is disruptive to the profession's dependence on empirical truth.

Rule II(c) effectively invokes II(b):

II (c). Engineers may coordinate an entire project provided that each design component is signed or sealed by the engineer responsible for that design component.

As long as each component of a project is satisfactorily approved as per Rule II(b), then it is permissible for an engineer to coordinate an entire project. This also underscores the importance of Rule II(b), as project managers have to rely heavily on the validity of the approvals for each prior part of a project.

As we saw in the earlier section on an engineer's obligations to society, engineers have duties to tell the truth, as in Rules I(d) and I(e). But there is a corollary:

II (d). Engineers shall not reveal professional information without the prior consent of the employer or client except as authorized or required by law.

Just as one must not lie or misinform, so also must one restrict to whom one reveals professionally relevant information. A professional engineer must be trustworthy as well as honest, and being trusted not to reveal confidential information is an important kind of trust.

Confidentiality is a central factor in assuring employers and clients that one's professional services for them are indeed for them alone, and that they can rely upon one's discretion in not revealing to others any private information without their full consent. This rule is also related to Rules II(e) through II(h) below, in that any revealing of information to others would probably create conflicts of interest or other serving-more-than-one-master problems of those kinds.

The next rule begins the first of four rules dealing with conflicts of interest:

II (e). Engineers shall not solicit or accept valuable considerations, financial or otherwise, directly or indirectly, from contractors, their agents, or other parties while performing work for employers or clients.

Conflicts of interest are cases where one has some primary professional interest or group of interests—generally, to carry out some project for an employer or client—but where other factors might enter into the picture that would activate other, nonprofessional interests one also has that would then conflict with the professional interests.

In the case of Rule II(e), the concern is that soliciting or accepting such things as gifts, hospitality, or suggestions of possible future job offers for oneself would activate non–job-related, personal interests (for additional pay, career advancement, etc.), which would then be in conflict with one's primary professional interests and duties concerning a current project.

Special attention should be paid to this and the other conflict-of-interest rules, and they should be strictly observed. People are sometimes tempted to think that breaking these rules is ethically harmless, on the grounds that if one has a strong enough character, then one will not actually be professionally influenced in a detrimental way by gifts, etc., and hence that accepting such inducements cannot do any harm.

However, it is important to realize that even the *appearance* of a conflict of interest (however careful one is to avoid actually undermining one's professional interests) can create serious ethical problems. One way this happens is through the potential for loss of trust among employers or clients. Just as the client needs to know that the engineer will keep his or her information confidential, as in Rule II(d), so the client also needs to know that the engineer is single-mindedly working with only the client's interests at heart. Any doubts raised because of the appearance of a conflict of interest could be very damaging to the professional relationship between the client and the engineer. This is why professional engineers must avoid doing anything that would create conflicts of interest, actual or merely apparent.

Further rules are necessary to deal with conflicts of interest, because unfortunately in some situations the appearance or possibility of conflicts of interest may be virtually unavoidable, no matter how ethically careful everyone is. Fortunately, however, at the same time there is a powerful method available for minimizing any ethically problematic effects of such situations. This is the method of full disclosure of potential conflicts to all interested parties, and it is addressed in the following two rules:

II (f). Engineers shall disclose to their employers or clients potential conflicts of interest or any other circumstances that could influence or appear to influence their professional judgment or the quality of their service.

428 **Studying Engineering**

II (g). An engineer shall not accept financial or other compensation from more than one party for services rendered on one project unless the details are fully disclosed and agreed to by all parties concerned.

These rules both address issues of full disclosure, or keeping all the relevant parties fully informed as to areas of potential conflict, or potentially undue external influences. The basic idea behind full disclosure is that it can maintain trust and confidence between all parties—something essential to the integrity of professional practice—in several important ways.

First, if engineer A informs other party B about a potential conflict or influence, then A has been honest with B about that matter, hence maintaining or reinforcing B's trust in A. Furthermore, if B is not further concerned about the matter once he or she knows about it, then the potential problem (namely, the apparent conflict or potentially bad influence on A's professional conduct) has been completely defused.

Suppose, on the other hand, that B is initially concerned about the issue even after it was honestly revealed to B by A. Even so, the problem is already lessened: at least A has honestly revealed the area of concern. Things would be much worse if B later discovered the problem for himself or herself, in a case when A had not fully disclosed it—that would be very destructive of trust between A and B.

Furthermore, now that B knows about the area of concern, and knows that A is fully cooperating with him or her in disclosing the potential problem, both of them can proceed to work out mutually acceptable ways of minimizing or disposing of the problem to their joint satisfaction. Thus even if the full disclosure leads to an initial problem that needs to be resolved, it won't be one that disintegrates trust between A and B. In fact it may even reinforce trust, in that A's willingness to fully disclose a potential conflict/influence, and even negotiate with B about it, will be taken by B as good evidence of A's professional honesty and sincerity.

Rule II(h) deals with a special case of potential conflicts of interest, namely when one of the interested parties is a governmental body:

II (h). To avoid conflicts of interest, engineers shall not solicit or accept a professional contract from a governmental body on which a principal or officer of their firm serves as a member. An engineer who is a principal or employee of a private firm and who serves as a member of a governmental body shall not participate in decisions relating to the professional services solicited or provided by the firm to the governmental body.

The idea is simpler than it sounds. Engineering firms have an obligation to avoid conflicts of interest when it comes to providing services that influence government decisions.

The rule is interesting because it shows that in a government-related case, a somewhat stricter rule is required than for the more usual cases involving only nongovernmental agencies. In nongovernmental cases, it is ethically sufficient to fully disclose potential conflicts of interests to all parties, and then to negotiate with the other parties as to how to deal with the potential conflicts. For example, if one's engineering firm has an official who was also on the board of directors of a bank, it would be ethically acceptable to accept a professional contract from that bank, as long as all parties were fully informed about the official's joint appointment prior to the agreement, and also as long as they could come to agree that the joint appointment was not an impediment to their signing a contract.

But in the governmental case, the stricter rule, II(h), is required when one of the parties is a governmental body. This is because in the case of agreements among private persons or businesses, they themselves are the only parties having a legitimate interest in the negotiations, and hence whatever they freely decide among themselves—assuming that no other ethical rules or laws are being broken—is acceptable. But in a case in which a governmental body is involved, there is another interested party that is not directly represented in negotiations, namely the electorate or citizens of the jurisdiction covered by that governmental body. The governmental body must act only in ways that fully respect the interests and concerns of the electorate, and engineers—as professionals with obligations to the public interest—must respect that.

In such a case, it is impossible to ensure that full disclosure to all of the citizens of the electorate of the potential conflicts of interest would be made in a case such as that envisaged in Rule II(h), and hence there is a need for the stricter rule that completely prohibits conflicts of interest. Only then can public trust both in the engineering profession and in governmental bodies be preserved.

The Engineer's Obligations to Other Engineers

The third subsection of the NCEES's Rules of Professional Conduct primarily addresses **the engineer's obligations to other engineers**. It might seem strange at first that codes of engineering ethics address the relationship between professional engineers. But, given the themes of this chapter, it's actually to be expected. The integrity of the profession depends upon cooperation and mutual concern among professional engineers for achieving the goals of the profession. And, as we shall see, such ethical obligations are grounded in considerations that are relevant not merely to registered engineers but also to those in training, preparing for careers in engineering. Let's begin with the first of three rules in this section:

III (a). Engineers shall not misrepresent or permit misrepresentation of their, or any of their associates', academic or professional qualifications. They shall not misrepresent their level of responsibility in prior assignments nor the complexity of those prior assignments. Pertinent facts relating to employers, employees, associates, joint ventures, or past accomplishment shall not be misrepresented when soliciting employment or business.

This rule requires that one not misrepresent one's own qualifications, or those of one's associates, which is a very important kind of truthfulness and objectivity. This obligation also applies to the representation of prior levels of responsibility and to the representation of the complexity of previous assignments. The rule concludes by requiring that all pertinent facts in one's previous history not be misrepresented.

But why does it really matter whether a particular engineer misrepresents his or her professional qualifications in subtle ways, or helps a peer to do the same? The now-familiar answer has to do with the abilities of engineers to maintain the status of the profession. But it also helps to consider what competing normative ethical theories like deontology and utilitarianism say about one's decision to misrepresent one's credentials for the purpose of advancing one's own career.

Applying utilitarianism, a quick assessment of the consequences shows moderate and largely short-term benefits to just one person at most. Perhaps I get a better job and an increase in salary to go with it. But I'm also now quite likely in a position for which I'm underqualified and potentially undertrained, and unlikely to enjoy that experience or last very long in it anyway. Moreover, a utilitarian analysis requires analyzing the consequences for *all* parties involved, not just for me. It quickly becomes obvious that my rather minor short-term gains are far outweighed not just by my own long-term dissatisfaction, but also by the significant long-term problems I cause for everyone else's well-being, including the person who promoted me to the position (who now looks incompetent), my coworkers (who are stuck compensating for my shortcomings), and even the public (who stand to be harmed in cases in which my incompetence jeopardizes the quality of technological projects, and at the very least stand to reap fewer benefits from my firm's engineering contributions to society).

The deontological analysis is particularly insightful in this case. It tells me to reflect on the reasons that guide or justify my decision. Could I will those reasons to be universal (i.e., reasons that others could also act from)? What the deontological analysis helpfully demonstrates is that what really drives my decision to misrepresent my own credentials is the hope that I get to be the rare exception to the rule that makes the profession possible. It's obvious that if all engineers, or even most, went around distorting their credentials, such credentials would quickly become meaningless, and decisions about who should be assigned to which positions and projects would soon have no basis in actual knowledge, talent, or qualification at all. It's hard to imagine a state of affairs more disruptive and contrary to the very idea of a profession. It's always tempting to make ourselves the *one* exception to the rule. But the problem with that line of thinking—readily exposed by the deontological insight—is that *everyone else* thinks the same thing too. And when people start acting on it, disaster ensues. The deontological analysis is rather helpful at reminding us that the whole idea behind living well is recognizing that you don't get to be the special exception to the rule without also taking on beliefs you can't justify. You don't want people to act in ways that treat themselves as exceptions to rules whose existence are necessary, and that goes for you the same as it does for everyone else. If it didn't, there could be no such rule.

It's also worth taking a moment to point out that all of these considerations apply equally to engineers in training. The same problematic consequences that arise for professional engineers who misrepresent credentials (like having underqualified people in important positions) and the same irrational thinking (about making oneself an exception to a well-justified rule) apply to engineering coursework and training too. Nobody wants to go to a doctor who cheated his way through school, or lied about his experiences in medical residency, and the same goes for the profession of engineering.

Rule III(b) continues the prohibitions against the conflicts of interest that were found in Rules II(e) through II(h):

III (b). Engineers shall not directly or indirectly give, solicit, or receive any gift or commission, or other valuable consideration, in order to obtain work, and shall not make a contribution to any political body with the intent of influencing the award of a contract by a governmental body.

The previous rules mainly covered cases where an engineer was already employed, while Rule III(b) here specifically applies to attempts to obtain future work, including the award of a contract by a governmental body. This rule also emphasizes that it is just as wrong to attempt to unduly influence someone else (e.g., a potential employer) as it would be to allow others to unduly influence oneself. Thus the rule underlines that it is just as ethically unacceptable to try to cause conflicts of interest in others as it is to allow oneself to be enmeshed in improper conflicts of interest.

The second part of Rule III(b), "and shall not make a contribution to any political body with the intent of influencing the award of a contract by a governmental body," specifically mentions the intent of the person making the contribution. Contributions are not prohibited, only contributions with the wrong intent.

This part of the rule could be difficult to apply or enforce in practice, because it may be very hard to establish what an engineer's actual intent was in making a political contribution. Also, the freedom to make political contributions to organizations of one's own choice is generally viewed as a basic right that should be limited as little as possible. So this part of the rule very much depends on and appeals to the ethical conscience of the individual engineer, who must judge his or her own intentions in such cases and avoid such contributions when his or her own intent would be self-interested in the manner prohibited by the rule. But since we now better understand what it means to say that the moral conclusions we derive from doing ethics are rational ones that we expect intelligent professionals to be capable of deriving for themselves, the fact that one must engage in this kind of self-regulation does not present any particular problem. It is rather a basic demand on professional competence.

Note also that the first part of Rule III(b), "Engineers shall not directly or indirectly give, solicit, or receive any gift or commission, or other valuable consideration, in order to obtain work" also mentions the reason or intention behind giving or receiving gifts, etc., in the phrase "in order to obtain work." However, in practice it is much easier to judge when gift giving is ethically unacceptable than when political contributions are unacceptable, since there are more behavioral and social tests for suspicious inducements to obtain work than there are for suspicious political support. So it is much easier to police and regulate infringements of this first part of Rule III(b) than it is for the second part. Nevertheless the ethical justification remains the same.

The final rule, III(c), is somewhat unclear, so it will require some discussion to get to its ethical core. The rule says:

III (c). Engineers shall not attempt to injure, maliciously or falsely, directly or indirectly, the professional reputations, prospects, practice or employment of other engineers, nor indiscriminately criticize the work of other engineers.

First, how should the subordinate phrases "maliciously or falsely" and "directly or indirectly" be interpreted? On one possible interpretation, Rule III(c) states that engineers should never attempt to injure in any way or for any reason the professional reputations of other engineers. On this interpretation those phrases just give examples of possible modes of injury that are prohibited, leaving unmentioned any other possible modes of injury, which are nevertheless also assumed to be prohibited.

However, another interpretation is possible, according to which it is only certain kinds of injury that are prohibited by Rule III(c), namely those spelled out by those same phrases interpreted so that the "directly or indirectly" clause modifies the "maliciously or falsely" clause. On this interpretation, Rule III(c) prohibits only malicious or false attempts (whether carried out directly or indirectly) to injure the reputations of other engineers. Notice that this second interpretation would ethically permit attempts to injure the reputations of other engineers, as long as the attempts were carried out in a non-malicious and honest, truthful way (and presumably with the public welfare in mind as well).

Some support for this second interpretation can be derived from the final clause of the rule, "nor indiscriminately criticize the work of other engineers." Clearly, in this case it is not all criticism of the work of other engineers that is being prohibited, but only indiscriminate criticism. Thus the last claim outlaws any criticism that is overly emotional, biased, not well reasoned or factually inaccurate, and so on, but it does not prohibit well-reasoned, careful, accurate, fact-based criticisms of other engineers.

Thus, in support of the second interpretation, it might be said that it is not all attempts to injure reputation that are being prohibited, but only those that are malicious, false, indiscriminate, or otherwise irresponsible in the methods they employ.

However, there is still something to be said in favor of the first interpretation as well (which, it will be recalled, involves a blanket condemnation of *all* attempts to injure reputations). Those defending it might do so as follows: it is arguable on general ethical grounds that any attempt to injure someone's reputation must be viewed as going too far and therefore becoming unethical. Even if one is convinced that someone else's work is shoddy, dishonest, and so on, at most (it could be argued) one has a duty to point out the problems and shortcomings in their work. It is a big leap from criticizing an engineer's actions, on the one hand, to condemning the engineer in a way designed to injure his or her reputation, on the other hand.

Thus, on this line of thinking, one should criticize an engineer only out of a disinterested desire to let the truth be known by all, not with the aim of injuring someone's reputation—it is for others to judge whether the truth of what one has pointed out will diminish the reputation and so on of the engineer in question.

Fortunately, it is not necessary to definitively decide between these different interpretations of Rule III(c). What is important is to become sensitive to the ethical issues involved in each interpretation. And for the purposes of conforming to the rule, both sides can agree that if criticism of other engineers ever becomes necessary, it should be done in a very cautious and objective manner, and with all due respect for the professional status of the person being criticized.

**EXAMPLE
14.7**

With respect to the Model Rules of Professional Conduct for Engineers:

a) The rules are a bad thing because they encourage engineers to spy on and betray their colleagues.

b) The rules are a useful legal defense in court, when engineers can demonstrate that they obeyed the rules.

c) The rules enhance the image of the profession and hence its economic benefits to its members.

d) The rules are important in providing a summary of what the public has a right to expect from responsible engineers.

Solution: Each answer has some truth to it, but only one has the most truth.

Choice A: It is true that the rules require those who have knowledge of violations of the rules to report such cases to the relevant state board. And this could involve one in collecting more information on the possible infringements, and hence exposing those involved to the scrutiny of the state board. However, the generally negative connotations of "spying" and "betraying colleagues" do not make the rules bad—the activities in question are a necessary part of responsible reporting of possible violations to the authorities, and hence are morally fully justified.

Choice B: It is true that proof in court that one has followed the rules may be a useful legal defense. However, this is only a secondary, indirect effect of the ethical value of the rules themselves, and so B does not provide the best answer.

Choice C: Again, it is true that the adoption of the rules by the engineering profession will enhance its image and economic benefits. But like B, this is only a secondary and derivative effect of the rules.

Choice D is the right answer. Since the basic function of the rules is to provide a guide for ethical conduct for engineers, the rules also provide a useful summary for the public of what they expect, and have a right to expect, from responsible engineers.

**EXAMPLE
14.8**

The Model Rules of Professional Conduct require registered engineers to conform to all but one of the following rules—which rule is not required?

a) Do not charge excessive fees.

b) Do not compete unfairly with others.

c) Perform services only in the areas of your competence.

d) Avoid conflicts of interest.

Solution: There may be one or two problems of this kind in the exam used to obtain a professional engineering license, which can be solved by memorizing the rules or by checking them directly to see which are or are not included. However, it is better to acquire a good understanding of the basic ethical concerns behind the rules, in which case the right answer here will be clear immediately.

Rule A is not required because fees are a matter of free negotiation between engineers and clients. Hence a fee that might seem excessive to some may be acceptable to others because of an interest in extra quality, or unusually quick delivery time, and so on. The other three rules are required.

EXAMPLE 14.9

You are a quality control engineer, supervising the completion of a product whose specification includes using only U.S.-made parts. However, at a very late stage you notice that one of your subcontractors has supplied you with a part having foreign-made bolts in it—but these aren't very noticeable, and would function identically to U.S.-made bolts. Your customer urgently needs delivery of the finished product. What should you do?

a) Say nothing and deliver the product with the foreign bolts included, hoping the customer won't notice.

b) Find (or, if necessary, invent) some roughly equivalent violation of the contract or specifications for which the customer (rather than your company) is responsible, and then tell them you'll ignore their violation if they ignore your company's violation.

c) Tell the customer about the problem, and let them decide what they want you to do next.

d) Put all your efforts into finding legal loopholes in the original specifications, or in the way they were negotiated, to avoid your company's appearing to have violated the specifications.

Solution: Choice A: This is wrong because it is dishonest—it violates the requirement of being objective and truthful in reports, etc.

Choice B: This is wrong because "two wrongs don't make a right." Negotiations with clients should always be done in an ethically acceptable manner.

Choice D: This would violate at least the spirit of the initial agreement. The ethical requirement of being objective and truthful includes an obligation not to distort the intent of any agreements with clients.

Choice C is the correct answer. Being honest with a client or customer about any production difficulties allows them to decide what is in their best interest given the new disclosures, and provides a basis for further good-faith negotiations between the parties.

EXAMPLE 14.10

You are the engineer of record on a building project that is behind schedule and urgently needed by the clients. Your boss wants you to certify some roofing construction as properly completed even though you know some questionable installation techniques were used. What should you do?

a) Certify it, and negotiate a raise from your boss as your price for doing so.

b) Refuse to certify it.

c) Tell the client about the problem, saying that you'll certify it if they want you to.

d) Certify it, but keep a close watch on the project in the future in case any problems develop with it.

Solution: There are some temptations and half-right elements in some of these, but they must be resisted as not being completely ethical.

Choice A: Even if you inform your boss of the problem and negotiate his or her consent to your certifying it, it is always wrong to certify work that does not measure up to the highest professional standards.

Choice C: Wrong for the same reason as A. Even if you honestly reveal the problem to the client and get their consent, nevertheless it is your professional duty as an engineer not to certify dubious work.

Choice D: The initial certification would be wrong, no matter how carefully you monitor future progress in the hope of minimizing any future problems.

Choice B is the right choice. Whether or not your boss is happy with this, it is your professional duty to refuse certification in such cases, even if as a result you are reassigned or fired.

EXAMPLE 14.11

You are an engineer and a manager at an aerospace company with an important government contract supplying parts for a space shuttle. As an engineer, you know that a projected launch would face unknown risks, because the equipment for which you are responsible would be operating outside its tested range of behaviors. However, since you are also a manager you know how important it is to your company that the launch be carried out promptly. What should you do?

a) Allow your judgment as a manager to override your judgment as an engineer, and so permit the launch.

b) Toss a coin to decide; your engineering and managerial roles are equally important, so neither should take precedence over the other.

c) Abstain from voting in any group decision in the matter, since as both a manager and an engineer you have a conflict of interest in this case.

d) Allow your judgment as an engineer to override your judgment as a manager, and so do not permit the launch.

Solution: A real-life case similar to this problem occurred with the Challenger space-shuttle disaster.

Choice A: Wrong, because engineers have special professional duties and ethical commitments that go beyond those of corporate managers.

Choice B: Wrong for the same reason as A; ethically, engineering responsibilities are more important than managerial responsibilities.

Choice C: Wrong, because your duties as a professional engineer require appropriate action even if other factors may seem to point in the opposite direction or toward abstention.

Choice D is the correct choice. Whatever other duties an engineer has, his or her professional engineering responsibilities must always be given first priority.

EXAMPLE 14.12

Your company buys large quantities of parts from various suppliers in a very competitive market sector. As a professional engineer, you often get to make critical decisions on which supplier should be used for which parts. A new supplier is very eager to get your company's business. Not only that, but you find they are very eager to provide you personally with many benefits—free meals at high-class restaurants and free vacation weekends for (supposedly) business meetings and demonstrations, and other more confidential things such as expensive gifts that arrive through the mail, club memberships, and so on. What should you do?

a) Do not accept any of the gifts that go beyond legitimate business entertaining, even if your company would allow you to accept such gifts.
b) Report all the gifts, etc., to your company, and let them decide whether or not you should accept them.
c) Accept the gifts without telling your company, because you know that your professional judgment about the supplier will not be biased by the gifts.
d) Tell other potential suppliers about the gifts, and ask them to provide you with similar benefits so you won't be biased in favor of any particular supplier.

Solution: Choice B: Wrong, because even if your company finds such gifts acceptable, it is still your duty as a professional engineer not to become involved in such conflicts of interest.

Choice C: Also wrong. It doesn't matter whether you believe you can remain unbiased, because you still have the conflict, and the possible perception by others that you might be biased by it also remains an ethical problem.

Choice D: Wrong. Remember, two wrongs don't make a right, and the same principle applies no matter how many wrong actions are involved.

Choice A is the right answer. As with the reasoning on choice B, it makes no difference whether others (having less demanding ethical standards than engineers) would find such things acceptable. You must not accept any gifts that would involve you in conflicts of interest.

EXERCISES AND ACTIVITIES

Introductory Note

As with any form of ethics, engineering ethics demands more than an impersonal, spectator point of view from us. It is important to consider ethical problems not simply as some other person's problem—to be handled with impartial advice or citing of ethical rules—but instead as a vital problem for yourself that must be resolved in terms that you can rationally live with, and expect others to live with.

To encourage this perspective, many of the cases are presented in the second person—that is, as problems for you rather than some abstract person. In answering them it is important to maintain this perspective and honestly explain how you would actually deal with each situation yourself.

14.1   Your company urgently needs a new manager in another division, and you have been invited to apply for the position. However, one of the job requirements is competence in the use of some sophisticated CAD (computer-aided design) software. Currently you do not have that competence, but you are enrolled in a CAD course, so eventually you will have that competence (but not before the deadline for filling the new job).

Is it ethically permissible to claim that you have (in order to get the job for which otherwise you are well qualified), even though strictly speaking this is not true?

Would your answer be any different if your current boss informally (off the record) advised you to misrepresent your qualifications in this way, giving as his or her reason that you are overall the most qualified person for the new managerial position?

14.2   You are a supervisor for a complex design project and are aware of the NCEES code rule according to which engineers may coordinate an entire project provided that each design component is signed or sealed by the engineer responsible for that design component.

You are very overworked and so have got into the habit of, in effect, rubber-stamping the design decisions of the subordinate engineers working on your project. You do no testing or checking of your own on their projects and routinely approve their plans (after they have signed or sealed them) with only the most superficial read-through of their work. In your mind this policy is justifiable because the NCEES rule says nothing about the supervisory responsibilities of someone in your position. Besides which, you trust the engineers to do a good job, and so see no need to check up on them.

Is your position ethically justifiable?

14.3   Newly hired as a production engineer, you find a potential problem on the shop floor: workers are routinely ignoring some of the government-mandated safety regulations governing the presses and stamping machines.

The workers override safety features such as guards designed to make it impossible to insert a hand or arm into a machine. Or they rig up convenience controls so they can operate a machine while close to it, instead of using approved safe switches, etc., which requires more movement or operational steps. Their reason (or excuse) is that if the safety features were strictly followed, then production would be very difficult, tiring, and inefficient. They feel that their shortcuts still provide adequately safe operation with improved efficiency and worker satisfaction.

Should you immediately insist on full compliance with all the safety regulations, or do the workers have enough of a case so that you would be tempted to ignore the safety violations? And if you were tempted to ignore the violations, how would you justify doing so to your boss?

Also, how much weight should you give to the workers' clear preference for not following the regulations? Ethically, can safety standards be relaxed if those to whom they apply want them to be relaxed?

14.4   You and an engineer colleague work closely on designing and implementing procedures for the proper disposal of various waste materials in an industrial plant. Your colleague is responsible for liquid wastes, which are discharged into local rivers.

During ongoing discussions with your colleague, you notice that he is habitually allowing levels of some toxic liquid waste chemicals that are slightly higher than the levels permitted by the law for those chemicals. You tell him that you have noticed this, but he replies that, since the levels are only slightly above the legal limits, any ethical or safety issues are trivial in this case, and not worth the trouble and expense to correct them.

Do you agree with your colleague? If not, should you attempt to get him to correct the excess levels, or is this none of your business since it is he rather than you who is responsible for liquid wastes?

If he refuses to correct the problem, should you report this to your boss or higher management? And if no one in your company will do anything about the problem, should you be prepared to go over their heads and report the problem directly to government inspectors or regulators? Or should you do that only in a case where a much more serious risk to public health and safety is involved?

14.5   Your automotive company is expanding into off-road and all-terrain vehicles (ATVs). As a design engineer you are considering two alternative design concepts for a new vehicle, one having three wheels and another having four.

Engineering research has shown that three-wheeled vehicles are considerably less stable and safe overall than four-wheeled ones. Nevertheless, both designs would satisfy the existing safety standards for the sale of each class of vehicles, so either design could be chosen without any legal problems for the company.

However, there is another factor to consider. Market research has shown that a three-wheeled version would be easier to design and produce, and would sell many more vehicles at higher profit margins than would a four-wheeled version.

Is it ethically acceptable for you as an engineer to recommend that your company should produce the three-wheeled version, in spite of its greater potential safety risks? Or do engineers have an ethical obligation always to recommend the safest possible design for any new product?

14.6   Your company has for some time supplied prefabricated wall sections, which you designed, to construction companies. Suddenly one day a new idea occurs to you about how these might be fabricated more cheaply using composites of recycled waste materials.

Pilot runs for the new fabrication technique are very successful, so the company decides to entirely switch over to the new technique on all future production runs for the prefabricated sections. Nevertheless, there are managerial debates about how, or even whether, to inform the customers about the fabrication changes.

The supply contracts were written with specifications in functional terms, so that the load-bearing capacities and longevity, etc., of the wall sections were specified but no specific materials or fabrication techniques were identified in the contracts. Thus it would be possible to make the changeover without violating the ongoing contracts with customers.

On the other hand, since there is a significant cost saving in the new fabrication method, does your company have an ethical obligation to inform the customers of this, and perhaps even to renegotiate supply at a reduced cost, so that the customers also share in the benefits of the new technique?

More specifically, do you have any special duty, as a professional engineer and designer of the new technique, to be an advocate in your company for the position that customers should be fully informed of the new technique and the associated cost savings?

14.7   Your company manufactures security systems. Up to now these have raised few ethical problems, since your products were confined to traditional forms of security, using armed guards, locks, reinforced alloys that are hard to cut or drill, and similar methods.

However, as a design engineer you realize that with modern technology much more comprehensive security packages could be provided to your customers. These could also include extensive video and audio surveillance equipment, along with bio-metric monitoring devices of employees or other personnel seeking entry to secure areas that would make use of highly personal data such as a person's fingerprints, or retinal or voice patterns.

But there is a potential problem to be considered. A literature search reveals that there are many ethical concerns about the collection and use of such personal data. For example, these high-tech forms of surveillance could easily become a form of spying, carried out without the knowledge of employees and violating their privacy. Or the data collected for security reasons could easily be sold or otherwise used out-side legitimate workplace contexts by unscrupulous customers of your surveillance systems.

Your boss wants you to include as much of this advanced technology as possible in future systems, because customers like these new features and are willing to pay well for them.

However, you are concerned about the ethical issues involved in making these new technologies available. As an engineer, do you have any ethical responsibility to not include any such ethically questionable technologies in products that you design and sell, or to include them only in forms that are difficult to misuse? Or is the misuse of such technologies an ethical problem only for the customers who are buying your equipment, rather than it being your ethical responsibility as an engineer?

14.8   Can you think of any issue that should be of ethical concern to engineers but is not adequately addressed by current codes of engineering ethics?

If so, should it be added to the current rules, or is it something that cannot ade-quately be summed up in an ethical rule?

14.9   Can you think of any rule included in current codes of engineering ethics that is unnecessary, too restrictive, or even ethically wrong? If so, should it simply be removed from the current rules, or is it something that serves a useful function even though it is not ethically required?

If you don't think there are any superfluous rules, instead pick one that you think is one of the least ethically important, and explain why you think it is not important. Does the justification for the rule survive additional scrutiny?

14.10   As an engineering expert on a state planning board, you have to decide which traffic safety projects (involving installation of traffic lights, road widening, etc.) should be funded by the state. Previously these matters were decided politically: each region received roughly equal funding, which tended to maximize voter satisfaction and ensure re-election for the politicians. But now nonelected engineering experts such as yourself will decide these matters.

On what grounds should decisions such as these be made? What would be the soundest and most ethically justifiable way for you to allocate the funds?

In particular, would a utilitarian approach (the greatest good for the greatest number) be the best way to allocate funds? But there is a potential problem with this approach. Since the greatest number of people live in cities and suburbs and very few in rural areas, it is likely that scarce funds would always tend to be allocated to cities with this approach, with almost none going to sparsely populated areas. This policy may tend to save the greatest number of lives or minimize accidents overall, but is it fair to those living in or traveling through rural areas?

On the other hand, how could you justify spending scarce resources in rural areas when you know that this would benefit far fewer people than if you spent the money in cities instead?

CHAPTER 15

Units and Conversions

15.1 History

Weights and measures may be ranked among the necessaries of life to every individual of human society. They enter into the economical arrangements and daily concerns of every family. They are necessary to every occupation of human industry; to the distribution and security of every species of property; to every transaction of trade and commerce; to the labors of the husbandman; to the ingenuity of the artificer; to the studies of the philosopher; to the researches of the antiquarian, to the navigation of the mariner, and the marches of the soldier; to all the exchanges of peace, and all the operations of war. The knowledge of them, as in established use, is among the first elements of education, and is often learned by those who learn nothing else, not even to read and write. This knowledge is riveted in the memory by the habitual application of it to the employments of men throughout life.

—Secretary of State John Quincy Adams
Report to Congress, 1821

Primitive societies used crude measures for many of their tasks. Length was measured with the forearm, the hand, or the finger. Time was measured with reference to the periods of the sun and moon. The volumes of containers such as gourds or clay vessels were measured with the use of plant seeds: a container would be filled with seeds and the number of seeds would determine the volume of the container. Seeds and stones also served as a measure of weight. The carat, used to measure the weight of a gem, was derived from the use of the carob seed.

With the evolution of society, it became more and more necessary to accurately measure various things. Commerce, land division, taxation, and scientific research made it mandatory that measurements could be reproduced with accuracy time after time and in various places. However, with limited international trade, different systems were developed for the same purposes in various parts of the world.

To ensure uniform standards for weights and measures in the United States, the Constitution gave power to Congress to establish the National Bureau of Standards, which ensures uniformity throughout the country. The British system was instituted because of the dominance of Great Britain throughout the world and the position that she had in the affairs of the United States.

But the need for a single, worldwide measurement system soon became obvious. Because the British system uses 12 (12 inches in a foot) and 16 (16 ounces in a pound) as bases, it was not selected as the preferred system. The system selected was the metric system because of its simplicity. It was first developed by the French Academy of Sciences in 1790. The unit of length was to be a fraction of the Earth's circumference, one ten-millionth of the distance from the North Pole to the Equator—the meter. The unit of measurement for volume was to be the cubic decimeter—the liter. The unit of mass was to be the mass of one cubic centimeter of water—the gram—at the temperature of maximum density. Larger and smaller values for each unit were to be found by multiplying or dividing by multiples of 10, thereby greatly simplifying calculations. In 1866, by an Act of Congress, the United States made it lawful to employ the weights and measures of the metric system in all contracts, dealings, or court proceedings.

In the late 19th century the Metric Convention was established, due to the need for better metric standards as required by the scientific community. This treaty was signed by 17 nations, including the United States. It set up well-defined standards for length and mass, and established a method to further refine the metric system when such refinements were required. Since 1893, the metric standards have served as the fundamental weights and measures standards of the United States. The metric system is now either obligatory or permissible throughout the world. In 1971 the Secretary of Commerce recommended that the U.S. move toward predominant use of the metric system through a coordinated program. This change is occurring and has been instituted in many sectors of the U.S., but several sectors continue using the U.S. customary system, most notably those sectors associated with the construction industry.

The International Bureau of Weights and Measures (BIPM), located in Sèvres, France, continues to serve as a permanent secretariat for the Metric Convention, coordinating the exchange of information about the use and refinement of the metric system. As scientific advances occur and the need for refinements arise, the General Conference of Weights and Measures—the diplomatic organization made up of members of the Convention—meets periodically to ratify changes in the system and standards.

In 1960 the General Conference on Weights and Measures ratified an extensive revision and simplification of the metric system. The French name *Le Système International d'Unités*—the International System of Units, abbreviated SI after the French name—was adopted as the modernized metric system.

15.2   The SI System of Units

The International System of Units—the SI system—is a modernized metric system adopted by the General Conference, a multinational organization that includes the

U.S. The SI system is built upon a foundation of seven base units, which are presented in Table 15.1. All other SI units are derived from these seven units. Multiples and sub-multiples are expressed using a decimal system. Table 15.1 also includes the supplementary units, radians, and steradians.

Table 15.1 The SI Base and Supplementary Units

Unit (Symbol)
Quantity	*Definition*
meter (m) length	1650763.73 wavelengths in a vacuum of the orange-red line of the spectrum of krypton-86
kilogram (kg) mass	A cylinder of platinum-iridium alloy kept by the International Bureau of Weights and Measures in Paris
second (s) time	The duration of 9192631770 periods of the radiation associated with a specified transition of the cesium-133 atom
ampere (A) electric current	That current which, if maintained in each of two parallel wires separated by one meter in free space, would produce a force of 2×10^{-7} N/m
kelvin (K) temperature	1/273.16 of the temperature of the triple point of water
mole (mol) amount of substance	The amount of substance of a system that contains as many elementary entities as there are atoms in 0.012 kg of carbon-12
candela (cd) luminous intensity	The luminous intensity of 1/600000 of a square meter of a blackbody at the temperature of freezing platinum
radian (rad) plane angle	The plane angle with its vertex at the center of a circle that is subtended by an arc equal in length to the radius
steradian (sr) solid angle	The solid angle with its vertex at the center of a sphere that is subtended by an area of the spherical surface equal to that of a square with sides equal in length to the radius

The symbols used for the base and supplementary units are included in Table 15.1. In general, the first letter of a symbol is capitalized if the name of a symbol is derived from a person's name; otherwise, it is lowercase. One exception is the symbol for the liter: "L" should be used so that the lowercase "l" is not confused with the numeral "1."

Names of units and prefixes are not capitalized except at the beginning of sentences and in titles and headings in which all main words are capitalized: "Temperature is measured in kelvins" or "Meters are longer than feet." Note that the temperature unit kelvin is not capitalized, but we do capitalize "Celsius." So we would write, "four degrees Celsius."

Names of units are often plural for numerical values equal to zero, greater than 1, or less than –1. The singular form is used for other values. We would say 1.1 meters, 0.1 meter, or 0 meters. We could, however, say 1.1 meter or –1 degree Celsius. In symbolic form, 100 m would be read "100 meters."

When writing units after a numerical value, note the following items:

- The product of two or more units is indicated by a dot, dash, or a space, although it is possible to omit these if no confusion results. We would write "meter gram" as m·g, m-g, and m g but never as mg because mg could be read as milligram.
- The division of units can be written as m/s, or m·s⁻¹, for example. In more complicated groupings of units, we could write m/s^2 or $m \cdot s^{-2}$, but not m/s/s. A more complicated example would be $m \cdot kg / (s^2 \cdot A)$ or $m \cdot kg \cdot s^{-2} \cdot A^{-1}$, but we would not write $m \cdot kg / s^2 / A$.
- No compound prefixes should be used. For example, kMm (kilo mega meter) is not allowed.

15.3 Derived Units

Derived units are expressed algebraically in terms of base and supplementary units using the three recommendations listed at the end of Section 15.2. Several derived units have been given special names and symbols, such as newton with symbol N. These special names and symbols may be used to express the units of a quantity in a simpler way than by using the base units. The following two tables list many such derived units. Table 15.2 lists dimensions whose derived units are expressed in terms of the base and supplementary units. Table 15.3 lists the dimensions whose derived units have special names.

Units of temperature are worth discussing in further detail. The unit of temperature in the SI system is kelvins. However, Celsius is often used in many engineering calculations. The kelvin scale is an absolute temperature scale since the zero point on this scale refers to the lowest possible temperature. There are no negative temperatures in

Table 15.2 Dimensions Whose Units Are Expressed in Terms of Base and Supplementary Units

Quantity	SI Unit	SI Symbol
Area	square meter	m^2
Volume	cubic meter	m^3
Speed, velocity	meter per second	m/s
Acceleration	meter per second squared	m/s^2
Density	kilogram per cubic meter	kg/m^3
Specific volume	cubic meter per kilogram	m^3/kg
Magnetic field strength	ampere per meter	A/m
Concentration	mole per cubic meter	mol/m^3
Luminance	candela per square meter	cd/m^2
Kinematic viscosity	square meter per second	m^2/s
Angular velocity	radian per second	rad/s
Angular acceleration	radian per second squared	rad/s^2

Table 15.3 Dimensions Whose Units Have Special Names

Quantity	SI Name	SI Symbol	Other SI Units
Frequency	hertz	Hz	cycle/s
Force	newton	N	kg·m/s^2
Pressure, stress	pascal	Pa	N/m^2
Energy, work	joule	J	N·m
Power	watt	W	J/s
Electric charge	coulomb	C	A·s
Electric potential	volt	V	W/A
Capacitance	farad	F	C/V
Electric resistance	ohm	Ω	V/A
Conductance	siemens	S	A/V
Magnetic flux	weber	Wb	V·s
Magnetic flux density	tesla	T	Wb/m^2
Inductance	henry	H	Wb/A
Luminous flux	lumen	lm	
Illuminance	lux	lx	

the kelvin scale. On the other hand, the Celsius scale is a relative scale since the zero point on this scale does not have any particular significance and negative temperatures are possible. The difference between the two scales is not in the actual scale but rather in the starting point of the scale. Therefore, the relationship is of the addition/subtraction type rather than multiplication/division, which is usually what we use to convert one unit to a different unit.

$$T[^\circ K] = T[^\circ C] + 273.15$$

One degree Celsius is equal to one kelvin, but temperature is read differently on these scales. For example, the two statements "water temperature was raised by 15 degrees Celsius" and "water temperature was raised by 15 kelvins" are equivalent. In cases where engineering units are presented in terms of per unit of temperature, we can use either Celsius or kelvin without changing the value. For example, the specific heat of water is 4.186 J/(kg K) or 4.186 J/(kg °C).

The Celsius temperature is expressed in degrees Celsius (i.e., 20°C) but we do not use the small degree symbol with K (i.e., 100 K).

Another relative temperature scale that was independently developed is the Fahrenheit scale. There are two noteworthy differences between the Celsius and the Fahrenheit scales. One difference is that

$$1^\circ C = 1.8^\circ F$$

The second difference is that their zero points correspond to different temperatures. Due to these two differences in the temperature scales, the conversion process from one scale to the other involves both addition/subtraction and multiplication/division. The two scales are related by the equation

$$T[°C] = 1.8 * (T[°F] - 32) \quad \text{or}$$

$$T[°F] = 1/1.8 * T[°C] + 32$$

15.4   Prefixes

Rather than write extremely large numbers or very small numbers, we use prefixes that have been defined for the SI system. Their symbols and pronunciations are listed in Table 15.4. Some are capitalized, while others are lowercase to avoid confusion. G stands for giga while g stands for gram, K for kelvin and k for kilo, M for mega and m for milli, and N for newton and n for nano. We do not leave a space between the prefix and the letters making up a symbol or a name. We write "milliliter" and "mL."

Hecto, deka, deci, and centi are metric prefixes that are to be avoided, except for unit multiples of area and volume and for the non-technical use of centimeter, as for body and clothing measurements.

Table 15.4 Prefixes for Metric Units

Multiplication Factor	Prefix	Symbol	Pronunciation	Term (USA)
1000 000 000 000 000 000 = 10^{18}	exa	E	Texas	one quintillion
1000 000 000 000 000 = 10^{15}	peta	P	petal	one quadrillion
1000 000 000 000 = 10^{12}	tera	T	terrace	one trillion
1000 000 000 = 10^9	giga	G	gig-ah	one billion
1000 000 = 10^6	mega	M	megaphone	one million
1000 = 10^3	kilo	k	kilowatt	one thousand
100 = 10^2	hecto	h	hector	one hundred
10	deka	da	D'Cartes	ten
0.1 = 10^{-1}	deci	d	decimal	one tenth
0.01 = 10^{-2}	centi	c	sentiment	one hundredth
0.001 = 10^{-3}	milli	m	military	one thousandth
0.000 001 = 10^{-6}	micro	m	microphone	one millionth
0.000 000 001 = 10^{-9}	nano	n	nan-o	one billionth
0.000 000 000 001 = 10^{-12}	pico	p	peek-o	one trillionth
0.000 000 000 000 001 = 10^{-15}	femto	f	fem-to	one quadrillionth
0.000 000 000 000 000 001 = 10^{-18}	atto	a	at-o	one quintillionth

In general, use prefixes that result in numerical values between 0.1 and 1000. However, use the same prefix for all items in a given context or for the entries of the same quantity in a table.

Notation in powers of 10 is often used rather than a prefix. Rather than writing 20 MJ, we often write 2×10^7 J. Either way is equally acceptable.

Avoid mixing prefixes. We would dimension an area as 40 mm wide and 1500 mm long, not 40 mm wide and 1.5 m long. However, a wire would be described as 1500 meters of 2-mm-diameter wire since the difference in size is extreme.

Never use two units for one quantity. We would say a board is 3.5 m long, not 3 m 50 cm long, or 3 m 500 mm long. A volume contains 13.58 L, not 13 L 580 mL.

15.5 Numerals

A space is always left between the numeral and the unit name or symbol, except when we write the degree symbol, as in 30°, or the symbol for temperature, as in 30°C. It would be incorrect to write 30 °C, with a space before the degree symbol. Likewise, it is incorrect to write 30m without a space before the meter symbol.

When a quantity is used in an adjectival sense, a hyphen is used between the number and the symbol or unit name (except for angle degree). An example would be: The length of the 20-mm-diameter pole is 30 m. When names are used rather than the symbols, the hyphens are still used.

In calculations a number must often be rounded off. If the number 8.3745 is rounded to three digits, it would be 8.37. If it is rounded to four digits, it would become 8.374 since the digit before the "5" is an even number. If the digit before 5 were odd, it would be rounded up. For example, the number 8.3755 would be rounded to 8.376.

Finally, a word is in order regarding significant digits. With the advent of the calculator and the computer, each calculation may be carried out to eight or more digits. In some engineering disciplines a material property such as density, viscosity, or conductivity may be used in the calculations. Material properties are either given in the problem statement or found in an appropriate table. A material property, which has been measured in some laboratory, is seldom known to more than four significant digits. Most properties are known to only three significant digits. It is impossible to calculate an answer to more significant digits than are present in the data used to arrive at that answer. We usually assume that information given in a problem statement, such as the number 6 in a 6-mm-diameter pipe, is known to four significant digits. To present an answer with six significant digits, when at most four significant digits are known in the data provided, is an error made by many students. Significant figures should not be confused with the number of decimal places, which is just the number of digits to the right of a decimal point.

For a given number, all non-zero digits are significant. The only digit that we need to note to determine the number of significant digits is zero.

Zeros *are significant* if they appear between two nonzero digits or if they appear as the last digits in a decimal number. In the following examples there are 5 significant digits in each of the given numbers:

40015

4.0015

41.500

4105.0

Zeros *are not significant* when they appear on the left side of a decimal number or on the right side of an integer number. Sometimes values are shown in the scientific notation to specify the exact number of significant digits. In the following examples there are 4 significant digits in each of the given numbers:

0.0001203	equivalent to	1.203×10^{-4}
0.001230	equivalent to	1.230×10^{-3}
254600	equivalent to	2546×10^{2}
20590000	equivalent to	2059×10^{4}

15.6   Unit Conversions

We often encounter dimensions that have units that are not appropriate for the problem at hand. Table 15.5 presents such units. For example, to convert 40 acres to square meters we simply multiply 40 by 4046.86 and obtain 161 874 square meters if we desire six significant digits, or 161 900 if we desire four significant digits.

Table 15.5 Conversion Factors

To convert from	To	Multiply by
Acres	Square meters	4046.86
Acres	Square feet	43 560
Acres	Square meters	100
Atmospheres	Bars	1.0132
Atmospheres	Feet of water	33.90
Atmospheres	Inches of mercury	29.92
Atmospheres	psi	14.700
Atmospheres	Torrs	760
Bars	psi	14.504
Btu	Foot-pounds	777.6
Btu	Joules	1054.4
Btu	kWh	0.000 292 9
Btu/min	Ton of refrigeration	0.004 997
Btu/lb	kJ/kg	2.324

To convert from	To	Multiply by
Btu/sec	Horsepower	1.4139
Btu/sec	kW	1.0544
Calories	Btu	0.003 968
Coulombs	Ampere-seconds	1
Cubic feet	Cubic meters	0.028 317
Cubic feet	Liters	28.317
Cubic feet	Gallons	7.4805
Cubic inches	Milliliters	16.387
Degrees	Radians	0.017 453
Dynes	Newtons	0.000 01
Electron volts	Ergs	1.60219×10^{-12}
Ergs	Joules	10^{-7}
Hectares	Acres	2.4711
Hectares	Square meters	10 000
Horsepower	Btu/hr	2545
Horsepower	Ft-lb/sec	550
Horsepower	kW	0.745 70
Inches	Centimeters	2.54
Inches of mercury	psf	70.7
Kilowatts	Horsepower	1.341 02
Kilowatt-hrs	Btu	3409.5
Knots	Feet/sec	1.6878
Knots	Meters per second	0.514 44
Maxwells	Webers	10^{-8}
Meters	Feet	3.2808
Miles	Feet	5280
Miles	Meters	1609.34
Miles/hr	Kilometers/hr	1.609 34
Newtons	Pound-force	0.224 81
Pound-mass	Kilograms	0.453 59
Pound-mass	Slugs	0.031 08
Pound-mass per cubic foot	kg/m³	16.018
Pound-force per square foot	N/m² (Pa)	47.88
Pound-force per square inch	N/m² (Pa)	6895
Radians	Revolutions	0.159 15
Radians	Degrees	57.296
Revolutions	Radians	6.2832
Siemens	Mhos	1
Slugs per cubic foot	kg/m³	515.38
Spans	Feet	0.75
Square feet	Square meters	0.092 90
Square kilometers	Acres	247.10

(continued)

Table 15.5 Conversion Factors *(continued)*

To convert from	To	Multiply by
Square meters	Square feet	10.764
Steres	Cubic meters	1
Tons (metric)	Kilograms	1000
Tons (short)	Pound-mass	2000
Tons of refrigeration	Btu/hr	12 000
Tons of refrigeration	Horsepower	4.716
Tons of refrigeration	lb of ice melted/day	2000
Torrs	mm of mercury	1
Watts	Horsepower	0.001 341
Webers	Maxwells	10^8

A number of examples for converting units are shown below.

EXAMPLE 15.1 Convert the following units to their SI equivalents: a) 10.0 lb (mass); b) 5.17 hp/in³; c) 25.6 Btu/(lb$_m$. °F).

a) In this example mass is converted from pounds to its SI equivalent, which is kg:

$$10.0 \ lb \ \times \frac{0.453 \ kg}{1 \ lb} = 4.53 \ kg$$

b) The dimensions in this example are power per unit volume. Therefore, hp needs to be converted to kg and in³ to m³:

$$5.17 \ \frac{hp}{in^3} \ \times \ \frac{746 \ W}{1 \ hp} \ \times \left(\frac{1 \ in}{0.0254 \ m} \right)^3 = 2.35 \ \times 10^8 \ \frac{W}{m^3}$$

c) A careful study of the units used in this example reveals that we are dealing with energy per unit mass per unit temperature. When converting temperature scale, care must be taken so as not to use the conversion equation used for reading the temperature. To convert F to K, we use the relationship 1 K = 1°C = 1.8 °F

$$25.6 \ Btu/(lb_m \ F) = \frac{1.055kJ}{1 \ Btu} \ \times \ \frac{1 \ lbm}{0.453 \ kg}s \ \times \frac{1.8°F}{1K} = 107.7 \ \frac{kJ}{kg \ K}$$

15.7   Dimensional Homogeneity and Dimensionless Numbers

If an equation expresses a proper relationship among variables in a physical process, then it should be **dimensionally homogeneous**. Such equations are correct for any system of units, and consequently each group of terms in the equation must have the same dimensional representation. This is also known as the law of **dimensional homogeneity**.

Consider Newton's universal gravitational law:

$$F = G \frac{m_1 m_2}{r^2}$$

In this equation, F has dimension of force, m has dimension of mass, and r has dimension of length. For this equation to be dimensionality homogeneous, G must have dimension of force * length^2/mass^2. And indeed, $G = 5.67 \times 10^{-11}$ N·m^2/kg^2.

Now let's consider another familiar equation that describes the position of a particle in a uniformly accelerated linear motion:

$$x = \frac{1}{2}at^2 + v_0 t + x_0$$

Here x denotes position, v denotes velocity, a is acceleration, and t is time. In this equation, there are three terms on the right side of the equation that are added together. To add these terms they *must* have the same dimensions. Here we claim that

$\frac{1}{2}at^2$, $v_0 t$, *and* x_0 have the same dimensions. Let's examine these terms using the

SI units for the given dimensions:

$$\frac{1}{2}at^2 \ [(m/s^2)(s^2) = m]$$

$$v_0 t \ [(m/s)(s)] = m$$

$$x_0 \ [m]$$

Note that each term has units of length; therefore, this equation is dimensionally homogeneous.

Dimensional homogeneity is necessary but not a sufficient condition for an equation to be valid. Two more conditions need to be satisfied for an equation to be valid. One condition is similarity of units. Consider:

$$A = B + C$$

where all three variables A, B, and C have dimensions of length. But if B is given as 5 meters and C is given as 3 inches, we cannot add these two numbers before performing some unit conversions.

The second condition that needs to be satisfied for an equation to be valid is that the equation should correctly model a physical process. For example, let us consider the following equation for calculating the area of a rectangle, where a and b denote the sides of the rectangle:

$$A = a^2 + b^2$$

Even though the equation is dimensionally homogeneous and the units may be correctly used, it does not constitute a valid equation, since for a rectangle $A = ab$.

Dimensionless Numbers

Dimensionless numbers fall into two major categories. First are those that are dimensionless by definition, such as counting numbers. A second category, which is more important to engineering, is combinations of variables where all the dimension/units have canceled out so that the net term has no dimension. These are often called "dimensionless groups" or "dimensionless numbers" and often have special names and meanings. Most of these have been found using techniques of "dimensional analysis," a way of examining physical phenomena by looking at the dimensions that occur in the problem without considering any numbers.

Some of the most commonly used dimensionless numbers are π, Mach number, and specific gravity, which are defined as follows:

$$\pi\pi = \frac{\text{Circumference of a circle}}{\text{diameter of a circle}}$$

$$M = \frac{\text{speed of an object}}{\text{speed of sound}}$$

$$SG = \frac{\text{density of a substance}}{\text{density of water}}$$

There are many other dimensionless numbers used in engineering, such as Reynolds number, Prandtl number, friction coefficient, drag coefficient, lift coefficient, Euler number, and Rayleigh number. Interestingly enough, trigonometric functions such as sin(x), cos(x), tan(x), and so forth are all dimensionless since they represent ratios of similar dimensions. For example, sine of an angle is the ratio of the opposite side to the hypotenuse in a right triangle.

A dimensionless quantity holds the same value regardless of the measurement system used. Such dimensions are independent of unit choice. There are certain computations that require dimensionless groups or else the computation will not make dimensional sense. For example, the quantity

$$c = a^b$$

can only have meaning if b is dimensionless. There are a few other situations where certain parts of an equation *must* be dimensionless. The following is a partial list of such situations. In each of these examples the variable x must be dimensionless:

- Logarithmic functions: $\log(x)$
- Exponential arguments: e^x
- Trigonometric functions: $\sin(x), \cos(x), \tan(x)$

One reason why the functions listed above are dimensionless is that these functions can be represented as a series of fractions of various power. The only time we can add quantities with different powers is when each term is dimensionless. For example, e^x can be represented by:

$$e^x = 1 + x + \frac{x^2}{2} + \frac{x^3}{6} + \frac{x^4}{24} + \dots$$

In the function above, if x were meters (m) then we could not add terms involving $m, m^2, m^3, m^4, \dots$ The only way we can add these terms is if each term is dimensionless, which requires x to be dimensionless.

REFERENCES

American National Metric Council, *Metric Editorial Guide*, 5th ed., 1993.

U.S. Department of Commerce/National Bureau of Standards, *Brief History of Measurement Systems*, Special Publication 304A, 1974.

U.S. Department of Commerce/National Bureau of Standards, *International System of Units (SI)*, 1974.

EXERCISES AND ACTIVITIES

15.1 State both the primitive and the modern definitions of time. Which do you find the more reasonable?

15.2 Select the angle that is approximately equal to a radian:

   a) 180 degrees
   b) 30 degrees
   c) 57 degrees
   d) 360 degrees

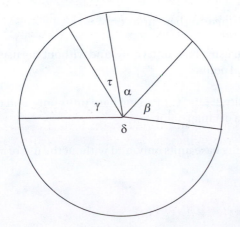

15.3   Select the correct statement:

   a) The temperature is 10°K.
   b) The temperature is 283°C.
   c) The temperature is 10 kelvins.
   d) The temperature is 283 Celsius.

15.4   Write each of the following in a correct form without using any exponents:

   a) 12300 Nmkg^{-1} K^{-1}
   b) 109.433 Cs m^{-1} K^{-1}
   c) 0.000433 W s m^{-3} K^{-1}

15.5   Write each of the following in a correct form without using any negative exponents:

   a) 0.004562 kNm m^{-3} K^{-1}
   b) 10^{-5} MNs^{-1} m^{-3} K^{-1}
   c) 0.000473001 mJs^{-1} m^{-3} kg^{-1}

15.6   Write each of the following in an acceptable form without using a decimal point:

   a) 0.0000546 kW/K
   b) 65.3207 kH/kg
   c) 54.6367 × 10^{-6} MF/kg

15.7   Write the viscosity $\mu = 0.001$ Pa.s using newtons and not pascals. Also, write the viscosity without using either newtons or pascals.

15.8   Write the heat flux density 300 W/m^2 with the units in the following forms:

   a) one that includes joules
   b) one that includes newtons
   c) one that includes kilograms

15.9   Write the permeability 0.231 H/m using base units only.

15.10   Write the permittivity 10^{-5} F/m using base units only, and without the use of a negative exponent or a decimal point.

15.11   Write the electric charge density 3 × 10^{-4} C/m^3 using base units only, and without the use of a negative exponent or a decimal point.

15.12   Write each of the following using base units only, and without the use of a prefix:

   a) 5643 kN/t.d
   b) 453000 μH/(kg.h)
   c) 654.0004 nS.m/(h.kg)

15.13 Write each of the following using base units without the use of a prefix:

a) 4×10^{-6} GF/m
b) 65.2 fF/m
c) 0.00347 pPa/kg

15.14 Write each of the following using base units without the use of an exponent:

a) 4×10^{-6} GF/m
b) 43.64×10^{6} mPa/mm^2
c) 0.000453×10^{-4} TT/mm^3

15.15 Viscosity is measured as follows. Write each in acceptable form using base units only:

a) $\dfrac{17}{9800}$ pa·s

b) $\dfrac{32}{8741}$ N·s / m^2

c) $\dfrac{61}{34\,982}$ kN·h/mm^2

15.16 Round off 6743.865 to:

a) six significant digits
b) five significant digits
c) four significant digits
d) three significant digits
e) two significant digits

15.17 Express each of the following to four significant digits:

a) $\dfrac{1}{3}$

b) $89\dfrac{2}{3}$

c) 2300.71

d) $\dfrac{3}{1000}$ s

e) $\dfrac{1}{3000}$

f) $\dfrac{9}{882}$

15.18   Express each of the following conversions to four significant digits:

a) 340 Btu/s to watts
b) 3458 degrees to radians
c) 25 hectares to square meters
d) 240 horsepower to watts
e) 67 851.22 kWh to joules
f) 2×10^{-7} miles to meters
g) 40 mph to km/s

15.19   In the study of flow characteristics in pipes, the relative effect of velocity to viscous effects is given by the Reynolds number, which is a dimensionless number. Reynolds number is given by:

$$Re = \frac{\rho VD}{\mu}$$

where ρ is the fluid density [kg/m³], V is velocity [m/s], and D is the pipe diameter [m]. What is the unit of absolute viscosity (μ), ?

15.20   Biot number, Bi, is a dimensionless number that describes the relative effect of convection to conduction heat transfer. This number is given as

$$Bi = \frac{hL}{k}$$

where h is the heat transfer coefficient [W/(m² °C)], and L is the characteristic length [m]. What is the unit of thermal conductivity, k ?

15.21   A jet plane is traveling at 900 km/h at an altitude where the temperature is such that the speed of sound is 600 km/h.

a) Determine the Mach number for the jet plane.
b) How will the answer in part (a) be different if the speeds were given in miles per hour?

CHAPTER 16

Mathematics Review

Mathematics is used as a tool to help solve the problems encountered in the analysis and design of physical systems. We will review those parts of mathematics that you may have already studied and that are used fairly often. The topics include algebra, trigonometry, analytic geometry, linear algebra (matrices), and calculus. The review here is intended to be brief and not exhaustive. There may be some subjects included in this chapter that you have not yet encountered. If that is the case, your instructor might not require that you review that material.

16.1 Algebra

It is assumed that you are familiar with most of the rules and laws of algebra as applied to both real and complex numbers. We will review some of the more important of these and illustrate several with examples. The three basic rules are:

Commulative Law: $\qquad\qquad a+b=b+a \qquad\qquad ab=ba$ $\qquad$ (16.1.1)

Distributive Law: $\qquad\qquad a(b+c)=ab+ac$ $\qquad\qquad\qquad$ (16.1.2)

Associative Law: $\qquad\quad a+(b+c)=(a+b)+c \qquad a(bc)=(ab)c$ $\quad$ (16.1.3)

Exponents

Laws of exponents are used in many manipulations. For positive x and y we use

$$x^{-a}=\frac{1}{x^a}$$
$$x^a x^b = x^{a+b}$$
$$(xy)^a = x^a y^a$$
$$x^{ab} = (x^a)^b$$

$\qquad$ (16.1.4)

457

Logarithms

Logarithms are actually exponents. For example if $b^x = y$ then $x = \log_b y$; that is, the exponent x is equal to the logarithm of y to the base b. Most applications involve common logs, which have a base of 10, written as $\log y$, or natural logs, which have a base of e ($e = 2.7183\ldots$), written as $\ln y$. If any other base is used it will be so designated, such as $\log_5 y$.

Remember, logarithms of numbers less than one are negative, the logarithm of one is zero, and logarithms of numbers greater than one are positive. The following identities are often useful when manipulating logarithms:

$$\ln x^a = a \ln x$$

$$\ln(xy) = \ln x + \ln y$$

$$\ln\left(\frac{x}{y}\right) = \ln x - \ln y$$

$$\ln x = 2.303 \log x \qquad (16.1.5)$$

$$\log_b b = 1$$

$$\ln 1 = 0$$

$$\ln e^a = a$$

$$\log_a y = x \text{ implies } a^x = y$$

$$\ln y = x \text{ implies } e^x = y$$

The Quadratic Formula and the Binomial Theorem

We often encounter the quadratic equation $ax^2 + bx + c = 0$ when solving problems. The **quadratic formula** provides its solution; it is

$$x = \frac{-b \pm \sqrt{b^2 - 4ac}}{2a} \qquad (16.1.6)$$

If $b^2 < 4ac$, the two roots are complex numbers. Cubic and higher-order equations are most often solved by trial and error.

The **binomial theorem** is used to expand an algebraic expression of the form $(a + x)^n$. It is

$$(a + x)^n = a^n + na^{n-1}x + \frac{n(n-1)}{2!}a^{n-2}x^2 + \cdots \qquad (16.1.7)$$

If n is a positive integer, the expansion contains $(n + 1)$ terms. If it is a negative integer or a fraction, an infinite series expansion results.

Partial Fractions

A rational fraction $P(x)/Q(x)$, where $P(x)$ and $Q(x)$ are polynomials, can be resolved into partial fractions for the following cases.

Case 1: $Q(x)$ factors into n different linear terms,

$$Q(x) = (x - a_1)(x - a_2)\cdots(x - a_n)$$

Then

$$\frac{P(x)}{Q(x)} = \sum_{i=1}^{n} \frac{A_i}{x - a_i}$$

(16.1.8)

Case 2: $Q(x)$ factors into n identical terms,

$$Q(x) = (x - a)^n$$

Then

$$\frac{P(x)}{Q(x)} = \sum_{i=1}^{n} \frac{A_i}{(x - a)^i}$$

(16.1.9)

Case 3: $Q(x)$ factors into n different quadratic terms,

$$Q(x) = (x^2 + a_1x + b_1)(x^2 + a_2x + b_2)\cdots(x^2 + a_nx + b_n)$$

Then

$$\frac{P(x)}{Q(x)} = \sum_{i=1}^{n} \frac{A_ix + B_i}{x^2 + a_ix + b_i}$$

(16.1.10)

Case 4: $Q(x)$ factors into n identical quadratic terms,

$$Q(x) = (x^2 + ax + b)^n$$

Then

$$\frac{P(x)}{Q(x)} = \sum_{i=1}^{n} \frac{A_ix + B_i}{(x^2 + ax + b)^i}$$

(16.1.11)

EXAMPLE 16.1   The temperature at a point in a body is given by $T(t) = 100e^{-0.02t}$. At what value of t does $T = 20$?

Solution: The equation takes the form

$$20 = 100e^{-0.02t} \text{ or } 0.2 = e^{-0.02t}$$

Take the natural logarithm of both sides and obtain

$$\ln 0.2 = \ln e^{-0.02t}$$

Using a calculator, and Eq. 16.1.5, we find

$$-1.6094 = -0.02t$$

$$\therefore t = 80.47$$

EXAMPLE 16.2

Solve for V if $3V^2 + 6V = 10$.

Solution: Use the quadratic formula, Eq. 16.1.6. In standard form, the equation is

$$3V^2 + 6V - 10 = 0$$

The solution is then

$$V = \frac{-6 \pm \sqrt{36 - 4 \times 3 \times (-10)}}{2 \times 3} = \frac{-6 \pm 12.49}{6}$$

$$= 1.082 \text{ or } -3.082$$

Note: Some physical situations may not permit a negative answer, so V = 1.082 would be the solution.

EXAMPLE 16.3

Resolve $\dfrac{3x-1}{x^2+x-6}$ into partial fractions.

Solution: The denominator is factored:

$$x^2 + x - 6 = (x+3)(x-2)$$

Using Case 1 there results

$$\frac{3x-1}{x^2+x-6} = \frac{A_1}{x+3} + \frac{A_2}{x-2}$$

This can be written as

$$\frac{3x-1}{x^2+x-6} = \frac{A_1(x-2) + A_2(x+3)}{(x+3)(x-2)}$$

$$= \frac{(A_1 + A_2)x - 2A_1 + 3A_2}{(x+3)(x-2)}$$

The numerators on both sides must be equal. Equating the coefficients of the various powers of x provides us with two equations:

$$A_1 + A_2 = 3$$

$$-2A_1 + 3A_2 = -1$$

These are solved to give $A_2 = 1$, $A_1 = 2$. Finally,

$$\frac{3x-1}{x^2+x-6} = \frac{2}{x+3} + \frac{1}{x-2}$$

16.2 Trigonometry

The primary functions in trigonometry involve the ratios between the sides of a right triangle. Referring to the right triangle in Figure 16.1, the functions are defined by

$$\sin\theta = \frac{y}{r}, \quad \cos\theta = \frac{x}{r}, \quad \tan\theta = \frac{y}{x} \tag{16.2.1}$$

In addition, there are three other functions that find occasional use, namely

$$\cot\theta = \frac{x}{y}, \quad \sec\theta = \frac{r}{x}, \quad \csc\theta = \frac{r}{y} \tag{16.2.2}$$

The trig functions $\sin\theta$ and $\cos\theta$ are periodic functions with a period of 2π. Figure 16.2 shows a plot of the three primary functions.

In the above relationships, the angle θ is usually given in radians for mathematical equations. It is possible, however, to express the angle in degrees; if that is done it may be necessary to relate degrees to radians. This can be done by remembering that there are 2π radians in $360°$. Hence, we multiply radians by $(180/\pi)$ to obtain degrees, or multiply degrees by $(\pi/180)$ to obtain radians. A calculator may use either degrees or radians for an input angle.

Most problems involving trigonometry can be solved using a few fundamental identities. They are

$$\sin^2\theta + \cos^2\theta = 1 \tag{16.2.3}$$

$$\sin 2\theta = 2\sin\theta\cos\theta \tag{16.2.4}$$

$$\cos 2\theta = \cos^2\theta - \sin^2\theta \tag{16.2.5}$$

$$\sin(\alpha \pm \beta) = \sin\alpha\cos\beta \pm \sin\beta\cos\alpha \tag{16.2.6}$$

$$\cos(\alpha \pm \beta) = \cos\alpha\cos\beta \pm \sin\alpha\sin\beta \tag{16.2.7}$$

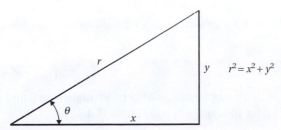

Figure 16.1 A right triangle

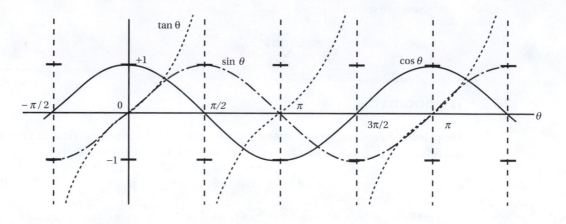

Figure 16.2 The trig functions.

A general triangle may be encountered such as that shown in Figure 16.3. For this triangle we may use the following equations:

$$\text{law of sines: } \frac{\sin\alpha}{a} = \frac{\sin\beta}{b} = \frac{\sin\gamma}{c} \tag{16.2.8}$$

$$\text{law of cosines: } a^2 = b^2 + c^2 - 2bc\cos\alpha \tag{16.2.9}$$

Note that if $\gamma = 90°$, the law of cosines becomes the *Pythagorean Theorem*

$$c^2 = a^2 + b^2 \tag{16.2.10}$$

The hyperbolic trig functions also find occasional use. They are defined by

$$\sinh x = \frac{e^x - e^{-x}}{2}, \cosh x = \frac{e^x + e^{-x}}{2}, \tanh x = \frac{\sinh x}{\cosh x} \tag{16.2.11}$$

Useful identities follow:

$$\cosh^2 x - \sinh^2 x = 1 \tag{16.2.12}$$

$$\sinh(x \pm y) = \sinh x \cosh y \pm \cosh x \sinh y \tag{16.2.13}$$

$$\cosh(x \pm y) = \cosh x \cosh y \pm \sinh x \sinh y \tag{16.2.14}$$

The values of the primary trig functions of certain angles are listed in Table 16.1. A unit circle is provided for your reference in Figure 16.4.

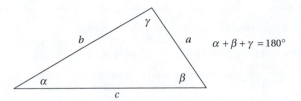

Figure 16.3 A general triangle

$$\alpha + \beta + \gamma = 180°$$

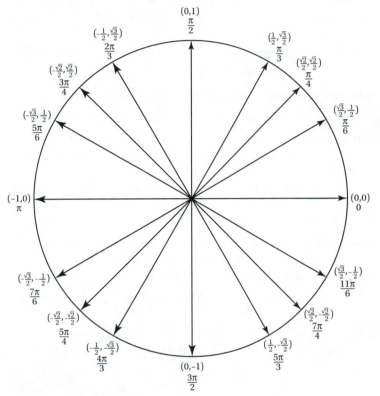

Figure 16.4 A unit circle

Table 16.1 Functions of Certain Angles

	0	30°	45°	60°	90°	135°	180°	270°	360°
$\sin q$	0	$\dfrac{1}{2}$	$\dfrac{\sqrt{2}}{2}$	$\dfrac{\sqrt{3}}{2}$	1	$\dfrac{\sqrt{2}}{2}$	0	-1	0
$\cos q$	1	$\dfrac{\sqrt{3}}{2}$	$\dfrac{\sqrt{2}}{2}$	$\dfrac{1}{2}$	0	$\dfrac{\sqrt{2}}{2}$	-1	0	1
$\tan q$	0	$\dfrac{1}{\sqrt{3}}$	1	$\sqrt{3}$	∞	-1	0	∞	0

EXAMPLE 16.4

Express $\cos^2\theta$ as a function of $\cos 2\theta$.

Solution: Substitute Eq. 16.2.3 into Eq. 16.2.5 and obtain

$$\cos 2\theta = \cos^2\theta - \left(1 - \cos^2\theta\right)$$

$$\cos 2\theta = 2\cos^2\theta - 1$$

There results

$$\cos^2\theta = (1 + \cos 2\theta)$$

EXAMPLE 16.5

If $\sin\theta = x$, what is $\tan\theta$?

Solution: Think of $x = x/1$. Thus, the hypotenuse of an imaginary right triangle is of length unity and the leg opposite θ is of length x. The other leg is of length $\sqrt{1 - x^2}$.
Hence, $\tan\theta = \dfrac{x}{\sqrt{1 - x^2}}$

EXAMPLE 16.6

An airplane leaves Lansing flying due southwest at 300 km/hr, and a second plane leaves Lansing at the same time flying due west at 500 km/hr. How far apart are the airplanes after 2 hours?

Solution: After 2 hours, the respective distances from Lansing are 600 km and 1000 km. A sketch is quite helpful (Fig. 16.5).The distance d that the two airplanes are apart is found using the law of cosines:

$$\text{Let } a = d = \text{distance}$$

$$b = 1000 \text{ km}$$

$$c = 600 \text{ km}$$

$$\alpha = 45°$$

$$a^2 = b^2 + c^2 - 2bc\,\cos\alpha$$

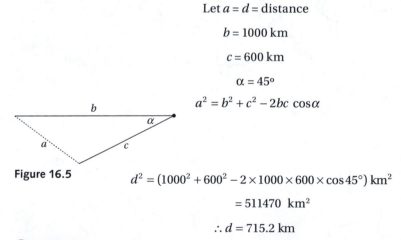

Figure 16.5

$$d^2 = (1000^2 + 600^2 - 2 \times 1000 \times 600 \times \cos 45°)\ \text{km}^2$$

$$= 511470\ \text{km}^2$$

$$\therefore d = 715.2 \text{ km}$$

16.3 Geometry

A regular polygon with n sides has a vertex angle (the central angle subtended by one side) of $2\pi/n$. The included angle between two successive sides is given by $\pi(n-2)/n$.
Some common geometric shapes are displayed in Figure 16.5.

The equation of a straight line can be written in the general form

$$Ax + By + C = 0 \tag{16.3.1}$$

There are three particular forms that this equation can take. They are:

$$\text{Point} - \text{slope} : y - y_1 = m\,(x - x_1) \tag{16.3.2}$$

$$\text{Slope} - \text{intercept} : y = m\,(x - x_1) \tag{16.3.3}$$

$$\text{Two} - \text{intercept} : \frac{x}{a} + \frac{y}{b} = 1 \tag{16.3.4}$$

In the above equations m is the slope, (x_1, y_1) a point on the line, "a" the x-intercept, and "b" the y-intercept (Fig. 16.6). The perpendicular distance d from the point (x_3, y_3) to the line $Ax + By + C = 0$ is given by (see Fig. 16.6)

$$d = \frac{|\,Ax_3 + By_3 + C\,|}{\sqrt{A^2 + B^2}} \tag{16.3.5}$$

The equation of a plane surface is given as

$$Ax + By + Cz + D = 0 \tag{16.3.6}$$

The general equation of second degree

$$Ax^2 + 2Bxy + Cy^2 + 2Dx + 2Ey + F = 0 \tag{16.3.7}$$

represents a set of geometric shapes called *conic sections* which are shown in Figure 16.8. They are classified as follows:

$$\begin{aligned}
&\text{Ellipse} : B^2 - AC < 0\,(\text{circle} : B = 0, A = C) \\
&\text{Parabola} : B^2 - AC = 0 \\
&\text{Hyperbola} : B^2 - AC > 0
\end{aligned} \tag{16.3.8}$$

If $A = B = C = 0$, the equation represents a line in the xy-plane, not a parabola. Let's consider each in detail.

Area of a triangle using lengths of three sides

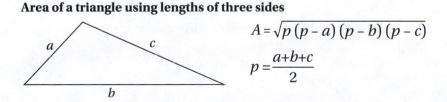

$$A = \sqrt{p\,(p - a)\,(p - b)\,(p - c)}$$

$$p = \frac{a+b+c}{2}$$

Figure 16.6 Area of a triangle.

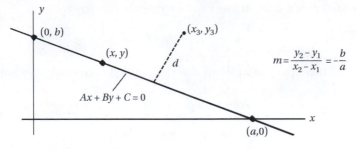

Figure 16.7 A straight line.

Circle: The circle is a special case of an ellipse with $A = C$. Its general form can be expressed as

$$(x - a)^2 + (y - b)^2 = r^2$$

(16.3.9)

where its center is at (a, b) and r is the radius.

Ellipse: The sum of the distances from the two foci, F, to any point on an ellipse is a constant. For an ellipse centered at the origin

$$\frac{x^2}{a^2} + \frac{y^2}{b^2} = 1$$

(16.3.10)

where a and b are the semi-major and semi-minor axes. The foci are at $(\pm c, 0)$ where $c^2 = a^2 - b^2$ (Fig. 16.7a).

Parabola: The locus of points on a parabola are equidistant from the focus and a line (the directrix). If the vertex is at the origin and the parabola opens to the right, it is written as

$$y^2 = 2px$$

(16.3.11)

where the focus is at $(p/2, 0)$ and the directrix is at $x = -p/2$ (Fig. 16.7b). For a parabola opening to the left, simply change the sign of p. For a parabola opening upward or downward, interchange x and y.

Hyperbola: The difference of the distances from the foci to any point on a hyperbola is a constant. For a hyperbola centered at the origin opening left and right, the equation can be written as

$$\frac{x^2}{a^2} - \frac{y^2}{b^2} = 1$$

(16.3.12)

The lines to which the hyperbola is asymptotic are asymptotes:

$$y = \pm\frac{b}{a}x$$

(16.3.13)

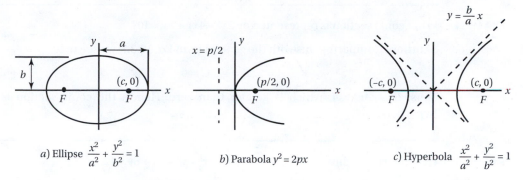

a) Ellipse $\dfrac{x^2}{a^2} + \dfrac{y^2}{b^2} = 1$ b) Parabola $y^2 = 2px$ c) Hyperbola $\dfrac{x^2}{a^2} + \dfrac{y^2}{b^2} = 1$

Figure 16.8 The three conic sections.

If the asymptotes are perpendicular, a rectangular hyperbola results. If the asymptotes are the x and y axes, the equation can be written as

$$xy = \pm k^2 \tag{16.3.14}$$

Finally, in our review of geometry, we will present three other coordinate systems often used in analysis. They are the polar (r, θ) coordinate system, the cylindrical (r, θ, z) coordinate system, and the spherical (r, θ, ϕ) coordinate system (Fig. 16.9). The polar coordinate system is restricted to a plane:

$$x = r \cos \theta, \quad y = r \sin \theta \tag{16.3.15}$$

For the cylindrical coordinate system

$$x = r \cos \theta, \quad y = r \sin \theta, \, z = z \tag{16.3.16}$$

And, for the spherical coordinate system

$$
\begin{aligned}
x &= r \sin \phi \cos \theta \\
y &= r \sin \phi \sin \theta \\
z &= r \cos \theta
\end{aligned}
\tag{16.3.17}
$$

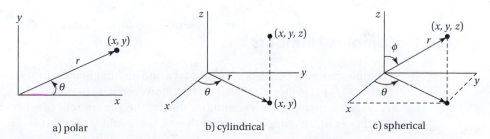

a) polar   b) cylindrical   c) spherical

Figure 16.9 The polar, cylindrical, and spherical coordinate systems.

EXAMPLE 16.7

What conic section is represented by $2x^2 - 4xy + 5x = 10$?

Solution: Comparing this with the general form Eq. 16.3.7, we see that

$$A = 2, \ B = -2, \ C = 0$$

Thus, $B^2 - AC = 4$, which is greater than zero. Hence, the conic section is a hyperbola.

EXAMPLE 16.8

Calculate the radius of the circle given by $x^2 + y^2 - 4x + 6y = 12$.

Solution: Write the equation in standard form (see Eq. 16.3.9):

$$x^2 + y^2 - 4x + 4 + 6y + 9 = 12 + 4 + 9$$

$$(x - 2)^2 + (y + 3)^2 = 25$$

The radius is $r = \sqrt{25} = 5$.

Note: The terms $4x$ and $6y$ demand that we write $(x - 2)^2$ and $(y + 3)^2$.

EXAMPLE 16.9

Write the general form of the equation of a parabola, vertex at (2, 4), opening upward, with directrix at $y = 2$.

Solution: The equation of the parabola (see Eq. 16.3.11) can be written as

$$(x - x_1)^2 = 2p(y - y_1)$$

where we have interchanged x and y so that the parabola opens upward. For this example, $x_1 = 2$, $y_1 = 4$, and $p = 4$ ($p/2$ is the distance from the vertex to the directrix). Hence, the equation is

$$(x - 2)^2 = 2(4)(y - 4)$$

or, in general form,

$$x^2 - 4x - 8y + 36 = 0$$

16.4 Complex Numbers

A complex number consists of a real part x and an imaginary part y, written as $x + iy$, where $i = \sqrt{-1}$. (In electrical engineering, however, it is common to let $j = \sqrt{-1}$ since i represents current.) In real number theory, the square root of a negative number does not exist; in complex number theory, we would write $\sqrt{-4} = \sqrt{4(-1)} = 2i$. The

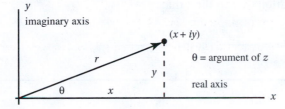

Figure 16.10 The complex number.

complex number may be plotted using the real x-axis and the imaginary y-axis, as shown in Figure 16.10.

It is often useful to express a complex number in polar form as

$$x + iy = re^{i\theta} \qquad (16.4.1)$$

where we use Euler's equation

$$e^{i\theta} = \cos\theta + i\sin\epsilon \qquad (16.4.2)$$

to verify the relations

$$x = r\cos\theta, \; y - r\sin\theta \qquad (16.4.3)$$

Note that $e^{i\theta} = e^{i(\theta + 2n\pi)}$ where n is an integer. This simply adds 360° (2π radians) to θ, and hence in Figure 16.10 $re^{i\theta}$ and $re^{i(\theta + 2n\pi)}$ represent the identical point.

Multiplication and division are accomplished with either form:

$$(a + ib)(c + id) = ac - ad + i(ad + bc) \qquad (16.4.4)$$

$$= r_1 e^{i\theta_1} r_2 e^{i\theta_2} = r_1 r_2 e^{i(\theta_1 + \theta_2)}$$

$$\frac{a + ib}{c + id} = \frac{a + ib}{c + id}\frac{c - id}{c - id} = \frac{(a + ib)(c - id)}{c^2 + d^2} \qquad (16.4.5)$$

$$= \frac{r_1}{r_2} e^{i(\theta_1 - \theta_2)}$$

The complex number introduced in Equation 16.4.5 "$c - id$" is a complex conjugate. It is formed by changing the sign of the imaginary part of the complex number.

It is usually easier to find powers and roots of complex numbers using the polar form:

$$(x + iy)^k = r^k e^{ik\theta}, (x + iy)^{1/k} = r^{1/k} e^{i\theta/k} \qquad (16.4.6)$$

When finding roots, more than one root results by using $e^{i\theta}$ and $e^{i(\theta + 2n\pi)}$. An example illustrates. Remember, in mathematical equations we usually express θ in radians.

Using Euler's equation we can show that

$$\sin\theta = \frac{e^{i\theta} - e^{i\theta}}{2i}, \quad \cos\theta = \frac{e^{i\theta} + e^{-i\theta}}{2} \tag{16.4.7}$$

EXAMPLE 16.10

Divide $(3 + 4i)$ by $(4 + 3i)$.

Solution: We perform the division as follows:

$$\frac{3 + 4i}{4 + 3i} = \frac{3 + 4i}{4 + 3i} \cdot \frac{4 - 3i}{4 - 3i} = \frac{12 + 16i - 9i + 12}{16 + 9} = \frac{24 + 7i}{25} = 0.96 + 0.28i$$

Note that we multiplied the numerator and the denominator by the complex conjugate of the denominator.

EXAMPLE 16.11

Find $(3 + 4i)^6$.

Solution: This can be done multiplying $(3 + 4i)$ six times, or using the polar form. Polar form is

$$r = \sqrt{x^2 + y^2}, \text{ and } \theta = \tan^{-1}\frac{y}{x}$$

$$r = \sqrt{3^2 + 4^2} = 5, \theta = \tan^{-1}(4/3) = 0.9273 \text{ rad}$$

We normally express θ in radians. The complex number, in polar form, is

$$3 + 4i = 5e^{0.9273i}$$

Thus,

$$(3 + 4i)^6 = 5^6(e^{0.9273i})^6 = 5^6 e^{5.564i}$$

Converting back to rectangular form we have

$$5^6 e^{5.564i} = 15625(\cos 5.564 + i\sin 5.564)$$

$$= 11755 - 10293i$$

EXAMPLE 16.12

Find the three roots of 1.

Solution: We express the complex number in polar form as

$$1 = 1e^{0i}$$

Since the trig functions are periodic, we know that

$$\sin\theta = \sin(\theta + 2\pi) = \sin(\theta + 4\pi)$$

$$\cos\theta = \cos(\theta + 2\pi) = \cos(\theta + 4\pi)$$

Thus, in addition to the first form, we have

$$1 = e^{2\pi i} = e^{4\pi i}$$

Taking the one-third root of each form, we find the three roots to be

$$1^{1/3} = 1e^{0i/3} = 1$$

$$1^{1/3} = 1e^{2\pi i/3}$$

$$= \cos 2\pi / 3 + i\sin 2\pi / 3 = -0.5 + 0.866i$$

$$1^{1/3} = 1e^{4\pi i/3}$$

$$= \cos 4\pi / 3 + i\sin 4\pi / 3 = -0.5 + 0.866i$$

If we added 6π to the angle we would be repeating the first root, so obviously this is not done.

16.5 **Linear Algebra**

The primary objective in linear algebra is to find the solution to a set of n linear algebraic equations for n unknowns. To do this we must learn how to manipulate a matrix, a rectangular array of quantities arranged into rows and columns.

An $m \times n$ matrix has m rows (the horizontal lines) and n columns (the vertical lines). An $m \times n$ matrix multiplied by an $n \times s$ matrix produces an $m \times s$ matrix. When multiplying two matrices the columns of the first matrix must equal the rows of the second. Their product is a third matrix:

$$[c_{ij}] = \sum_{k=1}^{n}[a_{ik}][b_{kj}] \tag{16.5.1}$$

We are primarily interested in square matrices since we usually have the same number of equations as unknowns, such as

$$a_{11}x_1 + a_{12}x_2 + a_{13}x_3 + a_{14}x_4 = r_1 \tag{16.5.2}$$

$$a_{21}x_1 + a_{22}x_2 + a_{22}x_3 + a_{24}x_4 = r_2$$

$$a_{31}x_1 + a_{32}x_2 + a_{33}x_3 + a_{34}x_4 = r_3$$

$$a_{41}x_1 + a_{42}x_2 + a_{43}x_3 + a_{44}x_4 = r_4$$

In matrix form this can be written as

$$[a_{ij}][x_j] = [r_i] \text{ or } \text{Ax} = \text{r} \tag{16.5.3}$$

where $[x_j]$ and $[r_i]$ are column matrices. (A column matrix is often referred to as a **vector**.) The coefficient matrix $[a_{ij}]$ and the column matrix $[r_i]$ are assumed to be known quantities. The solution $[x_j]$ is expressed as

$$[x_j] = [a_{ij}]^{-1}[r_i] \text{ or } \boldsymbol{x} = \boldsymbol{A}^{-1}\boldsymbol{r} \tag{16.5.4}$$

where $[a_{ij}]^{-1}$ is the **inverse** matrix of $[a_{ij}]$. It is defined as

$$[a_{ij}]^{-1} = \frac{[a_{ij}]^+}{|a_{ij}|} \text{ or } A^{-1} = \frac{A^+}{|A|} \tag{16.5.5}$$

where $[a_{ij}]^+$ is the **adjoint** matrix and $|a_{ij}|$ is the **determinant** of $[a_{ij}]$. Let us review how the determinant and the adjoint are evaluated.

In general, the determinant may be found using the **cofactor** A_{ij} of the element a_{ij}. The cofactor is defined to be $(-1)^{i+j}$ times the **minor**, the determinant obtained by deleting the ith row and the jth column. The determinant is then

$$|aij| = \sum_{j=1}^{n} a_{ij} A_{ij} \tag{16.5.6}$$

where i is any value from 1 to n. Recall that the third-order determinant can be evaluated by writing the first two columns after the determinant and then summing the products of the elements of the diagonals, using negative signs with the diagonals sloping upward.

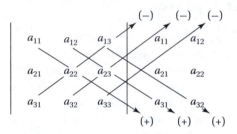

The elements of the adjoint $[a_{ij}]^+$ are the cofactors A_{ij} of the elements a_{ij}; for the matrix $[a_{ij}]$ of Eq. 16.5.2 we have

$$[a_{ij}]^+ = \begin{bmatrix} A_{11} & A_{21} & A_{31} & A_{41} \\ A_{12} & A_{22} & A_{32} & A_{42} \\ A_{13} & A_{23} & A_{33} & A_{43} \\ A_{14} & A_{24} & A_{34} & A_{44} \end{bmatrix} \tag{16.5.7}$$

Note that A_{ij} takes the position of a_{ji}.

Finally, the solution $[x_j]$ of Eq. 16.5.4 results if we multiply the square matrix $[a_{ij}]^{-1}$ by the column matrix $[r_i]$. In general, we multiply the elements in each left-hand matrix row by the elements in each right-hand matrix column, add the products, and place the sum at the location where the row and column intersect. The following examples will illustrate.

Before we work some examples, though, we should point out that the above matrix presentation can also be presented as **Cramer's rule**, which states that the solution element x_n can be expressed as

$$x_n = \frac{|b_{ij}|}{|a_{ij}|} \tag{16.5.8}$$

where $|b_{ij}|$ is formed by replacing the nth column of $|a_{ij}|$ with the elements of the column matrix $|r_j|$.

Note: If the system of equations is homogeneous (i.e., $r_i = 0$), a solution may exist if $|a_{ij}|$. If the determinant of a matrix is zero, that matrix is **singular** and its inverse does not exist.

In the solution of a system of first-order differential equations, we encounter the matrix equation

$$(\boldsymbol{A} - \lambda I)x = \boldsymbol{O} \qquad (16.5.9)$$

where "O" represents a matrix with all zero elements. The scalar λ is the **eigenvalue** and the vector **x** is the **eigenvector** associated with the eigenvalue. The matrix I is called the unit matrix, which for a 2×2 matrix is $\begin{bmatrix} 1 & 0 \\ 0 & 1 \end{bmatrix}$. If A is 2×2, λ has two distinct values, and if A is 3×3 it has three distinct values. Since $r_i = 0$ in Eq. 16.5.9, the scalar equation

$$|A - \lambda I| = 0 \qquad (16.5.10)$$

provides the equation that yields the eigenvalues. Then Eq. 16.5.9 is solved with each eigenvalue to give the eigenvectors. We often say that λ represents the eigenvalues of the matrix A.

EXAMPLE 16.13 Calculate the determinants of $\begin{bmatrix} 2 & -3 \\ 1 & 4 \end{bmatrix}$ and $\begin{bmatrix} 2 & 3 & 0 \\ 1 & 4 & -2 \\ 0 & 3 & 5 \end{bmatrix}$.

Solution: For the first matrix we have

$$\begin{vmatrix} 2 & -3 \\ 1 & 4 \end{vmatrix} = 2 \times 4 - 1(-3) = 11$$

The second matrix is set up as follows:

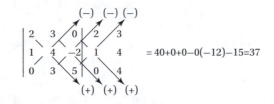

$$= 40 + 0 + 0 - 0(-12) - 15 = 37$$

EXAMPLE 16.14 Multiply the two matrices $\begin{bmatrix} 1 & 1 \\ -1 & -1 \end{bmatrix}$ and $\begin{bmatrix} 1 & 1 \\ -1 & -1 \end{bmatrix}$

Solution: Multiply the two matrices using Eq. 16.5.1. If we desire c_{12}, we use the first row of the first matrix and the second column of the second matrix so that

$$c_{12} = a_{11}b_{12} + a_{12}b_{22} = 1(1) + 1(-1) = 0$$

Doing this for all elements we find

$$\begin{bmatrix} 1 & 1 \\ -1 & -1 \end{bmatrix} \begin{bmatrix} 1 & 1 \\ -1 & -1 \end{bmatrix} = \begin{bmatrix} 0 & 0 \\ 0 & 0 \end{bmatrix}$$

Note: Even though a matrix has no zero elements, its square may be zero. Matrix multiplication is not like other forms of multiplication.

EXAMPLE 16.15

Find the adjoint of the matrix

$$\left[a_{ij} \right] = \begin{bmatrix} 1 & 0 & -2 \\ -1 & 2 & 0 \\ 1 & 2 & 1 \end{bmatrix}$$

Solution: The cofactor of each element of $[a_{ij}]$ must be determined. The cofactor is found by multiplying $(-1)^{i+j}$ times the determinant formed by deleting the ith row and the jth column. They are found to be

$$A_{11} = 2, \quad A_{12} = 1, \quad A_{13} = -4$$
$$A_{21} = -4, \quad A_{22} = 3, \quad A_{23} = -2$$
$$A_{31} = 4, \quad A_{32} = 2, \quad A_{33} = 2$$

The adjoint is then

$$\left[a_{ij} \right]^+ = \left[A_{ij} \right] = \begin{bmatrix} 2 & -4 & 4 \\ 1 & 3 & 2 \\ -4 & -2 & 2 \end{bmatrix}$$

Note: The matrix $[A_{ji}]$ is called the transpose of $[A_{ij}]$ (i.e., $[A_{ji}] = [A_{ij}]^T$).

EXAMPLE 16.16

Find the inverse of the matrix

$$a_{ij} = \begin{bmatrix} 1 & 0 & -2 \\ -1 & 2 & 0 \\ 1 & 2 & 1 \end{bmatrix}$$

Solution: The inverse is defined to be the adjoint matrix divided by the determinant $|a_{ij}|$. Hence, the inverse is (see Examples 16.13 and 16.15)

$$[a_{ij}]^{-1} = \frac{1}{10} \begin{bmatrix} 2 & -4 & 4 \\ 1 & 3 & 2 \\ -4 & -2 & 2 \end{bmatrix} = \begin{bmatrix} 0.2 & -0.4 & 0.4 \\ 0.1 & 0.3 & 0.2 \\ -0.4 & -0.2 & 0.2 \end{bmatrix}$$

EXAMPLE 16.17

Find the solution to

$$x_1 \qquad -2x_3 = 2$$
$$-x_1 + 2x_2 \qquad = 0$$
$$x_1 + 2x_2 \quad x_3 = -4$$

Solution: The solution matrix is (see Examples 16.13 and 16.15 or use $[a_{ij}]^{-1}$ from Example 16.16)

$$[x_j] = [a_{ij}]^{-1}[r_i] = \frac{\left[a_{ij}\right]^+}{|a_{ij}|}[r_i]$$

$$= \frac{1}{10}\begin{bmatrix} 2 & -4 & 4 \\ 1 & 3 & 2 \\ -4 & -2 & 2 \end{bmatrix}\begin{bmatrix} 2 \\ 0 \\ -4 \end{bmatrix}$$

First, let's multiply the two matrices; they are multiplied row by column as follows:

$$2\cdot2+(-4)\cdot0+4\cdot(-4)=-12$$
$$1\cdot2+3\cdot0+2\cdot(-4)=-6$$
$$-4\cdot2-2\cdot0+2\cdot(-4)=-16$$

The solution vector is then

$$[x_i] = \frac{1}{10}\begin{bmatrix} -12 \\ -6 \\ -16 \end{bmatrix} = \begin{bmatrix} -1.2 \\ -0.6 \\ -1.6 \end{bmatrix}$$

In component form, the solution is

$$x_1 = -1.2, \; x_2 = -0.6, \; x_3 = -1.6$$

EXAMPLE 16.18

Use Cramer's rule and solve

$$\begin{array}{rrrr} x_1 & & -2x_3 & = & 2 \\ -x_1 & +2x_2 & & = & 0 \\ x_1 & +2x_2 & +x_3 & = & -4 \end{array}$$

Solution: The solution is found (see Example 16.13) by evaluating the ratios as follows:

$$x_1 = \frac{\begin{vmatrix} 2 & 0 & -2 \\ 0 & 2 & 0 \\ -4 & 2 & 1 \end{vmatrix}}{D} = \frac{-12}{10} = -1.2 \qquad x_2 = \frac{\begin{vmatrix} 1 & 2 & -2 \\ -1 & 0 & 0 \\ 1 & -4 & 1 \end{vmatrix}}{D} = \frac{-6}{10} = -0.6$$

$$x_3 = \frac{\begin{vmatrix} 1 & 0 & 2 \\ -1 & 2 & 0 \\ 1 & 2 & -4 \end{vmatrix}}{D} = \frac{-16}{10} = -1.6$$

where

$$D = \begin{vmatrix} 1 & 0 & -2 \\ -1 & 2 & 0 \\ 1 & 2 & 1 \end{vmatrix} = 10$$

Note that the numerator is the determinant formed by replacing the ith column with right-hand side elements r_i when solving for x_i.

**EXAMPLE
16.19**   Find the eigenvalues of $A = \begin{bmatrix} 4 & 0 & 2 \\ 0 & 8 & 0 \\ 3 & 0 & 5 \end{bmatrix}$

Solution: To find the eigenvalues of the matrix, we form the equation

$$|A - \lambda I| = 0 \qquad \text{or} \qquad \begin{vmatrix} 4 - \lambda & 0 & 2 \\ 0 & 8 - \lambda & 0 \\ 3 & 0 & 5 - \lambda \end{vmatrix} = 0$$

Expanding the determinant using cofactors we have

$$(8 - \lambda) \begin{vmatrix} 4 - \lambda & 2 \\ 3 & 5 - \lambda \end{vmatrix} = (8 - \lambda)[(4 - \lambda)(5 - \lambda) - 6] = (8 - \lambda)[\lambda^2 - 9\lambda + 14] = 0$$

or

$$(8 - \lambda)(\lambda - 8)(\lambda - 2) = 0$$

The eigenvalues are then $\lambda = 8, 7, 2$.

16.6 Calculus

Differentiation

The slope of a curve $y = f(x)$ is the ratio of the change in y to the change in x as the change in x becomes infinitesimally small. This is the first derivative, written as

$$\frac{dy}{dx} = \lim_{\Delta x \to 0} \frac{\Delta y}{\Delta x} \tag{16.6.1}$$

This may be written using abbreviated notation as

$$\frac{dy}{dx} = Dy = y' = \dot{y} \tag{16.6.2}$$

The second derivative is written as

$$\frac{d^2 y}{dx^2} = D^2 y = y'' = \ddot{y} \tag{16.6.3}$$

and is defined by

$$\frac{d^2y}{dx^2} = \lim_{\Delta x \to 0} \frac{\Delta y'}{\Delta x}$$ (16.6.4)

Some derivative formulas, where f and g are functions of x, and k is constant, follow.

$$\frac{dk}{dx} = 0$$ (16.6.5)

$$\frac{d(kx^n)}{dx} = knx^{n-1}$$

$$\frac{d}{dx}(f+g) = f' + g'$$

$$\frac{df^n}{dx} = nf^{n-1}f'$$

$$\frac{d}{dx}(fg) = fg' + gf'$$

$$\frac{d}{dx}(\ln x) = \frac{1}{x}$$

$$\frac{d}{dx}(e^{kx}) = ke^{kx}$$

$$\frac{d}{dx}(\sin x) = \cos x$$

$$\frac{d}{dx}(\cos x) = -\sin x$$

Maxima and Minima

Derivatives are used to locate points of inflection, maxima, and minima. Note the following:

$f'(x) = 0$ at a maximum or a minimum.
$f''(x) = 0$ at an inflection point.
$f''(x) > 0$ at a minimum.
$f''(x) < 0$ at a maximum.

An inflection point always exists between a maximum and a minimum.

L'Hopital's Rule

Differentiation is also useful in establishing the limit of $f(x)/g(x)$ as $x \to a$ if $f(a)$ and $g(a)$ are both zero or $\pm\infty$. **L'Hopital's rule** (frequently written "L'Hospital's rule") is used in such cases and is as follows:

$$\lim_{x \to a} \frac{f(x)}{g(x)} = \lim_{x \to a} \frac{f'(x)}{g'(x)} = \lim_{x \to a} \frac{f''(x)}{g''(x)}$$ (16.6.6)

Integration

The inverse of differentiation is the process of **integration**. If a curve is given by $y = f(x)$, then the area under the curve from $x = a$ to $x = b$ is given by

$$A = \int_a^b y \, dx \tag{16.6.7}$$

The length of the curve between the two points is expressed as

$$L = \int_a^b (1 + y'^2)^{1/2} \, dx \tag{16.6.8}$$

Volumes of various objects are also found by an appropriate integration.

If the integral has limits, it is a **definite integral**; if it does not have limits, it is an **indefinite integral** and a constant is always added. Some common indefinite integrals follow:

$$
\begin{aligned}
&\int dx = x + C \\
&\int cy \, dx = c \int y \, dx \\
&\int x^n \, dx = \frac{x^{n+1}}{n+1} + C \quad n \neq -1 \\
&\int x^{-1} \, dx = \ln x + C \\
&\int e^{ax} \, dx = \frac{1}{a} e^{ax} + C \\
&\int \sin x \, dx = -\cos x + C \\
&\int \cos x \, dx = \sin x + C \\
&\int \cos^2 x \, dx = \frac{x}{2} + \frac{1}{4} \sin 2x + C \\
&\int u \, dv = uv - \int v \, du
\end{aligned} \tag{16.6.9}
$$

This last integral is often referred to as "integration by parts." If the integrand (the coefficients of the differential) is not one of the above, then in the last integral, $\int v \, du$ may in fact be integrable. An example will illustrate.

EXAMPLE 16.20

Find the slope of $y = x^2 + \sin x$ at $x = 0.5$.

Solution: The derivative is the slope:

$$y'(x) = 2x + \cos x$$

At $x = 0.5$ the slope is

$$y'(0.5) = 2 \cdot 0.5 + \cos 0.5 = 1.878$$

Note: Use 0.5 radians when evaluating cos 0.5.

**EXAMPLE
16.21**

Find $\dfrac{d}{dx}(\tan x)$

Solution: Writing $\tan x = \sin x \,/\, \cos x = f(x){\cdot}g(x)$ we find

$$\frac{d}{dx}(\tan x) = \frac{1}{\cos x}\frac{d}{dx}(\sin x) + \sin x \frac{d}{dx}(\cos x)^{-1}$$

$$= \frac{\cos x}{\cos x} + \frac{\sin^2 x}{\cos^2 x} = 1 + \tan^2 x$$

$$= \frac{\cos^2 x + \sin^2 x}{\cos^2 x} = \frac{1}{\cos^2 x} = \sec^2 x$$

Either expression is acceptable.

**EXAMPLE
16.22**

Locate the maximum and minimum points of the function $y(x) = x^3 - 12x - 9$ and evaluate y at those points.

Solution: The derivative is

$$y'(x) = 3x^2 - 12$$

$$y'(x) = 3(x^2 - 4)$$

$$y'(x) = 3(x + 2)(x - 2)$$

The points at which $y'(x) = 0$ are at

$$x = 2, -2$$

At these two points the extrema are

$$y_{\min} = (2)^3 - 12.2 - 9 = -25$$

$$y_{\max} = (-2)^3 - 12(-2) - 9 = 27$$

Let us check the second derivative. At the two points we have

$$y''(2) = 6.2 = 12$$

$$y''(-2) = 6 \cdot (-2) = -12$$

The point $x = 2$ is a minimum since its second derivative is positive there.

**EXAMPLE
16.23**

Find the limit as of $x \to 0$ of $\sin x \,/\, x$.

Solution: If we let $x = 0$ we are faced with the ratio of 0/0, an indeterminate quantity. Hence, we use L'Hopital's rule and differentiate both numerator and denominator to obtain

$$\lim_{x \to 0}\frac{\sin x}{x} = \lim_{x \to 0}\frac{\cos x}{1}$$

Now, we let $x = 0$ and find

$$\lim_{s \to 0} \frac{\sin x}{x} = \lim_{s \to 0} \frac{\cos x}{1} = 1$$

EXAMPLE 16.24 Find the area of the shaded area in Figure 16.11.

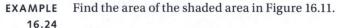
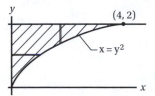

Figure16.11

Solution: We can find this area by using either a horizontal strip or a vertical strip. We will use both. First, for a horizontal strip:

$$A = \int_0^2 x \, dy$$

$$= \int_0^2 y^2 \, dy = \frac{y^3}{3} \Big|_0^2 = \frac{8}{3}$$

Using a vertical strip we have

$$A = \int_0^4 (2 - y) \, dx$$

$$= \int_0^4 (2 - x^{1/2}) \, dx = \left[2x - \frac{2}{3} x^{3/2} \right]_0^4 = 8 - \frac{2}{3} 8 = \frac{8}{3}$$

Either technique is acceptable. The first is the simpler one.

EXAMPLE 16.25 Find the volume enclosed by rotating the shaded area of Example 16.24 about the y-axis.

Solution: If we rotate the horizontal strip about the y-axis we will obtain a disc with volume

$$dv = \pi x^2 dy$$

This can be integrated to give the volume, which is

$$V = \int_0^2 \pi x^2 \, dy$$

$$= \pi \int_0^2 y^4 \, dy = \frac{\pi y^5}{d} \Big|_0^2 = \frac{32\pi}{5}$$

Now, let us rotate the vertical strip about the y-axis to form a cylinder with volume

$$dV = 2\pi x (2 - y) dx$$

This can be integrated to yield

$$V = \int_0^4 2\pi x(2-y)\,dx$$

$$= \int_0^4 2\pi x(2-x^{1/2})\,dx = 2\pi\left[x^2 - \frac{2x^{5/2}}{5}\right]_0^4 = \frac{32\pi}{5}$$

Again, using the horizontal strip results in a simpler solution.

EXAMPLE 16.26

Show that $\int xe^x\,dx = (x-1)e^x + C$

Solution: Let's attempt the last integral of (16.6.9). Define the following:

$$u = x, dv = e^x dx$$

Then,

$$du = dx, v = \int e^x\,dx = e^x$$

and we find that

$$\int xe^x\,dx = xe^x - \int e^x\,dx$$

$$= xe^x - e^x + C = (x-1)e^x + C$$

16.7 Probability and Statistics

Events are independent if the probability of occurrence of one event does not influence the probability of occurrence of other events. The number of permutations (a particular sequence) of n things taken r at a time is

$$P(n,r) = \frac{n!}{(n-r)!} \tag{16.7.1}$$

If the starting point is unknown, as in a ring, the *ring permutation* is

$$P(n,r) = \frac{(n-1)!}{(n-r)!} \tag{16.7.2}$$

The number of *combinations* (no order-conscious arrangement) of n things taken r at a time is given by

$$C(n,r) = \frac{n!}{r!(n-r)!} \tag{16.7.3}$$

Note that $0! = 1$.

For independent events of two sample groups A and B the following rules are necessary:

1. The probability of A or B occurring equals the sum of the probability of occurrence of A and the probability of occurrence of B; that is,

$$P(A \text{ or } B) = P(A) + P(B) \tag{16.7.4}$$

2. The probability of both A and B occurring is given by the product

$$P(A \text{ and } B) = P(A)P(B) \tag{16.7.5}$$

3. The probability of A not occurring is given as

$$P(\text{not } A) = 1 - P(A) \tag{16.7.6}$$

4. The probability of either A or B occurring is given by

$$P(A \text{ or } B) = P(A) + P(B) - P(A)P(B) \tag{16.7.7}$$

The probability of an event occurring is in the range of 0 to 1. An impossible event has a probability of 0 and an event that is certain to occur has a probability of 1.

The data gathered during an experiment can be analyzed using quantities defined by the following:

1. The arithmetic mean $\bar{x}$ is the average of the observations; that is,

$$\bar{x} = \frac{x_1 + x_2 + x_3 + \cdots + x_n}{n}$$

2. The median is the middle observation when all the data are ordered by magnitude; half the values are below the median. The median for an even number of data is the average of the two middle values.

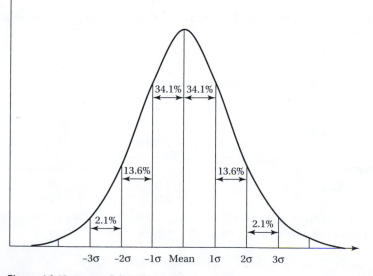

Figure 16.12 Normal distribution

3. The mode is the observed value that occurs most frequently.
4. The standard deviation σ of the sample is a measure of variability. It is defined as

$$\sigma = \left[\frac{(x_1 - \bar{x})^2 + (x_2 - \bar{x})^2 + \ldots + (x_n - \bar{x})^2}{n-1} \right]^{1/2} \tag{16.7.8}$$

$$= \left[\frac{x_1^2 + x_2^2 + \cdots + x_n^2 - n\bar{x}^2}{n-1} \right]^{1/2} \tag{16.7.9}$$

For large observations (over 50), it is customary to simply use *n*, rather than (*n* – 1), in the denominators of the above. In fact, if *n* is used in the above equations, σ is often referred to as the *standard deviation of the population*.

5. The *variance* is defined to be σ². Figure 16.12 shows what is called a "normal" distribution, also known as a bell curve. The distribution is considered "normal" when 68 percent of collected data fall within 1 standard deviation from the mean, $\bar{x}$, 95 percent fall within 2 standard deviations, and 99 percent are within 3 standard deviations from the mean.

EXAMPLE 16.27

How many different ways can seven people be arranged in a lineup? In a circle?

Solution: This is the number of permutations of seven things taken seven at a time. The answer is

$$P(7,7) = \frac{n!}{(n-r)!}$$

$$= \frac{7!}{(7-7)!} = 5040$$

In a circle we use the ring permutation:

$$P(7,7) = \frac{(n-1)!}{(n-r)!}$$

$$= \frac{(7-1)!}{(7-7)!} = 720$$

EXAMPLE 16.28

How many different collections of eight people can fit into a six-passenger vehicle? (Only six will fit at a time.)

Solution: The answer does not depend on the seating arrangement. If it did, it would be a permutation. Hence, we use the combination relationship and find

$$C(8,6) = \frac{n!}{(n-r)!r!}$$

$$= \frac{8!}{(8-6)!6!} = 28$$

**EXAMPLE
16.29**

A carnival booth offers $10 if you pick a red ball and then a white ball (the first ball is re-inserted) from a bin containing 60 red balls, 15 white balls, and 25 blue balls. If $1 is charged for an attempt, will the operator make money?

Solution: The probability of drawing a red ball on the first try is 0.6. If it is then re-inserted, the probability of drawing a white ball is 0.15. The probability of accomplishing both is then given by

$$P(\text{red and white}) = P(\text{red})P(\text{white})$$

$$= 0.6 \times 0.15 = 0.09$$

or 9 chances out of 100 attempts. Hence, the entrepreneur will pay out $90 for every $100 taken in and will thus make money.

**EXAMPLE
16.30**

If the operator of the bin of balls in Example 16.29 offers a $1.00 prize to contestants who pick either a red ball or a white ball from the bin on the first attempt, and charges $0.75 per attempt, will the operator make money?

Solution: The probability of selecting either a red ball or a white ball on the first attempt is

$$P(\text{red and white}) = P(\text{red})P(\text{white})$$

$$= 0.6 \times 0.15 = 0.75$$

Consequently, 75 out of 100 gamblers will win and the operator must pay out $75 for every $75 taken in. The operator would not make any money.

**EXAMPLE
16.31**

The operator of Example 16.30 has two identical bins and offers a $1.00 prize for withdrawing a red ball from the first bin or a white ball from the second bin. Will the operator make money if he charges $0.75 per attempt?

Solution: The probability of selecting a red ball from the first bin (sample group Ai) or a white ball from the second bin (sample group Bi) is

$$P(\text{red and white}) = P(\text{red}) + P(\text{white}) - P(\text{red})P(\text{white})$$

$$= 0.6 + 0.15 - 0.6 \times 0.15 = 0.66$$

For this situation the owner must pay $66 to every 100 gamblers who pay $75 to participate. His profit is $9.

**EXAMPLE
16.32**

The temperature at a given location in the South at 2 p.m. each August 10 for 25 consecutive years was measured, in degrees Celsius, to be 33, 38, 34, 26, 32, 31, 28,

39, 29, 36, 32, 29, 31, 24, 35, 34, 32, 30, 31, 32, 26, 40, 27, 33, 39. Calculate the arithmetic mean, the median, the mode, and the sample standard deviation.

Solution: Using the appropriate equations, we calculate the arithmetic mean:

$$\bar{T} = \frac{T_1 + T_2 + \cdots + T_{25}}{25}$$

$$= \frac{33 + 38 + 34 + \cdots + 39}{25} = \frac{801}{25} = 32.04\,°C$$

The median is found by first arranging the values in order. We have 24, 26, 26, 27, 28, 29, 29, 30, 31, 31, 31, 32, 32, 32, 32, 33, 33, 34, 34, 35, 36, 38, 39, 39, 40. Counting 12 values in from either end, the median is found to be 32°C.
The mode is the observation that occurs most often; it is 32°C.
The sample standard deviation is found to be

$$\sigma = \left[\frac{T_2^1 + T_2^2 + \cdots + T_{25}^2 - n\bar{T}^2}{n-1} \right]^{1/2}$$

$$= \left[\frac{33^2 + 38^2 + \cdots + 39^2 - 25 \times 32.04^2}{25-1} \right]^{1/2} = \sqrt{\frac{26,099 - 25,664}{24}} = 4.26$$

EXERCISES AND ACTIVITIES

Algebra

16.1 A growth curve is given by $A = 10e^{2t}$. At what value of t is $A = 100$?

a) 5.261　　　　b) 3.070　　　　c) 1.151　　　　d) 0.726

16.2 If $\ln(x) = 3.2$, what is x?

a) 18.65　　　　b) 24.53　　　　c) 31.83　　　　d) 64.58

16.3 If $\log_s(x) = -1.8$, find x.

a) 0.00483　　　　b) 0.0169　　　　c) 0.0552　　　　d) 0.0783

16.4 One root of the equation $3x^2 - 2x - 2 = 0$ is

a) 1.215　　　　b) 1.064　　　　c) 0.937　　　　d) 0.826

16.5 $\sqrt{4+x}$ can be written as the series

a) $2 - \dfrac{x}{4} + \dfrac{x^2}{64} + \cdots$ 　　　c) $2 - \dfrac{x^2}{4} - \dfrac{x^4}{64} + \cdots$

b) $2 + \dfrac{x}{8} - \dfrac{x^2}{128} + \cdots$ 　　　d) $2 + \dfrac{x}{4} - \dfrac{x^2}{64} + \cdots$

16.6 Resolve $\dfrac{2}{x(x^2-3x+2)}$ into partial fractions.

a) $\dfrac{1}{x}+\dfrac{1}{x-2}-\dfrac{2}{x-1}$

c) $\dfrac{1}{x}-\dfrac{2}{x-2}+\dfrac{1}{x-1}$

b) $\dfrac{1}{x}-\dfrac{2}{x-2}+\dfrac{1}{x-1}$

d) $\dfrac{1}{x}+\dfrac{1}{x-2}+\dfrac{1}{x-1}$

16.7 Express $\dfrac{4}{x^2(x^2-4x+4)}$ as the sum of fractions.

a) $\dfrac{1}{x}-\dfrac{1}{x-2}+\dfrac{1}{(x-2)^2}$

c) $\dfrac{1}{x^2}+\dfrac{1}{(x-2)^2}$

b) $\dfrac{1}{x}+\dfrac{1}{x^2}-\dfrac{1}{x-2}+\dfrac{1}{(x-2)^2}$

d) $\dfrac{1}{x}+\dfrac{1}{x^2}+\dfrac{1}{x-2}+\dfrac{1}{(x-2)^2}$

16.8 A germ population has a growth curve of $Ae^{0.4t}$. At what value of t does its original value double?

a) 9.682   b) 7.733   c) 4.672   d) 1.733

Trigonometry

16.9 If $\sin\theta = 0.7$, what is $\tan\theta$?

a) 0.98   b) 0.94   c) 0.88   d) 0.85

16.10 If the short leg of a right triangle is 5 units long and the long leg is 7 units long, what is the angle opposite the short leg, in degrees?

a) 26.3   b) 28.9   c) 31.2   d) 35.5

16.11 The expression $\tan\theta\,\sec\theta\,(1-\sin^2\theta)\,/\,\cos\theta$ simplifies to

a) $\sin\theta$ b) $\cos\theta$ c) $\tan\theta$ d) $\sec\theta$

16.12 A triangle has sides of length 2, 3, and 4. What angle, in radians, is opposite the side of length 3?

a) 0.55   b) 0.61   c) 0.76   d) 0.81

16.13 The length of a lake is to be determined. A distance of 850 m is measured from one end to a point x on the shore. A distance of 732 m is measured from x to the other end. If an angle of 154° is measured between the two lines connecting x, what is the length of the lake?

a) 1542   b) 1421   c) 1368   d) 1261

16.14   Express $2 \sin^2 \theta$ as a function of $\cos 2\theta$.

a) $\cos 2\theta - 1$ b) $\cos 2\theta + 1$ c) $\cos 2\theta + 2$ d) $1 - \cos 2\theta$

Geometry

16.15   The included angle between two successive sides of a regular eight-sided polygon is

a) $150°$ b) $135°$ c) $120°$ d) $75°$

16.16   A large 15-m-dia cylindrical tank that sits on the ground is to be painted. If one liter of paint covers 10 m², how many liters are required if it is 10 m high? (Include the top.)

a) 65   b) 53   c) 47   d) 38

16.17   The equation of a line that has a slope of –2 and intercepts the x-axis at $x = 2$ is

a) $y + 2x = 4$ b) $y - 2x = 4$ c) $y + 2x = -4$ d) $2y + x = 2$

16.18   The equation of a line that intercepts the x-axis at $x = 4$ and the y-axis at $y = -6$ is

a) $2x - 3y = 12$ b) $3x - 2y = 12$ c) $2x + 3y = 12$ d) $3x + 2y = 12$

16.19   The shortest distance from the line $3x - 4y = 3$ to the point $(6, 8)$ is

a) 4.8   b) 4.2   c) 3.8   d) 3.4

16.20   The equation $x^2 + 4xy + 4y^2 + 2x = 10$ represents which conic section?

a) circle   b) ellipse   c) parabola   d) hyperbola

16.21   The x- and y-axes are the asymptotes of a hyperbola that passes through the point $(2, 2)$. Its equation is

a) $x^2 - y^2 = 0$ b) $xy = 4$ c) $y^2 - x^2 = 0$ d) $x^2 + y^2 = 4$

16.22   A 100-m-long track is to be built 50 m wide. If it is to be elliptical, what equation could describe it if the 100-m length is along the x-axis?

a) $50x^2 + 100y^2 = 1000$ c) $4x^2 + y^2 = 2500$

b) $2x^2 + y^2 = 250$ d) $x^2 + 2y^2 = 250$

16.23   The cylindrical coordinates $(5, 30°, 12)$ are expressed in spherical coordinates as

a) $(13, 30°, 67.4°)$ c) $(15, 52.6°, 22.6°)$

b) $(13, 30°, 22.6°)$ d) $(15, 52.6°, -22.6°)$

16.24   The equation of a 4-m-radius sphere using cylindrical coordinates is

a) $x^2 + y^2 + z^2 = 16$ c) $r^2 + z^2 = 16$

b) $r^2 = 16$ d) $x^2 + y^2 = 16$

Complex Numbers

16.25   Divide $3 - i$ by $1 + i$.

a) $1 - 2i$ b) $1 + 2i$ c) $2 - i$ d) $2 + i$

16.26   Find $(1 + i)^6$.

a) $1 + i$ b) $1 - i$ c) $8i$ d) $-8i$

16.27   Find the root of $(1 + i)^{1/5}$ with the smallest argument.

a) $0.168 + 1.06i$ c) $1.06 - 0.168i$

b) $1.06 + 0.168i$ d) $0.168 - 1.06i$

16.28   Express $(3 + 2i)\, e^{2it} + (3 - 2i)\, e^{-2it}$ in terms of trigonometric functions.

a) $3 \cos 2t - 4 \sin 2t$ c) $6 \cos 2t - 4 \sin 2t$

b) $3 \cos 2t - 2 \sin 2t$ d) $3 \sin 2t + 2 \sin 2t$

16.29   Subtract $5e^{0.2i}$ from $6e^{2.3i}$.

a) $-0.903 + 3.481i$ c) $-8.898 - 5.468i$

b) $-8.898 + 3.481i$ d) $-0.903 - 5.468i$

Linear Algebra

16.30   Find the value of the determinant $\begin{vmatrix} 3 & 2 & 1 \\ 0 & -1 & -1 \\ 2 & 0 & 2 \end{vmatrix}$

a) 8   b) 4   c) -8 d) -4

16.31   Evaluate the determinant $\begin{vmatrix} 1 & 0 & 1 & 1 \\ 2 & -1 & 0 & 1 \\ 0 & 0 & 2 & 0 \\ 3 & 2 & 1 & 1 \end{vmatrix}$.

a) 8   b) 4   c) 0   d) -4

16.32   The cofactor A_{21} of the determinant of Prob. 16.30 is

a) -5 b) -4 c) 3   d) 4

16.33   The cofactor A_{34} of the determinant of Prob. 16.31 is

a) 4   b) 6   c) -6 d) -4

16.34 Find the adjoint matrix of $\begin{vmatrix} 1 & -4 \\ 0 & 2 \end{vmatrix}$.

a) $\begin{bmatrix} 4 & 2 \\ 0 & 1 \end{bmatrix}$

c) $\begin{bmatrix} 2 & 4 \\ 1 & 0 \end{bmatrix}$

b) $\begin{bmatrix} 1 & 0 \\ 4 & 2 \end{bmatrix}$

d) $\begin{bmatrix} 2 & 4 \\ 0 & 1 \end{bmatrix}$

16.35 The inverse matrix of $\begin{bmatrix} 2 & 3 \\ 1 & 1 \end{bmatrix}$ is

a) $\begin{bmatrix} -1 & 3 \\ 1 & -2 \end{bmatrix}$

c) $\begin{bmatrix} -1 & 1 \\ -3 & 2 \end{bmatrix}$

b) $\begin{bmatrix} 1 & -1 \\ -3 & 2 \end{bmatrix}$

d) $\begin{bmatrix} -2 & 3 \\ 1 & -1 \end{bmatrix}$

16.36 Calculate $\begin{bmatrix} 2 & -1 \\ 3 & 2 \end{bmatrix} \begin{bmatrix} 2 \\ 1 \end{bmatrix}$

a) $\begin{bmatrix} 8 \\ 3 \end{bmatrix}$

b) $\begin{bmatrix} 3 \\ 8 \end{bmatrix}$

c) $\begin{bmatrix} -3 \\ -8 \end{bmatrix}$

d) $[3,8]$

16.37 Determine $\begin{bmatrix} 1 & 2 \\ 2 & 1 \end{bmatrix} \begin{bmatrix} -1 & 0 \\ 1 & 2 \end{bmatrix}$.

a) $\begin{bmatrix} 1 & 4 \\ -1 & 2 \end{bmatrix}$

b) $\begin{bmatrix} 1 & -1 \\ 4 & 2 \end{bmatrix}$

c) $\begin{bmatrix} 1 \\ -1 \end{bmatrix}$

d) $\begin{bmatrix} 4 \\ 2 \end{bmatrix}$

16.38 Solve for $[x_i]$.

$$3x_1 + 2x_2 = -2$$

$$x_1 - x_2 + x_3 = 0$$

$$4x_1 + 2x_3 = 4$$

a) $\begin{bmatrix} 2 \\ 4 \\ -6 \end{bmatrix}$

b) $\begin{bmatrix} -2 \\ 4 \\ 12 \end{bmatrix}$

c) $\begin{bmatrix} 2 \\ 4 \\ 8 \end{bmatrix}$

d) $\begin{bmatrix} -6 \\ 8 \\ 14 \end{bmatrix}$

16.39 Find the eigenvalues of

a) 4, –1

b) 4, 1

c) 1, –4

d) 3, 2

Calculus

16.40   The slope of the curve $y = 2x^3 - 3x$ at $x = 1$ is

a) 3   b) 5   c) 6   d) 8

16.41   If $y = \ln x + e^x \sin x$, find dy/dx at $x = 1$.

a) 1.23   b) 3.68   c) 4.76   d) 6.12

16.42   At what value of x does a maximum of $y = x^3 - 3x$ occur?

a) 2   b) 1   c) 0   d) –1

16.43   Where does an inflection point occur for $y = x^3 - 3x$?

a) 2   b) 1   c) 0   d) –1

16.44   Evaluate $\lim\limits_{x \to \infty} \dfrac{2x^2 - x}{x^2 + x}$

a) 2   b) 1   c) 0   d) –1

16.45   Find the area between the y-axis and $y = x^2$ from $y = 4$ to $y = 9$.

a) $\dfrac{29}{3}$ b) $\dfrac{32}{3}$ c) $\dfrac{34}{3}$ d) $\dfrac{38}{3}$

16.46   The area contained between $4x = y^2$ and $4y = x^2$ is

a) $\dfrac{10}{3}$ b) $\dfrac{11}{3}$ c) $\dfrac{13}{3}$ d) $\dfrac{16}{3}$

16.47   Rotate the shaded area of Example 16.24 about the x-axis. What volume is formed?

a) 4π b) 6π c) 8π d) 10π

16.48   Evaluate $\int_0^2 (e^x + \sin x)\, dx$

a) 7.81   b) 6.21   c) 5.92   d) 5.61

16.49   Evaluate $2\int_0^1 e^x \sin x\, dx$

a) 1.82   b) 1.94   c) 2.05   d) 2.16

16.50   Derive an expression $\int x \cos x\, dx$

a) $x \cos x - \sin x + C$ c) $x \sin x - \cos x + C$

b) $x \sin x + \cos x + C$ d) $x \cos x + \sin x + C$

Probability and Statistics

16.51   You reach into a jelly bean bag and grab one bean. If the bag contains 30 red, 25 orange, 15 pink, 10 green, and 5 black beans, the probability that you will get a black bean or a red bean is nearest to

   a) 0.6   b) 0.5   c) 0.4   d) 0.3

16.52   Two jelly bean bags are identical to that of Prob. 16.51. One bean is to be selected from each bag. The probability of selecting a black bean from the first bag and a red bean from the second bag is nearest to

   a) $\dfrac{1}{100}$ b) $\dfrac{2}{100}$ c) $\dfrac{3}{100}$ d) $\dfrac{4}{100}$

16.53   From the original bag of Prob. 16.52, the probability of selecting five beans, the first three of which are red and the next two of which are orange, is nearest to

   a) $\dfrac{3}{1000}$ b) $\dfrac{4}{1000}$ c) $\dfrac{5}{1000}$ d) $\dfrac{6}{1000}$

16.54   Two bags each contain two black balls, one white ball, and one red ball. One ball is to be selected from each bag. What is the probability of selecting the white ball from the first bag or the red ball from the second bag?

   a) $\dfrac{1}{2}$ b) $\dfrac{7}{16}$ c) $\dfrac{1}{4}$ d) $\dfrac{3}{8}$

16.55   A professor gives the following scores to her students. What is the mode?

Frequency:	1	3	6	11	13	10	2
Score:	35	45	55	65	75	85	95

   a) 65   b) 75   c) 85   d) 11

16.56   For the data of Prob. 16.55, what is the arithmetic mean?

   a) 68.5   b) 68.9   c) 69.3   d) 70.2

16.57   Calculate the sample standard deviation for the data of Prob. 16.55.

   a) 9.27   b) 10.11   c) 11.56   d) 13.78

CHAPTER 17

Engineering Fundamentals

Most undergraduate engineering curricula begin with a first year of calculus-based math, foundational physical sciences, and some natural sciences depending on the major course of study. A math review was presented in the previous chapter, and this chapter reviews several of the topics students will continue to encounter as their studies progress.

Keeping in mind the ABET definition of engineering, which mentions the goal of economically using the "materials and forces of nature," this chapter will contain what will be a brief review for some, and an introduction for others, centered around this goal. Broadly defined, physics is the scientific study of matter, energy, space, time, and their interactions. A study of the forces of nature and their interactions with matter begins in physics and continues through further engineering courses in statics, dynamics, circuit analysis, power, electricity, thermodynamics, material sciences, solid mechanics, and fluid mechanics. Students pursuing chemical engineering and bioengineering fields will also build part of their foundation with appropriate chemistry and biology classes, which will further acquaint them with thermodynamics and provide a foundation for further study in mass and energy balances, chemical reaction kinetics, reactor design, and other biological systems analyses. These courses are typically taken during a student's sophomore and junior years, followed by a year of technical electives, many of which have most of these studies as prerequisites, and a capstone design experience, which draws upon all of the student's previous understanding of his or her studies to achieve a stated design goal.

We have selected several subjects to be reviewed in this chapter that you have undoubtedly studied or will study in your physics courses, along with a short presentation of economics. These topics are considered fundamental engineering sciences for all engineers, and so make up the foundational body of knowledge evaluated in the professional licensing process. We hope this short review will better prepare you for your future coursework and your future licensing endeavors.

17.1 Statics

Statics is concerned primarily with the equilibrium of bodies subjected to force systems. It has been introduced in physics courses and is reviewed here to refresh your memories of the major components that make up the subject. Forces and moments are the two entities that are of most interest in statics. Let's review them.

Forces, Moments, and Resultants

A force is the manifestation of the action of one body upon another. Forces arise from the direct action of two bodies in contact with one another, or from the action at a distance of one body upon another, as occurs with gravitational and magnetic forces. We classify forces as either **body forces**, which act (and are distributed) throughout the volume of the body, or **surface forces**, which act over a surface portion of the body. If the surface over which the force system acts is very small, we usually assume localization at a specific point in the surface and speak of a *concentrated* force at that point.

Mathematically, forces are represented by **vectors**. Geometrically, a vector is a directed line segment having a head and a tail (i.e., an arrow). Its orientation defines the line of action, and the direction of the arrow (tail to head) gives the sense of the force. Vectors can be manipulated mathematically. Two vectors can be added by placing the tail of one vector at the tip of the other. Similarly, one vector can be subtracted from another by adding its reverse (Fig. 17.1). The concepts are extendable to vectors in three or more dimensions. Vector addition is commutative and associative (refer to Section 16.1). Vectors can be multiplied by scalars. Given a Vector $\mathbf{A} = (a_1, a_2, a_3)$, and a real number m, $m\mathbf{A} = (ma_1, ma_2, ma_3)$.

Systems of concentrated forces are **concurrent** when all of the forces act, or could act, at a single point; otherwise they are **non-concurrent** systems. Parallel force systems are in this second group. Also, force systems are often described as two-dimensional (acting in a single plane) or three-dimensional (spatial systems).

In addition to the push–pull effect on the point at which it acts, a force creates a **moment** about axes passing through the body. Conceptually, a moment may be thought of as a tendency to rotate the body upon which it acts about a certain axis. The moment of a force R, acting at a point A sketched in Figure 17.2, about the *y*-axis is found by passing a plane normal to the *y*-axis through point A. The component of

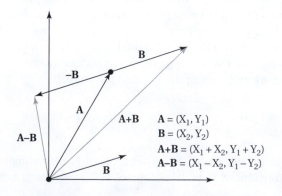

$\mathbf{A} = (X_1, Y_1)$
$\mathbf{B} = (X_2, Y_2)$
$\mathbf{A+B} = (X_1 + X_2, Y_1 + Y_2)$
$\mathbf{A-B} = (X_1 - X_2, Y_1 - Y_2)$

Figure 17.1 Vector addition and subtraction

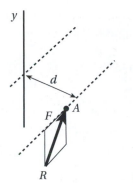

Figure 17.2 Moment of force R about the y-axis

R in this plane, which we will call F, multiplied by the perpendicular distance d from the line of action of F to the y-axis, equals the moment of R about the y-axis—that is,

$$M_y = F \times d \qquad (17.1.1)$$

The **resultant** force of a system of forces is the equivalent force of the total system. It is the vector sum of the individual forces.

EXAMPLE 17.1   Determine the magnitude of the resultant force R for the (a) plane and (b) space concurrent systems shown (the force B is in the yz-plane).

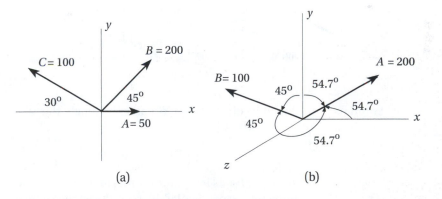

(a)   (b)

Solution: a)   Sum components in the x-direction:

$$\text{let } \theta_1 = 30°$$

$$\text{let } \theta_2 = 45°$$

$$R_x = -C\cos\theta_1 + B\cos\theta_2 + A$$

$$R_x = -100\cos 30° + 200\cos 45° + 50 = 104.8$$

Sum components in the y-direction:

$$R_y = C\sin\theta_1 + B\sin\theta_2$$

$$R_y = 100\sin 30° + 200\sin 45° = 191.4$$

$$\therefore R = \sqrt{104.8^2 + 191.42^2} = 218.2$$

b) Sum components in the x-, y-, and z-directions:

$$R_x = 200 \cos 54.7° + 0 = 115.5$$
$$R_y = 200 \cos 54.7° + 100 \cos 45° = 186.3$$
$$R_z = 200 \cos 54.7° + 100 \cos 45° = 186.3$$
$$\therefore R = \sqrt{115.5^2 + 186.3^2 + 186.2^2} = 287.6$$

EXAMPLE 17.2   Determine the moment of force F:

a) about the x-axis.
b) about the y-axis.

Distances are measured in meters.

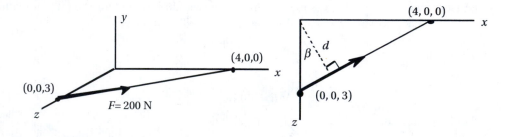

Solution: a) The line of action of F passes through the x-axis. Therefore, $d = 0$ and

$$M_x = 200 \times 0 = 0$$

b) The force F acts in a plane normal to the y-axis so we must determine the perpendicular distance d, sketched above:

$$d = 3\cos\beta = 3 \times 0.8 = 2.4 \text{m}$$

$$\therefore M_y = F \times d = 200 \times 2.4 = 480\text{N} \cdot \text{m}$$

Equilibrium

If the system of forces acting on a body is one whose resultant is absolutely zero (vector sum of all forces is zero, and the resultant moment of the forces about each of the axes is zero), the body is in **equilibrium**. Mathematically, equilibrium requires the six equations

$$\sum F_x = 0 \quad \sum M_x = 0$$
$$\sum F_y = 0 \quad \sum M_y = 0$$
$$\sum F_z = 0 \quad \sum M_z = 0 \tag{17.1.2}$$

which must hold for any orientation of the xyz-system. If the forces are concurrent and their vector sum is zero, the sum of moments will be satisfied automatically.

If all the forces act in a single plane, say the xy-plane, one of the above force equations and two of the moment equations are satisfied identically, so that equilibrium requires only

$$\sum F_x = 0, \quad \sum F_y = 0 \quad \sum M_z = 0 \tag{17.1.3}$$

In this case we can solve for only three unknowns instead of six as when Eqs. 17.1.2 are required.

The solution for unknown forces and moments in equilibrium problems rests firmly upon the construction of a good **free body diagram**, abbreviated FBD, from which the detailed Eqs. 17.1.2 or 17.1.3 may be obtained. An FBD is a neat sketch of the body (or of any appropriate portion of it) showing all forces and moments acting on the body, together with all important linear and angular dimensions.

A body in equilibrium under the action of two forces only is called a *two-force member*, and the two forces must be equal in magnitude and oppositely directed along the line joining their points of application. If a body is in equilibrium under the action of three forces (i.e., is a three-force member) those forces must be coplanar, and concurrent (unless they form a parallel system). Examples are shown in Figure 17.3.

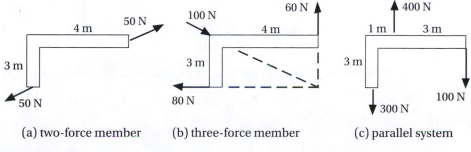

(a) two-force member   (b) three-force member   (c) parallel system

Figure 17.3 Plane force systems

A knowledge of the possible reaction forces and moments at various supports is essential in preparing a correct FBD. Several of the basic reactions are illustrated in Figure 17.4 showing a block of concrete subjected to a horizontal pull P (Fig. 17.4a) and a cantilever beam carrying both a distributed and concentrated load (Fig. 17.4b). The correct FBDs are on the right.

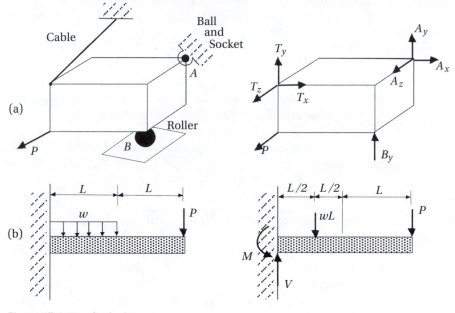

Figure 17.4 Free body diagrams

EXAMPLE 17.3 Determine the tension in the two cables supporting the 700 N block.

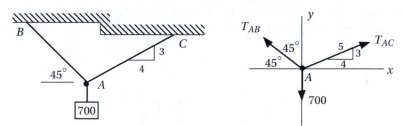

Solution: Construct the FBD of junction A of the cables. Sum forces in x and y directions:

$$\text{let } \theta_1 = 45°$$

$$\cos\theta_2 = \frac{4}{5}$$

$$\sin\theta_2 = \frac{3}{5}$$

$$\sum F_x = -T_{AB}\cos\theta_1 + T_{AC}\cos\theta_2 = 0$$

$$\sum F_y = T_{AB} \sin\theta_1 + T_{AC} \sin\theta_2 - 700 = 0$$

$$\sum F_x = -0.707 T_{AB} + 0.8 T_{AC} = 0$$

$$\sum F_y = 0.707 T_{AB} + 0.6 T_{AC} - 700 = 0$$

Solve, simultaneously, and find

$$T_{AC} = 700 / 1.4 = 500 \text{N}$$

$$T_{AB} = 0.8(500) / 0.707 = 565.8 \text{N}$$

EXAMPLE 17.4 A 12-m bar weighing 140 N is hinged to a vertical wall at A, and supported by the cable BC. Find the tension in the cable and the horizontal and vertical components of the force reaction at A.

Solution: Construct the FBD showing force components at A and B. Write the equilibrium equations and solve: $l = 12$ m, $\alpha = 50°$, $\beta = 60°$.

$$\sum M_A = d_x T_x + d_y T_y + d_W W$$

$$\sum M_A = l\sin(90 - \beta)T \sin\alpha + l\cos(90 - \beta)T \cos\alpha - \cos(90 - \beta)\frac{1}{2}lW = 0, \text{ or}$$

$$\sum M_A = l\cos\beta T \sin\alpha + l\sin\beta T \cos\alpha - \sin\beta\frac{1}{2}lW = 0$$

$$\sum M_A = 12 \times \frac{1}{2} \times T \sin 50° + 12 \times \frac{\sqrt{3}}{2}T \cos 50° - \frac{\sqrt{3}}{2} \times \frac{1}{2} \times 12 \times 140 = 0$$

$$\therefore T = 64.5 \text{ N}$$

$$\sum F_x = A_x - T \sin\alpha = A_x - 64.5(0.766) = 0$$

$$\therefore A_x = 49.4 \text{ N}$$

$$\sum R_y = A_y + T \cos\alpha - 140 = A_y + 64.5(0.643) = 0$$

$$\therefore A_y = 98.5 \text{ N}$$

$$\sum M_A = 6T \sin 50° + 6\sqrt{3}T \cos 50° - 140(6)\sqrt{3} / 2 = 0$$

$$\therefore T = 64.5 \text{N}$$

$$\sum F_x = A_x - T \sin 50° = A_x - 64.5(0.766) = 0$$

$$\therefore A_x = 49.4 \text{N}$$

$$\sum F_y = A_y + T \cos 50° - 140 = A_y + 64.5(0.643) - 140 = 0$$

$$\therefore A_y = 98.5 \text{N}$$

17.2 Dynamics

Dynamics is separated into two major divisions: **kinematics**, which is a study of motion without reference to the forces causing the motion, and **kinetics**, which relates the forces on bodies to their resulting motions. Newton's laws of motion are necessary in relating forces to motions; they are:

First law: A particle remains at rest or continues to move in a straight line with a constant velocity if no unbalanced force acts on it.

Second law: The acceleration of a particle is proportional to the force acting on it and inversely proportional to the particle mass; the direction of acceleration is the same as the force direction.

Third law: The forces of action and reaction between contacting bodies are equal in magnitude, opposite in direction, and colinear.

Law of gravitation: The force of attraction between two bodies is proportional to the product of their masses and inversely proportional to the square of the distance between their centers.

Kinematics

In rectilinear motion of a particle in which the particle moves in a straight line, the acceleration a, the velocity v, and the displacement s are related by

$$a = \frac{dv}{dt}, v = \frac{ds}{dt}, a = \frac{d^2s}{dt^2} = v\frac{dv}{ds} \qquad (17.2.1)$$

If the acceleration is a known function of time, the above can be integrated to give $v(t)$ and $s(t)$. For the important case of constant acceleration, integration yields

$$a\,dt = dv; v = v_o + at$$

$$v\,dt = ds; s = v_o t + a\frac{t^2}{2}$$

$$v^2 = v_o^2 + 2as \qquad (17.2.2)$$

where at $t = 0$, $v = v_o$ and $s_o = 0$.

Angular displacement is the angle θ that a line makes with a fixed axis, usually the positive x-axis. Counterclockwise motion is assumed to be positive, as shown in Figure 17.5. The angular acceleration α, the angular velocity ω, and θ are related by

$$\alpha = \frac{d\omega}{dt}, \ \omega = \frac{d\theta}{dt}, \ \alpha = \omega\frac{d\omega}{d\theta} = \frac{d^2\theta}{dt^2} \qquad (17.2.3)$$

If α is a constant, integration of these equations gives

$$\omega = \omega_0 + \alpha t \quad \theta = \omega_0 t + \alpha t^2 / 2 \quad \omega^2 = \omega_0^2 + 2\alpha\theta \qquad (17.2.4)$$

where we have assumed that $\omega = \omega_o$ and $\theta_o = 0$ at $t = 0$.

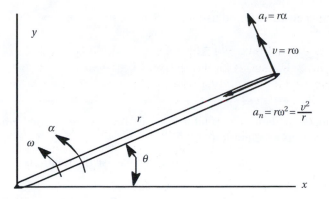

Figure 17.5 Angular motion

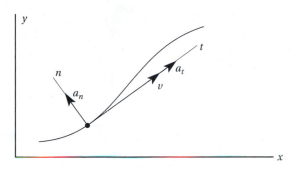

Figure 17.6 Motion on a plane curve

When a particle moves on a plane curve as shown in Figure 17.6, the motion may be described in terms of coordinates along the normal n and the tangent t to the curve at the instantaneous position of the particle.

The acceleration is the vector sum of the normal acceleration a_n and the tangential acceleration a_t. These components are given as

$$a_n = \frac{v^2}{r}, \quad a_t = \frac{dv}{dt} \tag{17.2.5}$$

where r is the radius of curvature and v is the magnitude of the velocity. The velocity is always tangential to the curve, so no subscript is necessary to identify the velocity.

It should be noted that a rigid body traveling without rotation can be treated as particle motion.

**EXAMPLE
17.5**

The velocity of a particle is $v(t) = 5 + 10t$ m/s. Find the acceleration and the displacement at t = 10 s, if $s_o = 0$ at t = 0.

Solution: The acceleration is found to be

$$a = \frac{dv}{dt} = 10 \text{ m/s}^2$$

The displacement is found using $v_o = 5$ m/s:

$$s = v_o t + at^2 / 2$$
$$= 5 \times 10 + 10 \times 10^2 / 2 = 550 \text{ m}$$

EXAMPLE 17.6   An automobile skids to a stop 60 m after its brakes are applied while traveling 25 m/s. What is its acceleration?

Solution: Since speed and distance are given we use the relationship

$$v^2 = v_o^2 + 2as$$

Letting $v = 0$, we find

$$a = -\frac{v_o^2}{2s} = -\frac{25^2}{2 \times 60} = -5.21 \text{ m/s}^2$$

EXAMPLE 17.7   A wheel, rotating at 100 rad/s ccw (counterclockwise), is subjected to an angular acceleration of 20 rad/s² cw. Find the total number of revolutions (cw plus ccw) through which the wheel rotates in 8 seconds.

Solution: The time at which the angular velocity is zero is found as follows:

$$\cancel{\omega}^{\,0} = \omega_o + \alpha t$$

$$\therefore t = -\frac{\omega_o}{\alpha} = -\frac{100}{-20} = 5 \text{ s}.$$

After 3 additional seconds the angular velocity is determined by

$$\omega = \cancel{\omega_o}^{\,0} + \alpha t$$

$$= -20 \times 3 = -60 \text{ rad/s}.$$

The angular displacement from 0 to 5 s is

$$\theta = \omega_o t + \alpha t^2 / 2$$

$$= 100 \times 5 - 20 \times 5^2 / 2 = 250 \text{ rad}$$

During the next 3 s, the angular displacement is

$$\theta = \alpha t^2 / 2$$

$$= -20 \times 3^2 / 2 = -90 \text{ rad}$$

The total number of revolutions rotated is

$$\theta = (250 + 90) / 2\pi = 54.1 \text{ rev}$$

EXAMPLE 17.8

Consider idealized projectile motion (no air drag) in which ax = 0 and ay = −g. Find expressions for the range R and the maximum height H in terms of v_0 and θ.

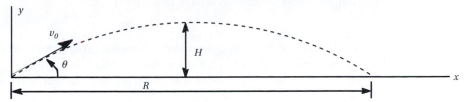

Solution: Using Eq. 17.2.2 for constant acceleration in the y-direction we have the point (R, 0):

$$\cancel{y}^0 = (v_o \sin\theta)t - gt^2/2.$$

$$\therefore t = 2v_o \sin\theta/g.$$

where $(vy) = v_o \sin\theta$. From the x-component equation recognizing that ax = 0:

$$x = (v_o \cos\theta)t - \cancel{a}^0_x t^2/2.$$

$$\therefore R = (v_o \cos\theta)2v_o \sin\theta/g = v_o^2 \sin 2\theta/g.$$

using $2\sin\theta \cos\theta = \sin 2\theta$. Obviously, the maximum height occurs when the time is one-half that which yields the range R. Hence, with ymax = H,

$$H = (v_o \sin\theta)\frac{v_o \sin\theta}{g} - \frac{g}{2}\left(\frac{v_o \sin\theta}{g}\right)^2$$

$$= \frac{v_o^2}{2g}\sin^2\theta$$

Note: The maximum R for a given v_o occurs when $\sin 2\theta = 1$, which means θ = 45° for Rmax.

EXAMPLE 17.9

It is desired that the normal acceleration of a satellite be 9.6m/s² at an elevation of 200 km. What should be the velocity be for a circular orbit? The radius of the Earth is 6400 km.

Solution: The normal acceleration, which points toward the center of the Earth, is

$$a_n = \frac{v^2}{r}$$

$$\therefore v = \sqrt{a_n r} = \sqrt{9.6 \times (6400 + 200) \times 1000} = 7960\,\text{m/s}$$

The normal acceleration is essentially the value of gravity near the Earth's surface. Gravity varies only slightly if the elevation is small with respect to the Earth's radius.

Kinetics

To relate the force acting on a body to the motion of that body we use Newton's laws of motion. Newton's second law is used in the form

$$\Sigma = F = ma \tag{17.2.6}$$

where the mass of the body is assumed to be constant and the vector is the acceleration of the center of mass (center of gravity) if the body is rotating.

The gravitational attractive force between one body and another is given by

$$F = K\frac{m_1 m_2}{r^2} \tag{17.2.7}$$

where $K = 6.67 \times 10^{-11}$ N · m²/kg². **Note:** Since metric units are used in the above relations, mass must be measured in kilograms. The weight is related to the mass by

$$W = mg \tag{17.2.8}$$

where we use $g = 9.8$ m/s², unless otherwise stated.

**EXAMPLE
17.10**

Find the tension in the rope and the distance the 300-kg mass moves in 3 seconds. The mass starts from rest and the mass of the pulley is negligible. The force due to friction is μN where N is the normal force and μ is the coefficient of friction. The pulley is frictionless.

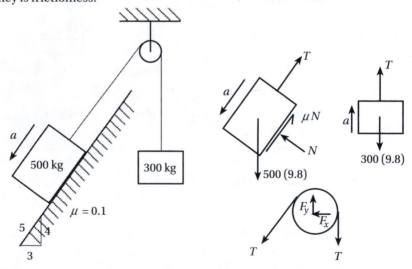

Solution: First, sketch the free bodies of the two masses and the pulley, as shown. The pulley does not change tension T in the rope. Next, sum forces normal to the surface:

$$\Sigma F_n = 0$$

$$N - 500 \times 9.8 \times 0.6 = 0. \therefore N = 2940\text{N}. \therefore F = \mu N = 294\text{N}$$

Applying Newton's second law to the 500-kg mass parallel to the surface gives

$$\sum F = ma$$

$$0.8 \times 500 \times 9.8 - 294 - T = 500a$$

By studying the pulley we observe the 300-kg mass to also be accelerating at a. Hence, we have

$$\sum F = ma$$

$$T - 300 \times 9.8 = 300 \times a$$

Solving the above equations simultaneously results in

$$a = 0.8575 \text{m} / \text{s}^2, \quad T = 3197\text{N}$$

The distance the 300-kg mass moves is

$$s = \frac{1}{2}at^2 = \frac{1}{2} \times 0.8575 \times 3^2 = 3.31\text{m}$$

EXAMPLE 17.11 Find the tension in the string if at the position shown, $v = 4$ m/s. Calculate the angular acceleration.

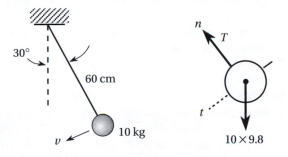

Solution: Sum forces in the normal direction and obtain

$$\sum F_n = ma_n = mv^2 / r$$

$$T - 10 \times 9.8\cos 30° = 10 \times 4^2 / 0.6$$

$$\therefore T = 352\text{N}$$

Sum forces in the tangential direction and find

$$\sum F_t = ma_t = mr\alpha$$

$$10 \times 9.8\sin 30° = 10 \times 0.6\alpha$$

$$\therefore \alpha = 8.17\,\text{rad/s}^2$$

17.3 Thermodynamics

Thermodynamics involves the storage, transformation, and transfer of energy. It is stored as internal energy, kinetic energy, and potential energy; it is transformed between these various forms; and it is transferred as work or heat transfer. We will present some introductory concepts as they apply primarily to ideal gases. These same concepts can be applied to substances that change phase, such as water and refrigerants. Tabulated thermodynamic properties are needed but are not included in this textbook.

Overview, Definitions, and Laws

Both a **system**, a fixed quantity of matter, and a **control volume**, a volume into which and/or from which a substance flows, can be used in thermodynamics. (A control volume may also be referred to as an "open system.") A system and its surroundings make up the **universe.** Some useful definitions follow:

Phase—matter that has the same composition throughout; it is homogeneous

Mixture—a quantity of matter that has more than one phase

Property—a quantity that serves to describe a system

Simple system—a system composed of a single phase, free of magnetic, electrical, and surface effects. Only two properties are needed to fix a simple system.

State—the condition of a system described by giving values to its properties at the given instant

Intensive property—a property that does not depend on the mass

Extensive property—a property that depends on the mass

Specific property—an extensive property divided by the mass

Thermodynamic equilibrium—when the properties do not vary from point to point in a system and there is no tendency for additional change

Process—the path of successive states through which a system passes

Quasi-equilibrium—if, in passing from one state to the next, the deviation from equilibrium is infinitesimal

Reversible process—a process that, when reversed, leaves no change in either the system or surroundings

Isothermal—temperature is constant

Isobaric—pressure is constant

Isometric/isochoric—volume is constant

Isentropic—entropy is constant

Adiabatic—no heat transfer (an insulated surface)

Experimental observations are organized into mathematical statements or laws. Some of those used in thermodynamics follow:

Zero-th law of thermodynamics—If two bodies are equal in temperature to a third, they are equal in temperature to each other.

First law of thermodynamics—During a given process, the net heat transfer minus the net work output equals the change in energy.

Second law of thermodynamics—During a process, the net entropy of the universe cannot decrease.

Boyle's law—The volume varies inversely with pressure for an ideal gas.

Charles' law—The volume varies directly with temperature for an ideal gas.

Avogadro's law—Equal volumes of different ideal gases with the same temperature and pressure contain an equal number of molecules.

Dalton's law of partial pressures—In a gas mixture, each separate gas exerts a pressure, called partial pressure, the sum of which for a mixture equals the total pressure of the gas mixture.

Density, Pressure, and Temperature

The density ρ is the mass divided by the volume,

$$\rho = \frac{m}{V} \tag{17.3.1}$$

The specific volume is the reciprocal of the density,

$$\upsilon = \frac{1}{\rho} = \frac{V}{m} \tag{17.3.2}$$

The pressure P is the normal force divided by the area upon which it acts. In thermodynamics, it is important to use **absolute pressure**, defined by

$$P_{abs} = P_{gauge} + P_{atmospheric} \tag{17.3.3}$$

where the atmospheric pressure is taken as 100 kPa (14.7 psi), unless otherwise stated. If the gauge pressure is negative, it is a **vacuum**.

The temperature scale is established by choosing a specified number of divisions, called degrees, between the ice point and the steam point, each at 101 kPa absolute. In the Celsius scale, the ice point is set at 0°C and the steam point at 100°C. *In thermodynamics, pressures are always assumed to be given as absolute pressures.* Expressions for the absolute temperature in kelvins and degrees Rankine are, respectively,

$$T_K = T_{celcius} + 273; \; T_R = T_{fahrenheit} + 460 \tag{17.3.4}$$

The temperature, pressure, and specific volume for an ideal (perfect) gas are related by the **ideal gas law**, which can be expressed in various forms.

$$P\upsilon = RT, \; P = \rho RT, \; \text{or } PV = mRT \quad R = \bar{R}/M \qquad (17.3.5)$$

$$PV = n\bar{R}T$$

where the **universal gas constant** $\bar{R} = 8.314 \text{ kJ}/\text{kmol} \cdot \text{K}$, n is the total number of moles, M is the molar mass, and R is the gas constant for the specific gas under consideration. The specific gas constant for air is 0.287 kJ/kg K. Keep in mind that other textbooks and reference materials may use slightly different symbols for the universal gas constant and the specific gas constant, such as R_u and R, respectively.

Where there is a mixture of more than one gas, each gas exerts a partial pressure, which contributes to the total pressure of the gas. Consider a mixture of gases A, B, and C. Each partial pressure can be calculated: $p_i = \dfrac{n_i \bar{R}T}{V}$

$$p_A + p_B + p_C = P_{total} = P$$

$$n_A + n_B + n_C = n_{total} = n$$

where n_i is the total number of moles of substance $_i$ and V is the total volume occupied by the gas mixture.

EXAMPLE 17.12   What mass of air is contained in a room 20 m $\times$ 40 m $\times$ 3 m at standard conditions?

Solution: Standard conditions are T = 25°C and P = 100 kPa. Using M = 28.97,

$$\rho = \frac{P}{RT}$$

$$= \frac{100}{0.287 \times 298} = 1.17 \text{kg}/\text{m}^3, \quad \text{where } R = \frac{8.314}{28.97} = 0.287 \text{kj}/\text{kg}\cdot\text{K}$$

$$\therefore m = \rho V$$

$$= 1.17 \times 20 \times 40 \times 3 = 2810 \text{kg}$$

Dimensional analysis for this problem confirms the following:

$$\frac{\text{kg}}{\text{m}^3} = \frac{\text{kPa}}{(\frac{\text{kJ}}{\text{kg} \cdot \text{K}}) \cdot \text{K}} \times \frac{\frac{\text{N}}{\text{m}^2}}{\text{Pa}} \times \frac{\text{J}}{\text{N} \cdot \text{m}}$$

The First Law of Thermodynamics for a System

The first law of thermodynamics, referred to as the "first law" or the "energy equation," is expressed for a cycle as

$$Q_{net} = W_{net} \tag{17.3.6}$$

and for a process as

$$Q - W = \Delta E \tag{17.3.7}$$

where Q is the heat transfer, W is the work, and E represents the energy (kinetic, potential, and internal) of the system. Heat transfer to the system is positive and work done by the system is positive. (Some authors define work done on the system as positive so that $Q + W = \Delta U$.) In thermodynamics attention is focused on internal energy with kinetic and potential energy changes neglected (unless otherwise stated) so that we have

$$Q - W = \Delta U \text{ or } q - w = \Delta u \tag{17.3.8}$$

where the specific internal energy, heat transfer per unit mass (not specific heat), and specific work are the following:

$$u = \frac{U}{m} \quad q = \frac{Q}{m} \quad w = \frac{W}{m} \tag{17.3.9}$$

Heat transfer may occur during any of the three following modes.

Conduction is heat transfer due to molecular activity. For steady-state heat transfer through a constant wall area **Fourier's law** states

$$\dot{Q} = kA\Delta T \, / \, L = A\Delta T \, / \, R \tag{17.3.10}$$

where k is the **conductivity** (dependent on the material), R is the **resistance factor**, and the length, L, is normal to the heat flow. A dot signifies a rate, so that $\dot{Q}$ has units of J/s.

Convection is heat transfer due to fluid motion. The mathematical expression used is

$$\dot{Q} = hA\Delta T = A\Delta T \, / \, R \tag{17.3.11}$$

where h is the **convective heat transfer coefficient** (dependent on the surface geometry, the fluid velocity, the fluid viscosity and density, and the temperature difference).

Radiation is heat transfer due to the transmission of waves. The heat transfer from body 1 is

$$\dot{Q} = \sigma \in A(T_1^4 - T_2^4)F_{1-2} \tag{17.3.12}$$

in which the **Stefan-Boltzmann constant** is $\sigma = 5.67 \times 10^{-11} \text{kJ/s} \cdot \text{m}^2 \cdot \text{K}^4$, ε is the **emissivity** ($\varepsilon = 1$ for a black body), and F_{1-2} is the **shape factor** ($F_{1-2} = 1$ if body 2 encloses body 1).

For a two-layer composite wall with an inner and outer convection layer we use the resistance factors and obtain

$$\dot{Q} = A\Delta T / (R_i + R_1 + R_2 + R_o) = UA\Delta T \qquad (17.3.13)$$

where U is the **overall heat transfer coefficient**. It is not to be confused with internal energy of Eq. 17.3.8.

Work can be accomplished mechanically by moving a boundary, resulting in a quasi-equilibrium work mode

$$W = \int P dV \qquad (17.3.14)$$

It can also be accomplished in non–quasi-equilibrium modes such as with paddle wheel or by electrical resistance, but then Eq. 17.3.14 cannot be used.

We introduce **enthalpy** for convenience, and define it to be

$$H = U + PV$$

$$h = u + Pv$$

$$h = \frac{H}{m} \qquad (17.3.15)$$

For ideal gases we assume constant specific heats and use

$$\Delta u = c_v \Delta T \qquad (17.3.16)$$

$$\Delta u = c_p \Delta T \qquad (17.3.17)$$

where c_v is the **constant volume specific heat** and c_p is the **constant pressure specific heat**. From the above we can find

$$c_p = c_v + R \qquad (17.3.18)$$

We also define the **ratio of specific heats** k to be

$$k = c_p / c_v \qquad (17.3.19)$$

For air $c_v = 0.716$ kJ/kg $\cdot$ K, $c_p = 1.00$ kJ/kg $\cdot$ K, and $k = 1.4$. For most solids and liquids we can find the heat transfer using

$$Q = mc_p \Delta T \qquad (17.3.20)$$

For water $c_p = 4.18$ kJ/kg $\cdot$ °K, and for ice $c_p \cong 2.1$ kJ/kg $\cdot$ K.

When a substance changes phase, **latent heat** is involved. The energy necessary to melt a unit mass of a solid is the **heat of fusion**; the energy necessary to vaporize a unit mass of liquid is the **heat of vaporization**; the energy necessary to vaporize a unit mass of solid is the **heat of sublimation**. For ice, the heat of fusion is approximately 320 kJ/kg and the heat of sublimation is about 2040 kJ/kg; the heat of vaporization of water varies from 2050 kJ/kg at 0°C to zero at the point where no vaporization occurs (at very high pressures).

We consider the preceding paragraphs and summarize as follows for quasi-equilibrium processes.

Constant Temperature (Isothermal)

First law:
$$Q - W = m\Delta u \text{ or } q - w = \Delta u \tag{17.3.21}$$

Ideal gas:
$$Q = W = mRT \ln \frac{\upsilon_2}{\upsilon_1} = mRT \ln \frac{P_1}{P_2} \tag{17.3.22}$$

$$P_2 = P_1 \upsilon_1 / \upsilon_2 \tag{17.3.23}$$

Constant Pressure (Isobaric)

First law:
$$Q = m\Delta u \text{ or } q = \Delta u \tag{17.3.24}$$

$$W = 0 \tag{17.3.25}$$

Ideal gas:
$$Q = mc_p \Delta T \tag{17.3.26}$$

$$T_2 = T_1 \upsilon_2 / \upsilon_1 \tag{17.3.27}$$

Constant Volume (Isometric)

First law:
$$Q = m\Delta u \text{ or } q = \Delta u \tag{17.3.28}$$

$$W = 0 \tag{17.3.29}$$

Ideal gas:
$$Q = mc_p \Delta T \tag{17.3.30}$$

$$T_2 = T_1 P_2 / P_1 \tag{17.3.31}$$

Both Adiabatic AND Reversible Process (Isentropic)

First law:
$$-W = m\Delta u \text{ or } -w = \Delta u \tag{17.3.32}$$

$$Q = 0 \tag{17.3.33}$$

Ideal gas:
$$-W = mc_\upsilon \Delta T \tag{17.3.34}$$

$$T_2 = T_1 \left(\upsilon_1 / \upsilon_2 \right)^{k-1} = T_1 \left(P_2 / P_1 \right)^{(k-1)/k} \tag{17.3.35}$$

$$P^2 = P_1 \left(t_1 / t_2 \right)^k \tag{17.3.36}$$

EXAMPLE 17.12

How much heat is needed to raise the temperature of 100 kg of ice at $T_1 = 10°C$ to 80°C?

Solution: The heat transfer is related to the enthalpy by

$$Q = m\Delta h$$
$$= m(c\Delta T_{ice} + \text{heat of fusion} + c\Delta T_{water})$$

Using the values given in this example,

$$Q = 100 \left(2.1 \times 10 + 320 + 4.18 \times 80 \right)$$
$$= 67540 \text{kj or } 67.54 \text{MJ}$$

EXAMPLE 17.13

Calculate the work done by a piston if the 2 m³ volume of air is tripled while the temperature is maintained at 40°C. The initial pressure is 400 kPa.

Solution: The mass is needed in order to use Eq. 17.3.22 to find the work; it is

$$m = \frac{PV}{RT} = \frac{400 \times 2}{0.287 \times 313} = 8.91 \text{kg}$$

The work is then found to be

$$W = mRT \ln \upsilon_2 / \upsilon_1$$
$$= 8.91 \times 0.287 \times 313 \ln 3 = 879 \text{kJ}$$

Note: The temperature is expressed as 40 + 273 = 313 K.

EXAMPLE 17.14

How much work is necessary to compress air in an insulated cylinder from 0.2 m³ to 0.01 m³? Use $T_1 = 20°C$ and $P_1 = 100$ kPa.

Solution: For an adiabatic process Q = 0 so that the first law is

$$-W = m\left(u_2 - u_1 \right)$$
$$= mc_\upsilon \left(T_2 - T_1 \right)$$

To find the mass m we use the ideal gas equation:

$$m = \frac{PV}{RT} = \frac{100 \times 0.2}{0.287 \times 293} = 0.2378 \text{kg}$$

The temperature T_2 is found to be

$$T_2 = T_1(\upsilon_1 / \upsilon_2)^{k-1}$$
$$= 293(0.2 / 0.01)^{0.4} = 971.1K$$

The work is then

$$W = 0.2378 \times 0.716(971.1 - 293) = 115.5\text{kj}$$

EXAMPLE 17.15

A 10-cm-thick wall made of pine wood is 3 m high and 10 m long. Calculate the heat transfer rate if the temperature is 25°C on the inside and –20°C on the outside. Neglect convection. Use $k = 0.15$ J/(s $\cdot$ m $\cdot$ °C).

Solution: The heat transfer occurs due to conduction. Using the given k Eq. 17.3.10 provides

$$\dot{Q} = kA\Delta T / L$$
$$= 0.15 \times (3 \times 10) \times [25 - (-20)] / 0.1 = 2025\text{J} / \text{s}$$

EXAMPLE 17.16

The surface of the glass in a 1.2-m $\times$ 0.8-m skylight is maintained at 20°C. If the air temperature is –20°C, estimate the rate of heat loss from the window. Use $h = 12$ J/ (s $\cdot$ m^2 $\cdot$ °C).

Solution: The convective heat transfer coefficient depends on several parameters so, as usual, it is specified. Using Eq. 17.3.11 the rate of heat loss is

$$\dot{Q} = hA\Delta T$$
$$= 12 \times (12. \times 08) \times [20 - (-20)] = 461\text{J} / \text{s}$$

EXAMPLE 17.17

A 2-cm-diameter heating oven is maintained at 1000°C and the oven walls are at 500°C. If the emissivity of the element is 0.85, estimate the rate of heat loss from the 2-m-long element.

Solution: The heat loss will be primarily due to radiation. Neglecting any convection loss, using $F_{1-2} = 1$ since the oven encloses the element, and Eq. 17.3.12 provides us with

$$\dot{Q} = \sigma \in A(T_1^4 - T_2^4)$$
$$= 5.67 \times 10^{-11} \times 0.85 \times (\pi \times 0.02 \times 2)(1273^4 - 773^4) = 13.7\text{kJ/s}$$

Note that absolute temperature must be used. Also, the area A is the surface area of the cylinder (i.e., πDL).

The First Law of Thermodynamics for a Control Volume

The continuity equation, which accounts for the conservation of mass, may be used in a certain situation involving a **control volume** (a volume into which and/or from which a fluid flows). It is stated as

$$\dot{m} = \rho_1 A_1 V_1 = \rho_2 A_2 V_2 \tag{17.3.37}$$

where, in control volume formulations, V is the velocity and $\dot{m}$ is called the "mass flux." In the above continuity equation, we assume **steady flow** (that is, the variables are independent of time) and one inlet and one outlet. For such steady-state flow situations, the first law takes the form

$$\frac{\dot{Q} - \dot{W}}{\dot{m}} = \frac{V_2^2 - V_1^2}{2} + h_2 - h_1 + g(z_2 - z_1) \tag{17.3.38}$$

where the dot signifies a rate so that Q and W_s have the units of kJ/s. In most devices the potential energy change is negligible. Also, the kinetic energy change can often be ignored (but if sufficient information is given, it should be included, as in a nozzle) so that the first law is most often used in the simplified form

$$\dot{Q} - \dot{W}_s = \dot{m}(h_2 - h_1) \tag{17.3.39}$$

Particular devices are of special interest. The energy equation for a **valve** or a **throttle plate** is simply

$$h_2 - h_1 \tag{17.3.40}$$

neglecting kinetic energy change.

For a **turbine** expanding a gas, the heat transfer is negligible so that

$$\dot{W}_T = \dot{m}(h_1 - h_2) \tag{17.3.41}$$

The work input to a **gas compressor** with negligible heat transfer is

$$\dot{W}_c = \dot{m}(h_2 - h_1) \tag{17.3.42}$$

A **boiler** and a **condenser** are simply heat transfer devices. The first law then simplifies to

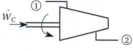

$$\dot{Q} = \dot{m}(h_2 - h_1) \tag{17.3.43}$$

For a nozzle or a **diffuser** there is no work or heat transfer; we must include, however, the kinetic energy change, resulting in

$$0 = \frac{V_2^2 - V_1^2}{2} + h_2 - h_1 \tag{17.3.44}$$

For a pump or a **hydroturbine** we take a slightly different approach. We return to Eq. 17.3.38 and write it using $v = 1/\rho$, and with Eq. 17.3.15, as

$$\frac{\dot{Q} - \dot{W}}{\dot{m}} = \frac{V_2^2 - V_1^2}{2} + u_2 - u_1 + \frac{P_2 - P_1}{\rho} + g(z_2 - z_1) \tag{17.3.45}$$

For most ideal situations we do not transfer heat and assume constant temperature so that $u_2 = u_1$. Often the kinetic and potential energy changes are negligible so

$$-\dot{W}_s = \dot{m}\frac{P_2 - P_1}{\rho} \tag{17.3.46}$$

This would provide the minimum pump power requirement or the maximum turbine power output. The inclusion of an efficiency would increase the pump power requirement or decrease the turbine output.

A gas turbine or compressor efficiency is based on an isentropic process as the ideal process. For a gas turbine or a compressor we have

$$\eta_T = \frac{\dot{W}_a}{\dot{W}_s} = \frac{w_a}{w_s} \quad \eta_C = \frac{\dot{W}_s}{\dot{W}_a} = \frac{w_s}{w_a} \tag{17.3.47}$$

where $\dot{W}_a$ is the actual power and W_s is the power assuming an isentropic process.

EXAMPLE 17.18

What is the minimum power requirement of a pump that is to increase the pressure from 2 kPa to 6 MPa for a mass flux of 10 kg/s of water?

Solution: With a liquid we let $h_2 - h_1 = u_2 - u_1 + (P_2 - P_1)\,v$ since $v =$ const. We let $u_2 - u_1 = 0$ so that the first law simplifies to

$$\dot{W}_P = \dot{m}\frac{P_2 - P_1}{\rho}$$

$$= 10\frac{6000 - 2}{1000} = 59.98\,\text{kW}$$

EXAMPLE
17.19

A nozzle accelerates air from 100 m/s, 400°C, and 400 kPa to a receiver where $P = 20$ kPa. Assuming an isentropic process, find V_2.

Solution: The energy equation takes the form

$$0 = \frac{V_2^2 - V_1}{2} + h_2 - h_1$$

Assuming air to be an ideal gas with constant cp we have

$$c_p(T_1 - T_2) = \frac{V_2^2 - V_1^2}{2}$$

We can find T_2 from Eq. 17.3.35 to be

$$T_2 = T_1(P_2 / P_1)^{k-1/k}$$
$$= 673(20 / 400)^{0.4/1.4} = 286\text{K}$$

The exiting velocity is found as follows:

$$1000(673 - 286) = \frac{V_2^2 - 100^2}{2}$$
$$\therefore V_2 = 885\text{m/s}$$

Note: cp must be used as 1000 J/kg· K so that the units are consistent.

17.4 **Electrical Circuits**

Circuits

Electric circuits are an interconnection of electrical components for the purpose of either generating and distributing electrical power; converting electrical power to some other useful form such as light, heat, or mechanical torque; or processing information contained in an electrical form (i.e., electrical signals). Most electrical circuits contain a source (or sources) of electrical power, passive components that store or dissipate energy, and possibly active components that change the electrical form of the energy or information being processed by the circuit.

Circuits may be classified as **direct current** (DC) circuits when the currents and voltages do not vary with time or as **alternating current** (AC) circuits when the currents and voltage vary sinusoidally with time. Both DC and AC circuits are said to be operating in the **steady state** when their respective current/voltage time variation is purely constant or purely sinusoidal with time. A **transient circuit**

Table 17.1 Quantities Used in Electric Circuits

Quantity	Symbol	Unit	Defining Equation	Definition
Charge	Q	Coulomb $C = $ Amperes	$Q = \int I \, dt$	
Current	I	A, ampere	$I = \dfrac{dQ}{dt}$	Time rate of flow of charge past a point in the circuit
Voltage	V	Volt $V = $ Watts/Ampere	$V = \dfrac{dW}{dQ}$	Energy per unit charge either gained or lost through a circuit element
Energy	W	Joule $J = $ N·m	$W = \int V \, dQ = \int P \, dt$	
Power	P	Watt $W = $ J/s	$P = \dfrac{dW}{dt} = IV$	Time rate of energy flow

condition occurs when a switch is thrown that turns a source either on or off. This review will cover the DC steady state. The primary quantities of interest in making circuit calculations are presented in Table 17.1. For more details regarding units, please refer to Chapter 15. The ampere is the only fundamental unit presented in Table 17.1, from which the coulomb and volt are derived. The Q symbol here is not to be confused with the symbol for heat transfer, which itself carries a unit of energy.

Circuit Components

The circuits reviewed in this section will contain one or more sources interconnected with passive components. These passive circuit components include resistors, inductors, and capacitors.

a) **Resistors** absorb energy and have a resistance value R measured in ohms:

$$I = \frac{V}{R}$$

$$\text{AMPERES} = \frac{\text{VOLTS}}{\text{OHMS}}$$

(17.4.1)

b) **Inductors** store energy and have an inductance value L measured in henries:

$$V = L\frac{dI}{dt}$$

$$\text{VOLTS} = \frac{\text{AMPERES} \cdot \text{HENRIES}}{\text{SECONDS}}$$

(17.4.2)

c) **Capacitors** store energy and have a capacitance value C measured in farads:

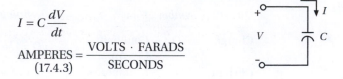

$$I = C\frac{dV}{dt}$$

$$\text{AMPERES} = \frac{\text{VOLTS} \cdot \text{FARADS}}{\text{SECONDS}}$$

$$(17.4.3)$$

Sources of Electrical Energy

Sources in electric circuits can be either independent of current and/or voltage values elsewhere in the circuit, or they can be dependent upon them. In this section only independent sources will be considered. Figure 17.7 shows both ideal and linear models for current and voltage sources.

Kirchhoff's Laws

Two laws of conservation govern the behavior of all electrical circuits:

a) **Kirchhoff's voltage law** (KVL), for the conservation of energy, states that the sum of voltage rises or drops around any closed path in an electrical circuit must be zero:

$$\sum_{DROPS}^{V} = 0 \quad \sum_{RISES}^{V} = 0 \,(\text{around closed path}) \tag{17.4.4}$$

b) **Kirchhoff's current law** (KCL), for the conservation of charge, states that the flow of charges either into (positive) or out of (negative) any node in a circuit must add to zero:

$$\sum_{IN}^{I} = 0 \quad \sum_{OUT}^{I} = 0 \,(\text{at node}) \tag{17.4.5}$$

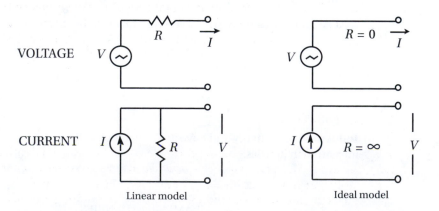

Figure 17.7 Ideal and linear models of current and voltage sources

Ohm's Law

Ohm's law is a statement of the relationship between the voltage across an electrical component and the current through the component. For DC circuits, where the components are resistors, Ohm's law is

$$V = IR \text{ or } I = V / R \tag{17.4.6}$$

For AC circuits, with resistors, capacitors, and inductors, Ohm's law, stated in terms of the component impedance Z, is

$$V = IZ \text{ or } I = V / Z \tag{17.4.7}$$

where all variables can be complex (real and imaginary parts).

Reference Voltage Polarity and Current Direction

Circuit analysis requires defining first a reference current direction with an arrow placed next to the circuit component. For each of the components a reference current direction is arbitrarily defined. Once the current reference direction is defined, the voltage reference polarity marks can be placed on each component. The polarity marks on passive components are always placed so that the current flows from the plus (+) mark to the minus (−) mark, the passive sign convention.

Current values can be either positive or negative. A positive current value shows that the current does in fact flow in the reference direction. A negative current value shows that the current flows opposite to the reference direction. Voltage values can be either positive or negative. A positive voltage value indicates a loss of energy or reduction in voltage when moving through the circuit from the plus polarity mark to the minus polarity mark. A negative voltage value indicates a gain of energy when moving through the circuit from the plus polarity mark to the minus polarity mark.

It is proper electrical terminology to talk about voltage drops and voltage rises. A voltage drop is experienced when moving through the circuit from the plus (+) polarity mark to the minus (−) polarity mark. A voltage rise is experienced when moving through the circuit from the minus (−) polarity mark to the plus (+) polarity mark.

Circuit Equations

When writing circuit equations, the current is assumed to have a positive value in the reference direction and the voltage is assumed to have a positive value as indicated by the polarity marks. To write the KVL circuit equation one must move around a closed path in the circuit and sum all the voltage rises or all the voltage drops. For example, for the circuit in Figure 17.8 begin at point a and move to b, then c, then d, and back to a. For $\sum_{RISES}^{V} = 0$ obtain

$$V_s - IR_1 - IR_2 - IR_3 = 0 \qquad\qquad (17.4.8)$$

For $\sum_{DROPS}^{V}$ obtain

$$-V_s + IR_1 + IR_2 + IR_3 = 0 \qquad\qquad (17.4.9)$$

Either of these equations can now be solved for the one unknown current

$$I = \frac{V_s}{R_1 + R_2 + R_3} = \frac{V_s}{R_{eq}} \qquad\qquad (17.4.10)$$

where R_{eq} is the equivalent resistance for the circuit.

Circuit Equations Using Branch Currents

The circuit in Figure 17.9 has meshes around which two voltage equations (KVL) can be written. There are, however, three branches in the circuit. An unknown current with a reference direction is assumed in each branch. The polarity marks are

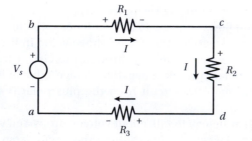

Figure 17.8 A simple circuit

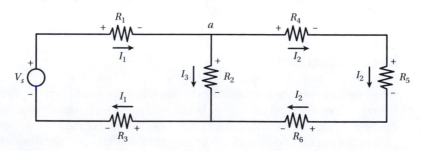

Figure 17.9 A circuit with two meshes and three branches

indicated for each resistor so the KVL can be written. Write two KVL equations, one around each mesh. Using $\Sigma_{DROPS}^V = 0$ obtain:

$$-V_s + I_1 R_1 + I_3 R_2 + I_1 R_3 = 0$$

$$-I_3 R_2 + I_2 R_4 + I_2 R_5 + I_2 R_6 = 0 \qquad (17.4.11)$$

Write one KCL equation at circuit node a. This additional equation is necessary since there are three unknown currents:

$$I_1 - I_2 - I_3 = 0 \qquad (17.4.12)$$

These three equations can be solved for I_1, I_2, and I_3. The current I_1 is (see Eq. 16.5.8)

$$
I_1 = \frac{\begin{vmatrix} V_s & 0 & R_2 \\ 0 & R_4 + R_5 + R_6 & -R_2 \\ 0 & -1 & -1 \end{vmatrix}}{\begin{vmatrix} R_1 + R_3 & 0 & R_2 \\ 0 & R_4 + R_5 + R_6 & -R_2 \\ 1 & -1 & -1 \end{vmatrix}} \qquad (17.4.13)
$$

Circuit Equations Using Mesh Currents

A simplification in writing the circuit equations for Figure 17.9 occurs if mesh currents are used. Note that

$$I_3 = I_1 - I_2 \qquad (17.4.14)$$

Redefine the reference currents in the network of Figure 17.9 as shown in Figure. 17.10. Now there are only two unknown currents to solve for instead of the three in Figure 17.9. The current through R_1 and R_3 is I_1. The current through R_4, R_5, and

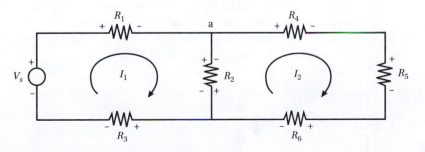

Figure 17.10 A circuit with two meshes

R_6 is I_2. The current through R_2 is $I_1 - I_2$, which is consistent with the KCL equation written for the network of Figure 17.9. Write two KVL equations, one around each mesh:

$$-V_s + I_1 \left(R_1 + R_2 + R_3 \right) - I_2 R_2 = 0$$
$$-I_1 R_2 + I_2 \left(R_2 + R_4 + R_5 + R_6 \right) = 0 \tag{17.4.15}$$

These two equations can be solved for I_1 and I_2. The current I_1 is equivalent to that of Eq. 17.4.13:

$$I_1 = \dfrac{\begin{vmatrix} V_s & -R_2 \\ 0 & R_2 + R_4 + R_5 + R_6 \end{vmatrix}}{\begin{vmatrix} R_1 + R_2 + R_3 & -R_2 \\ -R_2 & R_2 + R_4 + R_5 + R_6 \end{vmatrix}} \tag{17.4.16}$$

Circuit Simplification

It is possible to simplify a circuit by combining components of the same kind that are grouped together in the circuit. The formulas for combining several R's to a single R, several L's to a single L, and several C's to a single C are found using Kirchhoff's laws (see Eq.17.4.10). Consider two inductors in series as shown in Figure 17.11. We see that $L_{eq} = L_1 + L_2$. Combinations of circuit components are summarized in Table 17.2.

$$V - L_1 \frac{dI}{dt} - L_2 \frac{dI}{dt} = 0$$
$$V - \left(L_1 + L_2 \right) \frac{dI}{dt} = 0$$
$$V - L_{eq} \frac{dI}{dt} = 0$$

DC Circuits

In a DC circuit the only crucial components are resistors. Another component, the inductor, appears as a zero resistance connection (or short circuit) and a third component, a capacitor, appears as an infinite resistance, or open circuit. The three circuit components are summarized in Table 17.3.

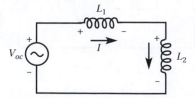

Figure 17.11 Circuit simplification

Table 17.2 Circuit Components in Series and Parallel

Component	Series	Parallel
R	$R_{eq} = R_1 + R_2 + \ldots + R_N$	$\dfrac{1}{R_{eq}} = \dfrac{1}{R_1} + \dfrac{1}{R_2} + \ldots + \dfrac{1}{R_N}$
L	$L_{eq} = L_1 + L_2 + \ldots + L_N$	$\dfrac{1}{L_{eq}} = \dfrac{1}{L_1} + \dfrac{1}{L_2} + \ldots + \dfrac{1}{L_N}$
C	$\dfrac{1}{C_{eq}} = \dfrac{1}{C_1} + \dfrac{1}{C_2} + \ldots + \dfrac{1}{C_N}$	$C_{eq} = C_1 + C_2 + \ldots + C_N$

Table 17.3 DC Circuit Components

Component		Impedance	Current	Power	Energy Stored
Resistor		R	$I = V / R$	$P = I^2 R = V^2 / R$	None stored
Inductor		(Short Circuit)	Unconstrained	None dissipated	$W_L = \dfrac{1}{2} L I^2$
Capacitor		Infinite (Open Circuit)	Zero	None dissipated	$W_C = \dfrac{1}{2} C V^2$

EXAMPLE 17.20

Compute the current in the 10Ω resistor.

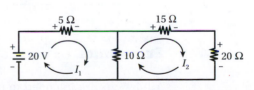

Solution: Assume loop currents I_1 and I_2. Write KVL around both meshes:

$$\sum_{DROPS}^{V} = 0$$

Mesh 1 : $-20 + 5I_1 + 10I_1 - 10I_2 = 0$

Mesh 2 : $-10I_1 + 10I_2 + 15I_2 + 20I_2 = 0$

These are arranged as

$$15I_1 - 10I_2 = 20$$
$$-10I_1 + 45I_2 = 0$$

The solution is

$$I_1 = 1.57\,A, I_2 = 0.35\,A$$

The current in the 10Ω resistor is

$$I = I_1 - I_2$$
$$= 1.57 - 0.35 = 1.22\,A$$

EXAMPLE 17.21

Compute the power delivered to the 6Ω resistor.

Solution: Only one current path exists since no DC current flows through a capacitor; the voltage drop across the inductor is zero. Write KVL:

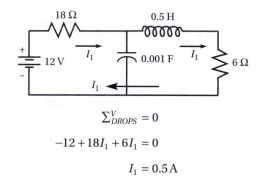

$$\sum_{DROPS}^{V} = 0$$

$$-12 + 18I_1 + 6I_1 = 0$$

$$I_1 = 0.5\,A$$

The power is then

$$P = I^2 R$$
$$= 0.5^2 \times 6 = 1.5W$$

17.5 Economics

All designs are intended to produce good results. In general, the good results are accompanied by unavoidable and/or undesirable effects, including the costs of manufacturing or construction. Selecting the best design requires the designer to anticipate and compare the good and bad outcomes. If outcomes are evaluated in dollars, and if "good" is defined as positive monetary value, then design decisions may be guided by an economic analysis.

Value and Interest

"Value" is not synonymous with "amount." The value of an amount of money depends on when the amount is received or spent. For example, the promise that you will be given a dollar one year from now is of less value to you than a dollar received today. The difference between the anticipated amount and its current value is called interest and is frequently expressed as a time rate. If an interest rate of 10 percent per year is used, the expectation of receiving $1.00 one year hence has a value now of about $0.91. In economics, interest usually is stated in percent per year. If no time unit is given, "per year" is assumed.

EXAMPLE 17.22   What amount must be paid in two years to settle a current debt of $1,000 if the interest rate is 6 percent?

Solution: Value after one year

$$= \$1000 + \$1000 \times 0.06$$
$$= \$1000(1 + 0.06)$$
$$= \$1060$$

Value after two years

$$= 1060 + 1060 \times 0.06$$
$$= 1000(1 + 0.06)^2$$
$$= \$1124$$

Hence, $1,124 must be paid in two years to settle the debt.

Cash Flow Diagrams

As an aid to analysis and communication, an economics problem may be represented graphically by a horizontal time axis and vertical vectors representing dollar amounts. The cash flow diagram for Example 17.22 is sketched in Figure 17.12. Income vectors point up and expenditure vectors point down. It is important to pick a point of view and stick with it. For example, the vectors in Figure 17.12 would have

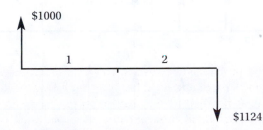

Figure 17.12 Cash flow diagram for Example 17.22

been reversed if the point of view of the lender had been adopted. It is a good idea to draw a cash flow diagram for every economics problem that involves amounts occurring at different times.

In economics, amounts are almost always assumed to occur at the ends of years. Consider, for example, the value today of the future operating expenses of a truck. The costs probably will be paid in varied amounts scattered throughout each year of operation, but for computational ease the expenses in each year are represented by their sum (computed without consideration of interest) occurring at the end of the year. The error introduced by neglecting interest for partial years is usually insignificant compared to uncertainties in the estimates of future amounts.

Cash Flow Patterns

Economics problems involve the following four patterns of cash flow, both separately and in combination (Fig. 17.13).

P-pattern:	A single amount *P* occurring at the beginning of *n* years. *P* frequently represents *present* amounts.
F-pattern:	A single amount *F* occurring at the end of *n* years. *F* frequently represents *future* amounts.
A-pattern:	Equal amounts *A* occurring at the ends of each of *n* years. The *A*-pattern frequently is used to represent *annual* amounts.
G-pattern:	End-of-year amounts increasing by an equal annual gradient *G*. Note that the first amount occurs at the end of the second year. *G* is the abbreviation of "gradient."

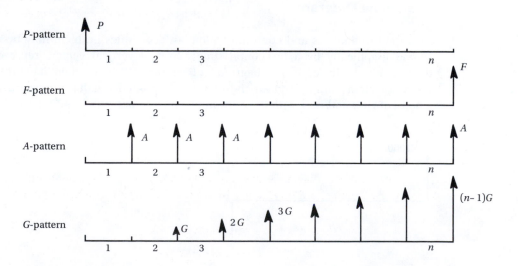

Figure 17.13 Four cash flow patterns

Equivalence of Cash Flow Patterns

Two cash flow patterns are said to be equivalent if they have the same value. Most of the computational effort in economics problems is directed at finding a cash flow pattern that is equivalent to a combination of other patterns. Example 17.22 can be thought of as finding the amount in an F-pattern that is equivalent to $1,000 in a P-pattern. The two amounts are proportional, and the factor of proportionality is a function of interest rate i and number of periods n. There is a different factor of proportionality for each possible pair of the cash flow patterns defined in the next section. To minimize the possibility of selecting the wrong factor, mnemonic symbols are assigned to the factors. For Example 17.23, the proportionality factor is written and solution is achieved by evaluating

$$F = \left(F / P\right)_n^i P.$$

To analysts familiar with the canceling operation of algebra, it is apparent that the correct factor has been chosen. However, the letters in the parentheses together with the subscripts and superscripts constitute a single symbol; therefore, the canceling operation is not actually performed. Table 17.4 lists symbols and formulas for commonly used factors. Table 17.5, located at the end of this chapter, presents a convenient way to find numerical values of interest factors. Those values are tabulated for selected interest rates i and number of interest periods n; linear interpolation for intermediate values of i and n is acceptable for most situations.

EXAMPLE 17.23   A new widget twister, with a life of six years, would save $2,000 in production costs each year. Using a 12 percent interest rate, determine the highest price that could be justified for the machine. Although the savings occur continuously throughout each year, follow the usual practice of lumping all amounts at the ends of years.

Solution: First, sketch the cash flow diagram.

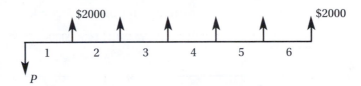

The cash flow diagram indicates that an amount in a P-pattern must be found that is equivalent to $2,000 in an A-pattern. The corresponding equation is

$$P = \left(P / A\right)_n^i A$$
$$= \left(P / A\right)_6^{12\%} 2000$$

Table 17.4 Formulas for Interest Factors

Symbol	To Find	Given	Formula
$(F/P)_n^i$	F	P	$(1+i)^n$
$(P/F)_n^i$	P	F	$\dfrac{1}{(1+i)^n}$
$(A/P)_n^i$	A	P	$\dfrac{i(1+i)^n}{(1+i)^n-1}$
$(P/A)_n^i$	P	A	$\dfrac{(1+i)^n-1}{i(1+i)^n}$
$(A/F)_n^i$	A	F	$\dfrac{i}{(1+i)^n-1}$
$(F/A)_n^i$	F	A	$\dfrac{(1+i)^n-1}{i}$
$(A/G)_n^i$	A	G	$\dfrac{1}{i}\left(\dfrac{n}{(1+i)^n-1}\right)$
$(F/G)_n^i$	F	G	$\dfrac{1}{i}\left[\dfrac{(1+i)^n-1}{i}-n\right]$
$(P/G)_n^i$	P	G	$\dfrac{1}{i}\left[\dfrac{(1+i)^n-1}{i(1+i)^n}-\dfrac{n}{(1+i)^n}\right]$

Table 17.5 is used to evaluate the interest factor for $i = 12$ percent and $n = 6$:

$$P = 4.1114 \times 2000$$
$$= \$8223$$

EXAMPLE 17.24

How soon does money double if it is invested at 8 percent interest?

Solution: Obviously, this is stated as

$$F = 2P.$$

Therefore,

$$(F/P)_n^{8\%} = 2$$

In the 8 percent interest table, the tabulated value for (F/P) that is closest to 2 corresponds to $n = 9$ years.

EXAMPLE
17.25

Find the value in 2002 of a bond described as "Acme 8% of 2015" if the rate of return set by the market for similar bonds is 10 percent.

Solution: The bond description means that the Acme Company has an outstanding debt that it will repay in the year 2015. Until then, the company will pay out interest on that debt at the 8 percent rate. Unless otherwise stated, the principal amount of a single bond is $1,000. If it is assumed that the debt is due December 31, 2015, interest is paid every December 31, and the bond is purchased January 1, 2002, then the cash flow diagram, with unknown purchase price P, is:

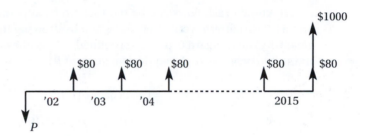

The corresponding equation is

$$P = (P/A)_{14}^{10\%} \, 80 + (P/F)_{14}^{10\%} \, 1000$$

$$= 7.3667 \times 80 + 0.2633 \times 1000$$

$$= \$853$$

That is, to earn 10 percent the investor must buy the 8 percent bond for $853, a discount of $147. Conversely, if the market interest rate is less than the nominal rate of the bond, the buyer will pay a premium over $1,000.

The solution is approximate because bonds usually pay interest semiannually, and $80 at the end of the year is not equivalent to $40 at the end of each half-year—but the error is small and is neglected.

EXAMPLE
17.26

You are buying a new appliance. From past experience you estimate future repair costs as:

First Year	$5
Second Year	$15
Third Year	$25
Fourth Year	$35

The dealer offers to sell you a four-year warranty for $60. You require at least a 6 percent interest rate on your investments. Should you invest in the warranty?

Solution: Sketch the cash flow diagram.

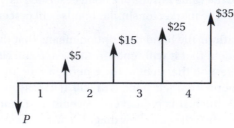

The known cash flows can be represented by superposition of a $5 *A*-pattern and a $10 *G*-pattern. Verify that statement by drawing the two patterns. Now it is clear why the standard *G*-pattern is defined to have the first cash flow at the end of the second year. Next, the equivalent amount *P* is computed:

$$P = (P/A)_4^{6\%} A + (P/G)_4^{6\%} G$$

$$= 3.4651 \times 5 + 4.9455 \times 10$$

$$= \$67$$

Since the warranty can be purchased for less than $67, the investment will earn a rate of return greater than the required 6 percent. Therefore, you should purchase the warranty.

If the required interest rate had been 12 percent, the decision would be reversed. This demonstrates the effect of a required interest rate on decision making: increasing the required rate reduces the number of acceptable investments.

EXAMPLE 17.27   Compute the annual equivalent maintenance costs over a 5-year life of a laser printer that is warranted for two years and has estimated maintenance costs of $100 annually. Use $i = 10$ percent.

Solution: The cash flow diagram appears as:

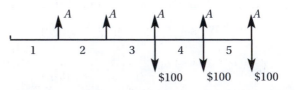

There are several ways to find the 5-year *A*-pattern equivalent to the given cash flow. One of the more efficient methods is to convert the given 3-year *A*-pattern to an *F*-pattern, and then find the 5-year *A*-pattern that is equivalent to that *F*-pattern. That is,

$$A = (A / F)_5^{10\%} (F / A)_3^{10\%} 100$$
$$= \$54$$

Unusual Cash Flows and Interest Periods

Occasionally an economics problem will deviate from the year-end cash flow and annual compounding norm. The examples in this section demonstrate how to handle these situations.

**EXAMPLE
17.28**

PAYMENTS AT BEGINNINGS OF YEARS

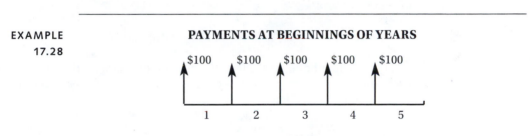

Using a 10 percent interest rate, find the future equivalent of:

Solution: Shift each payment forward one year. Therefore,

$$A = (F / P)_1^{10\%} 100 = \$110$$

This converts the series to the equivalent *A*-pattern:

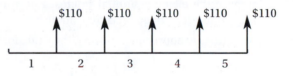

and the future equivalent is found to be

$$F = (F / A)_5^{10\%} 110 = \$672$$

Alternative Solution: Convert to a six-year series:

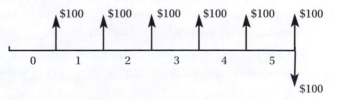

The future equivalent is

$$F = (F/A)_6^{10\%} 100 - 100 = \$672$$

EXAMPLE
17.29 **ANNUAL PAYMENTS WITH INTEREST COMPOUNDED m TIMES PER YEAR**
Compute the effective annual interest rate equivalent to 5 percent nominal annual interest compounded daily. (There are 365 days in a year.)

Solution: The legal definition of nominal annual interest is

$$i_n = mi$$

where i is the interest rate per compounding period. For the example,

$$i = i_{n/m}$$
$$= 0.05 / 365 = 0.000137 \text{ or } 0.0137 \text{ per day}$$

Because of compounding, the effective annual rate is greater than the nominal rate. By equating (F/P)-factors for one year and m periods, the effective annual rate i_e may be computed as follows:

$$(1 + i_e)^1 = (1 + i)^m$$
$$i_e = (1 + i)^m - 1$$
$$= (1.000137)^{365} - 1 = 0.05127 \text{ or } 5.127\%$$

EXAMPLE
17.30 **CONTINUOUS COMPOUNDING**
Compute the effective annual interest rate i_e equivalent to 5 percent nominal annual interest compounded continuously.

Solution: As m approaches infinity, the value for i_e is found as follows:

$$i_e = e^i - 1$$
$$= e^{0.05} - 1$$
$$= 0.051271 \text{ or } 5.1271\%$$

EXAMPLE
17.31 **ANNUAL COMPOUNDING WITH m PAYMENTS PER YEAR**
Compute the year-end amount equivalent to 12 end-of-month payments of $10 each. Annual interest rate is 6 percent.

Solution: The usual simplification in economics is to assume that all payments occur at the end of the year, giving an answer of $120. This approximation may not be acceptable for a precise analysis of a financial agreement. In such cases, the agreement's policy on interest for partial periods must be investigated.

EXAMPLE   **ANNUAL COMPOUNDING WITH PAYMENT EVERY m YEARS**
17.32   With interest at 10 percent, compute the present equivalent of

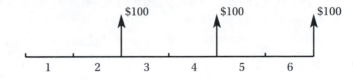

Solution: First convert each payment to an A-pattern for the m preceding years. That is,

$$A = (A/F)_2^{10\%} 100$$
$$= \$47.62$$

Then, convert the A-pattern to a P-pattern:

$$P = (P/A)_6^{10\%} 47.62$$
$$= \$207$$

EXERCISES AND ACTIVITIES

Statics

17.1   Find the component of $A = 200$ in the direction of $B = 100$.

a) 158
b) 143
c) 129
d) 115

17.2   Find the magnitude of the resultant of $A = 100$, $B = 50$, and $C = 120$.

a) 194
b) 202
c) 226
d) 275

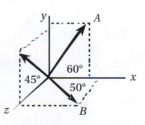

17.3   Determine the magnitude of the moment about the y-axis of the force $F = 500$ ($F_x = 300$, $F_y = 200$, $F_z = ?$) acting at $(4, -6, 4)$.

a) 186

b) 1385

c) 2580

d) 3185

17.4   Find the magnitude of the moment of the two forces F_1 ($F_x = 50$, $F_y = 0$, $F_z = 40$) and F_2 ($F_x = 0$, $F_y = 60$, $F_z = 80$) acting at $(2, 0, -4)$ and $(-4, 2, 0)$, respectively, about the x-axis.

a) 0

b) 80

c) 160

d) 240

17.5   The force system shown may be referred to as being

a) non-concurrent, non-coplanar

b) coplanar

c) parallel

d) two-dimensional

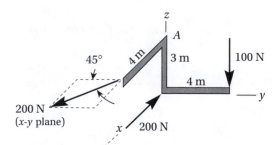

17.6   Equilibrium exists due to a rigid support at A in the figure of Prob. 17.5. Find the magnitude of the force that exists at A.

a) 153 N

b) 173 N

c) 257 N

d) 382 N

17.7   Equilibrium exists on the object in Prob. 17.5. Find the magnitude of the reactive moment that exists at support A.

a) 915 N · m

b) 691 N · m

c) 862 N · m

d) 721 N · m

17.8   If three nonparallel forces hold a rigid body in equilibrium, they must

a) be equal in magnitude

b) be concurrent

c) be non-concurrent

d) form an equilateral triangle

17.9   Find the magnitude of the reaction at support B.

a) 400 N

b) 500 N

c) 600 N

d) 700 N

17.10 What moment M exists at support A?

a) 5600 N · m
b) 5000 N · m
c) 4400 N · m
d) 4000 N · m

17.11 Calculate the reactive force at support A.

a) 250 N
b) 350 N
c) 450 N
d) 550 N

17.12 Find the support moment at A.

a) 66 N · m
b) 77 N · m
c) 88 N · m
d) 99 N · m

17.13 Calculate the magnitude of the equilibrating force at A for the three-force body shown.

a) 217 N
b) 287 N
c) 343 N
d) 385 N

17.14 Find the magnitude of the equilibrating force at point A.

a) 187 N
b) 142 N
c) 114 N
d) 84 N

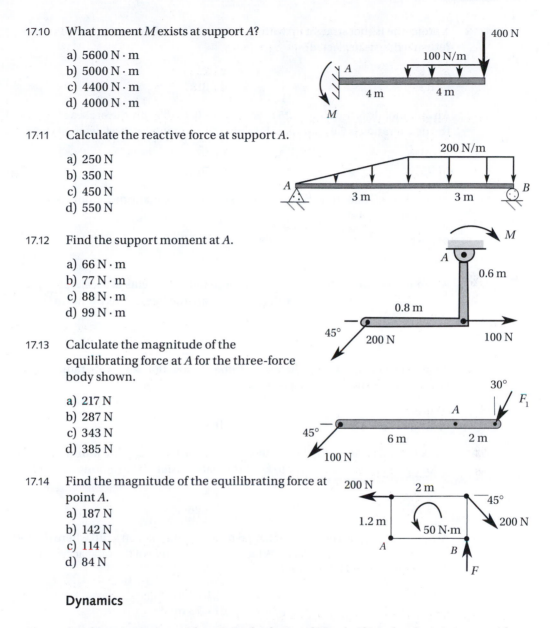

Dynamics

17.15 An object is moving with an initial velocity of 20 m/s. If it is decelerating at 5 m/s² how far does it travel before it stops?

a) 10 m
b) 20 m

c) 30 m
d) 40 m

17.16   A projectile is shot straight up with v_o = 40 m/s. After how many seconds will it return if drag is neglected?

a) 4 s
b) 6 s

c) 8 s
d) 10 s

17.17   An automobile is traveling at 25 m/s². It takes 0.3 s to apply the brakes after which the deceleration is 6.0 m/s². How far does the automobile travel before it stops?

a) 40 m
b) 45 m

c) 50 m
d) 60 m

17.18   A wheel accelerates from rest with α = 6 rad/s². How many revolutions are experienced in 4 s?

a) 7.64
b) 9.82

c) 12.36
d) 25.6

17.19   A 2-m-long shaft rotates about one end at 20 rad/s. It begins to accelerate with α = 10 rad/s². After how long will the velocity of the free end reach 100 m/s?

a) 3 s
b) 4 s

c) 5 s
d) 6 s

17.20   A roller-coaster reaches a velocity of 20 m/s at a location where the radius of curvature is 40 m. Calculate the acceleration, in m/s².

a) 8
b) 9

c) 10
d) 12

17.21   A bucket full of water is to be rotated in the vertical plane. What minimum angular velocity, in rad/s, is necessary to keep the water inside if the rotating arm is 120 cm?

a) 2.86
b) 3.15

c) 3.86
d) 4.26

17.22   An automobile is accelerating at 5 m/s² on a straight road on a hill where the radius of curvature of the hill is 200 m. What is the magnitude of the total acceleration when the car's speed is 30 m/s?

a) 5 m/s²
b) 5.46 m/s²

c) 6.04 m/s²
d) 6.73 m/s²

17.23   A particle experiences the displacement shown. What is its velocity at t = 1 s?

a) −0.58 m/s
b) −0.76 m/s
c) −0.92 m/s
d) −1.0 m/s

17.24 Neglecting the change of gravity with elevation, estimate the speed a satellite must have to orbit the earth at an elevation of 100 km. Earth's radius = 6400 km.

a) 4000 m/s

b) 6000 m/s

c) 8000 m/s

d) 10 000 m/s

17.25 Find an expression for the maximum range of a projectile with initial velocity v_o at angle θ with the horizontal.

a) $\dfrac{v_o^2}{g}\sin 2\theta$

b) $\dfrac{v_o^2}{2g}\sin^2\theta$

c) $\dfrac{v_o^2}{2g}\sin\theta$

d) $\dfrac{v_o^2}{g}\cos\theta$

17.26 Find the maximum height, in meters.

a) 295

b) 275

c) 255

d) 235

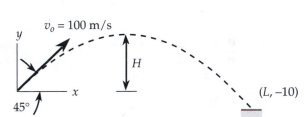

17.27 Calculate the time, in seconds, it takes the projectile of Problem 17.26 to reach the low point.

a) 14.6

b) 12.2

c) 11.0

d) 10.2

17.28 What is the distance L, in meters, in Problem 17.26?

a) 530

b) 730

c) 930

d) 1030

17.29 What is a_A?

a) 2.09 m/s²

b) 1.85 m/s²

c) 1.63 m/s²

d) 1.47 m/s²

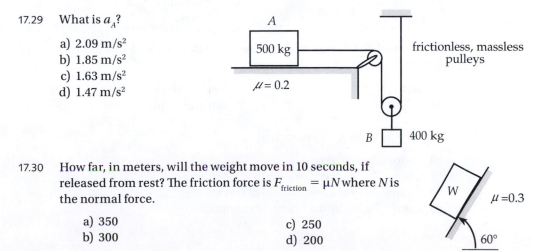

17.30 How far, in meters, will the weight move in 10 seconds, if released from rest? The friction force is $F_{friction} = \mu N$ where N is the normal force.

a) 350

b) 300

c) 250

d) 200

17.31   At what angle, in degrees, should a road be slanted to prevent an automobile traveling at 25 m/s from tending to slip? The radius of curvature is 200 m.

a) 22   c) 18
b) 20   d) 16

17.32   A satellite orbits the Earth 200 km above the surface. What speed, in m/s, is necessary for a circular orbit? The radius of the Earth is 6400 km and $g = 9.2$ m/s^2

a) 7800   c) 6600
b) 7200   d) 6000

17.33   Determine the mass of the Earth, in kg, if the radius of the Earth is 6400 km.

a) 6×10^{22} c) $6 3 \, 10^{24}$
b) 6×1023 d) 6×10^{25}

17.34   The coefficient of sliding friction between rubber and asphalt is about 0.6. What minimum distance, in meters, can an automobile slide on a horizontal surface if it is traveling at 25 m/s?

a) 38   c) 48
b) 43   d) 53

Thermodynamics

17.35   Which would be considered a system rather than a control volume?

a) a pump   c) a pressure cooker
b) a tire   d) a turbine

17.36   Which of the following is an extensive property?

a) temperature   c) pressure
b) velocity   d) mass

17.37   An automobile heats up while sitting in a parking lot on a sunny day. The process can be assumed to be

a) isothermal   c) isometric
b) isobaric   d) isentropic

17.38   A scientific law is a statement that

a) we postulate to be true
b) is generally observed to be true
c) is a summary of experimental observation
d) is agreed upon by the scientific community

17.39   In a quasi-equilibrium process involving a system:

a) The pressure remains constant.
b) The pressure varies with temperature.
c) The pressure is constant at an instant at each point in the system.
d) The pressure force results in a positive work input.

17.40   The density of air at vacuum of 40 kPa and –40°C is, in kg/m³,

a) 0.598   c) 0.897
b) 0.638   d) 0.753

17.41   A cold tire has a volume of 0.03 m³ at –10°C and 180 kPa gauge. If the pressure and temperature increase to 210 kPa gauge and 30°C, find the final volume in m³.

a) 0.0304   c) 0.0312
b) 0.0308   d) 0.0316

17.42   A 300-watt light bulb provides energy in a 10-m-diameter spherical space. If the outside temperature is 20°C, find the inside steady-state temperature, in °C, if $R = 1.5\ \mathrm{hr} \cdot \mathrm{m}^2 \cdot °C/kj$ for the wall.

a) 40.5   c) 32.6
b) 35.5   d) 25.2

17.43   Select a correct statement of the first law.

a) Heat transfer equals the work done for a process.
b) Net heat transfer equals the net work for a cycle.
c) Net heat transfer equals net work plus internal energy change for a cycle.
d) Heat transfer minus work equals change in enthalpy.

17.44   A cycle undergoes the following processes. All units are kJ. Find E_{after} for the process 1→2.

	Q	W	ΔE	E_{before}	E_{after}
1→2	20	5		10	
2→3		–5	5		
3→1	30			30	

a) 10   c) 20
b) 15   d) 25

17.45   For the cycle of Problem 17.44 find W_{3-1}.

a) 50   c) 70
b) 60   d) 80

17.46 Ten kilograms of –10°C ice is added to 100 kg of 20°C water. What is the eventual temperature, in °C, of the water? Assume an insulated container.

a) 9.2   c) 11.4
b) 10.8   d) 12.6

17.47 Select the incorrect statement. $(Q - W = \Delta U)$

a) Work and heat transfer represent energy crossing a boundary.
b) The differentials of work and heat transfer are exact.
c) Work and heat transfer are path integrals.
d) Net work and net heat transfer are equal for a cycle.

17.48 Determine the work, in kJ, necessary to compress 2 kg of air from 100 kPa to 4000 kPa if the temperature is held constant at 300°C.

a) –1210   c) –932
b) –1105   d) –812

17.49 There are 200 people in a 2000-m² room, lighted with 30 W/m². Estimate the maximum temperature increase, in °C, if the ventilation system fails for 20 min. Each person generates 400 kJ/h. The room is 3 m high.

a) 5.6   c) 8.6
b) 6.8   d) 13.4

17.50 Estimate the average c_p value, in kJ/kg · K, of a gas if 522 kJ/kg of heat are necessary to raise the temperature from 300 K to 800 K holding the pressure constant.

a) 1.000   c) 1.038
b) 1.026   d) 1.044

17.51 How much heat, in kJ, must be transferred to 10 kg of air to increase the temperature from 10°C to 230°C if the pressure is maintained constant?

a) 2200   c) 1890
b) 2090   d) 1620

17.52 A tire is pressurized to 100 kPa gauge in Michigan where $T = 0$°C. In Arizona the tire is at 70°C. Assuming a rigid tire, estimate the pressure in kPa gauge.

a) 120   c) 140
b) 130   d) 150

17.53 Air expands in an insulated cylinder from 200°C and 400 kPa to 20 kPa. Find T_2 in °C.

a) –24   c) –51
b) –28   d) –72

17.54   During an isentropic expansion of air, the volume triples. If the initial temperature is 200°C, find T_2 in °C.

a) 32   c) 16
b) 28   d) 8

17.55   Find the work, in kJ/kg, needed to compress air isentropically from 20°C and 100 kPa to 6 MPa.

a) −523   c) −423
b) −466   d) −392

17.56   What is the energy requirement, in kW, for a pump that is 75 percent efficient if it increases the pressure of 10 kg/s of water from 10 kPa to 6 MPa?

a) 60   c) 80
b) 70   d) 90

17.57   A river 60 m wide and 2 m deep flows at 2 m/s. A hydro plant develops a pressure of 300 kPa gage just before the turbine. What maximum power, in MW, is possible?

a) 72   c) 56
b) 64   d) 48

Electrical Circuits

17.58   For the circuit below, with voltage polarities as shown, KVL in equation form is

a) $\upsilon_1 + \upsilon_2 + \upsilon_3 - \upsilon_4 + \upsilon_5 = 0$
b) $-\upsilon_1 + \upsilon_2 + \upsilon_3 - \upsilon_4 + \upsilon_5 = 0$
c) $\upsilon_1 + \upsilon_2 - \upsilon_3 - \upsilon_4 + \upsilon_5 = 0$
d) $-\upsilon_1 - \upsilon_2 - \upsilon_3 + \upsilon_4 + \upsilon_5 = 0$

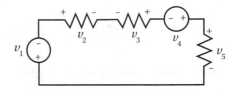

17.59   Find I_1 in amps.

a) 12
b) 15
c) 18
d) 21

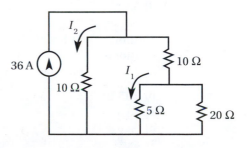

17.60   Find the magnitude and sign of the power, in watts, absorbed by the circuit element in the box.

a) −20
b) −8
c) 8
d) 12

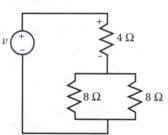

17.61   For the circuit shown with $v = 1$ V, the voltage across the 4-ohm resistor is

a) 1/4
b) 1/2
c) 2/3
d) 2

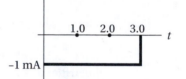

17.62   A 100µF capacitor has $I_c(t)$. The capacitor voltage $V_c(t)$ at $t = 2.5$ seconds ($V(0) = 1.0$V) is most nearly

a) −24
b) −25
c) 25
d) 26

17.63   The voltage across the 5-ohm resistor in the circuit shown is

a) 1.0
b) 2.5
c) 3.0
d) 5.83

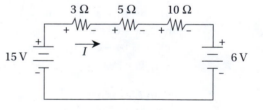

17.64   The power delivered to the 5-ohm resistor is

a) 1.5
b) 2.15
c) 2.85
d) 3.2

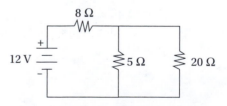

17.65 The voltage across a 10μF capacitor is $50t^2$ V. The time, in seconds, it will take to store 200 J of energy is most nearly

a) 0.15

b) 0.21

c) 1.38

d) 11.25

17.66 The equivalent resistance, in ohms, between points a and b in the circuit below is

a) 3

b) 5

c) 7

d) 8

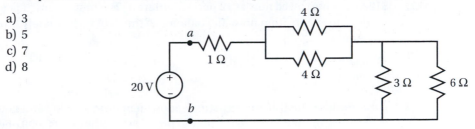

17.67 The voltage V_2 is

a) 6.4

b) 4.0

c) 2.0

d) 5.6

17.68 Find I_1 in amperes.

a) 4.0

b) 2.0

c) 4.11

d) 2.11

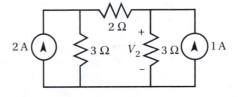

Economics

17.69 Which of the following would be most difficult to monetize?

a) maintenance cost

b) selling price

c) fuel cost

d) prestige

17.70 If $1,000 is deposited in a savings account that pays 6 percent annual interest and all the interest is left in the account, what is the account balance after three years?

a) $840

b) $1,000

c) $1,180

d) $1,191

17.71 Your perfectly reliable friend, Frank, asks for a loan and promises to pay back $150 two years from now. If the minimum interest rate you will accept is 8 percent, what is the maximum amount you will loan him?

a) $119   c) $129
b) $126   d) $139

17.72 $12,000 is borrowed now at 12 percent interest. The first payment is $4000 and is made three years from now. The balance of the debt immediately after the payment is

a) $4000   c) $12,000
b) $8000   d) $12,860

17.73 An alumnus establishes a perpetual endowment fund to help Saint Louis University. What amount must be invested now to produce income of $100,000 one year from now and at one-year intervals forever? Interest rate is 8 percent.

a) $8000   c) $1,250,000
b) $100,000   d) $10,000,000

17.74 The annual amount of a series of payments to be made at the end of each of the next 12 years is $500. What is the present worth of the payments at 8 percent interest compounded annually?

a) $500   c) $6,000
b) $3,768   d) $6,480

17.75 Consider a prospective investment in a project having a first cost of $300,000, operating and maintenance costs of $35,000 per year, and an estimated net disposal value of $50,000 at the end of 30 years. Assume an interest rate of 8 percent.
What is the present equivalent cost of the investment if the planning horizon is 30 years?

a) $670,000   c) $720,000
b) $689,000   d) $791,000

If the project replacement will have the same first cost, life, salvage value, and operating and maintenance costs as the original, what is the capitalized cost of perpetual service?

a) $670,000   c) $720,000
b) $689,000   d) $765,000

17.76 Maintenance expenditures for a structure with a 20-year life will come as periodic outlays of $1,000 at the end of the 5th year, $2,000 at the end of the 10th year, and $3,500 at the end of the 15th year. With interest at 10 percent, what is the equivalent uniform annual cost of maintenance for the 20-year period?

a) $200   c) $300
b) $262   d) $325

17.77 An alumnus has given Michigan State University $10 million to build and operate a laboratory. Annual operating cost is estimated to be $100,000. The endowment will earn 6 percent interest. Assume an infinite life for the laboratory and determine how much money may be used for its construction.

a) 5.00×10^6
b) $8.33 3 106
c) 8.72×10^6
d) 9.90×10^6

17.78 An investment pays $6000 at the end of the first year, $4000 at the end of the second year, and $2000 at the end of the third year. Compute the present value of the investment if a 10 percent rate of return is required.

a) $8333
b) $9667
c) $10,300
d) $12,000

17.79 An amount F is accumulated by investing a single amount P for n compounding periods with interest rate of i. Select the formula that relates P to F.

a) $P = F(1 + i)^{-n}$
b) $P 5 F (1 1 i)–n$
c) $P = F(1 + n)^{-i}$
d) $P = F(1 + ni)^{-1}$

17.80 At the end of each of the next 10 years, a payment of $200 is due. At an interest rate of 6 percent, what is the present worth of the payments?

a) $27
b) $200
c) $1472
d) $2000

17.81 The purchase price of an instrument is $12,000 and its estimated maintenance costs are $500 for the first year, $1500 for the second, and $2500 for the third. After three years of use the instrument is replaced; it has no salvage value. Compute the present equivalent cost of the instrument using 10 percent interest.

a) $14,070
b) $15,570
c) $15,730
d) $16,500

17.82 If an amount invested five years ago has doubled, what is the annual interest rate?

a) 15%
b) 12%
c) 10%
d) 6%

17.83 After a factory has been built near a stream, it is learned that the stream occasionally overflows its banks. A hydrologic study indicates that the probability of flooding is about 1 in 8 in any one year. A flood would cause about $20,000 in damage to the factory. A levee can be constructed to prevent flood damage. Its cost will be $54,000 and its useful life is 30 years. Money can be borrowed at 8 percent interest. If the annual equivalent cost of the levee is less than the annual expectation of flood damage, the levee should be built. The annual expectation of flood damage is $(1/8) \times 20,000 = $2,500$. Compute the annual equivalent cost of the levee.

a) $1,261
b) $1,800
c) $4,320
d) $4,800

17.84   If $10,000 is borrowed now at 6 percent interest, how much will remain to be paid after a $3,000 payment is made four years from now?

a) $7,000

b) $9,400

c) $9,625

d) $9,725

17.85   The maintenance costs associated with a machine are $2,000 per year for the first 10 years and $1,000 per year thereafter. The machine has an infinite life. If interest is 10 percent, what is the present worth of the annual disbursements?

a) $16,145

b) $19,678

c) $21,300

d) $92,136

17.86   A bank currently charges 10 percent interest compounded annually on business loans. If the bank were to change to continuous compounding, what would be the effective annual interest rate?

a) 10%

b) 10.517%

c) 12.5%

d) 12.649%

Table 17.5 Interest Tables (Compound Interest Factors)

i = 6.00%

n	(P/F)	(P/A)	(P/G)	(F/P)	(F/A)	(A/P)	(A/F)	(A/G)	n
1	0.9434	0.9434	−0.0000	1.0600	1.0000	1.0600	1.0000	−0.0000	1
2	0.8900	1.8334	0.8900	1.1236	2.0600	0.5454	0.4854	0.4854	2
3	0.8396	2.6730	2.5692	1.1910	3.1836	0.3741	0.3141	0.9612	3
4	0.7921	3.4651	4.9455	1.2625	4.3746	0.2886	0.2286	1.4272	4
5	0.7473	4.2124	7.9345	1.3382	5.6371	0.2374	0.1774	1.8836	5
6	0.7050	4.9173	11.4594	1.4185	6.9753	0.2034	0.1434	2.3304	6
7	0.6651	5.5824	15.4497	1.5036	8.3938	0.1791	0.1191	2.7676	7
8	0.6274	6.2098	19.8416	1.5938	9.8975	0.1610	0.1010	3.1952	8
9	0.5919	6.8017	24.5768	1.6895	11.4913	0.1470	0.0870	3.6133	9
10	0.5584	7.3601	29.6023	1.7908	13.1808	0.1359	0.0759	4.0220	10
11	0.5268	7.8869	34.8702	1.8983	14.9716	0.1268	0.0668	4.4213	11
12	0.4970	8.3838	40.3369	2.0122	16.8699	0.1193	0.0593	4.8113	12
13	0.4688	8.8527	45.9629	2.1329	18.8821	0.1130	0.0530	5.1920	13
14	0.4423	9.2950	51.7128	2.2609	21.0151	0.1076	0.0476	5.5635	14
15	0.4173	9.7122	57.5546	2.3966	23.2760	0.1030	0.0430	5.9260	15
16	0.3936	10.1059	63.4592	2.5404	25.6725	0.0990	0.0390	6.2794	16
17	0.3714	10.4773	69.4011	2.6928	28.2129	0.0954	0.0354	6.6240	17
18	0.3503	10.8276	75.3569	2.8543	30.9057	0.0924	0.0324	6.9597	18
19	0.3305	11.1581	81.3062	3.0256	33.7600	0.0896	0.0296	7.2867	19
20	0.3118	11.4699	87.2304	3.2071	36.7856	0.0872	0.0272	7.6051	20
21	0.2942	11.7641	93.1136	3.3996	39.9927	0.0850	0.0250	7.9151	21
22	0.2775	12.0416	98.9412	3.6035	43.3923	0.0830	0.0230	8.2166	22
23	0.2618	12.3034	104.7007	3.8197	46.9958	0.0813	0.0213	8.5099	23
24	0.2470	12.5504	110.3812	4.0489	50.8156	0.0797	0.0197	1.87951	24
25	0.2330	12.7834	115.9732	4.2919	54.8645	0.0782	0.0182	9.0722	25
26	0.2198	13.0032	121.4684	4.5494	59.1564	0.0769	0.0169	9.3414	26
28	0.1956	13.4062	132.1420	5.1117	68.5281	0.0746	0.0146	9.8568	28
30	0.1741	13.7648	142.3588	5.7435	79.0582	0.0726	0.0126	10.3422	30
∞	0.0000	16.6667	277.7778	∞	∞	0.0600	0.0000	16.667	∞

i = 8.00%

n	(P/F)	(P/A)	(P/G)	(F/P)	(F/A)	(A/P)	(A/F)	(A/G)	n
1	0.9259	0.9259	−0.0000	1.0800	1.0000	1.0800	1.0000	−0.0000	1
2	0.8573	1.7833	0.8573	1.1664	2.0800	0.5608	0.4808	0.4808	2
3	0.7938	2.5771	2.4450	1.2597	3.2464	0.3880	0.3080	0.9487	3
4	0.7350	3.3121	4.6501	1.3605	4.5061	0.3019	0.2219	1.4040	4
5	0.6806	3.9927	7.3724	1.4693	5.8666	0.2505	0.1705	1.8465	5
6	0.6302	4.6229	10.5233	1.5869	7.3359	0.2163	0.1363	2.2763	6
7	0.5835	5.2064	14.0242	1.7138	8.9228	0.1921	0.1121	2.6937	7
8	0.5403	5.7466	17.8061	1.8509	10.6366	0.1740	0.0940	3.0985	8
9	0.5002	6.2469	21.8081	1.9990	12.4876	0.1601	0.0801	3.4910	9
10	0.4632	6.7101	25.9768	2.1589	14.4866	0.1490	0.0690	3.8713	10
11	0.4289	7.1390	30.2657	2.3316	16.6455	0.1401	0.0601	4.2395	11
12	0.3971	7.5361	34.6339	2.5182	18.9771	0.1327	0.0527	4.5957	12
13	0.3677	7.9038	39.0463	2.7196	21.4953	0.1265	0.0465	4.9402	13
14	0.3405	8.2442	43.4723	2.9372	24.2149	0.1213	0.0413	5.2731	14
15	0.3152	8.5595	47.8857	3.1722	27.1521	0.1168	0.0368	5.5945	15
16	0.2919	8.8514	52.2640	3.4259	30.3243	0.1130	0.0330	5.9046	16
17	0.2703	9.1216	56.5883	3.7000	33.7502	0.1096	0.0296	6.2037	17
18	0.2502	9.3719	60.8426	3.9960	37.4502	0.1067	0.0267	6.4920	18
19	0.2317	9.6036	65.0134	4.3157	41.4463	0.1041	0.0241	6.7697	19
20	0.2145	9.8181	69.0898	4.6610	45.7620	0.1019	0.0219	7.0369	20
21	0.1987	10.0168	73.0629	5.0338	50.4229	0.0998	0.0198	7.2940	21
22	0.1839	10.2007	76.9257	5.4365	55.4568	0.0980	0.0180	7.5412	22
23	0.1703	10.3711	80.6726	5.8715	60.8933	0.0964	0.0164	7.7786	23
24	0.1577	10.5288	84.2997	6.3412	66.7648	0.0950	0.0150	8.0066	24
25	0.1460	10.6748	87.8041	6.8485	73.1059	0.0937	0.0137	8.2254	25
26	0.1352	10.8100	91.1842	7.3964	79.9544	0.0925	0.0125	8.4352	26
28	0.1159	11.0511	97.5687	8.6271	95.3388	0.0905	0.0105	8.8289	28
30	0.0994	11.2578	103.4558	10.0627	113.2832	0.0888	0.0088	9.1897	30
∞	0.0000	12.500	156.2500	∞	∞	0.0800	0.0000	12.5000	∞

i = 10.00%

n	(P/F)	(P/A)	(P/G)	(F/P)	(F/A)	(A/P)	(A/F)	(A/G)	n
1	0.9091	0.9091	−0.0000	1.1000	1.0000	1.1000	1.0000	−0.0000	1
2	0.8264	1.7355	0.8264	1.2100	2.1000	0.5762	0.4762	0.4762	2
3	0.7513	2.4869	2.3291	1.3310	3.3100	0.4021	0.3021	0.9366	3
4	0.6830	3.1699	4.3781	1.4641	4.6410	0.3155	0.2155	1.3812	4
5	0.6209	3.7908	6.8618	1.6105	6.1051	0.2638	0.1638	1.8101	5
6	0.5645	4.3553	9.6842	1.7716	7.7156	0.2296	0.1296	2.2236	6
7	0.5132	4.8684	12.7631	1.9487	9.4872	0.2054	0.1054	2.6216	7
8	0.4665	5.3349	16.0287	2.1436	11.4359	0.1874	0.0874	3.0045	8
9	0.4241	5.7590	19.4215	2.3579	13.5795	0.1736	0.0736	3.3724	9
10	0.3855	6.1446	22.8913	2.5937	15.9374	0.1627	0.0627	3.7255	10
11	0.3505	6.4951	26.3963	2.8531	18.5312	0.1540	0.0540	4.0641	11
I2	0.3186	6.8137	29.9012	3.1384	21.3843	0.1468	0.0468	4.3884	12
13	0.2897	7.1034	33.3772	3.4523	24.5227	0.1408	0.0408	4.6988	13
14	0.2633	7.3667	36.8005	3.7975	27.9750	0.1357	0.0357	4.9955	14
15	0.2394	7.6061	40.1520	4.1772	31.7725	0.1315	0.0315	5.2789	15
16	0.2176	7.8237	43.4164	4.5950	35.9497	0.1278	0.0278	5.5493	16
17	0.1978	8.0216	46.5819	5.0545	40.5447	0.1247	0.0247	5.8071	17
18	0.1799	8.2014	49.6395	5.5599	45.5992	0.1219	0.0219	6.0526	18
19	0.1635	8.3649	52.5827	6.1159	51.1591	0.1195	0.0195	6.2861	19
20	0.1486	8.5136	55.4069	6.7275	57.2750	0.1175	0.0175	6.5081	20
21	0.1351	8.6487	58.1095	7.4002	64.0025	0.1156	0.0156	6.7189	21
22	0.1228	8.7715	60.6893	8.1403	71.4027	0.1140	0.0140	6.9189	22
23	0.1117	8.8832	63.1462	8.9543	79.5430	0.1126	0.0126	7.1085	23
24	0.1015	8.9847	65.4813	9.8497	88.4973	0.1113	0.0113	7.2881	24
25	0.0923	9.0770	67.6964	10.8347	98.3471	0.1102	0.0102	7.4580	25
26	0.0839	9.1609	69.7940	11.9182	109.1818	0.1092	0.0092	7.6186	26
28	0.0693	9.3066	73.6495	14.4210	134.2099	0.1075	0.0075	7.9137	28
30	0.0573	9.4269	77.0766	17.4494	164.4940	0.1061	0.0061	8.1762	30
∞	0.0000	10.0000	100.0000	∞	∞	0.1000	0.0000	10.0000	∞

i = 12.00%

n	(P/F)	(P/A)	(P/G)	(F/P)	(F/A)	(A/P)	(A/F)	(A/G)	n
1	0.8929	0.8929	−0.0000	1.1200	1.0000	1.1200	1.0000	−0.0000	1
2	0.7972	1.6901	0.7972	1.2544	2.1200	0.5917	0.4717	0.4717	2
3	0.7118	2.4018	2.2208	1.4049	3.3744	0.4163	0.2963	0.9246	3
4	0.6355	3.073	4.1273	1.5735	4.7793	0.3292	0.2092	1.3589	4
5	0.5674	3.6048	6.3970	1.7623	6.3528	0.2774	0.1574	1.7746	5
6	0.5066	4.1114	8.9302	1.9738	8.1152	0.2432	0.1232	2.1720	6
7	0.4523	4.5638	11.6443	2.2107	10.0890	0.2191	0.0991	2.5515	7
8	0.4039	4.9676	14.4714	2.4760	12.2997	0.2013	0.0813	2.9131	8
9	0.3606	5.3282	17.3563	2.7731	14.7757	0.1877	0.0677	3.2574	9
10	0.3220	5.6502	20.2541	3.1058	17.5487	0.1770	0.0570	3.5847	10
11	0.2875	5.9377	23.1288	3.4785	20.6546	0.1684	0.0484	3.8953	11
12	0.2567	6.1944	25.9523	3.8960	24.1331	0.1614	0.0414	4.1897	12
13	0.2292	6.4235	28.7024	4.3635	28.0291	0.1557	0.0357	4.4683	13
14	0.2046	6.6282	31.3624	4.8871	32.3926	0.1509	0.0309	4.7317	14
15	0.1827	6.8109	33.9202	5.4736	37.2797	0.1468	0.0268	4.9803	15
16	0.1631	6.9740	36.3670	6.1304	42.7533	0.1434	0.0234	5.2147	16
17	0.1456	7.1196	38.6973	6.8660	48.8837	0.1405	0.0205	5.4353	17
18	0.1300	7.2497	40.9080	7.6900	55.7497	0.1379	0.0179	5.6427	18
19	0.1161	7.3658	42.9979	8.6128	63.4397	0.1358	0.0158	5.8375	19
20	0.1037	7.4694	44.9676	9.6463	72.0524	0.1339	0.0139	6.0202	20
21	0.0926	7.5620	46.8188	10.8038	81.6987	0.1322	0.0122	6.1913	21
22	0.0826	7.6446	48.5543	12.1003	92.5026	0.1308	0.0108	6.3514	22
23	0.0738	7.7184	50.1776	13.5523	104.6029	0.1296	0.0096	6.5010	23
24	0.0659	7.7843	51.6929	15.1786	118.1552	0.1285	0.0085	6.6406	24
25	0.0588	7.8431	53.1046	17.0001	133.3339	0.1275	0.0075	6.7708	25
26	0.0525	7.8957	54.4177	19.0401	150.3339	0.1267	0.0067	6.8921	26
28	0.0419	7.9844	56.7674	23.8839	190.6989	0.1252	0.0052	7.1098	28
30	0.0334	8.0552	58.7821	29.9599	241.3327	0.1241	0.0041	7.2974	30
∞	0.0000	8.333	69.4444	∞	∞	0.1200	0.0000	8.3333	∞

CHAPTER 18

The Campus Experience

18.1 Orienting Yourself to Your Campus

Webster's New World Dictionary defines "orienting" as "facing in any specific direction" or "adjusting to a situation." Both of these definitions describe the moment of truth when you enter the college campus of your choice for the first time. You have set your sights in a specific direction, and in most cases you will have to adjust to many situations that you have never before experienced. The road ahead can be fairly bumpy, but you can develop some very workable tools that will make the experience enjoyable and valuable.

This chapter introduces some of the things you can do to adjust to the college/ university experience and, more specifically, to your life as an engineer on campus. This is a transition period in your life and much will be asked of you. You will need to make numerous decisions relating to your future career. This chapter will provide you with a number of exercises that will encourage you to explore your surroundings. Hopefully, you will be creative and find many other ways to acclimate yourself to your new campus. So let's get started.

18.2 Exploring Your New Home Away from Home

At first, you might be reminded of your Scouting days when you had to learn how to read a map of some unknown territory. Well, come to think of it, that might be a good analogy. When a member of the Scouts is asked to perform an orienteering activity, it involves following a map with some particular set of instructions. Scouts must quickly acclimate themselves to their surroundings. How many seniors, or even campus employees, know all that much about their surroundings? If you make the effort to familiarize yourself with various locations on campus from the beginning, you will feel much more at home. Locating the proper offices for assistance will then be effortless. On the next page is a sample map of the campus at Michigan State University.

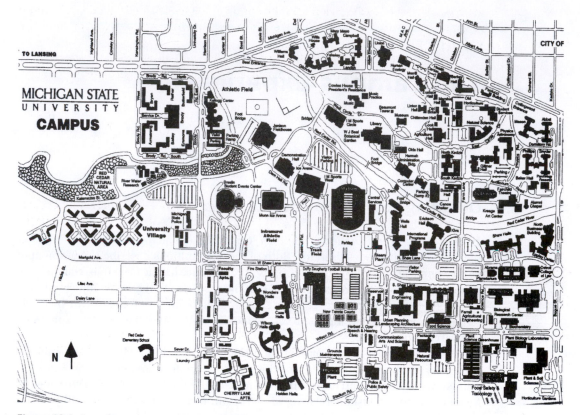

Figure 18.1 Sample campus map

18.3   Determining and Planning Your Major

There are many sources on campus to help you with the big decisions ahead of you. We are concerned here with determining your major, so we will limit our discussion to those related sources. You will need to begin making it a practice to ask as many questions as you can of those who can provide you with insightful answers. This is vital. Learn all you can from everyone about the fields that you think you might find worthwhile.

18.4   Get into the Habit of Asking Questions

Asking a multitude of questions will make life a lot easier. In life the active questioners are the ones who learn the most, who understand the course in a useful, permanent way, and who always seem to get assignments done correctly. It's not just luck! As the saying goes: "The harder I work, the luckier I get." Questioning is good work! These students make sure they understand everything, and then they move on to complete the task.

One of the most important things you can do as an engineer is to clearly convey information. You can practice this by verifying information you get from others.

18.5 The "People Issue"

Meeting People

It is sometimes very difficult to strike up a conversation with a perfect stranger, but this can be vitally important to your present coursework and to your career. You will find that working with others is one of the most important elements of the learning process. People are necessary for your success. It is only the very few who survive without constant collaboration with others. Make every effort to talk to other students in your classes. Find out what their interests are, where they're from, their abilities, and where they might be struggling. By uncovering such information you will discover opportunities for further conversation, opportunities where you can learn from them, and opportunities in which you can help them. A little pleasant conversation at the beginning of the term may lead to opportunities that spell the difference between getting a passing grade or acing the course. Making friends can also help you refine your perspective on engineering (and life) as you converse with peers, and with those who've already been through what you're just beginning (professors included).

Your Academic Advisor

When you enrolled in college you most likely found that you were assigned an academic advisor. In the early orientation period you might have had contact with this individual, but many students fail to speak with their advisor again until they prepare for graduation. Some students even find that their graduation is held up because they failed to meet one or two requirements. Advisors can help guide you in the right direction—to take specific courses and to avoid others that you're not qualified to take. They make sure that your prerequisites are fulfilled and that you take the necessary courses to adequately prepare for your career. Advisors also know a tremendous amount about the university itself. They can guide you to the right people to discuss financial aid or administrative problems. Advisors know what each course requires, and they know which instructors are good for you and which may not be.

The problem that many people encounter when entering college is a fear of asking questions—simple questions whose answers, in the long run, could greatly simplify the educational process. Questioning requires constant practice to fine-tune. Start by practicing on your advisor!

Investigating Instructors

Many students go through their academic years without engaging their instructors. Courses are taken, tests and assignments are graded, and that's it. With a little effort you can discover a great deal about your instructors and their methods of teaching.

Before blindly signing up for courses, take the time to ask questions of students who have taken the course with those instructors. Ask pertinent questions that will help you make decisions about selecting a particular instructor or opting for someone else:

- Does the instructor challenge students to really learn the material?
- Is the material covered logically?
- Can students ask questions?
- Is there an evident organization to the course?

Another method you can try is to ask to sit in on a class, or even simply sit outside a classroom and listen to what is going on. It's quite easy in such ways to get valuable info about the approach of the instructor and the setup of the course.

Networking

One of the most important things you can do early on is networking—building contacts that may be useful to you for the rest of your engineering career. Friendships can help you grow as a person, help you keep perspective in hard times, and help you avoid pitfalls. Friendships can also be helpful in the classroom. They can help with projects and homework, and can give access to a multitude of activities and opportunities that you would have never known about otherwise. This group includes the instructors in your courses, your classmates, and your neighbors in dormitories and apartments. By keeping contact with them, you will always have a group of people who can offer assistance and support, and with whom you can practice your skills by helping in return!

18.6   Searching for Campus Resources

Finding Out What's Available on Campus

Every school should have a document of some sort that lists the activities and opportunities on campus. It is your responsibility to find this document and use it. Nowadays, there may be a website that ties you into services available to you as a student. Michigan State University's website, for example, lists the following items to help a student start a search for helpful resources.

Things to Do, Places to Go

Abrams Planetarium
W. J. Beal Botanical Garden
Kellogg Hotel and Conference Center
Kresge Art Museum
MSU Museum
Wharton Center for Performing Arts
Breslin Center
Horticultural Gardens
MSU Butterfly House and Bug House
MSU Union
Michigan State University Celebrity Lecture Series
Pavilion for Agriculture and Livestock Education

What's Happening

The Academic Calendar (kept by the Office of the Registrar)
Calendar of Events (MSU News Bulletin)

Diversity Resources News and Information

The State News
MSU News Bulletin
WKAR Radio and TV
MSU Press
Common Phone Numbers
Mailing to Campus Addresses
Campus Security and Crime Awareness

Collections

W. J. Beal Botanical Garden
Michigan State University Herbarium
Botany and Plant Pathology Live Plant Collection
Campus Wood Plant Collection
Entomology Museum
Human Environment and Design Collections
MSU Museum
Kresge Art Museum
MSU Libraries Special Collections
University Archives and Historical Collections
G. Robert Vincent Voice Library

The website also informs you about:

- Library location and hours—a good place to study
- Services that will benefit you—legal aid, counseling, financial aid, advisors
- Extracurricular activities
  — Sports programs
  — Leisure time activities
  — Focused clubs: Formula Car, Hybrid Vehicle, Bridge Building, and more

18.7   Other Important Issues

If you haven't got the time to do it right, when will you have the time to do it over?

—Jeffrey Mayer

It is important that you take the initiative to look into areas where you may be slightly uncomfortable and seek information that will help you make your career preparation worthwhile. Asking questions and seeking help is truly something some people never seem to learn. For the list of topics below, we suggest some resources of which you may want to take advantage. They are simply some of the sources that you can check. The list is endless when it comes to books that you can access for help. Always keep your eyes open for resources that will help, and save any information that you find especially valuable.

Managing Time

If you have no idea where your time goes, you need to discover ways to get that time element under control.
Recommended reading: *Making Time, Making Money*—Rita Davenport; *7 Habits of Highly Successful People*—Steven Covey

The Usefulness of Reading

Engineering involves a lot of reading. You will be evaluating reports and consulting a wide range of technical texts in order to prepare projects and documents. This may be the first time you are required to digest large amounts of technical information. To prepare you for this, there are entire books devoted to teaching the art of reading. If you feel that you are a slow reader or that you have difficulty with reading comprehension, Student Services can steer you toward offices on campus that can provide information to improve your abilities. A few things to help get you started:

- Is there a synopsis at the beginning of the chapter? If there is, read it carefully.
- Are there questions at the end of the chapter? If so, read them first. This will focus your mind on what is important in the reading.

- Are there headings at the beginning of each section? If so, use these as the main points of your outline for taking notes.
- Each paragraph should have a central point and the rest of the paragraph should support that topic. Make sure that you understand each paragraph's central point. Recommended reading: *College Reading and Study Skills*—Kathleen McWhorter

Fulfilling Duties

Be proud of your abilities and aware of your obligations. It is important that you realize that you have much to offer your school. It is also vital that you realize that the years of training and the financial breaks (underwritten by tax dollars) that our society offers students oblige you to hold up your end of the deal in every way. You ought to fulfill the civic duty of your training just as soldiers fulfill the military duty of their training. It's not just a matter of money; other members of our society look up to and expect to rely on those whom it frees up to gain high-level training. So if you start to feel cynical about college, realize that in truth it is a high calling. It truly is an occasion to rise to. Use this awareness as a motivation that lasts longer than just getting an "A"! By giving, you receive—your abilities greatly improve and become clear to others. Work hard to provide quality feedback when it is asked for. You'll find your life to be more worthwhile when you create worth. To develop such helpfulness and creativity, be assertive. You have to be willing to work against inertia, to raise your hand, to volunteer, if you want to be your best. When you contribute, that's when you get what you need. So don't ever feel that you have nothing to offer. Contributing brings its own reward. When things are not going the way you think they should be, ask questions. One answer may solve a vast number of problems.

Using the Web

There's more to the Web than Google and your favorite social media sites. The World Wide Web is an invaluable tool that allows you to travel the world to discover information in seconds. Spend some time looking for specific information, not just wandering haphazardly. Just log onto a browser and travel. You'll soon see how this tool can be vital to your success, both academically and professionally.

Test-Taking Skills

Preparing outlines of the information that you have been presented in class or in books that you have been required to read for your courses will help you to focus on the course content. Team up with other members of your classes to discuss the information. Listen carefully to what the group is saying. Ask questions. The questions that you ask may clarify points that would have hindered you from doing the best job you could on a test.

Check with Student Services on campus and find out when they provide seminars or meetings on test-taking techniques. Most campuses offer these kinds of activities, and you will find them beneficial.

Recommended reading: *Proven Strategies for Successful Test Taking*—Thomas Sherman; *Effective Study Skills*—James Semone

Taking Notes

You will be required to take a mountain of notes in your four- or five-year undergraduate career. You will need to know how to organize the information and sort out what is essential from what is not. This requires specialized skill, effort, and good help.

Recommended reading: *Note-Taking Made Easy*—Kesselman-Turkel/Peterson; *Achieving Academic Success*—Kendall Hunt Publishing

Study Skills

We all need to study efficiently. Time is valuable, so you have to evaluate the effectiveness with which you study. Do you take initiative and use your time well, or do you rush to get the studying in at the last minute? Do you prepare an outline and notes to study with, or do you just pile up your books and hope to hit the right points as your fingers pass over the pages? Do you set yourself up for interruptions—or do you carefully protect your set-aside study time? Do you find studying to be tiring and stressful?

You should learn to make studying a calm, structured, routine part of your everyday life as soon as possible because you have plenty of it ahead of you! It's essential to learning, so if you find that you begrudge the time for it, try to look at it a new way. Have you thought about rewarding yourself for studying? Don't study in a way that gets you tense. Take periodic breaks for a change of pace. It's recommended to stand up, move around, and stretch a few times an hour. In fact, it is necessary for optimal study health. Find some fun things to do as a reward after you have devoted a proper amount of time to getting that studying done.

Recommended reading: *Improving Study Skills* —Conrad Lashley; *Orientation to College Learning*—Dianna Van Blerkom

Teaching Styles

You will encounter a variety of instructors in college—from graduate assistants to distinguished full professors. Your experience will run the gamut. Many will be excellent teachers and communicators, while some will only know how to monotonously present stale facts. Some will have accents that are difficult to understand, and some will downright dislike being in the classroom. It is your job to deal with these varying personalities and gain as much information as you can. You may have to rely

more on background information for one professor; you may have to actively participate in another professor's course to draw out the information you need most through
aggressive questioning; and you may be pushed to your limits just to keep up to gain
the full benefits of another professor's lectures. You must fully engage in the learning
process, not by just sitting there but by learning how to adjust to the teaching style of
each instructor. By knowing how they differ, you can adjust the way in which you
react to their particular style.

Learning Styles

We all learn in different ways. You have your own particular style for assimilating
information, so make sure you understand your learning style and incorporate effective methods of learning. Do you memorize everything you read or hear the first time?
Do you have to write out facts several times in order to remember them? Do you study
only at the last hour, or only when others force you to? Do you have particular places
you go when you need to study? Do you have any idea how you study best? If not, you
will need to learn.

Perspectives of Others

You will find that being able to keep an open mind about the thoughts, feelings, and
beliefs of others will make your college career more valuable. You don't need to go
along with everything that goes on around campus, but you do need to be able to
listen and rationally discuss your beliefs and the beliefs of others, and to work together
to put them all to equal tests. Your view might pass your own tests just fine, but maybe
you're missing some angles. Utilizing other people is the best way to get help to do
this. If you find it hard to be empathetic in certain discussions, then it is better to seek
help from campus resources that might at least help you widen your ability to listen,
or not block the rights of others to speak. You also have a right to be talked to at a level
that shows respect to you. No one should badger or assault you verbally with their
ideas.

Recommended reading: *Managing Diversity*—Norma Carr-Rufino

Listening Skills

You must listen with greater intensity in college than you have previously. Courses in
subjects that you are unfamiliar with will require that you concentrate harder than
you ever have before. You will need to focus your attention on what instructors are
saying and how it relates to other sources of information presented in the course. You
will find that your new environment consists of students, faculty, and staff from
myriad backgrounds. You do not need to agree with everything you hear, but you
must allow the other individual to speak so that you can in turn discuss your beliefs

and explain your conclusions. Dialog does not have to be confrontational if both parties are willing to listen respectfully.

Recommended reading: *Building Active Listening Skills*—Judi Brownell

Handling Stress

Develop study skills that are as agreeable to you as possible, but always take some time to relax. Do all things in moderation. This doesn't mean you should overdo one thing, then overdo another to offset the first and supposedly *balance* it. Such a way of handling stress creates more stress.

A common way to create unhealthy stress is to take a class for the wrong reason. You should study to learn and for no other reason. Your grade will, in the end, reflect what you learn. And the way you arrange your daily life will be set up around study, for the most part. But you shouldn't simply study for a grade and you shouldn't take a class just because you have to. Some required classes might seem pointless to you, but instead of getting a bad attitude, you should rise to the occasion, trying to see why the class was required so you can get more out of it than you thought possible. If you resent your studies, you're in for trouble.

You shouldn't even necessarily resent your weak points, since everyone has plenty of features that require improvement. Inevitable failures will roll off you like water off a duck if you approach them positively. After all, as the saying goes, "Success is but a series of failures handled properly." You must go beyond what you know, and are already good at, to learn and grow.

"Grades, sports, and partying" is an approach to school that some students *devolve* to, even as they think they're being clever. A results-oriented bias that overemphasizes a balance of study, exercise, and fun can put students in this mindset even without them hardly noticing a change. To whatever degree we take any type of performance more seriously than actual learning, we might tend to overdo one or more aspects, then try to balance it out against the others, so that more study requires more fun, on and on. . . . This can start a stressful trend, even a downward spiral. If you find yourself flopping between extremes, work to develop a better initial understanding of what your real purpose is in life and in school. This should defuse the stress, even if it does seem to make you a bit more boring to others or make your study seem to take more effort.

If you notice these things, that's good! Paying attention is a great way to beat stress. Notice the results of your increased effort. See how steady study really doesn't hurt you. Notice how, even when things aren't exciting or you can't seem to see any obvious improvement, if you have a good attitude—including patience!—things work out better in the end.

A "heads up" approach that lets you see the *benefits* of stress beyond its sometimes gloomy appearance will give you better endurance, toleration, and perspective. You can't avoid stress. But you can *grow* through it if you use it right. To do so, you have to assert yourself. You have to go against the grain as you work to fulfill your obligations. Even if some of your fellow students are negative about school sometimes, it doesn't mean you have to be. This sort of approach can protect you from the harmful effects of stress.

Of course, even with a good attitude and lifestyle you will have many challenges and problems as you grow through your college career. Even the best study system can break down. You will have unpleasant surprises and consequences. But if you handle them right, you can turn them to your favor even if things look bad outwardly. If handled improperly, even success can cause damage to your system and to your progress in learning.

Talking it out can help, so develop a varied group of people to whom you can direct your concerns over stressful situations (and who know they can let you know if they see you getting into trouble). This group should include friends and peers, but also professors, older students, mentors, role models, and anyone you look up to. Even your parents! (Oh yes, they'll soon start to seem wiser and more intelligent.)

There are also a number of agencies that provide listeners who can offer a sympathetic shoulder for you to unburden your anxieties. Operations like the Listening Ear let you talk to anonymous people on the phone, giving you a chance to start the venting process with no embarrassment. Never let the pressure build. Release the steam before it creates a damaging situation. But be aware that the buildup itself is a sign that you might need to change your daily habits and expectations.

Recommended reading: *Career Success/Personal Stress*—Christine Ann Leatz; *Why Zebras Don't Get Ulcers*—Robert Spolsky

18.8 Final Thoughts

The ideas expressed in this chapter should help you settle in with greater confidence at college. Efforts will be made on the part of many to make you a part of that life, but it still will depend on your curiosity and steadfastness to investigate every element that makes up the total package we call the college experience. You can do that if you get involved.

REFERENCES

Brownell, Judi, *Building Active Listening Skills*, Prentice Hall, 1986.

Carr-Ruffino, Norma, *Managing Diversity: People Skills For A Multicultural Workplace*, Pearson, 1998.

Cherney, Elaine, *Achieving Academic Success: A Learning Skills Handbook,* 2nd ed., Kendall Hunt Publishing, 1996.

Covey, Stephen, *The Seven Habits Of Highly Successful People.* Fireside/Simon & Schuster, 1989.

Davenport, Rita, *Making Time, Making Money: A Step-by-step Program to Set Your Goals and Achieve Success*, Davenport, 1982.

Kesselman-Turkel, Judi, and Franklynn Peterson, *Note-Taking Made Easy*, Madison, WI, University of Wisconsin Press, 2003.

Lashley, Conrad, *Improving Study Skills: A Competence Approach.* Cassell, 1995.

Leatz, Christine A., and Mark W. Stolar, *Career Success/Personal Stress: How to Stay Healthy in a High-Stress Environment.* McGraw-Hill Companies, 1992.

McWhorter, Kathleen T., *College Reading & Study Skills*, Addison-Wesley, 1998.

Semones, James K., *Effective Study Skills: A Step-by-Step System for Achieving Student Success*, Wadsworth Publishing Company, 1991.

Sherman, Thomas M., and Terry M. Wildman, *Proven Strategies for Successful Test Taking*, Merrill Publishing Company, 1982.

Spolsky, Robert, *Why Zebras Don't Get Ulcers*, 3rd ed., Redwood City, CA, Stanford University Press, 2004.

Van Blerkom, Dianna L., *Orientation To College Learning*, Cengage Learning, 2013.

EXERCISES AND ACTIVITIES

18.1　With a map of the university in front of you, find the following:

　　a) a research unit on your campus
　　b) a place you can go if you are feeling ill
　　c) the offices dealing with financial aid
　　d) a dormitory with an odd or interesting name

　　After locating these sites, plan the shortest route to visit each of the buildings and then do so. Write a brief paragraph that describes the journey and the buildings.

18.2　Find a building on campus that seems relatively mysterious to you. Visit the building and ask questions about the activities within it. Report your findings.

18.3　On your way to your classes, walk through buildings that you have never been in. What goes on in those buildings?

18.4　Begin a walking tour and enter every building in a chosen section of campus. Take time to look at what is in the buildings—departments, faculty, services—and make notes of these findings.

18.5　Stop in one of the offices along your way, identify yourself, and ask for information concerning what goes on in the office or building. Write a memo to your instructor concerning the information that you discovered.

18.6　Find at least five buildings on campus that you have never entered. Investigate what goes on in them, investigate their history, and find out what is planned for them in the future. Write a brief report on your experience.

18.7　Find the oldest building on campus. What is its history?

18.8　Are there any buildings on campus that have served more than one purpose in their existence?

18.9 How much office space is allotted to each faculty member in the engineering department?

18.10 Do the athletic facilities have areas that only certain students can use?

18.11 Find a bridge on campus and document when it was built. What was happening on campus at that time?

18.12 Find someone who graduated from your school at least 10 years ago. What seems different to them about today's campus?

18.13 When high school students visit your campus, what do they want to see?

18.14 After a chosen lecture, pose a set of questions that you could ask in order to clarify information from that lecture. Make a point of asking those questions during the next class period, or send an e-mail to the instructor focusing on those questions.

18.15 Suggest to your instructor that he/she have the class turn in a set of questions that relate to the previous lecture or lectures. Ask that the common questions be addressed during the next class period or placed on a website for easy viewing by the class.

18.16 Find out the following from your classmates:

a) who has traveled to the most countries of the world
b) who has gone to a recent concert
c) who wants to follow your same career path
d) who thinks that communication is critical to success in engineering
e) has any home-state student never traveled out of the state within which the college is located

18.17 As the semester goes on, find out who is getting a 4.0 in your courses. Ask them if they incorporate any unique methods in their study that help them obtain those high marks.

18.18 Make an effort to help someone who is having problems in one of your courses. Through helping him or her, you may well come to understand the material more fully.

18.19 Make an appointment with your academic advisor to discuss your future courses and the requirements for entering the engineering profession. Report on your conversation.

18.20 Collect information from other students about instructors in your major department.

18.21   When you have free time with no classes, ask if you can sit in on courses that you might take in the future. Write up your evaluations of the experience.

18.22   Make an effort to get to know at least two of your instructors. Interview them and ask about their educational experience. What suggestions can they give you about your own education?

18.23   In one of your courses, create a diagram of those areas where you feel you will need help in understanding the material. Find students who can provide you with the information that will help you succeed.

CHAPTER 19

Engineering Work Experience

19.1 A Job and Experience

"How do you get experience without a job, and how do you get a job without experience?"

This is the job-seeker's first, natural question. It was used in a national publicity campaign by the National Commission for Cooperative Education a number of years ago. It's one that all engineering students have considered or will consider at some point during their college career. Career-related experience, commonly called "experiential learning," has become an important supplement to a student's engineering education. Throughout the years, and despite the changing times of up-and-down job markets, numerous employers of engineering graduates and engineering graduate schools have continued to stress the fact that they seek to fill their annual openings by selecting only from those candidates who have done out-standing things during their college years to distinguish themselves from their peers. In most cases, those students who have participated in some form of experi-ential learning have gained the types of skills and competencies that prospective employers and graduate schools are seeking.

A college education has evolved over the past century to be quite a different experience from that of our predecessors. For most engineering students during the first half of the 20th century, college was completed in four years. Since studying engineering has always been a rigorous experience, most students chose to devote their time and efforts to experiences in the classroom, library, or laboratory. Those who chose to work while going to school were viewed by the conventional student population as "nontraditional," representing the working class who were forced to work for pay solely as a means of affording their schooling. However, in the latter part of the 20th century the costs of a college education soared. It is now the rule, rather than the exception, that most students are forced to work for pay in order to keep up with steadily increasing tuition, room and board, and associated expenses such as books, computers, and transportation. (Again, some countries fully subsidize college education—with one of the benefits being that students can concentrate fully on their studies, with less need for extracurricular employment during school.)

For many students, the idea of combining their academic studies with a career-related employment experience is something they had not previously considered. The thought of stretching their college education to five or even six years can seem to be an unnecessary burden. To their parents who grew up with the notion that a college education is a four-year plan with a direct path into the world of employment, lengthening the time spent in school is also an uncomfortable concept.

Consider the change in employment trends in recent years. The current generation of college students was born, for the most part, in the mid-1990s. During the relatively short span of their lifetime, this group has witnessed firsthand several major shifts within corporate America. Prior to the years when many of these students were born, a major recession was experienced in some of the largest manufacturing sectors. Especially hard hit was the automobile industry and the surrounding Midwest region of the country. Dubbed the "Rust Belt," this area saw a significant reduction in the workforce, especially in the traditional blue-collar jobs. Jobs that had traditionally been viewed as secure for a worker's lifetime were now eroding and in many cases disappearing. During this period, the employment of professionals remained fairly stable, and new engineering graduates were having only minor problems gaining employment at graduation time.

The period 1983–1986 saw a revival in the U.S. economy, and this was reflected by an increase in hiring of new engineers, with salaries well above the rate of inflation.

The period 1988–1994 was one of major restructuring within corporate America. Many major firms instituted a series of layoffs and experienced periods of downsizing. This period was notable for how the white-collar work force was affected. Professionals either were being laid off or provided with incentives to retire early. The net effect was a significant restructuring, reflected by one of the weakest labor markets ever for engineers.

The Collegiate Employment Research Institute at Michigan State University conducts an annual hiring survey of employers and is able to track the labor market for new college graduates. From 1994 to mid-2001, the job market for graduating engineers was excellent; experts agree that it was probably the hottest job market in over 30 years. Unfortunately, overall hiring contracted nearly 50 percent in 2002–2003, due in part to the aftershock of 9/11. Hiring trends slowly recovered each year from 2004 to 2008, but then the economic downturn of 2008 began to negatively influence employment opportunities for new college graduates, including engineers, as available positions became highly competitive. It was not until 2010–2011 when a gradual shift in hiring patterns began to emerge. Students were once again finding that employers were returning to college campuses, job prospects were slowly improving, and more graduates were finding employment.

The most recent survey of over 4,300 employers for 2012–2013 shows a slowly growing economy and a continued tight labor market for new college graduates. If this trend continues, opportunities for engineers will still be available, but students will need to devote increased efforts to finding the right opportunity. However, employment opportunities and starting salaries for engineers have traditionally surpassed those of most other majors on college campuses, and this trend continues.

Truly, the current generation of college students has witnessed a roller-coaster type of economy. Many realize that the rules of the game have changed and that lifelong job security with a single employer is a thing of the past. However, one fact remains clear from employers in good times and bad: the need for college graduates with experience is a constant. Many employers expect today's college graduates to have significant work experience. Why? Because engineering graduates with career-related work experience require less training and can produce results more quickly than the typical recent college graduate. To an employer, this is sound economic policy.

There are many forms of experiential learning available to engineering students. This chapter will discuss several of the most popular options, including on- and off-campus jobs, summer work, volunteer experiences, academic internships, research assistantships, and perhaps the most popular and beneficial program for students, cooperative education. Each of these has its own advantages and disadvantages, depending on a student's particular situation and background. These issues will be examined as part of this chapter.

19.2 Summer Jobs and On- and Off-Campus Work Experiences

Most engineering students arrive at college with some type of work experience. Some have mowed lawns, done babysitting, or worked in fast-food franchises, while others may have had the opportunity to actually work in a business or industry that was involved in some type of engineering activity. Whatever the experience, it can be helpful in preparing you for your future. Career-related work experience is clearly preferred by engineering employers, but one has to start someplace in developing a work history.

There is value in working in the residence hall cafeteria, or delivering pizzas, or helping out in a campus office. These experiences begin to build a foundation that your classroom learning and other career-related experiences will supplement. On- and off-campus jobs and summer work experiences may not seem highly relevant, but they can help develop important skills. Through such experiences, many students have strengthened their communication skills, learned to work as part of a team, and developed problem-solving abilities and an organizational style that can be applied to engineering courses and future employment positions. Jobs that you hold as a student will influence your perceptions about the things you like to do, the conditions and environments in which you like to do them, and the types of people and things with which you like to work. These jobs can be useful in helping you identify your personal strengths and weaknesses, interests, personal priorities, and the degree of challenge you are willing to accept in a future position.

Real-world experience can come in many forms. While career-related experience is important, do not underestimate the importance of learning some basic on-the-job skills. If you are seeking to develop some basic job skills or merely trying to earn some extra money to assist with your college expenses, you should not be afraid to explore some of the many employment options that are currently available on and off campus during the school year and in the summer.

19.3 Volunteer or Community Service Experiences

Another popular form of experiential learning is found in the many opportunities associated with volunteer or community service experiences. While these programs are available to engineering students of all class levels, they can be particularly beneficial to freshman or sophomore students who are trying to gain some practical experience. Generally, volunteer or community service projects can be short- or long-term arrangements. Typically, these positions are with nonprofit organizations such as human service groups, educational institutions at all levels, a variety of local, state, and federal government agencies, or small businesses. These groups are particularly interested in engineering students for their strong backgrounds in math, science, and computer skills, which can be applied to a variety of projects and activities.

Even though these positions usually are not paid, students can gain significant experience and develop important skills. Students have found that working with different groups can enhance their communication abilities, as they are often interacting with people much different from those in their collegiate environment. Through the various projects and activities, students can develop organizational skills, initiative, and the desire for independent learning that can be useful in many settings, including the classroom and other employment situations. Students involved in volunteer and community service experiences find that these opportunities are helpful with career decisions, while building confidence with the satisfaction of helping others and providing needed services.

For those students interested in developing a basic set of skills that can be useful throughout their professional engineering career, volunteer and community service work is worth exploring. For more information, contact your campus placement center or a particular firm or agency in your area. A faculty member or academic advisor may be able to provide assistance in this very valuable endeavor.

19.4 Supervised Independent Study or Research Assistantship

The **supervised independent study** or **research assistantship** is a form of experiential learning designed primarily for the advanced undergraduate engineering student who is considering graduate school and/or a career in research and development. Generally, a supervised independent study or research assistantship is a planned program of study or research that involves careful advance planning between the instructor and student, with the goals, the scope of the project, and the evaluation method specified in writing. Some of these opportunities are paid, while others involve the award of credit, and possibly both.

For students considering graduate school or a career in research and development, these opportunities can provide experience working in an environment that can be quite different from other engineering functions. Graduate school and research work involve the application of technical skills at a very theoretical level. Students should enjoy the challenges of advanced-level problem solving that will require a very in-depth use of their engineering, math, and science background.

To learn more about supervised independent study programs or research assistantships, talk with some of your professors with whom you enjoy working and with whom you share similar engineering interests.

19.5 Internships

What is an internship? Of all forms of experiential learning, some think that this is the most difficult to describe, as it can have different meanings, uses, and applications depending on a variety of circumstances. Generally, internships can be either paid or unpaid work experience that is arranged for a set period of time. This usually occurs during the summer so there is the least disruption to a student's academic schedule. For most students and employers, an internship is a one-time arrangement with no obligations by either party for future employment. Objectives and practices will vary greatly depending on the employer and the student's class level in school. Some internships involve observing practicing engineers and professionals to see what working in the field is really like. The intern's actual work often involves menial tasks designed to support the needs of the office. However, other internships can be structured as a capstone experience, which permits students to apply the principles and theory taught on campus to some highly technical and challenging real-world engineering problems.

At most engineering schools, internships are treated as an informal arrangement between the student and the employer. Therefore, this activity is usually under the general supervision of an experienced professional in the field in a job situation that places a high degree of responsibility on the student for success or achievement of desired outcomes. While many campus placement centers provide facilities and assist with arrangements for internship interviews, there generally is not a formal evaluation of the internship job description by any engineering faculty member. As a result, it is often difficult for students to understand how, and whether, this internship will be related to their engineering field of study. Since the typical engineering internship does not involve faculty input, there is usually not any formal evaluation or monitoring of the experience by any college officials. For these reasons, college credit is usually not awarded for these experiences.

For many students, however, an internship can be a very valuable part of their college education. Under the best of circumstances, the internship will be structured by the employer so it will be directly related to the student's field of study. It is advantageous for the student to try to obtain, in advance, a description of the duties and responsibilities of the position. The student may wish to review this material with an academic advisor or faculty member to gain an opinion on the relevance to the student's selected field. These individuals may be able to make suggestions and comments on ways that the student can maximize the benefits of the experience. The most productive internships are obtained by upper-level undergraduates who have completed a significant amount of engineering coursework. It is often easier for employers to match these students and their qualifications to specific projects and activities that will take advantage of the student's academic experiences. Students gain some practical real-world experience that will provide them with a better perspective of the engineering field they have selected. For the employer, internships provide a mechanism to supplement their workforce in order to complete many short-term engineering projects. The employer can also evaluate students in a variety of work-related situations. Many employers will use internship programs to screen candidates for possible full-time employment after graduation.

For students with restrictions related to time constraints, curriculum flexibility, scheduling difficulties, or location or geographical preferences, the one-time internship program can serve as a valuable supplement to their engineering education. While internships do not provide the depth of experience found in other forms of experiential learning, they can help students gain an appreciation for the engineering profession and perhaps provide additional opportunities for full-time employment after graduation.

19.6   Cooperative Education

At most engineering schools, cooperative education programs are viewed and promoted as the preferred form of experiential learning.

> *Engineering cooperative education (co-op) programs integrate theory and practice by combining academic study with work experiences related to the student's academic program. Employing organizations are invited to participate as partners in the learning process and should provide experiences that are an extension of and complement to classroom learning.*

This background statement, endorsed by the Cooperative Education Division of the American Society of Engineering Education, describes the basic foundation of **cooperative education** and some of the critical components of a quality engineering co-op program. Perhaps the most important part of any cooperative education program is its linkage to a college's engineering programs. Co-op is considered to be an academic program, and in fact at many colleges of engineering it is administered centrally, by the college itself.

As an academic program, co-op is structured so students' assignments are directly related to their major field of study. Typically, participating employers, who have been previously approved by the program administrators, will forward detailed job descriptions, including salary information of available positions, to co-op faculty and staff. These descriptions are reviewed to evaluate their correlation to the academic content and objectives of the educational program. Employers also provide a set of qualifications for the candidates they are seeking, to provide for appropriate matching of students with the positions they have available.

The integration of school and work is another important component of cooperative education programs. This is typically achieved by alternating several periods of increasingly complex co-op assignments, almost always with the same employer, with periods of study at school. This program, called the "alternating-term co-op," is the most common form of co-op scheduling. For example, a student might be at the XYZ Company during the fall semester, back at school for the spring semester, then return to XYZ for the summer semester, back to school for the fall, etc. This type of schedule allows students to apply their classroom learning to real-world problems in an engineering setting. Upon returning to school, students can relate the learning in the classroom to the experience they gained on the job. After another semester of schoolwork, students return to the employer with an even greater knowledge of engineering and scientific principles, which can be used for even more complex assignments in the workplace. As students mature academically and professionally they become stronger students and valuable employees. It is truly a winning combination for the employers, the students, and the schools.

Another form of cooperative education scheduling popular at some colleges and universities is called "parallel co-op." Under this arrangement, students usually enroll for a partial load of classwork while engaging in co-op assignments for 20 to 25 hours per week. This type of co-op program allows students to continue to make progress toward their degrees while gaining valuable work experience. The disadvantage of this arrangement is that students may encounter difficulties if the demands of the co-op position begin to interfere with the time needed for school assignments, or if the time devoted to schoolwork begins to hamper the performance on the co-op assignments. Great care and judgment should be exercised when students and employers consider undertaking this co-op option.

Under another arrangement, co-op assignments are scheduled consecutively, during back-to-back semesters. This option is particularly useful for co-op projects that require a longer commitment than a single semester, or where the activity is seasonal in nature. For example, many civil engineering students are employed by highway construction firms for assignments that typically last for a summer and fall semester. Students involved in this type of co-op may work two consecutive semesters, followed by a semester or two in which they attend school, and then they return to the employer for one or two additional co-op assignments. This pattern can provide many of the same benefits of the alternating plan if it is scheduled in a way that ensures that the academic coursework is still a major component of the overall plan.

Some co-op schedules are actually a combination of these different patterns. Due to the sequential scheduling of a typical engineering curriculum, it can be difficult to adhere to a true alternating co-op program. Therefore, some students and employers develop a series of assignments that may begin as an alternating plan for the first two or three rotations but may then require a back-to-back schedule or a parallel semester to complete the full schedule plan due to a student's particular academic schedule.

Cooperative education is rarely a summer-only program. Interrupting work for two semesters every year makes it difficult to maintain continuity in the integration of academics and the employment experience. In addition, most participating employers have needs that require them to have co-op students available for these positions throughout the year. In fact, most co-op employers plan their hiring so that while one student is away at school, another student is onsite working on a given co-op assignment.

For both an employer and a student to gain maximum benefit from the co-op experience, most employers and schools require that a student make a commitment for three or four semesters. This guarantees stability for the employers while ensuring the student at least a full year of job-related experience. Most engineering schools typically will recognize the student's participation in co-op after completion of the equivalent of one year of work experience by placing an official designation on the student's transcript and/or diploma.

Advantages of Cooperative Education Programs

Advantages for Students

Cooperative education programs provide students with an advantage as they seek employment or graduate school admission. Many employers desire to fill permanent positions with students who have had significant work experience. These employers use the cooperative education program to evaluate candidates on the job. For many students this leads to an offer of permanent employment upon graduation. Students are under no obligation to accept full-time employment from their co-op employer, but studies have demonstrated that many accept these offers. Students who choose to work for their co-op employer often begin with higher seniority and benefits, including higher starting salaries, increased vacation time, and other fringe benefits. Those students who elect to interview with other firms usually find that their co-op experience places them far ahead of their peers, as they are sought more often for interviews and plant trips, receive more offers, and earn higher starting salaries than their classmates.

Students who have participated in co-op programs usually comment on how they have been able to develop their technical skills. The nature of co-op assignments allows students to use the skills they have gained in the classroom and apply them to actual engineering problems. In the classroom, students' assignments and laboratory experiments are graded based on how close they come to achieving the correct answer. As they quickly learn on the job, real-life assignments and projects often are directed toward not a specific answer but a variety of potential solutions. This is where the student's technical background is developed and applied. Much of the actual technical learning occurs when students are forced to apply various principles to these unique situations. The learning that is achieved as part of the co-op experience provides students with new knowledge that often will make them better prepared for classroom challenges. The opportunity to learn by doing has been an important component of human development throughout time.

Many students who participate in cooperative education programs also find that their experiences can help solidify career decisions. Many engineering students are not fully aware of the spectrum of options and opportunities available to them. By engaging in the daily responsibilities of engineers, interacting with both the technical and non-technical individuals who work with engineers, and applying their academic background to their positions, students begin to gain a better understanding of the engineering profession. As a result, many find that their selection of a particular discipline has been confirmed, reinforced, and even strengthened. Others are exposed to new fields and technologies of which they were not previously aware. Some change their majors to reflect their new interests, while others modify their course schedule to include new courses to incorporate some of the exciting new areas to which they have been exposed.

All cooperative education assignments are, by definition, paid positions. Therefore, students find that the salaries they earn as co-op participants can greatly assist them in meeting the high costs of education. Recent studies have demonstrated that average salaries paid to engineering co-op students are in the neighborhood of $2,400 per month. For students who are working between four and seven months per year, co-ops can provide significant resources to meet their education expenses. It should be noted that average salaries can vary greatly depending on such factors as the employer, geographical location (pay is often higher in areas with a high cost of living), and the student's major, class level, and amount of experience.

Advantages for Employers

Many students often wonder why employers become involved in cooperative education programs. Some of the most common reasons given by participating employers are as follows:

- The cost of recruiting co-op students averages 16 times less than recruiting college graduates.
- Almost 50 percent of co-op students eventually accept permanent positions with their co-op employers. The retention of college graduates after five years of employment is 30 percent greater for co-op graduates.
- Typically, co-op students receive lower salaries and fewer fringe benefits than permanent employees. Total wages average 40 percent less for co-op students. In addition, employers are not required to pay unemployment compensation taxes on the wages of co-op students if they are enrolled in a qualified program.
- The percentage of minority members hired is twice as high among co-op students as among other college graduates, thus assisting co-op employers in meeting EEO objectives.
- Co-op programs provide an opportunity to evaluate employees prior to offering them full-time employment.

- The co-op graduate's work performance is often superior to that of a college graduate without co-op experience. Co-op students are typically more flexible and easily adapt to a professional environment.
- Regular staff members are freed up from more rudimentary aspects of their jobs to focus on more complex or profitable assignments.
- Co-op programs often supply students who have fresh ideas and approaches and who bring state-of-the-art technical knowledge to their work assignments.
- Co-op graduates are capable of being promoted sooner (and farther) than other graduates.
- Co-op programs build positive relationships between businesses and schools, which in turn helps employers with their recruiting.

Advantages for Schools

There also are many benefits to educational institutions that offer cooperative education programs to their engineering students:

- Cooperative work experiences provide for an extension of classroom experience, thus integrating theory and practice.
- Cooperative education keeps faculty members better informed of current trends in business and industry.
- Co-op programs build positive relationships between schools and businesses and provide faculty members with access to knowledgeable people working in their fields.
- Co-op programs enhance the institution's reputation and attract students interested in the co-op plan to their school.
- Cooperative education provides schools with additional business and industry training facilities that would otherwise be unaffordable.
- Cooperative education lowers placement costs for graduates.

More Benefits of Co-op

Co-op students have found that many professional development skills critical for success in today's workplace can be enhanced through their co-op experiences.

1. **Written and oral communication skills:** The ability to communicate ideas, both in written form and orally, is a critical skill. Most co-op assignments require students to document their findings and report them to others. This may take the form of an e-mail to a supervisor, a technical memo, a presentation to a group of engineers and technicians, or even a formal presentation to a group of directors or other executives. While communications and presentations may not be a favorite activity for many engineering students, co-ops provide students the opportunity to improve and develop these critical skills. For those who find this communication a liability, cooperative education assignments can help turn it into an asset.

2. **Networking:** Most co-op students find the opportunity to develop new contacts and work with people from a variety of backgrounds to be a great asset to their short- and long-term career objectives. Co-op students come in contact with many individuals who are able to provide advice, lend support with projects, and help get things done more efficiently. These same people can recommend new assignments to co-op students, and maybe even write an important letter of recommendation.

3. **Self-discipline:** Most co-op students quickly discover that the transition from school to work requires that they develop an organizational style that may be superior to the one they used in college. They find that within the co-op structure they are given a great amount of responsibility and freedom to succeed (or fail) with their projects. Time management, punctuality, adequate preparation, organization, and specific protocol and procedures are learned on the job. As students acquire and develop these skills, they find that they are able to apply them back at school, which in turn makes them more productive and successful as students.

4. **Interactions with a variety of people and groups:** Co-op students deal with a wide range of individuals as they complete their tasks. Typical individuals may include engineers, technicians, scientists, production workers, labor union representatives, clerical staff, managers, directors, and perhaps even CEOs. Each of these will contribute to the students' learning as they experience the successes, challenges, frustrations, and friendships that will evolve from these interactions. These individuals can help the co-op student evolve from a student to a successful engineer.

5. **Supervisory and management experience:** Many cooperative education assignments require that students take responsibility for other individuals or groups in order to accomplish their tasks. These opportunities to manage the work of others can provide invaluable experience to students, especially those who aspire to eventual management or leadership positions.

A Note of Caution

As discussed above, cooperative education programs can provide many positive benefits to students, employers, and schools. However, a co-op program may not be the appropriate form of experiential education for every engineering student. There are many aspects to the co-op structure that may not be consistent with a particular student's overall goals and objectives.

The biggest drawback to co-op participation is the intermingling of work and study. During each of the semesters that students are off campus working on their co-op assignments, they are not making progress toward their degree requirements. This means that typical co-op students add an extra year to the duration of their academic programs. For many, this means that co-op participation will extend their engineering degree program to five or even six years. The overwhelming majority of co-op students, schools, and employers feel that this sacrifice is minor compared to the outstanding benefits received from this type of education. However, students and parents should carefully weigh this factor.

Other factors also should be considered. The continual relocation between school and the work site is a minor inconvenience, but it does take time and effort—especially if the two sites are far apart.

The logistics of course scheduling also can present a difficulty because certain classes are offered sequentially and have prerequisites. Students should work out a long-term schedule with their academic advisor and co-op coordinator so that any class scheduling problems can be handled.

To find out more about engineering cooperative education programs at your school, contact your co-op office or placement center, or talk with your academic advisor or a faculty member. If your school doesn't have a co-op program, you may want to write or talk with a corporation of interest to you. You may be able to work out a co-op assignment with the help of an advisor, a faculty member, or the campus career center.

19.7   Which Is Best for You?

Many forms of experiential learning have been discussed in this chapter. Each of these opportunities has advantages and disadvantages, and only you can determine which program is best suited to your particular situation. Begin by asking yourself these types of questions:

- Is obtaining career-related work experience a high priority in my educational planning?
- Am I willing to sacrifice personal convenience to gain the best possible experience?
- Will I be flexible in considering all available work opportunities, or do I have special personal circumstances (class schedules, time constraints, geographical or location restrictions) that limit my choices?
- Do I have the drive and commitment necessary to succeed in a co-op program?

Your answers to such questions will be important as you move forward in your engineering career. Work with your academic advisor, a professor, or your campus placement office to get as much information as possible so you can make an informed choice in determining which is right for you. The resources are available to you—it is up to you to take advantage of them.

EXERCISES AND ACTIVITIES

19.1   Develop a list of your skills and abilities. Determine which could be improved with work experience and which will be developed in the classroom.

19.2 Interview one of your professors who is involved with a research project. Discuss the advantages and disadvantages of a career in research.

19.3 Talk with other students who have done volunteer and community service projects. List the benefits they have gained that have helped them in their engineering classes.

19.4 Visit your engineering co-op office or campus career planning or placement center and read the recruiting brochures of some of the employers in which you have interest. Make a list of the types of qualifications these firms are seeking for full-time and co-op or intern positions.

19.5 Attend a co-op orientation session. List some of the advantages and disadvantages as they relate to your personal priorities.

19.6 Attend your campus career/job fair. Talk with at least three recruiters from different firms. Write a report comparing their views of cooperative education experiences, internships, and other forms of experiential learning.

19.7 Visit with some upper-class engineering students who have been co-ops or interns. Write an essay that compares and contrasts their experiences. Which students do you feel gained the most from their experience, and why?

19.8 Visit the campus placement center. Gather information concerning the starting salaries and benefits of those students who have participated in cooperative education compared to those who have done internships or some other form of experiential learning.

19.9 Visit the engineering co-op office and read some of the job descriptions for current openings. Select one of these descriptions and write an essay discussing the benefits you feel would be gained from this position.

19.10 Meet with your academic advisor and develop a long-range course schedule. Modify this plan so it shows how a co-op and an internship would affect this schedule. Write a short paper discussing the advantages and disadvantages of extending your education to include extra time for work experience.

19.11 Develop a list of your personal career priorities. How can this list be enhanced by some form of experiential learning?

19.12 Develop a list of your personal and educational abilities that you think need to be strengthened. Write an essay that discusses how work experience could improve these skills for you.

19.13 Interview an engineering professor or graduate student who had been a co-op student as an undergraduate. Discuss how the co-op experience influenced him or her to pursue an advanced degree in engineering.

19.14 Visit http://epics.ecn.purdue.edu and review the volunteer student team projects that have been developed at this university. Prepare a short paper that discusses the benefits gained by these students and how these experiences can be an advantage in academic studies.

19.15 Attend a campus career fair. Talk with at least three recruiters about their opinions concerning career-related work experience. Develop a report that discusses the reasons that employers prefer to hire students for permanent positions who have had significant work experience as part of their undergraduate education.

19.16 Talk with an advanced engineering student who changed his or her engineering field of study as a result of a co-op or intern work experience. What factors in this student's experience caused him or her to do so? Does he or she view this work experience as a disadvantage or as an advantage to his or her overall career planning?

CHAPTER 20

Connections: Liberal Arts and Engineering

20.1 What Are Connections?

The connections that we will discuss in this chapter refer to the connections that exist between engineering and all those areas that are commonly referred to as liberal arts. Yes, liberal arts! You remember those courses you took in high school and may have referred to as the "soft courses"—courses that were not based on what engineers typically think of as science. Can you name a few? How about:

Literature	History
Music	Art
Social Studies	Philosophy

Did you, along with many of your science-oriented friends, feel that such courses would be of no significance to you as an engineer? Can you hear yourself saying, "I will never read another book, write another paper, or ever think about why some characters on stage are doing what they're doing"? Have you heard those sentiments echoed by fellow engineers? Well, if so, the purpose of this chapter is to get you to rethink those opinions.

It is important that you look closely at what engineers really are and what they really do. The root skills needed for art are similar to those needed for science. In the end, they're both about the quest for truth, as well as for what works. Art is amazingly rational and rigorous. It involves thinking about all aspects of physical objects, just like engineering, but it extends thinking past the usual limits of engineering. In creating their work, artists often have to think like engineers. But the flip side is also true: to do their best work engineers have to think like artists. Working to extend your mental horizons makes you a better engineer—and a better person. And it lets you see clearly that our human faculties do not stand alone, nor is one more important than another. Thought, emotion, willpower—all these need full development if we're to be the best we can be, on the job or anywhere else.

For a more specific definition of liberal arts, the dictionary says "liberal" comes from *liberty*, so that liberal arts means "works befitting a free man." Its goal is "a quest for truth, goodness, and beauty." Those are qualities that are difficult to put into equations, but all good equations do reflect them! Such a quest has everything, in the end, to do with engineering, but it's probably necessary to investigate it in more detail to show why this is so.

As for liberal education, it is half of the focus of many university degrees. It signifies a general big-picture education as distinct from technical training. The liberal aspect of even technical degrees was developed because people have a need for a strong, open mind in addition to a specialty in order to be well rounded—so they can have true liberty, and not be trapped by cultural blind spots. Technical training needs to be grounded in a general liberal education if it's going to have a healthy perspective.

It is often comforting to engineering students when they realize that a liberal education is not a vague project. There is a set of skills to learn and hone, and a modest amount of knowledge to acquire. There are definite ways to develop those skills. You learn them like you learn science skills. In fact, your talent in science will serve you well throughout your liberal education! There are also straightforward road maps to guide you so you get the most from the arts. You need to be taught how to use them, like anything else worth learning. When we say the arts are natural, we mean they're what all humans can acquire, or draw out of themselves, through education. (In fact, the word "education" comes from the Latin "to draw out.") Let's look further into why this is so.

20.2   Why Study Liberal Arts?

As an engineer, you will be considered an educated person. Your knowledge of science and mathematics will lead you to be held in esteem simply because of the fact that those two areas require a rigorous amount of study. As an engineer, you won't be sitting at a desk putting square pegs into square holes and round pegs into round holes. You will be investigating the world around you and creatively imagining how you will attack the problems with which you are confronted. You will not be a machine. You will be a living, breathing problem solver. Problem solving requires a creative and active mind. What do the liberal arts have to do with this?

Liberal arts help you improve your intellectual competence and expand your creative powers.

The Liberal Arts Help Improve . . . Your Broadness

Liberal arts give you practice at looking in many directions at once. They force you to ask questions about areas that do not have preset answers, and to evaluate why you have chosen the answers you have. This vibrant activity will keep your mind from becoming stagnant. It will allow you to grow and become more qualified to look at the world around you—a world of people who may not always act in obviously predictable ways.

The liberal arts also give you skills that are required in the real world, beyond school. As engineers in the 21st century you will be expected to be leaders. You will not be confined to laboratories. You must communicate with a wide variety of individuals, from corporate executives to the custodian in your plant, from the stockholder to the individual bringing a liability suit against your company. All of these people will expect you to listen to their concerns, evaluate the problems they raise, and speak to each of them in a way that tells them that you care and understand. Can the study of science and mathematics provide you with that training? It is possible, but improbable.

The liberal arts help you to develop the qualities of character and personality that will truly make you a leader in your field.

The Arts Improve . . . Your Perspective

As engineers, it is important for us to see the big picture of life so we can learn to use our skills in a wholesome way for everyday living. Otherwise, it's all too easy to let our lifestyles become fragmented, where what we do at work has little bearing on the rest of our life. The effort toward integrity, on the other hand, will always let you keep a good perspective and will give new insights into your engineering.

There's a strong tendency to become one-sided in modern life and in college. We then balance this one-sidedness with equally intense recreation of some type. This creates dangerous blind spots and stresses. Today, many students turn their university education into a vocational degree, like a shop mechanic would get at a trade school. We tend to focus on just getting that good job after we graduate. Our difficult studies make it even easier to become one-sided. A few years after we graduate, though, we often wish we'd taken the time for broader study. We finally see that money or a job isn't the goal of life, that they're just *means* to an end. We wish we'd studied the *ends* more seriously. They're a tougher nut to crack than even the toughest job. We easily overlook the fact that college is the one time in life—the best time in life—to develop the judgment that adult life requires. Thankfully, this is why engineers have graduation requirements that include the liberal arts! Even though they might seem annoying and beside the point at first, if we are to grow past immaturity we need to be fully integrated humans, with no part overlooked. High school didn't do it, work life doesn't allow the time—college is just right!

Here are some examples of how it works. We need philosophy to teach us a wide range of reasoning, and to appreciate the seriousness of our duty to use it right. Psychology, social science, and anthropology help us to see that there are two sides to everything, that we don't always see our own motivations clearly, and that imbalances create disease in society and ourselves. Engineers in particular need to learn that technology is a double-edged sword, the use of which requires great judgment. When we study history, we learn a perspective that keeps us humble despite the powerful techniques that we master in science. History provides us with many role models and shows how risky life can be. In general, the integrating, freeing benefits of liberal studies help us cover all the bases for a disciplined life, where we have the strength, skill, and insight to do the right thing, while keeping to the Golden Mean—avoiding excess in anything.

These skills give us the ability to do and to be our best in ways that transcend fads, trends, or temporary excitements. They give us integrity for a lifetime.

Perhaps this is a good time to make your own practical investigation into liberal arts and education, to help you see what we mean.

ACTIVITY 20.1

a) Look carefully at your world. What part does engineering play in the lives of the people around you? Collect six examples that show the ways that engineering is an integral part of the everyday lives of people.

b) Now think about those items from the standpoint of the engineers who created them. What did those engineers have to think about as they were working on perfecting the item? Did they only consider the science and mathematics of the object, or was there more?

c) Look at the latest models of automobiles. Are these automobiles designed only for gas efficiency and aerodynamics? Do we really need rich Corinthian leather? Heated seats? Thermometers in the rearview mirror? Why is there so much chrome? Why don't we just build an automobile that gets 80 miles to the gallon and gets us from point A to point B efficiently? Why is it necessary to try to produce beautiful automobiles? Why are they appealing?

The Arts Improve . . . Your Balance

The cultivation of an appreciation of art and its importance to the engineering field will allow you to see how artists work within their media and how the public is affected by their efforts. Because artists reflect their culture's view of the world, an artistic education will help you to see more clearly how the general public views the world. The insights that you gain will tell you that people do not always look at life with a scientific eye; many times they look with their hearts. They look with their whole person. And they need to keep this perspective, even when they work at specialized jobs, including engineering. This outlook perhaps can be thought of as a balance of views—energy, thought, and emotion, all working together, in their proper ways. There are many ways to express it, but if one is one-sided, it is hard to create work that has integrity or lasting value. A study of art can be beneficial in explaining this process. You likely will find a medium or style that best speaks to you, but all the disciplines and styles have something to contribute. Altogether, they've brought us to where we are now. Don't worry if some forms seem obscure: art and its meanings need to be taught as much as any other skill. With a good teacher, you'll soon understand. In this way, we introduce another benefit of liberal studies:

Liberal arts enrich our personal lives with new knowledge, insights, and a keener appreciation of beauty.

A well-rounded education that gives you a sense of what is happening within other areas of study cannot help but increase your ability to develop and present engineering ideas to the world.

A liberal education also provides you with practice in dealing with a variety of diverse ideas. Philosophy courses ask you to compare and contrast the varied interpretations of experience of individuals who are sometimes pitted in violent struggles against each other. Consider the difficulties in Northern Ireland or the Middle East or in the inner cities of the United States. How, as an engineer, do you explain the problems that have existed there for hundreds of years? Do you simply say, "$x + y = 3/z$"? Can mathematical thinking explain pain and suffering? Since math strives to realize harmony, in a certain way the answer is "yes." But you need to develop the skill to rigorously extend reasoning in new directions. Philosophy is not a soft science, after all. You need practice in deciphering and evaluating the misunderstandings of experience, laziness of reason, and fear of truth that lie at the source of so many troubles of life . . . and that can easily derail engineering projects.

The Arts Improve . . . Your People Skills

The people factors are always involved in the answers to these questions, and engineers need to know about people and how they act and react. Liberal studies of these factors will help us grow as humans and as such can't help but have ancillary benefits for our careers. For instance, we are living in a global society where the introduction of a car called the "Nova" in a Spanish-speaking country raises the eyebrows of customers. The automobile may be the most fantastic vehicle ever built, but since in Spanish "nova" means "it does not go," those customers may never spend their money on it. Learning to see all sides of a situation, including factors that are not at all obvious, will help you avoid such trouble!

The liberal education that you receive in college will provide you with the tools to understand these *people* issues, issues that may not have been considered noteworthy 50 years ago. Additionally, it will help you be aware of things that modern tendencies avoid or neglect. It may be that some colleges don't have many liberal requirements anymore, so you might find that you have to go against the grain to get the education you need. Be sure to let your advisor know your goals! He or she can help you avoid wasting your time or spinning your wheels. Not all liberal arts courses are equal. Some may not serve your needs. Some may be as technically specialized as any engineering course, which is not what you want. This is why it's best that a school plans a basic liberal curriculum to help fulfill your electives. Fifty years ago college was a place where even if you were getting a technical degree much of your effort went to developing yourself and your moral sense, as well as your awareness of the world. Today your liberal education could be neglected as you pursue a degree. To quote a commercial, "This isn't your father's Oldsmobile!" Times have changed. You, the engineer, will have to make sure you are a well-rounded, integrated individual. You'll be working in a global marketplace and you'll need to be ready to understand all sorts of cultural factors that may positively or negatively affect you and your company.

If you have had training in the liberal arts, you can meet those circumstances head on. If not, your shortsightedness will hurt you.

The Arts Improve . . . Your Sense of Duty and Responsibility

People the world over look to engineers to provide them with clean water, safe homes and transportation systems, warm places to sleep, work environments that will not injure or kill them, neighborhoods that are free of toxic fumes, and lakes uncongested with algae—and with technology experts who will not mislead them into putting too much focus on material goods in the first place. As you work to achieve these goals, the engineering profession will improve, but it will only improve with the human element as its core concern. Thus:

> *The liberal arts elevate, integrate, and unify the standards of the profession, focus them on the human, and create greater awareness and esteem for the engineer.*

A people orientation provides you with important tools to use for society:

> *You are better able to fulfill your duty in life, so society respects you more. Thus you are able to convince the public of issues that need to be resolved through effective engineering.*
> *By understanding people and their overall needs, you can apply your engineering expertise properly to any problem.*
> *By demonstrating the submission of science to the basic needs of mankind, you secure the dignity of the engineering profession.*

You are on the threshold of a great, yet humble, undertaking. You will be able to investigate the laws of science, examine data that will prove your assumptions, learn the ways of engineering, *and* contribute to making a better world. As a complete engineer and as a whole person, you will not only understand scientific parameters, but you will understand the people and the world in which you will provide your services.

EXERCISES AND ACTIVITIES

20.1   Assume that the department chairman within your major has selected you to explain to incoming freshmen why non-engineering courses are important to engineers. Discuss the role non-engineering courses can play in one's engineering career.

20.2   Evaluate a textbook you are using in a non-engineering course. Compare it to an engineering textbook that you are using. Do you see similarities and differences? Discuss them in a brief report.

APPENDIX A

Nine Excel Skills Every Engineering Student Should Know

Yeow K. Siow, PhD
University of Illinois at Chicago

A.1 Why Use Spreadsheets?

In life, you will most likely encounter situations where large amounts of information and data require your attention. In engineering, these situations are even more common. When they arise, your options are as follows:

A) Ignore it, in which case nothing gets done.
B) Manage the data using pen and paper. With this strategy you risk losing your work, not to mention the amount of paper and time required if you are dealing with a huge amount of data.
C) Use a calculator. This is effective only when handling small amounts of data.
D) Use spreadsheets.

Options A through C not only can be ineffective, but they can also cause headaches in your studies, job, and life. Spreadsheets, on the other hand, can help you organize, present, and share information quickly and consistently.

Using a spreadsheet is the digital equivalent of constructing a data table on a piece of paper, but spreadsheets are more powerful. In spreadsheets, you can automate calculations and generate plots with only a few clicks or taps. You can also update numbers instantly when changes need to be made. In other words, spreadsheets can save you time and resources and improve the quality of your life.

Figure A.1 shows the pressure inside the cylinder of an internal combustion engine. The graph represents more than 2,000 data points. Try managing these data and generating the plot by hand!

From tracking your spending to analyzing the performance of a nuclear facility, spreadsheets are often the ideal tool.

	CA	Cyl Press (bar)	P adjusted (Pa, abs)	(bar)	smooth p (b=11)	again	(bar)	Isentropic: cyl. Vol.
3	−180	1.459	281000	2.81				1656090
4	−179.9	1.46	281100	2.811				1656089
5	−179.8	1.46	281100	2.811				1656087
6	−179.7	1.46	281100	2.811				1656083
7	−179.6	1.46	281100	2.811				1656077
8	−179.5	1.46	281100	2.811				1656070
9	−179.4	1.459	281000	2.81				1656061
10	−179.3	1.459	281000	2.81				1656051
11	−179.2	1.458	280900	2.809				1656039
12	−179.1	1.457	280800	2.808				1656026
13	−179	1.457	280800	2.808	280766.9			1656011
14	−178.9	1.456	280700	2.807	280714.9			1655994
15	−178.8	1.456	280700	2.807	280660.3			1655976
16	−178.7	1.455	280600	2.806	280603.3			1655956
17	−178.6	1.454	280500	2.805	280545.5			1655935
18	−178.5	1.454	280500	2.805	280488.4			1655912
19	−178.4	1.453	280400	2.804	280431.4			1655887
20	−178.3	1.453	280400	2.804	280376			1655861
21	−178.2	1.452	280300	2.803	280321.5			1655833
22	−178.1	1.452	280300	2.803	280269.4			1655804
23	−178	1.451	280200	2.802	280218.2	280239.2	2.802392	1655773
24	−177.9	1.45	280100	2.801	280168.6	280193.3	2.801933	1655740
25	−177.8	1.45	280100	2.801	280122.3	280149.6	2.801496	1655706
26	−177.7	1.449	280000	2.8	280078.5	280108.4	2.801084	1655671
27	−177.6	1.449	280000	2.8	280038.8	280069.9	2.800699	1655633
28	−177.5	1.449	280000	2.8	280002.5	280034.1	2.800341	1655595
29	−177.4	1.448	279900	2.799	279968.6	280001.2	2.800012	1655554
30	−177.3	1.448	279900	2.799	279938.8	279971.3	2.799713	1655512
31	−177.2	1.448	279900	2.799	279912.4	279944.3	2.799443	1655469
32	−177.1	1.448	279900	2.799	279889.3	279920.2	2.799202	1655423
33	−177	1.447	279800	2.798	279868.6	279899	2.79899	1655377
34	−176.9	1.447	279800	2.798	279852.1	279880.5	2.798805	1655328
35	−176.8	1.447	279800	2.798	279838.8	279864.6	2.798646	1655278
36	−176.7	1.447	279800	2.798	279828.1	279851.1	2.798511	1655227
37	−176.6	1.447	279800	2.798	279819.8	279839.8	2.798398	1655174
38	−176.5	1.447	279800	2.798	279813.2	279830.5	2.798305	1655119
39	−176.4	1.447	279800	2.798	279808.3	279822.9	2.798229	1655063
40	−176.3	1.447	279800	2.798	279805	279816.9	2.798169	1655005
41	−176.2	1.447	279800	2.798	279802.5	279812.2	2.798122	1654946

evo −596   evc −336   ivo −364.5   ivc −134.5

	mm		mm^3	R	3.005272
bore	116.59	bowl vol.	73487.6	k	1.37
stroke	146.05	squish vol.	8406.35119		
conn rod	219.46	swept vol.	1559242.56		
clearance	0.7874	ring+valve recess	14953.4128		
C.R.	17.1	so, TDC vol.	96847.364		
		and, BDC vol.	1656089.92		

V = Vc * {1 + 0.5*(rc-1)*[R + 1 − cos T − sqrt(R^2−sin^2 T)]}

isentropic:
p2=p1*(v1/v2)^k

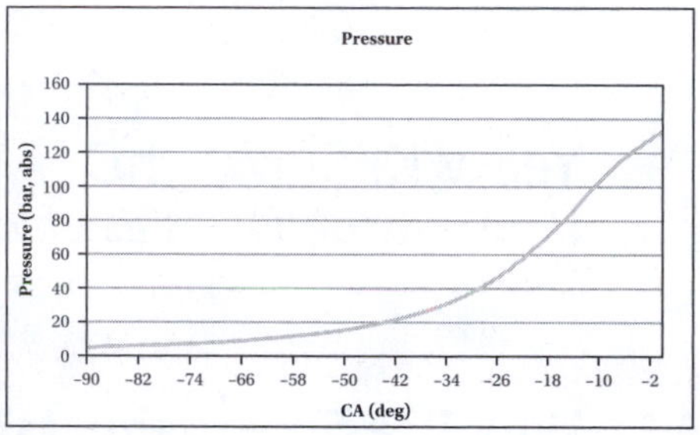

Figure A.1 A mountain of numbers: do not try this by hand!

A.2   Software Choices

While the proprietary Microsoft Excel is the *de facto* spreadsheet software of choice for many users and organizations, it is important to be aware of, if not proficient with, the alternatives. Open-source software such as OpenOffice Calc and LibreOffice Calc are popular and powerful choices for major operating systems including Windows, Mac, and Linux. Cloud-based programs such as Google Sheets can be handy for on-the-go productivity using any device running any operating system.

Cross-platform compatibility is essential in this age of computing ubiquity. Many of the Excel-alternatives *excel* in the area of compatibility. Excel can run natively on Mac, although its graphical user interface (GUI) can be somewhat different from the Windows version. While Excel does not run on Linux, you can install Windows on Linux by using either a parallel program such as Wine or a virtual machine such as VMware. Like Excel, the alternatives have strengths and weaknesses, but all of them are capable of satisfying the basic spreadsheet needs.

While Excel may be fee-free for many university students, open-source alternatives are, for the most part, free to use on any number of devices.

A.3 The Nine Skills

Below you will find a step-by-step guide to nine spreadsheet skills that are not only essential for engineers, but also beneficial in day-to-day life. These nine skills were developed based on the assumption that you are already familiar with the basic layout of a spreadsheet: cell, row, column.

Keep in mind that technical skills like these can only be gained through practice and repetition. This is why you will find downloadable examples in Section A.4. You're strongly encouraged to peruse the file contents and use them to help you complete the practice projects.

Although the title of this appendix refers to Excel, most, if not all, of the nine skills also pertain to the alternatives mentioned in the previous section. In fact, in some cases it may be more convenient and productive if you use Google Sheets, since it enables you collaborate with others in real time. It also allows you to enter data on the go and on one device, and then resume work at a later time on another device. Furthermore, since software can evolve quickly and the GUI can change drastically with every new release, these nine skills are more about *what you need to know to present useful data* than *which button to click to perform a task.*

Most examples used to illustrate the nine skills below are non-engineering in nature. In particular, you will be looking at *spending habits* and *GPA tracking*—two things you can, and should, easily relate to. Once you become familiar with the nine skills, you will be able to apply them to most engineering analyses, and the next section, A.4, includes practice projects based on real-world data.

Let's begin.

Skill 1: Organize Your Thoughts

The very first step in creating a spreadsheet is to form some *idea*, no matter how vague, regarding what you would like to achieve in the end. Do not worry about *how* you are going to achieve it—yet.

This idea of yours may be graphic in nature, such as those in Figure A.2. Would you like to *track* your GPA so far? Or perhaps *analyze* your grades, identify your weaknesses, and make adjustments to your study habits for the next semester? Would you like to see a *trend* of your spending? Or find out the *category* you spend most money on?

Once you have identified some goal(s), start thinking about what you will need to achieve this goal. Do you need to dig up past spending data (sales receipts, etc.) or start collecting new ones? Do you have an existing GPA to begin with? How do you calculate GPA anyway?

This last question alludes to an extremely important element in spreadsheets: formulas. Throughout the examples below, you will learn how to use the "formula" function to automate calculations, such as GPA, for you. In fact, formula building is so important that it transcends the nine skills.

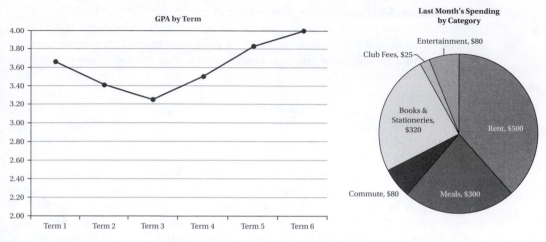

Figure A.2 Some everyday examples

Skill 2: Follow Best Practices

There are countless ways one may learn a new skillset, such as a sport, music, machine tools, and software, particularly when it is self-taught. More often than not, "best practices" exist whereby you can learn the new skillset effectively and efficiently. These best practices, typically established by experts in the field or experienced users, are intended to save you time, minimize mistakes, and "future-proof" your work.

Filename Convention

Always give your file a descriptive name, without spaces or special characters. Under-scores and dashes are the only "spacers" you should use. A good example:

john_GPA_purdue_2015-18.xlsx

A bad example:

john's G.P.A. @purdue 2015~18.xlsx

Following a good naming convention will ensure successful file saving and searching, as well as cross-platform compatibility.

File Location

Save your file in a convenient, dedicated location. A good *temporary* place to save it to is the desktop (when using standalone software on Linux, Mac, or Windows) or the "home directory" (when using cloud-based program like Google Sheets), since it is readily visible. However, once you have finished editing the file, you should create a dedicated folder/directory and move the spreadsheet there. Otherwise, the desktop space would become a crowded mess and you would not be able to locate your spreadsheet when you needed it.

Notes and Comments

Save the first few rows for adding notes, instructions, etc. for yourself or others. If you (or someone else) need to revisit the spreadsheet months later, having some explanations near the top of the spreadsheet will make life easy.

Escape Key

When in doubt, hit the Escape [esc] key.

Save Often

Software can and will crash; it is only a matter of *when*. Save every few minutes, and immediately after a major effort.

Skill 3: Populate Data

The fact that you choose (or are required) to use a spreadsheet means that you are dealing with more than just a few numbers; otherwise, a simple hand calculation or a calculator would have sufficed. This large amount of data needs to be entered—"populated"—into the spreadsheet. There are many ways to populate data, depending on where the data come from: Are those numbers in your head? Are you copying and pasting data from another document? Or do you need to generate new data based on equations?

One major and fundamental rule of entering data is to *never* mix data types (number vs. text) in a cell, and to enter each value in its own cell.

Let's explore.

Entering Data Manually (and Smartly)

Say you would like to enter your daily spending from the previous month. You would need two columns: Day and Dollars.

The "Day" data are straightforward, since they are already in your head and the numbers can be entered quickly—or can they? You may type "1," "2," "3," etc., one at a time, down Column A, until it reaches the last day of the month. Now imagine entering not days of the month, but all 365 days of the year. It would be an incredible waste of time trying to enter one number followed by the next. There is a much faster way to do this: **formulas**.

Begin by entering the first number, "1," in Cell A4. Below it, in Cell A5, type this formula:

$$=A4+1$$

and hit Enter. Then copy and paste this formula by either dragging the "fill handle" (tiny square at the bottom right corner) of Cell A5 downward (Figure A.3):

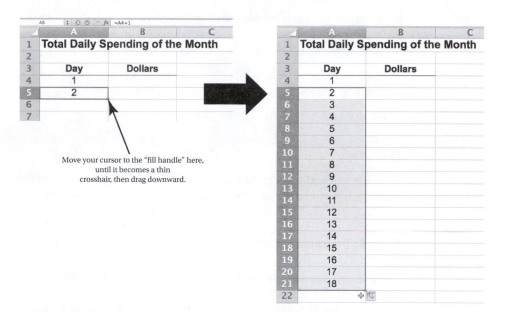

Figure A.3 Populate data: copy and paste formulas

or hit CTRL-C to copy Cell A5; then, using your mouse to highlight the next 29 cells, hit CTRL-V.

Next, simply type your spending data under the Dollars column (Column B). This step is labor-intensive, unless your numbers have already been typed up elsewhere, in which case you may simply import or copy the data.

Inputting Preexisting Data

If your data already exist and reside in another document, such as a web page, a text file, or another spreadsheet, you can either import or copy/paste the values, depending on the type of document the data are on. These source data may likely contain redundant values or may need to be reformatted. Fortunately, most software can allow you to trim and reformat the values so that only useful data are imported (or pasted).

The most common way of transferring data into your spreadsheet is a simple copy/paste process. On the external document, first copy the source data (which may be a mix of numbers, text, and symbols) and then, on your spreadsheet, hit either "Paste" or "Paste Special." The latter is often necessary, especially when the source document is a web page. Choosing "Paste Special" will most likely bring up this dialog box (Fig. A.4):

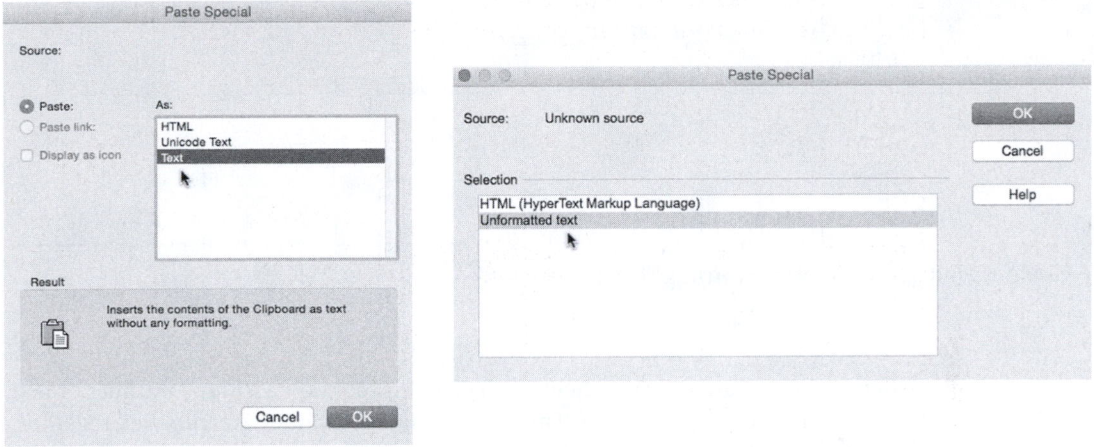

Excel   OpenOffice Calc

Figure A.4 Import data from an external source: "Paste Special"

Choose one of the options listed—for example, Text (in Excel) or Unformatted Text (in OpenOffice). In Excel, you may sometimes notice that multiple numbers are grouped together in a single cell while in fact they need to be separated (Fig. A.5):

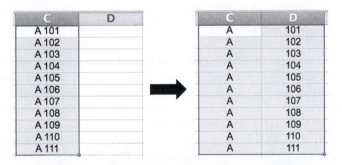

Figure A.5 Mixed data types should be split up.

When this occurs, highlight those cells, and use the "Text to Columns" feature (most likely found under the "Data" menu) to bring up a dialog similar to Figure A.6: This feature allows you to separate numbers (or text) by specifying the separation criteria (comma, space, tab, etc.).

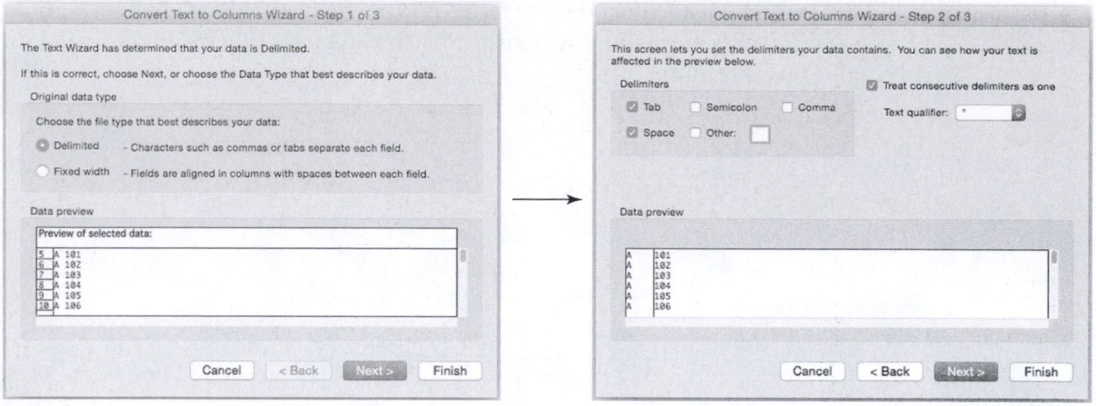

Figure A.6 Split up mixed data types using "Text to Columns."

Remember that in a spreadsheet, each cell must contain a unique number or text, so the above steps are necessary if any of the imported values are bunched together.

Another method of inputting data is when the source document is a CSV (comma-separated values) file. Simply go to File → Import and the rest is straightforward.

Skill 4: Manipulate Data

Once your data have been entered, it is often necessary to modify the values to make them presentable or graph-ready. Reasons for modifications include converting units, adding numbers, calculating new quantities based on existing numbers, etc. In any case, the use of *formulas* is necessary. Let's look at some examples.

Unit Conversion

From engineering quantities (e.g., pressure, temperature, force, stress) to currencies, we often need to convert units before the data will make sense. Let's assume that you studied abroad in Sweden for a month and would now like to tally up all your spending for that month. All expenses—rent, food, commute, gifts, etc.—were in Swedish krona. Converting the spending into U.S. dollars requires a simple formula (Fig. A.7):

Then simply drag Cell C6 down, or use the CTRL-C and CTRL-V procedure as previously discussed. Notice that the conversion rate of 0.12 Krona-to-USD is isolated and placed somewhere near the top of the spreadsheet, in Cell C3. This way, you can easily update the exchange rate only once instead of having to go into each and every formula.

The formula then references this cell, using the "$" symbol (i.e., "C$3" instead of "C3") in order to "protect" this C3 cell and prevent it from getting automatically

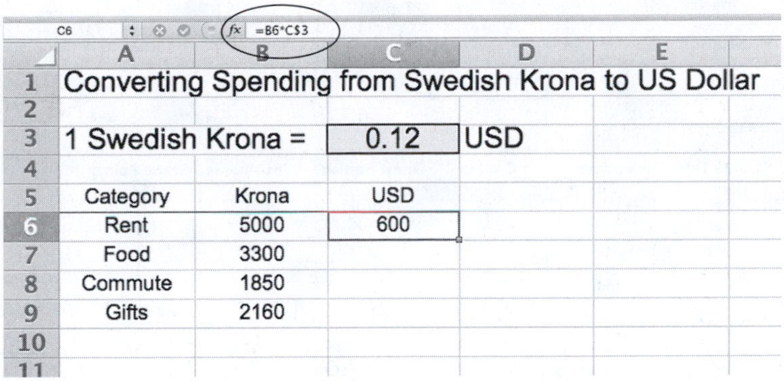

Figure A.7 Use multiplication in formulas to convert currency.

incremented (or "relatively referenced") when you drag the first formula down (or copy/paste the formula). Adding a "$" symbol before a column letter or row number in a formula changes the default, *relative* referencing to *absolute*. Here, only the number "3" needs to be made absolute (i.e., "$3"). However, there is no harm for using "C3" here instead, since the end result is the same. If, instead of populating Cell C6 downward, you decided to drag it to the right, then either "$C3" or "$C$3" in the formula would be necessary.

New Quantities Using Equations

Often we need to create a new quantity based on the input values, using mathematical relations. The idea is the same as the example above. Let's use GPA calculation as an example.

The equation for calculating GPA for a given term is

$$GPA = \frac{\text{Total Grade Points}}{\text{Total Credits}} \text{ where}$$

$$\text{Total Grade Points} = \Sigma(\text{Point received for each course} \times \frac{\text{Number of credits}}{\text{for that course}})$$

Figure A.8 shows a sample calculation, where the overall GPA is calculated in Cell F16 using the formula

$$=F14/C14$$

Total Credits is calculated in Cell C14 using the formula utilizing one of the most useful built-in functions, "sum":

$$=sum(C8:C13)$$

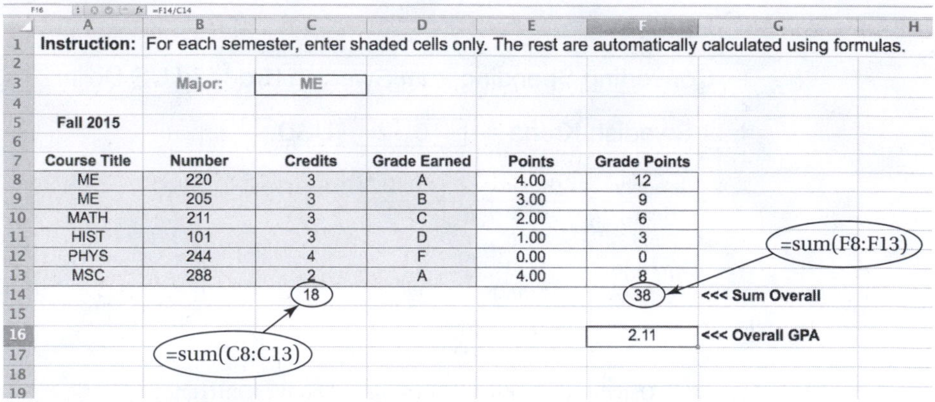

Figure A.8 GPA calculation

whereas Cell F14 contains the formula to calculate Total Grade Points:

$$=sum(F8{:}F13)$$

Note that the calculation of Total Grade Points requires first calculating the points received for each course (Column E) based on the grades earned (Column D)—that is, an "A" grade yields four points, "B" three, "C" two, etc. To automate the calculation of points, the use of "conditional quantities" is necessary.

Conditional Quantities: "If"

In the preceding paragraph, a *logic* exists whereby you may conduct a *test*: if the grade earned (Column D) is an "A," then return a number of "4" (in Column E). Otherwise, if the grade is a "B," then "3." Keep testing until you reach the minimum grade, "F," in which case returns "0." In other words, use the "if" function. Understanding, and mastering, the "if" function will unequivocally empower you as a spreadsheet user. It takes only a second to get used to it, so let's dive right in.

In Figure A.8, Column E is where you will use the "if" function. By the way, you *could* manually enter the points for each course and for each term, but to save time and prevent human error, automating this process using an "if" formula is *much* preferred.

The formula in Cell E8 is

$$=if(D8{=}"A",4,if(D8{=}"B",3,if(D8{=}"C",2,if(D8{=}"D",1,0))))$$

Let's break down this seemingly complex formula. The "format," or *syntax*, of the "if" function is

$$=if(condition,\ value\ if\ true,\ value\ if\ false)$$

where

condition: A logical test involving a comparison of x and y (i.e., x>y, x<y, x=y, x>=y, or x<=y)

value if true: The result if the above logical test is true

value if false: The result if the above logical test is false

Each of these three "fields" is separated by a comma (or by a semicolon in OpenOffice Calc). Putting them together, we have the formula in Cell E8, repeated in Figure A.9:

=if (D8="A" , 4, if (D8="B" , 3, if (D8="C" , 2, if (D8="D" , 1, 0))))

$\underbrace{}_{\text{Condition}}$ $\underbrace{}_{}$ $\underbrace{}_{\text{Value if false}}$

Value if true

Figure A.9 The powerful "if" function

Notice how the "value if false" contains yet another "if" function. This is called a "*nested* if" where an "if" function resides within another. In the example above, the "if" function is nested three times.

The remaining steps to calculate the overall GPA are a breeze by comparison.

In Figure A.8, notice that the course title and course number columns (A and B) are kept separate. Instead of typing "ME 220" in a single cell, "ME" and "220" are kept separate in two adjacent cells (A8 and B8). This is crucial in calculating the *Major* GPA (i.e., GPA based only on courses in your major field of study).

To accomplish this, another neat function called "sumif" is used (Fig. A.10). The formula used in Cell C15 is

=sumif($A8:$A13,C3,C8:C13)

	A	B	C	D	E	F	G	H
	\multicolumn{8}{l}{C16 =SUMIF($A8:$A13,C3,C8:C13)}							
1	**Instruction:** For each semester, enter shaded cells only. The rest are automatically calculated using formulas.							
2								
3		Major:	ME					
4								
5	Term 1							
6								
7	Course Title	Number	Credits	Grade Earned	Points	Grade Points		
8	ME	220	3	A	4.00	12		
9	ME	205	3	B	3.00	9		
10	MATH	211	3	C	2.00	6		
11	HIST	101	3	D	1.00	3		
12	PHYS	244	4	F	0.00	0		
13	MSC	288	2	A	4.00	8		
14			18			38	<<< Sum Overall	
15			6			21	<<< Sum Major Only	
16								
17						2.11	<<< Overall GPA	
18						3.50	<<< Major GPA	
19								

Figure A.10 Use "sumif" to perform conditional sum: Major-Only GPA.

The idea is similar to the "if" function in that it allows you to sum up numbers only if a certain criterion is satisfied. Here, to calculate the Major GPA, you will need to add the number of credits (Column C) and grade points (Column F) *pertaining to your major courses only*. So the criterion used here is Column A, the course title. Perform the sum if, and only if, it matches your major (Cell C3).

The remaining steps are straightforward.

Formulas are essential in creating and using your spreadsheet, not just for engineering but for everyday applications as well. In Section A.4, you will explore more advanced formulas in a variety of projects and exercises.

Skill 5: Format Data

The appearance of your data can significantly boost the effectiveness of the spreadsheet. Typeface and font choices, cell alignment and size, background color, and cell bordering can make a complex spreadsheet legible and easily comprehensible.

Figure A.11 shows an example of engineering optimization of a truss-type bridge design (included in Section A.4 as a project). Without formatting, the numerous data appear unorganized and incomprehensible. Applying some editing to the cells can drastically improve legibility.

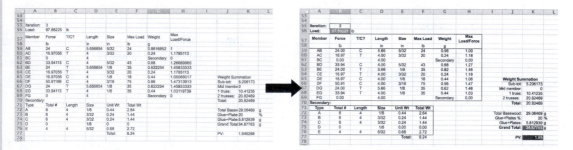

Figure A.11 Simple formatting can go a long way.

Excel, like any software, comes with its own "factory default" settings, including the formatting of cells. More often than not, these default behaviors are less than ideal. Simple efforts such as center-aligning the entire column, adding borders to header cells, and adding background colors to important data will go a long way.

Skill 6: Plot Data

It is often necessary to visually represent your data in 2D space, especially when you have a large amount of values or would like to analyze the trend. There are a number of ways to graph your data, depending on the type of your data values. Typically two types of graphs are used: Number versus Category (Fig. A.12) or Number versus Number (Fig. A.13).

Number Versus Category

If you need to visualize the trend of something descriptive or qualitative, then this type of graph is well suited for the purpose. To plot your spending (dollar amount) for each category (food, travel, rent, etc.), you may use a bar chart or a pie chart. To observe the pattern or trend of your past GPA, use an *xy* plot.

Number Versus Number

If your data involve at least one pair (or "set") of *related* numbers, often via some mathematical function (e.g., $y = x^2$), or if you would like to *quantify* (instead of simply observe) the data trend, then use scatter plot.

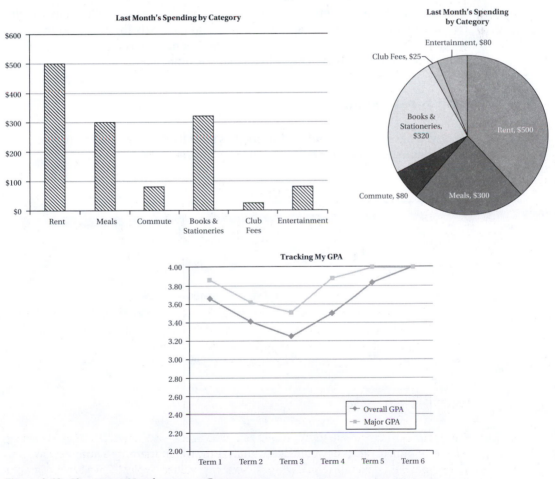

Figure A.12 Chart type: Number versus Category

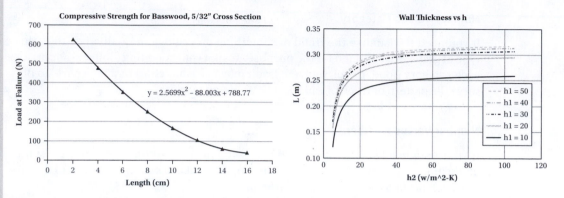

Figure A.13 Chart type: Number versus Number

If you have more than one set of data (e.g., spending category, dollar amount, *plus* frequency), then you can always add data sets to your existing plot. You may also specify a "secondary" *y*-axis to represent an additional dataset, especially when its range is in a different order of magnitude than the other data sets.

Skill 7: Format Plot

Once you have plotted your data, the graph is most likely in its "raw" form (Fig. A.14) and needs to be polished to make it legible and presentable.

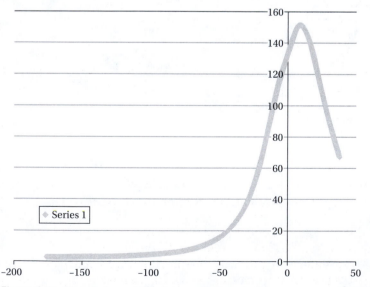

Figure A.14 Raw, unformatted plot

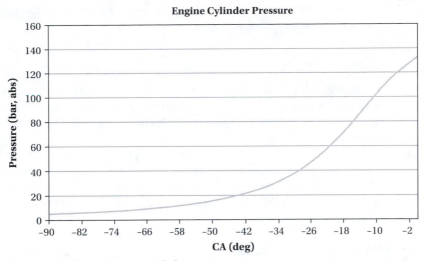

Figure A.15 Properly formatted plot

A typical formatted plot should contain the following elements: chart title, axis assignment and titles, axis numbers/text, data representation, legend, grid, chart background, and annotation (Fig. A.15).

Chart Title

Every plot must contain a concise title informing the audience about the big picture (e.g., Adam's GPA 2016–19, My Monthly Spending 2015).

Axis Assignment and Titles

A spreadsheet plot typically contains two axes: x and y. While you have full control over how your data are presented on this x-y space, take care when assigning x and y variables. Once you have decided which axis will represent a variable, give a clear and concise title to each axis.

Axis Numbers/Text

If your data along an axis contain numbers, carefully choose the spacing and decimals. Often the default behavior of the software will attempt to "borrow" the number format of the source data cells, which may not be suitable on the plot (e.g., too many decimal points). Make adjustments so that the axes are not crowded and the minutiae of the data trend are easily discernible by the human eye.

Also pay attention to the axis scale (i.e., the range of your data you want presented). If your data (e.g., GPA) are meaningful only in the range of, say, 3.2 to 3.6, and if your goal is to analyze the trend of your GPA, then consider limiting the axis scale to 3.0 to 3.8.

Another useful feature is "grid." Gridlines along an axis allow the reader to quickly capture the location of a data point.

Overall Appearance

The focus of a plot is the data itself. The rest—background color, borders, etc.—should not distract the reader from understanding your graph. Keep the plot background neutral; a white or gray background typically works well. Do not use a dark background, to save ink if printing becomes necessary.

Since your plots may not always be viewed on a monitor or printed with a color printer, consider using not only colors but also line types or symbols to represent your data, particularly when your plot contains multiple data sets. This will also allow readers who are colorblind to appreciate and understand your graphs.

Skill 8: Interpret Data

Now that you have created a well-polished plot for your data, let's try to make sense of it by analyzing the trend (i.e., the shape of the graph) and arriving at some conclusion and perhaps implication for the future.

If you noticed a significant jump on your "Overall Spending vs. Time" chart Fig. A.16), then an alarm should sound. It would also warrant an investigation

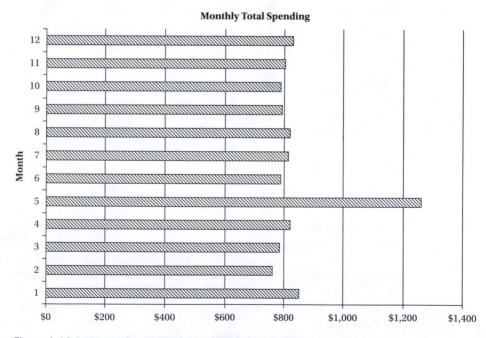

Figure A.16 Interpret data: Why the high spending during Month 5?

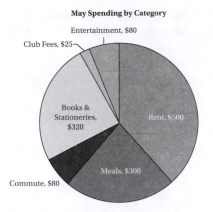

May Spending by Category

Entertainment, $80

Club Fees, $25

Books & Stationeries, $320

Rent, $500

Commute, $80

Meals, $300

Figure A.17 A pie chart may lend some insights.

regarding, say, what categories contribute most to the increase, by analyzing a particular "Spending Category" plot (Fig. A.17).

Visual inspection like this is sometimes insufficient, however, particularly in engineering analyses. Instead, a more quantitative description of the trend is desired, employing a technique called "curve-fitting" (Fig. A.18). Other common names include "regression analysis" and "trendline." Essentially, this technique attempts to generate an equation to best represent your discrete data.

The example in Figure A.18 shows experimentally observed compression loads that caused failure for various lengths of basswood. During the experiment, eight different basswood lengths, incremented by 2 cm, were tested. The test result for each

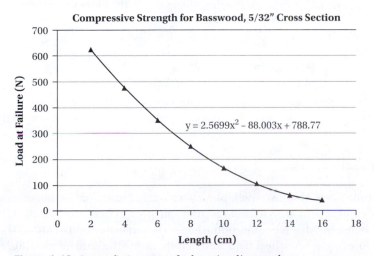

Compressive Strength for Basswood, 5/32" Cross Section

$y = 2.5699x^2 - 88.003x + 788.77$

Load at Failure (N)

Length (cm)

Figure A.18 Curve-fitting a set of otherwise discrete data

length is represented by a triangle. A curve-fit, or trendline, is now added, using the "second-order polynomial" option (among others). The curve-fit equation, as determined by Excel, is

$$y = 2.5699x^2 - 88.003x + 788.77$$

You may then use this equation to help predict failure at a length of, say, 4.25 cm:

$$\text{Failure Load at 4.25 cm} = 2.5699 \cdot (4.25)^2 - 88.003 \cdot (4.25) + 788.77 = 461.18N$$

Skill 9: Preserving Data

All types of software evolve, and some do not survive the test of time. File types such as Excel's "xls" or "xlsx" may have been around a long time, but this does not guarantee their existence in the future. Saving your spreadsheet in PDF or JPG format may remove some of the uncertainty as long as future software continues to support them, even though you would not be able to modify the sheet with these formats.

If a major corporation is about to go under, its proprietary software may very well go with it. Similarly, open-source programs can, and do, go through life cycles during which a particular file type may lose future support.

What this means for you is simply this: be aware of the status quo of computing advancement. If a file type is approaching extinction, then take steps toward converting your files.

Another way of preserving your work is by physical archiving—that is, printing your plots and data. Be mindful, however, of the amount of materials required if you have an enormous data range. Before printing, first carefully define the "print range," perform a print preview, and perhaps save as a PDF before you print.

A.4   Takeaway Projects

In addition to constructing and maintaining your GPA and spending spreadsheets, next you will find a variety of scenarios in which to practice the nine skills.

Gym Schedule

A sample Excel file, downloadable at www.oup.com/us/oakes, illustrates an analysis of the weekly activities in the large, multipurpose gym at a local YMCA. It is based on a publicly available schedule, published on the website of this YMCA. Translating the schedule into spreadsheet data, and manipulating the data for plotting, results in the graphs you see in the Excel file.

Your task now is to identify a local fitness facility, look up its published weekly schedule, and perform a similar analysis. Alternatively, you may also analyze your own weekly calendar schedule, and perhaps compare "target values" and "actual values." In other words, on a calendar (paper or electronic such as Google Calendar), allocate activities (e.g., "Do homework for ME 220, Tuesday 3 pm–5 pm") for a given week. Observe the actual occurrence, and enter these actual data on your spreadsheet. Make several charts to show your weekly activities, and analyze the accuracy of your targeted time allocation. If the actual time usage data is quite far off, then, based on the graphs you make, attempt to identify the root cause and make adjustments for the following week.

Diesel Engine

You may sometimes encounter data that are irregular or fluctuating, especially if those data have been collected by laboratory instruments during an experiment. Before plotting and analyzing the data, it may be necessary to "smooth out" the fluctuations. The file at www.oup.com/us/oakes contains real-life data for a diesel engine. Specifically, real-time cylinder pressure data, collected by pressure transducers, have been reported for each "crank angle," which, essentially, represents time. These experimental data are fluctuating due to the accuracy limitations of the instruments as well as the minute disturbances that occur in the natural world.

This Excel example presents a particular smoothing technique whereby you can define the degree (or extent) to filter out those little disturbances or "noise." The higher the degree, the more neighboring data points you will "borrow" to arrive at some average. The filtering-technique equation is included in the file, under the "Equations" tab. Study the "Raw Engine Data" tab, and compare the smoothing equation to the Excel formula in Columns E and F.

Your task is to smooth out the raw pressure data (Column C) by specifying a filtering degree—that is, choose a "b" value different from what the example uses (currently $b = 11$). When done, plot and compare both pre- and post-smooth pressure data. Zoom in, by adjusting the scale of the x- and y-axes, to appreciate what smoothing has done to the otherwise fluctuating data.

However, remember that if you are dealing with only a handful of data points, what appears to be erratic may in fact contain statistical significance. In this case you should not smooth the data.

Truss Bridge

In designing truss bridges, design load and bridge weight are among the primary considerations. The Excel file www.oup.com/us/oakes contains a tutorial for optimizing the load-to-weight ratio, dubbed the "performance value" (PV), for a model truss bridge, made of basswood, spanning 24 inches end to end. Only five sizes of basswood are allowed: 3/32", 1/8", 5/32", 3/16", and 1/4". The optimization of PV is an iterative process, and this spreadsheet includes a tutorial for coming up with the best PV. All property data for basswood are also given, particularly its compressive and tensile strengths (Abramowitz, H., *Basswood Bridges*, Paper presented in June 2008 at Annual Conference & Exposition, Pittsburgh, Pennsylvania. https://peer. asee.org/4111).

Impress Them: How to Make Presentations Effective

Yeow K. Siow, PhD
University of Illinois at Chicago

B.1 Why Bother?

Driven by workforce needs, today's engineering education emphasizes producing graduates who are not only technically savvy but also well rounded. This means employers are looking for individuals who exhibit many non-technical traits, including being able to communicate effectively. In other words, your ability to make an impression may very well land you a great career, or win you a scholarship, or help you earn an A.

Throughout your college years, you will need to make more than a few individual or group presentations. This need becomes even more pronounced in the work environment or graduate school. Learning what it takes to make an enticing and powerful presentation—and doing so as early as possible—can, and will, open many doors for you in the years to come.

This appendix intends to show you the makeup of a good presentation, some best practices to adopt, what *not* to do, and advice on confidence building. It does not, however, teach you the step-by-step process of using any particular type of software. We assume you already know your way around the modern computing environment.

B.2 What Does It Take?

A presentation does not necessarily always require software and a projector, but in technical situations you will most likely need them to help you get your points across. All in all, the tools you will need to make a strong presentation are: Vision, Audience, Visuals, Practice, and Intangibles.

A Clear Vision of Your Points

Having a vision seems like a no-brainer, but presenters can easily falter or struggle to stay on track if they have no goals to guide them in the first place. A vision will also help streamline the preparation of your presentation.

An Understanding of Your Audience

Knowing your audience and their background will dictate your presentation content as well as style. Are you presenting to a group of sixth-graders? Then keep it simple, colorful, and fun. Speaking to professors and your classmates? Turn up the level of math and technical drawings.

A Visual Representation of All Your Ideas

As mentioned, you will most likely need to rely on software and hardware to visually convey your points. No particular software is preferred or advocated here; use what is convenient or feels comfortable for you. Popular choices include Google Slides, Prezi, Apple Keynote, and Microsoft PowerPoint. Each has its pros and cons, but do try to familiarize yourself with at least a couple of software choices. Google Slides, being cloud-based, is an excellent tool for collaboration or editing using different devices. Prezi uses "nonlinear" or "no-slide" presentation, resulting in a dynamic—and sometimes dizzying—show. Apple Keynote has strong visual capabilities while suffering from cross-platform compatibility. Microsoft PowerPoint is simply ubiquitous.

Whatever you choose, chances are that presentation templates exist. Some templates, especially simple ones, can be easy to adopt and customize. Some others, however, may be clumsy and counterproductive. On the other hand, starting from scratch may provide you with the flexibility and opportunity to be creative.

Regardless of which software you use, keep in mind a few things:

- Go beyond the default settings, including typeface, font style, background, and transition.
- Find out, if possible, where and on what device you will be making the presentation. Standalone software uses the fonts available on the device you use to create the presentation.
- Experiment and be bold!

Visual effectiveness will be discussed in detail in Sections B.3 and B.4.

Ample Preparation and Practice

Always prepare and take full control of your own presentation. In the case of a team presentation, each presenter should peruse the entire document, and treat it as if he or she is the only person presenting.

Practice in front of a mirror or a mock audience, and, if possible, in the actual presentation room. Be so familiar with your entire presentation that, with only a quick glance, you will be able to elaborate on your points contained in a slide and make a smooth transition to the next.

The most effective speaker faces the audience almost 100 percent of the time. With enough practice, you will not need any cue cards or notes. Cue cards are not only distracting, but they also undermine the audience's perception of your confidence. If your slides are well made, and if you spend time practicing, a quick glance at a slide should instantly remind you of the points to elaborate on.

Intangibles: Eloquence and Charisma

After you have prepared all your presentation materials, it is time to put everything into action. Your words and overall *presence*—not appearance—are crucial for enticing the audience. Outward appearance, granted, can give an important first impression, but a first impression can only last so long. Get plenty of sleep the night before so your eyes do not become puffy (or your speech slurred). Dress comfortably, and do not be concerned if you should wear a tie or a jacket. Instead, focus on the more important thing: your delivery.

Smile, and your audience will smile back. Relax, and they will relax with you. Keep your head up high, scan the entire room, and always make eye contact—particularly with those who really pay attention or give you positive energy. Use a wireless clicker and laser pointer, if possible, to free yourself up so you can move around the floor and get closer to the audience. Keep smiling. Speak in your natural tone and pace; do not force any artificial style. No cue cards should be necessary; your slides *are* your cues. Do not recite a prewritten speech or, worse yet, read from the screen verbatim. A little humor helps, but only if you are comfortable with it. Do not stop smiling. Always face the audience. Look behind you only during slide transition or when pointing out interesting facts on the screen. Spend no more than a minute or two on each slide, and keep it flowing. Are you still smiling?

All in all, if you have a clear vision, know who your audience is, create great visuals, and practice over and over, you will eventually acquire the eloquence, confidence, and charisma to deliver an impressive presentation time and time again.

B.3 Some Samples

Let's take a tour of some sample slides. Judge for yourself if each one makes a strong, positive impression or makes your head spin.

Words Versus Graphics

Would you prefer to see an entire presentation that consists of only words (Fig. B.1) or one that contains strong visuals (Fig. B.2)?

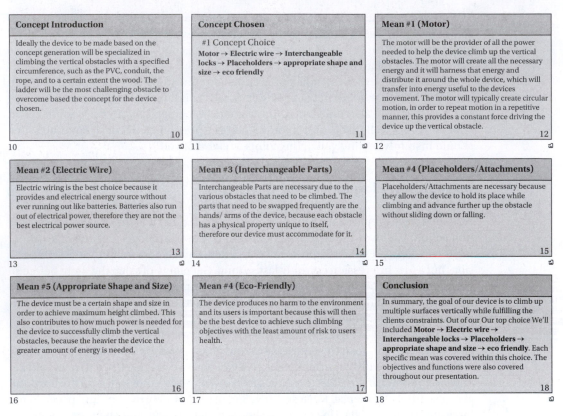

Figure B.1 Nothing but words in the entire presentation

First Slide

Which of the slides in Figure B.3 creates a better first impression?

Too Much Information?

In Figure B.4, one slide is simple and legible; the other forces the audience to squint and is simply stuffy.

Listen or Read?

Both slides in Figure B.5 attempt to present to a panel of judges a proposed engineering design for scaling a vertical surface. The first slide offers impactful visuals, the details of which are to be elaborated by the presenter. The second slide, while containing a sketch, has an accompanying essay that should be spoken words instead.

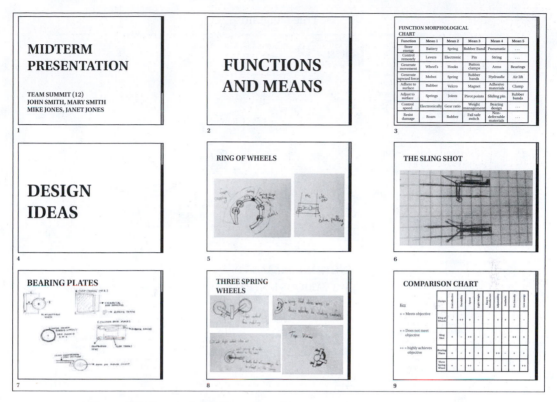

Figure B.2 Strong visuals throughout the presentation

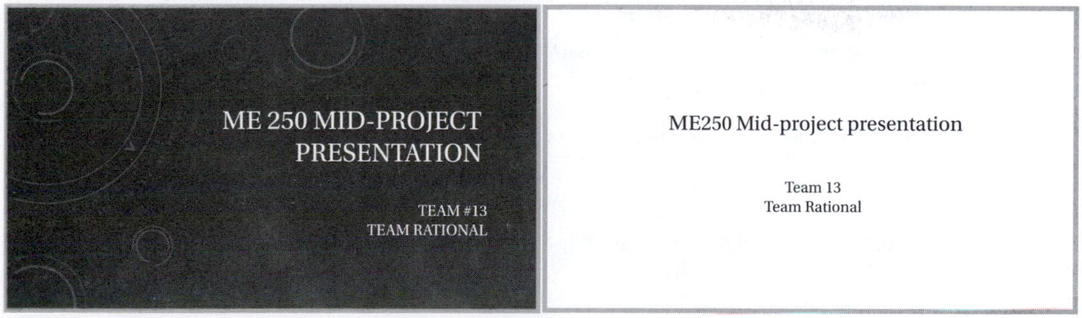

Figure B.3 First impression matters: fresh and modern (left) versus bland and bare (right)

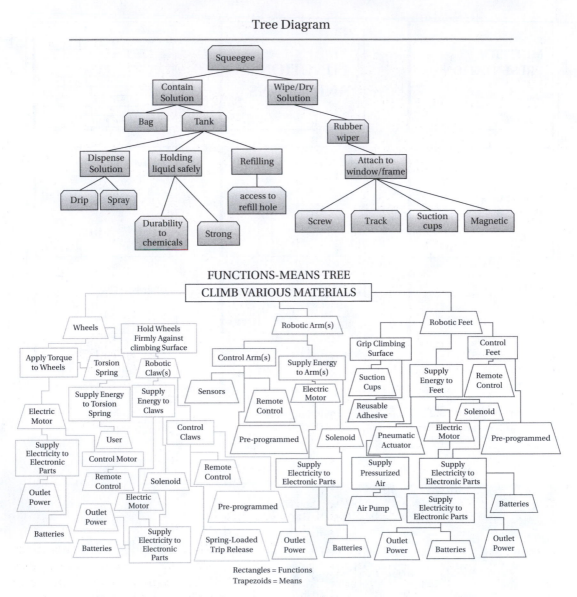

Figure B.4 How much information is too much?

B.4   **Break It Down: Elements of a Slide**

Having seen a sampling of slides here, we'll now dissect one and assess how each element contributes to the effectiveness.

Final Design

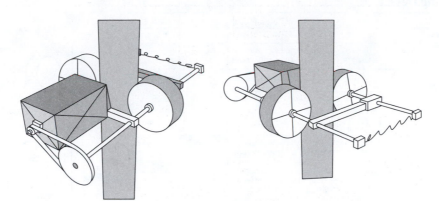

Final Design

The main function of this device is to act as a vice, in the sense of easily tightening around whatever is between the two sides. By having 2 wheels on the sides of the device rotating in opposite directions towards the material, the device will self-tighten around the material until the torque and friction forces are greater than that of gravity. The motor, when combined with the gears, is responsible for turning the wheels towards each other, and in combination with the pulleys, results in a self-tightening device. Powered by a battery, once the switch is set to on, the device will begin to tighten around the material. When the device is tight enough around the material, the friction from the rubber wheels and the torque from the motor will be a greater force than gravity moving the device vertically up the material.

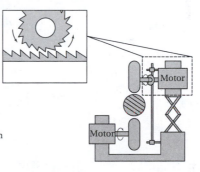

Figure B.5 Should the audience listen (left) or read (right)?

Element 1: Canvas and Background

A slide begins with a blank space, or canvas, on which to build your content. Typically you can choose between 4:3 or 16:9 aspect ratios for the canvas dimensions. Most newer projectors, screens, or monitors are flexible enough to handle these aspect ratios. In effect, it comes down to preference. If you feel that your contents fit better in, say, the 16:9 format, then go for it, and use it throughout your entire presentation. As shown in Figure B.6, the wider aspect ratio provides more horizontal real estate for side-by-side content.

While it may add value to incorporate a splash of style, it is typically best to keep the background clean and simple. You may add textures or use graphics for the background, as long as they do not distract the audience from the main points. Whatever

Rolling Device

Means	1	2	3	4	5
Functions					
Power Generation	Human Energy	Combustion	Air Compression	Battery	Hydro Powered
Mechanism to Attach to Obstacles	Suction	Clamps	Friction Wheel	Concave Wheel	VELCRO™
Gripping Force	Torsion Spring	Motor	Pneumatic Pump	Magnets	Linear Springs
Way to turn energy to motion	Torsion Spring	Gears	Motors	Pistons	Linear Spring
Chassis Material	Aluminum	PLA - Plastic	Wood	Cardboard	Steel
Fasten Power Generator to Body	Screws	Glue	Rubberbands	Rivets	Weld
Initial Power Input	Winding	Spark Plugs	On/Off Switch	Remote	Valve Release

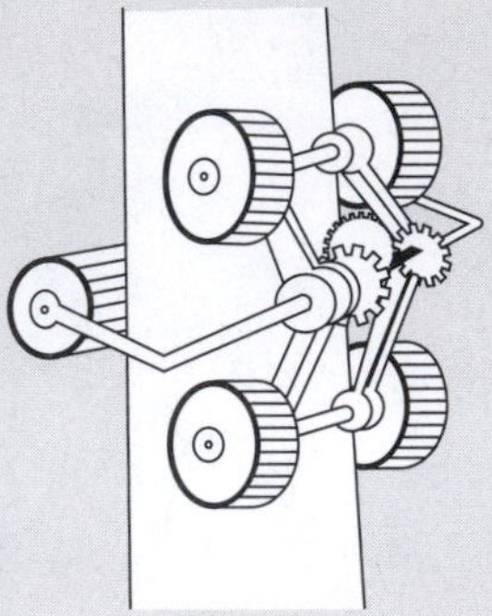

Rolling Device

Means	1	2	3	4	5
Functions					
Power Generation	Human Energy	Combustion	Air Compression	Battery	Hydro Powered
Mechanism to Attach to Obstacles	Suction	Clamps	Friction Wheel	Concave Wheel	VELCRO™
Gripping Force	Torsion Spring	Motor	Pneumatic Pump	Magnets	Linear Springs
Way to turn energy to motion	Torsion Spring	Gears	Motors	Pistons	Linear Spring
Chassis Material	Aluminum	PLA - Plastic	Wood	Cardboard	Steel
Fasten Power Generator to Body	Screws	Glue	Rubberbands	Rivets	Weld
Initial Power Input	Winding	Spark Plugs	On/Off Switch	Remote	Valve Release

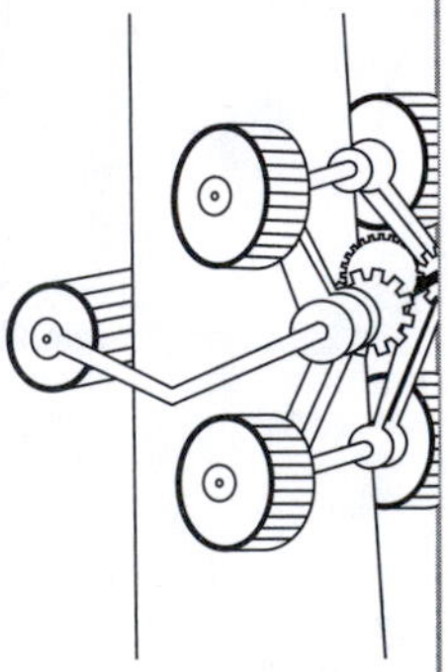

Figure B.6 Canvas aspect ratios: 16:9 (top) versus 4:3 (bottom)

background you choose, make sure it creates a coherent and pleasing contrast to the content (elements 3 and 4). Figure B.7 shows two background choices. One is modern and soothing, while the other is busy and distracting.

Element 2: Borders and Zones

Generally no borders are necessary, except for defining "zones" separating, say, the heading from the content. Typically each slide contains some title or header, the location and orientation of which are completely up to you.

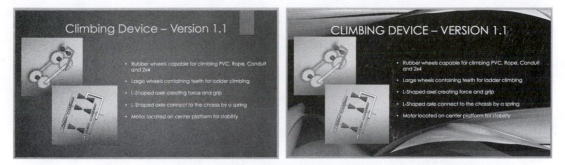

Figure B.7 Background choices: clean (left) or fussy (right)

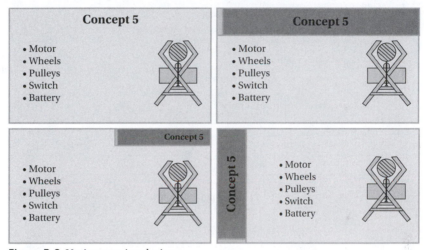

Figure B.8 Various zoning designs

Figure B.8 shows different styles of bordering (or the lack thereof). Each one is unique. Generally, even if you use a template, try to experiment with the "theme" or "slide master" until you find a bordering style that feels best for you. Then apply the same style to your entire presentation instead of mixing different zoning styles.

Element 3: Graphics

The sole purpose of using electronic slides in a presentation is to add *visual impact*. While Elements 3 and 4 go hand in hand, graphics should be the first thing you think of when it comes to disseminating your ideas. Can you use an image or chart to represent a 50-word statement, as shown in Figure B.9? Can you include a relevant graphic in each slide? Can your entire slide contain only images?

Do not, however, include random or irrelevant imagery simply for the sake of including graphics. Anything you put up on the screen must have a meaning and a purpose. Otherwise, it is nothing more than a distraction.

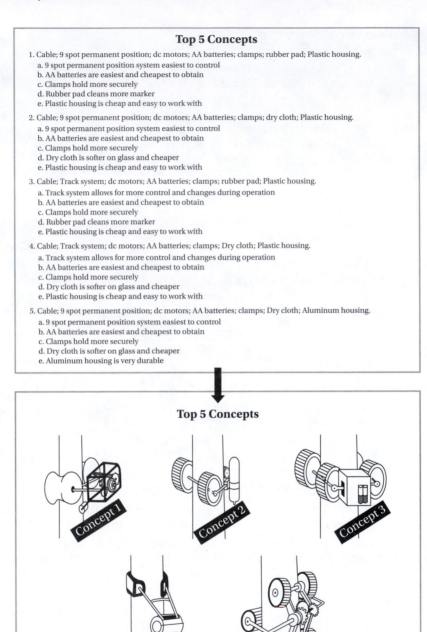

Figure B.9 Turn boring words into strong visuals.

Element 4: Text

Written words, of course, are necessary, but never make your audience read an essay! Any explanation or discussion should be done by you, using spoken words. Each slide should contain *just* enough text to achieve the following goals:

1. Instantly remind yourself of the details of a point.
2. Convey an instantly recognizable idea to the audience.

The audience members should split their attention between you and your slide. If they are fixated on the screen, your slide may contain too many words or too much information to digest.

To make words legible and quickly digestible, you should do the following:

- Use a good typeface (e.g., a common sans serif font like Arial, Calibri, or Helvetica).
- Do not mix more than two typefaces.
- Choose a good font size.
- Use a good font color that contrasts with the background. Avoid an overly saturated color palette.
- Use bullet points.
- Do not use complete sentences; instead, treat words as if you are writing news headlines.
- When you are done typing, read the slide a few times. If you find yourself having to reread some words or spend more than a few minutes reading, then consider rewriting or splitting up the slide.

Compare the two slides in Figure B.10. One uses mixed typefaces (Arial and Times New Roman), has low contrast, and carries two full paragraphs. The other slide has much better coherence and legibility, employing bullet points, an impactful sans serif typeface, and good font size.

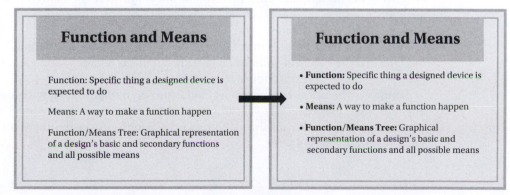

Figure B.10 Impact of typeface and font style

Element 5: Videos and Links (Optional)

Generally, moving images and audio can do more harm than good in a technical presentation, especially when the video is externally sourced (i.e., someone else's creation). Any unoriginal content is better avoided, because if someone else is explaining the topic for you, then why are you even here?

Of course, if your content inherently includes motion (e.g., research in combustion inside a car engine with moving pistons), then it may be beneficial to incorporate video clips in your presentation.

A major challenge in including a video is the need for a hyperlink or embedding of the clip, so that it can either auto-play or play on demand. Each software program handles videos differently, so be sure to experiment and conduct multiple tests using the exact hardware of your actual presentation. Another challenge with videos is the sound—whether it will sync with the video or play at all. For moving imagery without audio, "animated GIF" may be a reliable format and is easy to embed. Similarly, hyperlinks to websites are best avoided unless the purpose of your presentation is to, say, evaluate the effectiveness of today's web design.

Keep in mind that a successful test on your home computer does not guarantee that it will work the same way on the presentation hardware. Therefore, an *in situ* test—at the actual presentation facility using the actual hardware—is optimal if at all possible.

Element 6: Flow: Animation and Transition

After you have created all of the above elements, think about how much information, and in what sequence, you want your audience to see at any given time. Do you want your audience to see everything at once in a slide, or do you prefer showing one bullet point or image at a time? Most software lets you define and control content "animation" behavior, including special effects.

Figure B.11 shows a finalized slide containing a header, an image, and text. Instead of putting all these items on the screen at once, however, you may animate the sequence of appearance, as illustrated in Figure B.12. Simply hit a key or use a wireless clicker to advance to the next item.

Some software programs may allow you to predefine an elapsed time between items so that the slide will animate on its own, but this is not recommended. You, the presenter, should have full control of the pace.

Similarly, when advancing to the next slide, you may be able to specify a transitional effect as well. However, any special effect that is extravagantly designed may be distracting or even cause dizziness. Refrain from using any over-the-top features and focus instead on the contents.

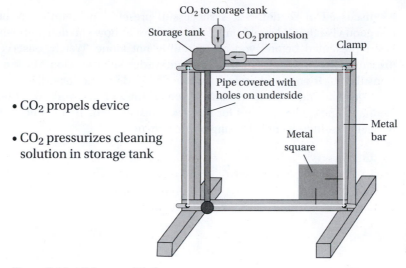

Figure B.11 All-in-one slide

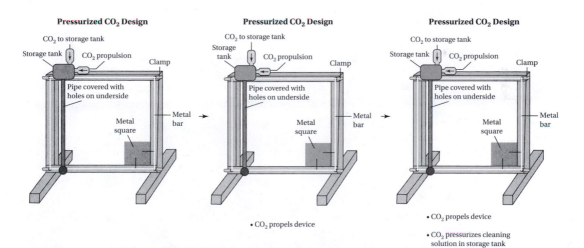

Figure B.12 Add animations for better flow.

B.5   **It's Much More than Slides**

As discussed in Section B.2, a successful presentation requires a lot more than just good visuals. Slides can only be as effective as how you deliver them. If you have never presented before, you are certainly not alone. Watch, observe, and study presentations—recorded or live—by luminaries such as Elon Musk or Steve Jobs. Consult your friends and professors. Practice, and practice again.

Do not hesitate to experiment. If something doesn't work out, then take note, evaluate, and make changes for the next presentation. There is no such thing as a perfect presentation, only lifelong improvement.

APPENDIX C

An Introduction to MATLAB

Yeow K. Siow, PhD
University of Illinois at Chicago

C.1 Introduction

Facing the world's challenges and finding solutions to problems, today's engineers and engineering students often need to handle large amounts of data and solve complex equations. Doing calculations by hand or a calculator may suffice for the simplest of situations, but in most cases a computer is necessary. Knowing how to rely on modern computing power to help solve problems is a skill every engineer must have. Being able to instruct a machine to perform calculations for you will empower you as an engineer as well as a human being.

The purpose of this appendix is to provide a brief introduction to the MATLAB® computing environment (MATLAB is a registered trademark of The MathWorks, Inc.). There are numerous texts written about MATLAB and its use. Our intent here is not to provide a thorough understanding of the MATLAB environment, as there are entire texts that do this well; our favorite is *Introduction to Technical Problem Solving with MATLAB* by Sticklen and Eskil. Rather, it is to begin exploring the possibilities of this powerful computer tool that you will be using throughout your engineering studies and career. As a way of introducing MATLAB, we will use examples of calculations that may be required in your freshman-level engineering courses.

The MATLAB program was originally written as a matrix manipulator named MATRIX LABORATORY. It has proven to be a very powerful tool in performing engineering computations and has grown beyond its original intent. Some of the first applications of MATLAB were in the areas of image and signal processing, which are part of electrical and computer engineering. It has since become widely accepted in a broad range of other engineering areas and programs. Many engineering curricula have moved to making MATLAB the primary computing tool in their undergraduate program, replacing traditional computer programming.

One reason for its popularity is that it is relatively easy to use and has the power of a computer language. In a traditional computer language, such as Fortran, C, or C++, a program is created. It then must be compiled, which means that it is translated into a language that the computer can understand. This compiled version of the program is then run by the computer. MATLAB provides the flexibility of programming in the traditional sense of a computer language but also has the functionality of a calculator. MATLAB removes the compiling step for the user.

MATLAB can run on many different platforms, including UNIX/Linux, PC, and Macintosh. The programs, data, and results are the same across platforms, making it easy to use and easy to work collaboratively on larger projects. The examples for this appendix have been created on a PC using the student version of MATLAB that is widely available for purchase from campus bookstores, although some colleges offer it for free. This student edition contains many of the capabilities of the full professional version but with the flexibility of being able to install it on your own computer. MATLAB has numerous functions that can be added through extra software packages called toolboxes. These toolboxes contain built-in functions that provide special capabilities in areas such as financial analysis, image and signal processing, fuzzy logic, control systems design, and symbolic calculations.

MATLAB commands can run in either **interactive mode** or **batch mode**. Sections C.2 through C.9 explain some basic features of MATLAB in interactive mode. In Section C.10, you will learn about batch mode, in which a collection of commands are executed all at once by the computer. In other words, you will learn about the basics of programming.

Although MATLAB is the prevalent software, a few popular alternatives exist and are worth investigating. The most highly regarded alternative is called Octave, from GNU (GNU's Not Unix). It is an open-source free software and can run on UNIX/Linux, Windows, and Mac. Its functionality and syntax are similar to MATLAB, with some minor but noted differences. You are encouraged to explore and tinker with software alternatives—the reward may be tremendous.

C.2   **MATLAB Environment**

Starting MATLAB is very simple. On a UNIX platform, simply type MATLAB at the UNIX prompt. On a PC or Mac, double-click on the MATLAB icon. A window will appear. This window consists of three components, your Command window, the Command History window, and the Launchpad window. The Command window is what you will use to run your programs and to see the results as shown in Figure C.1. A prompt should appear that looks like ">>" as in the figure. This indicates that MATLAB is ready for a command.

The Launchpad window allows you to start applications and demonstrations by clicking the icons in the window. This window can also show the variables that are active in the current session. To view these, click on the tab labeled "workspace." The Command History window shows a history of the commands that have been entered into the command window.

This window is one of three types of MATLAB windows. There are also graphics windows, which will be discussed in Section C.9 on plotting, and editor windows, which will be discussed with writing programs in Section C.10. A great way to begin

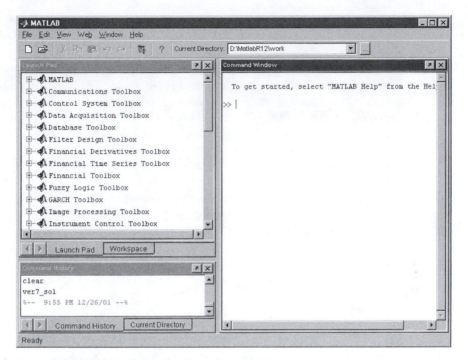

Figure C.1 MATLAB window for the student edition

to explore the possibilities of MATLAB is to run the demonstration programs, or demos. Type "demo" at the MATLAB prompt and then hit the Enter key. Alternatively, click on the "+" symbol next to MATLAB in the Launchpad and then double-click on the Demos icon. This will result in a graphical window (Fig. C.2). Select from the MATLAB topics on the left column to explore the capabilities of MATLAB. New MATLAB users should spend time exploring the demos. They also provide a great way to brush up on MATLAB commands if you have not used them for a while.

MATLAB has built-in help functions that are very useful. As with any software, understanding how to find and use the help functions will allow users to expand their abilities to use this powerful tool. The two main help commands are **help** and **lookfor**. To use the help command, you need to specify a MATLAB command (e.g., "help [command name]"). The help function can be accessed in the workspace using the help command followed by the MATLAB command name or in a separate help window using **helpwin**. The helpwin command will create a separate window in which the functions will be described. The advantage of this separate window is that the workspace window is not taken up with the help topic descriptions. This saves you from having to scroll past the help descriptions to find your earlier work.

The **lookfor** function searches keywords with MATLAB functions and commands. This is very helpful if you don't know the command you are looking for but know the topic. MATLAB uses keywords contained in its functions, so if a word does not yield the desired result using the lookfor command, try synonyms for that word.

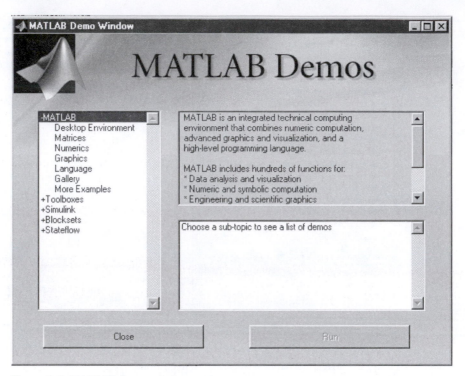

Figure C.2 MATLAB demo window

EXAMPLE
C.1

Use the lookfor command to find the MATLAB commands associated with logarithms. In the MATLAB workspace type the following:

```
» lookfor logarithm
```

The result is that MATLAB displays the following:

```
LOGSPACE Logarithmically spaced vector.
LOG Natural logarithm.
LOG10 Common (base 10) logarithm.
LOG2 Base 2 logarithm and dissect floating point number.
BETALN Logarithm of beta function.
GAMMALN Logarithm of gamma function.
LOGM Matrix logarithm.
```

Notice that the LOG function in MATLAB is for natural logarithms and that base 10 logarithms are calculated by log10.

MATLAB can be used to perform simple operations. The following example shows operations that could be performed on a calculator. In Example C.2, we can see that MATLAB performs simple arithmetic functions that are on most calculators. MATLAB also has features that calculators do not have. For example, MATLAB stores the variables and the results that you have produced in memory in the workspace. Values can be assigned to variables, as is done in Example C.2. At any time, the user can see the variables currently in memory by typing the **who** command. To erase the memory, type the **clear** command at the MATLAB prompt. To erase a variable, type the clear command followed by the variable name. For example, in Example C.2, we could remove the variable "apples" by entering "clear apples" at the MATLAB prompt.

Variable names in MATLAB must begin with a letter and can comprise up to 19 characters that include letters, numbers, and underscores (_). MATLAB is case-sensitive, so the variable "Apple" is different from *APPLE*," which is different from "apple," which is different from "aPPLe," which is different from "ApPlE," and so

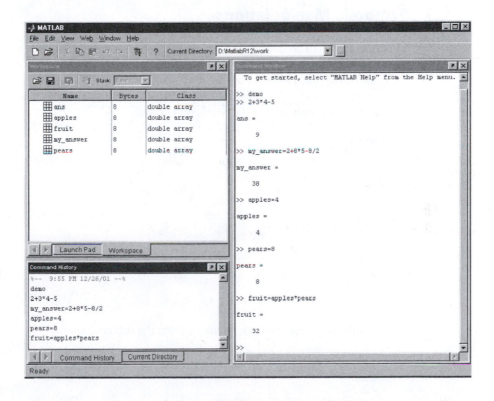

EXAMPLE C.2 This example shows some simple arithmetic using MATLAB. (Note that the workspace tab in the Launchpad window has been selected and shows the current variables for this session.)

---

on. It is advisable to use a single case when naming variables; having two variables with the same name but different cases can easily lead to confusion and errors.

When naming variables, avoid names that MATLAB already uses. If you use a name that MATLAB already uses for a function, you can disable that MATLAB function and create errors. Some of the names to avoid are **ans**, **Inf**, **NaN**, **i**, **j**, **eps**, **flops**, **nargin**, **narout**, **realmin**, and **realmax**, along with names of MATLAB functions. To determine if a name is in use by MATLAB, simply type that name at the prompt. If it is not being used by MATLAB, an error message will be returned "??? Undefined function or variable." Alternatively, you can type "help" followed by the name you wish to use. If it is not being used as a MATLAB function, the response "not found" will be returned. A suggestion is to use underscores in the variable names, as no MATLAB function contains an underscore (e.g., "good_name").

C.3 **Symbolic Manipulations**

MATLAB can also be used as a symbolic solver. To allow MATLAB to do symbolic manipulations, variables must be declared symbols. This can be done using the syms command. Variables listed after the syms command are designated as symbols, as shown in Example C.3. Symbolic manipulations can be done using algebra, integral, or differential calculus, as shown in the example.

EXAMPLE C.3   MATLAB can be used to perform symbolic manipulations. First, the variables x and y are declared to be symbols:

```
» syms x y
```

Algebraic expressions can be solved using the solve command:

```
» solve (x^2-4)

ans =
[  2]
[ -2]
```

Symbolic derivatives can be taken using the diff command:

```
» diff (y^3)

ans =
3*y^2
```

Symbolic integrals can also be taken using the int command:

```
» int(sin(x))

ans =
-cos(x)
```

C.4 **Saving and Loading Files**

It is possible to save your workspace in MATLAB for future use. This is done using the *save* command at the MATLAB prompt. On a PC or Mac, this can also be done under the file menu using the option for saving the workspace.

When you save a file, you need to know where the file is saved. As a default, MATLAB will put files in your present working directory. To save files to another directory, you must so specify. To find out the identity of your working directory, type "**pwd**" (print working directory) at the MATLAB prompt. This will tell you the address of your current directory. To change the working directory, you may use the **cd** (change directory) command (e.g., "cd c:\matlab\mystuff" would change the directory on a PC to the folder named "mystuff"), or use the directory window above the command window.

To save a file to a different directory, include the path name as part of the file name. On a PC or Mac, a window will appear that lets you select the directory in which you save the MATLAB files. Remember to use a .mat extension when saving your workspace. If you use the save and return commands with no file name, MATLAB will save the file under the name "matlab.mat." When choosing a name, similar restrictions apply as with variables. Remember to avoid file names that MATLAB already uses for functions.

To retrieve your file and return to your workspace, use the command *load* followed by the file name. You do not have to use the .mat extension with the load command. For example, if the workspace is saved under the name "my_workspace. mat," it can be reloaded by typing "load my_workspace" at the MATLAB prompt.

When loading files in MATLAB, make sure that MATLAB can find the files. MATLAB will look for files in your current directory. Then, it will use the paths that it has saved. To see a listing of the paths that MATLAB is currently using, type "**path**" at the MATLAB prompt. What will follow is a list of the directories that MATLAB will search for files. If files are stored in a directory other than your current working directory, you can add this location to the path list using the addpath command followed by the location. To check to see that the path has been added, type "path" again. The new path will appear at the beginning of the new list so you may need to scroll back to the beginning.

C.5 **Vectors**

MATLAB treats all numbers and variables as matrices; recall the original name of Matrix Laboratory. Earlier, we did simple arithmetic with scalars, or single numbers. In MATLAB, we can also use vectors and matrices.

A vector is simply a row or column of numbers. To create a vector, enter a variable name along with an equals sign just as was done with a single variable. The difference for a vector is that the numbers are enclosed in square brackets. To make a column vector, the numbers are separated by semicolons. To make a row vector, the numbers are separated by spaces or commas.

EXAMPLE C.4   Create a row vector and a column vector.

```
» row _ vector=[1 3 6 9 12]
row _ vector =
1 3 6 9 12
» col _ vector=[2;4;6;8;10]
col _ vector =
  2
  4
  6
  8
 10
```

Vectors are commonly encountered in engineering applications. For vectors to be added and subtracted, the vectors must be of the same type (row or column). The MATLAB **transpose** command can change a column vector into a row vector and vice versa. The vectors also must be the same size (or length) as MATLAB will add/subtract the first numbers of each, then the second, and so on. If the vectors have different sizes, there will be numbers left over, which will result in an error.

If we want to multiply or divide the numbers of one vector with the other, similarly to how we added and subtracted, we must use special MATLAB symbols. To multiply the elements of one vector by the elements of the other, use the command ".*". For division use "./". These symbols are referred to as element-by-element multiplication and division, respectively.

Vectors can be multiplied using the * operation and divided using the / operation. However, these are matrix algebra operations and follow matrix algebra rules. They are described along with matrices in the following section.

EXAMPLE C.5   Create two vectors and multiply and divide the elements.

```
» clear
» vec _ 1=[1 2 3 4 5]
vec _ 1 =
1 2 3 4 5
» vec _ 2=[2 4 6 8 10]
vec _ 2 =
2 4 6 8 10
» X = vec _ 1 .* vec _ 2
X =
2 8 18 32 50
» Y=vec _ 2 ./ vec _ 1
Y =
2 2 2 2 2
```

C.6 **Matrices**

Matrices, like vectors, are very common to engineering applications, and many engineering curricula require entire courses on matrices (or "linear algebra," as it might be called). For this brief introduction, we will provide some simple examples to show some of the uses of MATLAB for manipulating matrices.

A matrix is simply a group of numbers arranged in columns and rows. A vector is a special type of a matrix with a single row or column. An example of a matrix is

$$A = \begin{bmatrix} 1 & 2 & 3 \\ 4 & 5 & 6 \\ 7 & 8 & 9 \end{bmatrix}$$

Every matrix has a size. The size tells the numbers of rows and columns, respectively. The size of A is 3×3, three rows and three columns. The vectors in Example C.5 have sizes of 1×5 (one row by five columns). The **size** command in MATLAB gives the size of a matrix. For example, "size(A)" in MATLAB would return "3 3" for three rows and three columns.

Each element of a matrix is identified by the use of a numbering system called indexing. Each element has two numbers or indices associated with it. The first index is the row number and the second is the column number. For matrix A, each element and its index is shown:

$$A = \begin{bmatrix} a_{1,1} & a_{1,2} & a_{1,3} \\ a_{2,1} & a_{2,2} & a_{2,3} \\ a_{3,1} & a_{3,2} & a_{3,3} \end{bmatrix}$$

So, element $a_{1,1}$ (first row and first column of matrix A) = 1, $a_{3,2} = 8$, etc. In this way, individual elements or sections of matrices can be identified and used in calculations or graphing. Often in engineering applications, data are stored as a matrix, and pieces of the data are extracted to be analyzed or plotted. MATLAB can extract an entire row, say the first row, if we enter "A(1,:)," which tells MATLAB to take the first row and all columns. Similarly, we can ask for all rows and the first column of A by indicating "A(:,1)." If we wanted to take elements $a_{1,1}, a_{1,2}, a_{2,1}$ and $a_{2,2}$, we could do so by entering "A(1:2,1:2)." By separating the row and column numbers, we have required MATLAB to take rows one and two and columns one and two.

EXAMPLE C.6   Create a matrix and extract some elements.

```
» A = [1 2 3;4 5 6;7 8 9]
A =
1 2 3
4 5 6
7 8 9
» matrix_size = size(A)
matrix_size =
3 3
```

```
» ele _ 21 = A(2,1)
ele _ 21 =
4
» ele _ 32 = A(3,2)
ele _ 32 =
8
» vec _ row1 = A(1,:)
vec _ row1 =
1 2 3
» vec _ col2 = A(:,2)
vec _ col2 =
2
5
8
» upper _ square = A(1:2,1:2)
upper _ square =
1 2
4 5
```

A matrix can be any size, and the number of rows does not have to equal the number of columns as in the previous example. Arithmetic operations with matrices follow special rules. As with vectors, we can add, subtract, multiply, and divide two matrices, element by element, using the $+, -, .*,$ and $./$ symbols, respectively.

EXAMPLE C.7   Element-by-element operations can be performed using MATLAB. Several examples follow.

For addition:

$$\begin{bmatrix} a_{1,1} & a_{1,2} \\ a_{2,1} & a_{2,2} \end{bmatrix} + \begin{bmatrix} b_{1,1} & b_{1,2} \\ b_{2,1} & b_{2,2} \end{bmatrix} = \begin{bmatrix} a_{1,1} + b_{1,1} & a_{1,2} + b_{1,2} \\ a_{2,1} + b_{2,1} & a_{2,2} + b_{2,2} \end{bmatrix}$$

For subtraction:

$$\begin{bmatrix} a_{1,1} & a_{1,2} \\ a_{2,1} & a_{2,2} \end{bmatrix} - \begin{bmatrix} b_{1,1} & b_{1,2} \\ b_{2,1} & b_{2,2} \end{bmatrix} = \begin{bmatrix} a_{1,1} - b_{1,1} & a_{1,2} - b_{1,2} \\ a_{2,1} - b_{2,1} & a_{2,2} - b_{2,2} \end{bmatrix}$$

For multiplication:

$$\begin{bmatrix} a_{1,1} & a_{1,2} \\ a_{2,1} & a_{2,2} \end{bmatrix} .* \begin{bmatrix} b_{1,1} & b_{1,2} \\ b_{2,1} & b_{2,2} \end{bmatrix} = \begin{bmatrix} a_{1,1} \times b_{1,1} & a_{1,2} \times b_{1,2} \\ a_{2,1} \times b_{2,1} & a_{2,2} \times b_{2,2} \end{bmatrix}$$

For division:

$$\begin{bmatrix} a_{1,1} & a_{1,2} \\ a_{2,1} & a_{2,2} \end{bmatrix} ./ \begin{bmatrix} b_{1,1} & b_{1,2} \\ b_{2,1} & b_{2,2} \end{bmatrix} = \begin{bmatrix} a_{1,1} \div b_{1,1} & a_{1,2} \div b_{1,2} \\ a_{2,1} \div b_{2,1} & a_{2,2} \div b_{2,2} \end{bmatrix}$$

Matrices can be multiplied using the .* command (e.g., A.*B), but the operation follows linear algebra rules. Under these rules, the respective elements of the first row of matrix A are multiplied by the respective elements of the first column of matrix B and added. The sum becomes the element in the upper left corner (1,1) of the product matrix. The second row of matrix A is multiplied by the first row of matrix B, and so on. To illustrate, use the example of two 2 × 2 matrices called "A" and "B" where

$$A = \begin{bmatrix} a_{1,1} & a_{1,2} \\ a_{2,1} & a_{2,2} \end{bmatrix} \qquad B = \begin{bmatrix} b_{1,1} & b_{1,2} \\ b_{2,1} & b_{2,2} \end{bmatrix}$$

A.*B becomes

$$\begin{bmatrix} a_{1,1} & a_{1,2} \\ a_{2,1} & a_{2,2} \end{bmatrix} . * \begin{bmatrix} b_{1,1} & b_{1,2} \\ b_{2,1} & b_{2,2} \end{bmatrix} = \begin{bmatrix} (a_{1,1} \times b_{1,1} + a_{1,2} \times b_{2,1}) & (a_{1,1} \times b_{1,2} + a_{1,2} \times b_{2,2}) \\ (a_{2,1} \times b_{1,1} + a_{2,2} \times b_{2,1}) & (a_{2,1} \times b_{1,2} + a_{2,2} \times b_{2,2}) \end{bmatrix}$$

Note that, in general, these rules do not have the same commutative property as scalar multiplication: A.*B ≠ B.*A. For two matrices to be multiplied together, the number of columns of the first matrix must equal the number of rows in the second. For example, for A.*B, the number of columns in A must equal the number of rows in B. This is illustrated in Example C.8.

EXAMPLE C.8

Multiply the matrices of different sizes: A*x, where A and x are defined as

$$A = \begin{bmatrix} 1 & 2 \\ 3 & 4 \end{bmatrix} \text{ and } x = \begin{bmatrix} 5 \\ 6 \end{bmatrix}$$

A is a 2 × 2 and x is a 2 × 1 matrix. So A*x becomes (2 × 2)*(2 × 1). For the operation to be possible, the underlined numbers (middle numbers) must be the same. The resultant matrix becomes a 2 × 1, which are the remaining or outside numbers. The result of the multiplication is

$$A * x = \begin{bmatrix} 1 \times 5 + 2 \times 6 \\ 3 \times 5 + 4 \times 6 \end{bmatrix} = \begin{bmatrix} 17 \\ 39 \end{bmatrix}$$

C.7 Simultaneous Equations

One of the most helpful applications of using MATLAB to manipulate matrices is the ability to quickly solve simultaneous equations. To illustrate, consider three equations:

$$8Z + 3X + 6Y = 25$$

$$5Y + 7Z - 36 = 0$$

$$4X - 6Y - 8Z = -18$$

While we could solve these equations using algebra, it would take a lot of time and effort. It would take even longer if there were more equations and unknowns. With MATLAB, as the number of equations increases, the only time that significantly increases is the time it takes to input the elements. MATLAB uses matrix algebra to solve the equations, so the equations must be written in matrix form. There will be three matrices. One is a vector of the variables (x, y, and z) that we will call "x." The second matrix contains the coefficients of the variables and is called "B." The third vector, b, contains the constant terms of the equations.

The first step in writing the equations in matrix form is to put them in standard form:

$$3X + 6Y + 8Z = 25$$

$$5Y + 7Z = 36$$

$$4X - 6Y - 8Z = -18$$

Now the matrix form of Ax = b can be written as

$$\begin{bmatrix} 3 & 6 & 8 \\ 0 & 5 & 7 \\ 4 & -6 & -8 \end{bmatrix} \begin{bmatrix} X \\ Y \\ Z \end{bmatrix} = \begin{bmatrix} 25 \\ 36 \\ -18 \end{bmatrix}$$

If we follow the rules of matrix multiplication, we get our original three equations back. If the equation of Ax = b were an algebraic one, we would simply divide both sides by A and solve for x as x = b/A. With MATLAB, we use the left division operator as x = A\b.

EXAMPLE C.9   Solve the following simultaneous equations using MATLAB:

$$3X + 6Y + 8Z = 25$$

$$5Y + 7Z = 36$$

$$4X - 6Y - 8Z = -18$$

In MATLAB, enter:

```
» A = [3 6 8;0 5 7;4 -6 -8]
A =
3   6   8
0   5   7
4  -6  -8
```

```
» b = [25; 36; −18]
b =
25
36
−18
» x = A\b
x =
1.0000
−67.0000
53.0000
```

C.9 **Plotting**

MATLAB is very useful for plotting data. For the purposes of this appendix, we will discuss 2D plotting on linear, semi-log, and log scales. However, there are other types of plots, including 3D plotting, available in MATLAB. By using the **demo** command, the demonstrations, which were referred to earlier, will provide a good introduction.

The basic function in graphing is **plot** and is used to generate linear xy plots. The plot function requires:

$$\text{plot}(x \text{ axis values}, y \text{ axis values}, \text{ 'symbol or line type'})$$

The x- and y-axis values must be stored in MATLAB as vectors and they must have the same number of elements. The symbol or line type allows the user to specify symbols, type of line, and color. Note that the line type and symbol are enclosed in single quotes. The MATLAB names for different symbols, line types, and colors are shown in Table C.1. Symbols and line types can be combined, but they must be in the same color.

To plot multiple datasets, we can use the plot command by listing two different pairs of data. For example, if we want to plot data stored in vectors x and y as well as w and z, we can enter:

$$\text{plot}(x,y,'-k',w,z,':g')$$

The result will be a plot of x and y with a solid black line and w and z with a dotted green line.

A second way to plot multiple data sets is to use the **hold** command. By entering "hold on," subsequent plotting commands will be plotted on the same plot. Entering "hold off" can turn off the hold command. Example C.10 shows how this can be used to plot data multiple times. In this example, the axes are labeled using the commands **xlabel**, **ylabel**, and **title**. Notice that the descriptive text is enclosed in parentheses and single quotes. The axes can be scaled to different values using the *axis* command.

Table C.1 Line and Symbol Options

Color		Symbol		Line type	
y	yellow	.	point	-	solid
m	magenta	o	circle	:	dotted
c	cyan	x	x-mark	-.	dashdot
r	red	+	plus	--	dashed
g	green	*	star		
b	blue	s	square		
w	white	d	diamond		
k	black	v	triangle (down)		
		^	triangle (up)		
		>	triangle (right)		
		p	pentagram		
		h	hexagram		

If you wish to open multiple figure windows, MATLAB has the command **figure(n),** where n is the figure number. For instance, figure(2) would open a second figure window and label it "figure 2." This will make it active, and any subsequent plotting will be done in that window. To switch back from figure 2 to the first figure, simply type "figure(1)" or click on figure 1 with the mouse.

The command **subplot** is used to generate multiple plots on a single figure. The command is used as "subplot(i,j,k)", where i and j indicate the number of rows and columns of figures and k represents the graph number. The graph numbers begin in the upper left-hand corner and move left to right across a row and then down to the left side of the next row. For example, if we wish to put six plots on a single figure, we would enter "subplot(3,2,1)" to plot the figure in the upper left-hand corner. To activate the plot on the right side of the second row, enter "subplot(3,2,4)". Example C.11 demonstrates the use of the subplot command. In this example, four types of 2D plots are demonstrated.

Semi-log plots are also shown in Example C.11. In a semi-log plot, one axis has a log scale while the other has a linear scale. MATLAB has **semilogx** and **semilogy** commands to make the *x*- and *y*-axis the log scale, respectively. To make both scales log, use the plotting command **loglog**.

EXAMPLE C.10 Create a plot of x and y values by entering the data into vectors named x and y. Then plot the data and label the graph as shown below.

```
» x = [1 2 3 4 5 6 7 8 9]
x =
  1 2 3 4 5 6 7 8 9
» y = [1 4 9 16 25 36 49 64 81]
```

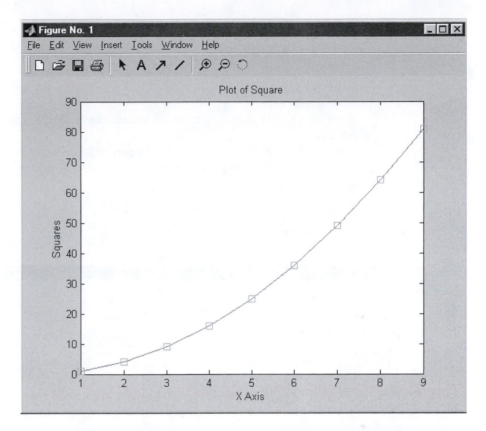

```
y =

    1 4 9 16 25 36 49 64 81
» plot(x,y,'-b')
» hold on
» plot(x,y,'rs')
» title('Plot of Square')
» xlabel('X Axis')
» ylabel('Squares')
```

**EXAMPLE
C.11**

Plot the function y = ex on linear, semi-log, and log-log plots. Place four plots on a single figure.

MATLAB commands:

```
»figure(2) % generates a new figure window
» x=[1 2 3 4 5 6 7 8 9] % enter the x data
x =
  1 2 3 4 5 6 7 8 9
```

```
» y=exp(x) % generating y data using the x data
y =
Columns 1 through 6
2.7183 7.3891 20.086 54.598 148.41 403.43
Columns 7 through 9
1096.6 2981 8103.1
» subplot(2,2,1) % selects the upper left-hand corner plot
» plot(x,y)
» subplot(2,2,2) % selects the upper right-hand corner plot
» loglog(x,y) % plots log scales on both x and y axes
» subplot(2,2,3)
» semilogx(x,y)
» subplot(2,2,4)
» semilogy(x,y)
```

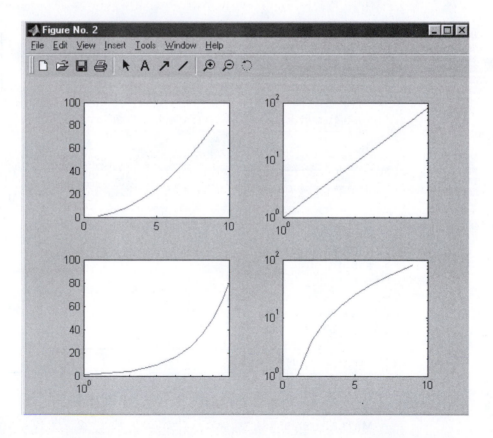

C.10 Programming

In MATLAB we can write programs similar to a programming language. These programs, also called scripts, consist of a series of MATLAB commands that can be saved to run in "batch mode" later.

To create a MATLAB program, we need to use a text editor (Fig. C.3). In UNIX, open any editor of your choice, such as *vi* or *emacs*. On a PC or Mac, there is a MATLAB editor that can be accessed through the file menu. Under "new," select "M-file" to open the programming editor.

In the editor, you can enter MATLAB commands just like you would type them into the workspace. You can also add comment statements by using the % symbol. Anything entered after the % will be ignored by MATLAB, so this is where you can enter comments for your own use. It is good programming practice, regardless of the programming language, to use comment statements to document your program. They will help you understand what the program does when you come back to it later.

Once you have entered the commands and comment lines into the file, you must save it in a file with a .m extension. The naming rules are the same as with the .mat (workspace) files described earlier in this appendix. Remember to avoid file names that MATLAB already uses. Also, similar to saving the .mat (workspace) files, you must also save the file in a directory that MATLAB will access. This may require using the path and addpath commands outlined earlier.

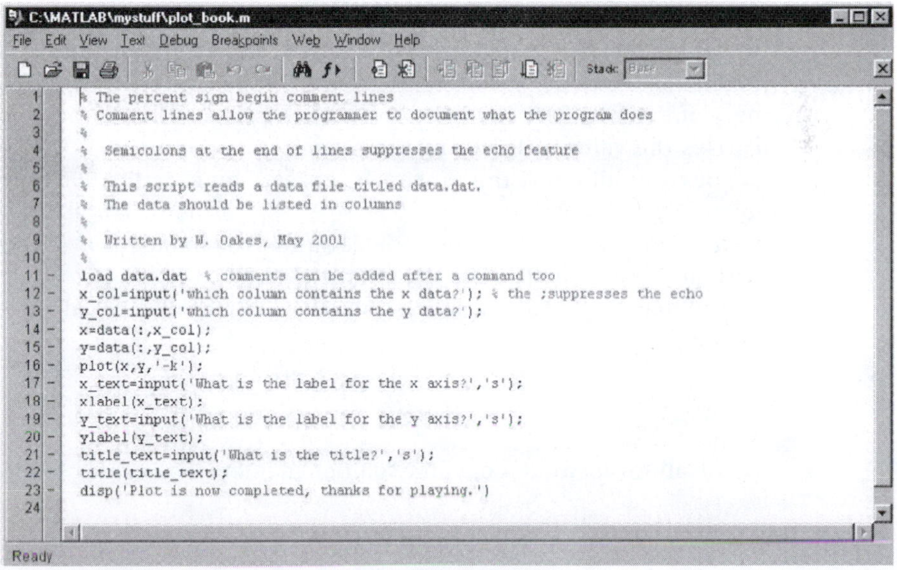

C:\MATLAB\mystuff\plot_book.m
File Edit View Text Debug Breakpoints Web Window Help
Stack:
% The percent sign begin comment lines
% Comment lines allow the programmer to document what the program does
%
% Semicolons at the end of lines suppresses the echo feature
%
% This script reads a data file titled data.dat.
% The data should be listed in columns
%
% Written by W. Oakes, May 2001
%
load data.dat % comments can be added after a command too
x_col=input('which column contains the x data?'); % the ;suppresses the echo
y_col=input('which column contains the y data?');
x=data(:,x_col);
y=data(:,y_col);
plot(x,y,'-k');
x_text=input('What is the label for the x axis?','s');
xlabel(x_text);
y_text=input('What is the label for the y axis?','s');
ylabel(y_text);
title_text=input('What is the title?','s');
title(title_text);
disp('Plot is now completed, thanks for playing.')
Ready

Figure C.3 Editor window

After the file is saved, it can be executed by typing the file name (without the .m extension) at the MATLAB prompt. For example, if the script name was "example_script.m," you would type "example script" at the MATLAB prompt.

Let's delve into the *content* of a MATLAB script, or code. You may, of course, include any or all of the commands introduced in the sections above, and once executed, the computer will perform each command one line at a time, top to bottom. However, running MATLAB in batch mode can be extremely powerful when additional features are used. In particular, the ***for*** and ***if*** functionalities can be used to perform—and automate—complex and/or repetitious computations.

The for Loop

Say you need to calculate the sum of whole numbers between 1 and 1,000—that is, find 1+2+3+...+1,000. Doing this calculation in interactive mode by typing the exact symbolic equation is both time-consuming and impractical. A much better and more efficient way is the use of a for loop in your script:

```
sum1=0;
for j=1:1000
    sum1=sum1+j;
end
```

In the simple code above, the variable sum1 is first zeroed out, a common and good practice in any programming language. The for loop then begins, where the variable sum1 is repeatedly growing in size, and each time the amount to be added to it is the "counter" j, which goes from 1 to 1,000, increasing by 1. During the first pass of the loop, what happens to sum1 is "sum1 = 0 + 1," at which point it now carries the value of 1. During the second pass, "sum1 = 1 + 2," and now sum1 carries the value of 3. This process is repeated for exactly 1,000 times. The solution for sum1 is 500,500.

It should be noted that the variable sum1 is so named to avoid potential conflict with the built-in MATLAB function called "sum."

You may have realized that this simple sum may also be calculated by using the relationship

$$\sum_{k=1}^{n} k = \frac{n(n+1)}{2}$$

which allows for a quick one-line solution in interactive mode:

» sum2=0.5*1000*(1000+1)

However, most real-life engineering problems do not have exact solutions such as the equation above. The for loop provides the flexibility necessary for carrying out complicated tasks.

The syntax of a basic for loop is straightforward, as illustrated above. If you would now like to calculate the sum of whole numbers between 1 and 1,000, *skipping every other number* (i.e., find 1+3+5+. . .+999), then the code may look like:

```
sum3=0;
for j=1:2:1000
    sum3=sum3+j;
end
```

The only difference here is the number 2 sandwiched between 1 and 1,000 in the first line of the for loop. This middle number defines the step size or step width. The absence of a middle number automatically assumes that the step size is 1.

The if **Statement**

In computing of any capacity, from calculating your GPA to a full-blown 3D transient analysis of fluid dynamics, we often need the code to make *conditional* decisions—that is, "if this is true, then do that; otherwise do something else (or nothing) instead."

As a quick example, the code below assigns a value of A to the variable Y if X equals 4, a value of B if X equals 3, and a value of C if X equals any number other than 3 or 4.

```
if X==4
    Y='A';
elseif X==3
    Y='B';
else
    Y= 'C';
end
```

The == is called a relational operator. Other choices are < (less than), > (greater than), <= (less than or equal), >= (greater than or equal), and ~= (not equal). Note that since the variable Y contains only letters (i.e., A, B, or C) instead of numbers, quotation marks must be used for each letter assigned to Y.

Let's now illustrate the use of if and for by calculating your GPA for a particular semester (Example C.12).

EXAMPLE C.12   Let's assume that the grades you earned from the last semester are:

Course	Grade Received	Number of Credits
ME 204	A	3
PHYS 222	C	4
MA 330	B	5
HIST 101	A	3

What is your semester GPA? The sample code below presents one way of automating the calculation for you, semester after semester. The only modifications you need to make are the first two lines containing the course grades and number of credits for each course.

The definition of GPA is

$$GPA = \frac{\text{Total Grade Points}}{\text{Total Credits}}$$

where Total Grade Points is the sum of "grade points" earned in each course based on the grade received for that course. A common grade point value scale is 4 for A, 3 for B, 2 for C, 1 for D, and 0 for F. This means for each credit of a course, you will earn 4 points if you receive an A in that course. So for a three-credit course, you will gain 12 points if you have earned an A. Total Credits is simply the sum of credit hours for the semester.

```matlab
% Enter course grades and # credits for each course
% according to the table above; must follow
% exact order
% Create an 'array' for the variable 'grades'
% and 'credits.' Notice curly braces are used for
% letters, square brackets for numbers.
grades = {'A','C','B','A'};
credits = [3,4,5,3];

% Do not modify from here on out.

% Define grade range and points for each grade
% based on the common scale
gp _ array1 = {'A','B','C','D','F'}
gp _ array2 = [4,3,2,1,0];

% Use 'for' and 'if' to calculate grade points

% Count number of courses from 'grades' above
% The function 'numel' counts how many elements
% are in an array
total _ courses = numel(grades);

for j=1:total _ courses
  for k=1:5
   if (grades{j}==gp _ array1{k})
    gradepoints(j)=gp _ array2(k)*credits(j);
   end
  end
end

% List the calculated grade points on the screen
gradepoints

% Calculate total grade points using 'sum' function
sum _ gp = sum(gradepoints);
sum _ gp
```

```
% Calculate total credits
sum _ cr = sum(credits);
sum _ cr

% Finally, calculate GPA
gpa = sum _ gp / sum _ cr;
gpa
```

The sample code above, while functional, is rather crude. A number of improvements can be made to help the code run more efficiently (in terms of RAM usage and number of operations), become more user-friendly, and be more fail-safe.

One possible modification to the code is the way course grades and number of credits are entered. Instead of hard-coding these items into the code, you may choose to prompt the user to enter this information interactively. The **input** command is very useful for this purpose. This command allows the user to enter data in response to a MATLAB prompt. For example, if we want the user to enter a number that the program might use, we could write

W = input('Enter a number to be used by the code')

The result is that MATLAB will display the prompt

Enter a number to be used by the code

and will wait for the user to enter a number and then press return. Once the number is entered, it is stored as the value for the variable W and can be used in the program as W.

If a word or letters are required rather than numbers, we must tell MATLAB to expect letters, which are called strings. This requires adding a " ,'s' " to the end of the input line. To ask the user for a string of letters would require the following:

my_word = input('Enter a word: ','s')

Example C.13 uses string inputs to label the plots.

If data are to be displayed, the function **disp** can be used. This function will display the contents of a variable. Disp(W) would display the value of W while disp(my_word) would display the contents of the variable my_word. Text can be displayed as it is in Example C.13 using single quotes. To wish the user good luck, enter "disp('good luck')" and "good luck" will be displayed on the screen.

EXAMPLE   In an editor, enter and save the following code under the name 'plot_example.m'
C.13   in the directory c:\matlab\mystuff.

```
% The percent sign (%) begins comment lines
% Comment lines allow the programmer to
% document the program
```

```
%
% Semicolons at the end of lines suppresses the echo
% feature in MATLAB
%
% This script reads a data file titled data.dat.
% The data to be plotted is listed in columns
%
% Written by W. Oakes
%
load data.dat % comments can also be added after a command
x _ col=input('which column contains the x data?');
y _ col=input('which column contains the y data?');
x=data(:,x _ col);
y=data(:,y _ col);
plot(x,y,'-k');
x _ text=input('What is the label for the x axis?','s');
xlabel(x _ text);
y _ text=input('What is the label for the y axis?','s');
ylabel(y _ text);
title _ text=input('What is the title?','s');
title(title _ text);
disp('Plot is now completed, thanks for playing.')
```

To run the program, the following was typed in MATLAB:

```
» addpath c:\matlab\mystuff
» plot _ example
» which column contains the x data? 1
which column contains the y data? 2
What is the label for the x axis? x axis
What is the label for the y axis? y axis
What is the title? Title of a Wonderful Graph
Plot is now completed, thanks for playing.
```

The result is a plot with a title and *x*- and *y*-axes labeled.

In conclusion, this appendix has introduced the idea of using the computer to perform calculations, with MATLAB being one of the many tools available out there. There is much more to MATLAB than what has been shown here, and you are urged to delve deeper into the full capabilities of MATLAB as well as other programming languages such as C, Java, Perl, Python, and C++. Once you have mastered a language or two, the skills you acquire can have a lifelong impact on the lives of many—including yours.

For MATLAB product information, please contact: The MathWorks, Inc., 3 Apple Hill Dr., Natick, MA 01760-2098, ph: 508-647-7000, fax: 508-647-7001, email: info@math-works.com, web: www.mathworks.com.

EXERCISES

C.1   Use the lookfor command to find the MATLAB commands that deal with:

a) sine   b) plot   c) logarithm
d) variable   e) path   f) exponential

C.2   Use the help function in MATLAB to find out what five functions do. Summarize these functions in a short report.

C.3   Run the MATLAB demo by typing "demo" at the MATLAB prompt. Explore at least two of the functions and prepare a one-page report on how MATLAB will be helpful to you as an engineering student.

C.4   Run the MATLAB demo by clicking on the demos icon in the Launchpad window to explore the graphics capabilities of MATLAB, and prepare a one-page report on how MATLAB's graphics capabilities could be used in engineering applications.

C.5   Use MATLAB to do simple arithmetic using variables. Print the screen you are working with to show the variables you created.

C.6   Solve y = 0.5x – w, where x = [2 64 82 106 94] and w = [5 6 7 8 9], using MATLAB.

C.7   Write the following equations in matrix form (Ax = b):

a) $x - 14 = -2y - 3z$
b) $40 - 4x = 6y + 8z$
c) $x + 3y + 7z - 28 = 0$

C.8   Solve Exercise C.7 using MATLAB.

C.9   Write a brief description of the following MATLAB commands:

a) round   b) format   c) ceil
d) floor   e) fix   f) abs
g) rem   h) exp

C.10 Create a y vector such that y = ex where x = [1,2,3,4,5,6,7,8,9]. Plot the result on an xy graph. Label the x- and y-axes and place a title on the graph.

C.11 Create a graph of x = [2 4 6 8 10 12] and y = [8 32 72 128 200 288]. Label the graph using red squares for the symbols connected with a dashed blue line.

C.12 Replot the data from Exercise C.10. Is there a graph type (e.g., linear, semi-log, or log-log) where the graph becomes a straight line?

C.13   Replot the data from Exercise C.11. Is there a graph type (e.g., linear, semi-log, or log-log) where the graph becomes a straight line?

C.14   Create symbol x and find the roots of:

a) $x^2 - 4$
b) $x^2 - 2x - 195$
c) $x^3 + 2x^2 - 13x + 10$

C.15   Use the *simple* command to multiply $(x + 4)(x - 5)(x + 6)(x - 2)$.

C.16   Use MATLAB to integrate the following:

a) $3x^2 + 2x + 1$ c) $\tan(x)$
b) $\sin(x)$ d) $\sin^2(x)$

C.17   Use MATLAB to differentiate the following:

a) $3x^3 + 4x^2 - 2x - 8$
b) $\cos(x)$
c) $\tan^2(x)$

C.18   Create a workspace with variables and save the file in a directory named "matlab_stuff." Clear the workspace and reload the file.

C.19   Write a MATLAB script that will plot data that the user inputs. Select some data and show a plot.

C.20   Write a MATLAB script that will ask the user his or her name and age and calculate the number of days the person has been living. Demonstrate the script with an example.

C.21   For x = 2 and y = 5, compute the following using MATLAB:

a) $\dfrac{yx^3}{x - y}$ b) $\dfrac{3x}{2y}$ c) $\dfrac{3xy}{2}$ d) $\dfrac{x^5}{\left(x^5 - 1\right)}$

C.22   For x = 3 and y = 4, use MATLAB to solve the following

a) $\left(1 - \left(\dfrac{1}{x^5}\right)\right) - 1$ b) $3\rho x^2$ c) $\dfrac{3y}{\left(4x - 8\right)}$ d) $\dfrac{4(y - 5)}{3x - 6}$

C.23   Suppose x takes on the values of x = 1, 1.2, 1.4, . . . , 5. Use MATLAB to compute the array y that results from the function $y = 7\sin(4x)$. Plot the results.

C.24 Create, save, and run a script that solves the following set of equations for given values of a, b, and c. Check your script for the values of a $= 95$, b $= -50$, and c $= 145$.

$$7x + 14y - 6z = a$$

$$12x - 5y + 9z = b$$

$$-5x + 7y + 15z = c$$

C.25 The volume of a sphere is given by $V = 4\pi r^3/3$, where r is the radius of the sphere. Make a plot of volume versus radius for radii of 0.1 to 10.

C.26 Repeat the previous problem using four subplots, one with linear axes, one with a semi-log on the x axis, one with a semi-log on the y axis, and one with both axes having a log scale (log-log). Label each plot properly.

C.27 Given the matrices

$$A = \begin{bmatrix} -7 & 16 \\ 4 & 9 \end{bmatrix} B = \begin{bmatrix} 6 & -5 \\ 12 & -2 \end{bmatrix} C = \begin{bmatrix} -3 & -9 \\ 6 & 8 \end{bmatrix}$$

use MATLAB to find

a) $A + B + C$
b) $A - B + C$
c) Verify the associative law $(A + B) + C = A + (B + C)$.
d) Verify the commutative law $A + B + C = A + C + B = B + C + A$.

C.28 Use *MATLAB* to solve (for x$= 2$):

$$(6x + 4x - 5) / (12x - 7x + 3x + 9)$$

C.29 ... for x $= 1, 1.1, 1.2, \ldots, 2$ and y $= 3x^3 + 4x^2 - 5x - 7$, plot x versus y using a blue dashed line with green circles for the data points. Provide a title and labels for the x and y axes.

C.30 ... for x $= 1, 1.1, 1.2, \ldots, 2$ and y $= x^2$, plot x and y using:

a) Red squares for the data points
b) Red solid line and blue squares for the data points
c) Yellow dotted line with green triangles for the data points
d) Green dashed line with black circles for the data points

Index

Page references for tables are indicated by *t* and for figures are indicated by *f*.